PHYSIOLOGICAL ECOLOGY

A Series of Monographs, Texts, and Treatises

A complete list of titles in this series appears at the end of this volume.

Allelopathy

Second Edition

ELROY L. RICE

Department of Botany and Microbiology
The University of Oklahoma
Norman, Oklahoma

 ACADEMIC PRESS, INC.

(Harcourt Brace Jovanovich, Publishers)

Orlando   San Diego   San Francisco   New York   London
Toronto   Montreal   Sydney   Tokyo   São Paulo

ACADEMIC PRESS, INC.
Orlando, Florida 32887

United Kingdom Edition published by
ACADEMIC PRESS, INC. (LONDON) LTD.
24/28 Oval Road, London NW1 7DX

Library of Congress Cataloging in Publication Data

Rice, Elroy Leon, Date
    Allelopathy.

    (Physiological ecology)
    Bibliography: p.
    Includes index.
    1. Allelopathy. I. Title. II. Series.
QK911.R5   1983   581.5'24   83-11782
ISBN 0-12-587055-8 (acid free paper)

PRINTED IN THE UNITED STATES OF AMERICA

 84 85 86   9 8 7 6 5 4 3 2 1

CONTENTS

PREFACE

This monograph still remains the only one on allelopathy published in English and the only comprehensive one on the subject in any language. The rate of publication of papers concerning allelopathy has increased markedly in the past decade. Thus, there has been a growing need for an updating of the first edition. That edition covered developments in allelopathy through 1972; this edition covers investigations through the first quarter of 1983.

I have completely reorganized the monograph to reflect the changes in research emphasis in the various areas of allelopathy and to make the sequence of topics more logical. There has been a tremendous increase in research related to agriculture and forestry, and exciting applications of the research findings are on the horizon. All chapters have been rewritten and updated, but there are some areas in which research activity has been slow.

Chapters 2, 5, 6, 11, and 15 in the first edition have been eliminated as separate chapters and pertinent material has been included in other chapters. Four new chapters were added to reflect more accurately the current directions of research in allelopathy, to discuss important topics that were covered inadequately or not at all in the first edition, or to emphasize areas in which research activity is critically needed: "Roles of Allelopathy in Plant Pathology" (Chapter 4), "Allelopathy and the Nitrogen Cycle" (Chapter 9), "Evidence for Movement of Allelopathic Compounds from Plants and Absorption and Translocation by Other Plants" (Chapter 12), and "Factors Determining Effectiveness of Allelopathic Agents after Egression from Producing Organisms" (Chapter 14).

The scientific names of all agricultural and horticultural species follow "Hortus Third" (Bailey et al., 1976). I have attempted to follow currently accepted scientific nomenclature for other species. Common names follow various U.S. Department of Agriculture standardized lists of names or names used by the authors for species not included in U.S.D.A. lists. I have attempted to be consistent in the use of both common and scientific names, and this necessitated the use of names in several instances other than those used by the authors.

I acknowledge with thanks the permissions granted by authors and publishers to use materials from their publications.

Elroy L. Rice

PREFACE TO THE FIRST EDITION

No general monograph on the subject of allelopathy has been published previously in the English language, and none in any language since Grodzinsky's in 1965, in Russian. His book has not been translated and has had limited distribution outside the USSR. Moreover, much of the research that has established the field of allelopathy has been published since that time. The wide acceptance by ecologists of allelopathy as an important ecological phenomenon has occurred only within the past ten years. Thus, there appears to be a need for a general reference source in this field, both for researchers in the discipline and as an overview for those who desire to learn something about the subject.

Most significant contributions in the field, available at the time of writing, have been discussed; but no attempt has been made to include all publications that are in some way related to allelopathy. In fact, I have deliberately refrained from discussing the antibiotics involved primarily in medicine and most of the research concerned with biochemical interactions involved in plant diseases. My primary goal has been to discuss the broad ecological roles of allelopathy.

I have used the term allelopathy in the broad sense of Molisch (1937) to include biochemical interactions among plants of all levels of complexity, including microorganisms. Any restriction of this use does not make practical sense, as a perusal of this monograph will confirm. All levels of interaction are inextricably interwoven in ecological phenomena.

Most of my own research and that of my students reported here was supported by The National Science Foundation, for which I am grateful. I deeply appreciate the enthusiastic contributions of my graduate students, without whose help this monograph would not have been possible. I acknowledge with thanks the permissions granted by numerous authors and publishers to use previously published materials. The support and help of Dr. T. T. Kozlowski (editor of the Physiological Ecology Series) and of the staff of Academic Press are gratefully acknowledged.

Elroy L. Rice

1

Introduction

I. ORIGIN AND MEANING OF ALLELOPATHY

Molisch (1937) coined the term *allelopathy* to refer to biochemical interactions between all types of plants including microorganisms. His discussion indicated that he meant the term to cover both inhibitory and stimulatory reciprocal biochemical interactions.

The first edition of this book (Rice, 1974) deviated from Molisch's use of the term and defined allelopathy as any direct or indirect harmful effect by one plant (including microorganisms) on another through production of chemical compounds that escape into the environment. Additional experimental research and review of the literature have convinced me that elimination of stimulatory effects from the definition is artificial. Apparently most, if not all, organic compounds that are inhibitory at some concentrations are stimulatory to the same processes in very small concentrations. Most researchers have consistently followed Molisch's definition of the term, except for a relatively small number in North America. Many extremely important ecological roles of allelopathy have been overlooked because of the concern with only the detrimental effects of the added chemicals. Thus, the term *allelopathy* will be used in this review according to Molisch's use of the term.

A very important point concerning allelopathy is that its effect depends on a chemical compound being added to the environment. It is thus separated from *competition,* which involves the removal or reduction of some factor from the environment that is required by some other plant sharing the habitat. Factors that may be reduced include water, minerals, food, and light.

Confusion has arisen in the literature because some biologists consider allelopathy to be part of competition. This confusion could be lessened by using the term *interference* to refer to the overall influence of one plant on another, as suggested by Muller (1969). Interference would thus encompass both allelopathy and competition, and I will use the term interference in this sense. There are no known techniques to eliminate possible allelopathic effects in "competition" experiments. Therefore, all such studies could be more accurately entitled "interference" studies.

II. SUGGESTED TERMINOLOGY FOR CHEMICAL INTERACTIONS BETWEEN PLANTS OF DIFFERENT LEVELS OF COMPLEXITY

Grümmer (1955) suggested that special terms be adopted for the chemical agents involved in allelopathy based on the type of plant producing the agent and the type of plant affected. He recommended the commonly used term *antibiotic* for a chemical produced by a microorganism and effective against a microorganism and Waksman's term *phytoncide* for an agent produced by a higher plant and effective against a microorganism. Gaumann's term *marasmin* was suggested for a compound produced by a microorganism and active against a higher plant, and Grümmer coined the term *kolines* for chemicals produced by higher plants and effective against higher plants.

A specific allelopathic substance may have a sharply limited scope of action such that it is not effective against a higher plant if it is an antibiotic. On the other hand, it may have a broad scope of action like the antibiotic patulin, which also has marked toxicity for higher plants (Grümmer, 1955). This same principle applies to certain kolines, phytoncides, and marasmins.

III. EARLY HISTORY OF ALLELOPATHY

Theophrastus (ca. 300 B.C.) stated that chick pea (*Cicer arietinum*) does not reinvigorate the ground as other related plants (legumes) do but "exhausts" it instead. He pointed out also that chick pea destroys weeds and "above all and soonest caltrop" (*Tribulus terrestris*).

Pliny (Plinius Secundus, 1 A.D.) reported that chick pea, barley (*Hordeum vulgare*), fenugreek (*Trigonella foenum-graecum*), and bitter vetch (*Vicia ervilia*) all "scorch up" cornland. Pliny stated that the "shade" of walnut (apparently *Juglans regia*) is "heavy, and even causes headache in man and injury to anything planted in the vicinity; and that of the pine tree also kills grass; . . ." He also stated that "at all events for the shadow of a walnut tree or a stone pine or a spruce or a silver fir to touch any plant whatever is undoubtedly poison." His discussion indicated that he was using the term shade in a broad sense to include the usual concept of partial exclusion of light, plus effects on nutrition, plus chemicals that escape from the plants into the environment. Pliny's awareness of release of chemicals by plants was indicated by his statement that "The nature of some plants though not actually deadly is injurious owing to its blend of scents or of juice . . . for instance the radish and laurel are harmful to the vine (grape); for the vine can be inferred to possess a sense of smell, and to be affected by odors in a marvellous degree. . . ." Pliny claimed also that "cytisus and the plant called halimon by the Greeks kill trees." He stated further that the best way to kill bracken fern (*Pteridium aquilinum*) is to knock off the stalk with a stick when the stalk is budding "as the juice trickling down out of the fern itself kills the roots."

Culpeper (1633) stated that basil (*Ocimum*) and rue never grow together nor near one another. He claimed also that there is such an antipathy between grape and cabbage plants that one will die where the other grows. Browne, in his "Garden of Cyrus" published in 1658, reported that "the good and bad effluviums of vegetables promote or debilitate each other" (Keynes, 1929). It is notable that this statement is still acceptable in terms of our present understanding of allelopathy.

Lee and Monsi (1963) found a report by Banzan Kumazawa in a Japanese document some 300 years old that rain or dew washing the leaves of the Japanese red pine (*Pinus densiflora*) was harmful to crops growing under the pine. This was substantiated by Lee and Monsi in a series of experiments.

Young (1804) claimed that clover was extremely apt to fail in districts where it has been cultivated constantly because the soil becomes sick of clover. T. A. (1845) pointed out that clover sickness can be prevented by allowing an interval of 7 or 8 years between clover crops.

DeCandolle (1832) suggested that the soil sickness problem in agriculture might be due to exudates of crop plants and that rotation of crops could help alleviate the problem. He observed that thistles in fields injured oats; euphorbe and *Scabiosa* injured flax; and rye plants injured wheat.

Beobachter (1845) stated that the heath plant (probably *Erica* sp.) has a unique property of forming a hard block stratum a few inches below the soil surface. This stratum is impervious to water and equally impervious to the roots of trees. Unless this stratum is penetrated, it is useless to plant trees because most will die and those that survive will not thrive. Beobachter pointed out that this stratum appears to result from excrements by the roots of the heath plants.

In 1881, Stickney and Hoy observed that vegetation under the black walnut tree (*Juglans nigra*) was very sparse compared with that under most other commonly used shade trees. They also pointed out that crops did not grow under or very near it. Stickney stated that there was a question as to whether this condition was caused by water dripping from the tree or by the tree's high requirement for minerals, thereby exhausting the soil. Hoy claimed, however, that the main reason vegetation does not thrive under these trees is the poisonous character of the drip. He said that the juice of the leaf was poisonous, and a solution made from it kept flies off horses.

It is evident from this brief history that many botanists, farmers, and gardeners have observed and suggested many cases of allelopathy for over 2000 years. It is equally clear that controlled scientific experiments on this phenomenon were not conducted until after the year 1900. It is significant, however, that most of the species suggested to have pronounced chemical effects on themselves or other species have been demonstrated subsequently to have such effects. It is noteworthy that many species of plants, which have been widely used in medicine and are known to have powerful medicinal effects, have pronounced allelopathic effects also. Selected research in allelopathy done after 1900 will be discussed in appropriate places in subsequent chapters.

IV. PHYLA OF PLANTS DEMONSTRATED TO HAVE ALLELOPATHIC SPECIES

Bold's (1957) system of classification of the plant kingdom is used in this brief outline of the known scope of allelopathy. Bold has designated more phyla (divisions) than most morphologists, and this permits a better overview of the groups of plants known to have species with allelopathic potential.

Phylum 1. Cyanophyta (Blue-green algae). Many species have been demonstrated to cause allelopathic effects (see Chapter 6 for references).

Phylum 2. Chlorophyta (Green algae). Widespread effects (see Chapter 6).

Phylum 3. Charophyta (Stoneworts). Hayes (1947) tested aqueous extracts of *Chara foetida* against four species of bacteria and found little or no activity. More research certainly needs to be done concerning this phylum.

Phylum 4. Euglenophyta (Euglenoids). No references on allelopathic effects by members of this phylum were found.

Phylum 5. Pyrrophyta (Cryptomonads and Dinoflagellates). Several species have been demonstrated to have allelopathic effects (see Chapter 6).

Phylum 6. Chrysophyta (Yellow-green and golden-brown algae and diatoms). Numerous species have allelopathic effects (see Chapter 6).

Phylum 7. Phaeophyta (Brown algae). Several species have allelopathic potential (see Chapter 6).

Phylum 8. Rhodophyta (Red algae). Numerous species have been found to be allelopathic (see Chapter 6).

Phylum 9. Schizomycota (Bacteria). Many species are known to produce antibiotics or marasmins (Chapters 4, 5, 9; Agnihothrudu, 1955; Hattingh and Louw, 1969a,b; Konishi, 1931; Stewart and Brown, 1969; Waksman, 1937,

1947). Anderson *et al.* (1980) found that a bacterium (*Pseudomonas* sp.) which they repeatedly isolated from the caryopses of the grass *Tripsacum dactyloides* inhibited growth of several species of fungi, including *Penicillium chrysogenum, Rhizopus stolonifer,* and *Trichoderma viride.* This inhibition may prevent fungal decomposition of the caryopses of *T. dactyloides,* and this may explain why many caryopses of this species can remain on or in the soil for 18 months or longer before germination without decaying. Extensive literature is available on this subject in relation to medicine, especially in connection with actinomycetes.

Phylum 10. Myxomycota (Slime molds). I know of no references on allelopathic effects of these organisms in relation to agriculture, ecology, or forestry. There may very well be some references, however, relating to medicine.

Phylum 11. Phycomycota (Alga-like fungi). Many of these cause plant diseases and, thus, produce marasmins that have marked effects on host plants (see Chapter 4). Tests of pure cultures of several species of this group have indicated that they are generally much less active in producing antibiotics than other phyla of fungi (Waksman, 1947). Obviously, much more research is needed on this group.

Phylum 12. Ascomycota (Sac fungi). Several genera of this phylum are well known for their production of antibiotics, for example, *Aspergillus* and *Penicillium.* Chaumont and Simeray (1982) tested acqueous extracts of ascocarps of seven species of Ascomycetes against growth of seven species of pathogenic fungi; extracts of two species inhibited growth of one or more test organisms. Many species of Ascomycetes are pathogenic also and therefore probably produce marasmins.

Phylum 13. Basidiomycota (Club-fungi). Marx (1972) listed over 100 species of ectomycorrhizal Basidiomycetes known to produce antibacterial or antifungal antibiotics either in pure culture or in basidiocarps. Wilkins and Harris (1944) made extracts from basidiocarps of more than 700 Basidiomycetes and found that over 24% exhibited antibacterial activity. Sevilla-Santos *et al.* (1964) tested aqueous extracts of sporophores of 587 samples of Philippine Basidiomycetes against several genera and species of bacteria. Tests of 506 extracts inhibited at least one of the test bacteria. Chaumont and Simeray (1982) tested water extracts of sporophores of 218 species of Basidiomycetes against growth of seven species of pathogenic fungi. Extracts of 75 species reduced growth of one or more of the pathogenic fungi. See Chapter 4 for additional information on this phylum.

Bold did not list a phylum for the Imperfect Fungi, but several species of Imperfects product potent antibiotics (Bell, 1977; Stillwell, 1966). Stillwell isolated an Imperfect fungus, *Cryptosporiopsis* sp. from *Betula alleghaniensis* and found in plate culture that this fungus inhibited growth of 31 Basidiomycetes isolated from decaying material in deciduous and coniferous forests, an ascomycete, the Dutch elm disease fungus (*Ceratocystis ulmi*), and a phycomycete (*Phytophthora infestans*). He found also that growth of *Fomes fomentarius,* the fungus most commonly associated with decay in living branches of *Betula alleghaniensis*, was inhibited markedly by the presence of *Cryptosporiopsis* in sterile yellow birch wood.

Phylum 14. Hepatophyta (Liverworts). Hayes (1947) tested water extracts of numerous kinds of plants, including liverworts, against four species of bacteria. Many of the extracts were inhibitory to one or more of the bacteria. Spencer (1979) listed 350 compounds that have been reported from 200 taxa of liverworts. Among these compounds were a great many known to be effective allelopathic agents: phenolics; mono-, di-, sesqui-, and tetraterpenes; quinones; fatty acids; flavonoids; and carboxylic acids.

Phylum 15. Bryophyta (Mosses). Some Arizona State University researchers were intrigued by the fact that mosses did not appear to be attacked by fungi generally, and subsequent research demonstrated that three local mosses (not named) inhibited growth of several taxa of bacteria and fungi (Anonymous, 1963). Schlatterer and Tisdale (1969) reported that litter of *Tortula ruralis* retarded germination and early growth of *Stipa thurberiana, Sitanion hystrix,* and *Agropyron spicatum*. Four weeks after germination, however, growth of the same species was stimulated by the litter. Brown (1967) reported that *Sphagnum capillaceum* inhibited germination of jack pine (*Pinus banksiana*) seeds, and Livingston (1905) stated that water from sphagnum bogs was inhibitory to growth of the alga *Stigeoclonium*. Boiko (1973) reported that water extracts of four of ten moss species inhibited radish seed germination, and extracts of the other six stimulated germination slightly.

Phylum 16. Psilophyta (Whisk ferns). Little research has been reported on allelopathic effects of *Psilotum* and none on *Tmesipteris*. Hayes (1947) tested water extracts of *Psilotum nudum* against four species of bacteria and found little or no effect.

Phylum 17. Microphyllophyta (Club mosses). Hayes (1947) tested water extracts of one species of *Lycopodium* and two of *Selaginella* against four species of bacteria and found little or no effect. Horsley (1977a) reported that the presence of a club moss (*Lycopodium obscurum*) was correlated with reduced numbers of black cherry (*Prunus serotina*) seedlings. Moreover, foliage extracts of club moss inhibited growth of black cherry seedlings that had exhausted cotyledonary reserves.

Phylum 18. Arthrophyta (Horsetails). Aqueous extracts of *Equisetum arvense* had little or no effect against four species of bacteria (Hayes, 1947). Szczepańska (1971) demonstrated that *E. limosum* decreased productivity of *Phragmites communis* and stimulated productivity of bulrush (*Schoenoplectus lacustris*) when cultivated together in lake mud. Stimulation of bulrush suggested a chemical action.

Phylum 19. Pterophyta (Ferns). An appreciable amount of research has been done on allelopathic effects of ferns in forestry (see Chapter 3). Little research has been done on allelopathic effects of ferns in relation to life cycles of these organisms. Davidonis and Ruddat (1973) reported that growth of *Thelypteris normalis* gametophytes was inhibited under *T. normalis* sporophytes. They experimentally eliminated competition, change in pH, and microbial inhibitors as causes of the inhibition, and they subsequently isolated two inhibitors from root exudates of the sporophyte. The inhibitors were designated thelypterin A and B; thelypterin A gave an Ehrlich-positive reaction indicative of a secondary aromatic amine and an ultraviolet absorption spectrum indicative of a heterocyclic ring. Both compounds inhibited growth of gametophytes of *Pteris, Phlebodium,* and *Thelpteris.* Subsequent research indicated that the thelypterin A noncompetitively inhibited auxin-enhanced elongation of *Avena* coleoptiles (Davidonis and Ruddat, 1974). Growth of *T. normalis* root segments was not affected by the thelypterins, but growth of young sporophytes of *T. normalis* was reduced by mature sporophytes.

Bell and Klikoff (1979) found that leachates of rhizomes and roots of *Polypodium vulgare, Polystichum acrostichoides,* and *Onoclea sensibilis* inhibited growth of *O. sensibilis* gametophytes and leachates of *P. vulgare* and *P. acrostichoides* slightly stimulated growth of *P. vulgare* gametophytes. Results in greenhouse tests of gametophytes growing on soil under the sporophytes "agreed fairly closely."

Lellinger (1976) reported that root exudates of the tropical fern *Phlebodium aureum* cv mandianum completely inhibited growth of *Bryophyllum* seedlings when leaves were present on the fern. When the leaves dehisced, however, bryophyllum tripled its height in about 2 weeks. As soon as new leaves appeared on the fern, the bryophyllum plants stopped growing again.

Phylum 20. Cycadophyta (Cycads). No publications were found that relate allelopathic effects to members of this phylum.

Phylum 21. Ginkgophyta (*Ginkgo*). Hayes (1947) tested aqueous extracts of *Ginkgo biloba* against four species of bacteria and reported little or no effect.

Phylum 22. Coniferophyta (Conifers). Allelopathy is very common among many species in this phylum (see Chapter 3).

Phlyum 23. Gnetophyta (*Ephedra, Gnetum, Tumboa*). No publications on allelopathy in this phylum were found.

Phylum 24. Anthophyta (Flowering plants). Most of the research on allelopathy, not related to the production of antibiotics for use in medicine, has involved flowering plants. See most of the other chapters in this book for references.

There are obviously numerous phyla of the plant kingdom about which we know very little or nothing concerning possible allelopathic effects. Hopefully, this brief summary will stimulate further research.

2

Manipulated Ecosystems: Roles of Allelopathy in Agriculture

I. EFFECTS OF WEED INTERFERENCE ON CROP YIELDS

Most of the reported effects of weeds on crop yields have been labeled "competitive," although no evidence was provided generally to indicate whether the effects were due to competition, allelopathy, or both (interference). First, the discussion will concentrate on the interference on crop yields and the allelopathic effects of several important weeds that have been demonstrated to have allelopathic potential and second on the demonstrated allelopathic potential of many additional weeds for which specific data on effects on crop yields are not readily available. Many publications on effects of weeds on crop yields have involved several weeds simultaneously, but I have selected only examples based on individual species of weeds.

Velvetleaf (*Abutilon theophrasti*) has become a serious threat to several crops in the United States, such as cotton and soybeans. Holm *et al.* (1979) listed it as a principal weed in Canada and a common weed in the United States. Several scientists have determined the specific effects of this weed on the yield of cotton (*Gossypium hirsutum*) and soybean (*Glycine max*) (Table 1). Average yield reductions of soybeans under a variety of velvetleaf densities, placements, and durations of interference ranged from 14 to 41%. Reductions in cotton yields ranged from 44 to 100%. All researchers cited in Table 1 attributed reductions in crop yields to competition, although none performed experiments to determine whether the effects were due to competition, as defined in Chapter 1. The term competition may have been used in a broad sense, equivalent to the term interference as defined by Muller (1969), but this was not indicated in any of these papers or others cited elsewhere, which attributed yield reductions to competition.

TABLE 1. Interference of Velvetleaf on Crop Yields

Years (no.)	State	Crop	Weed placement	Yield (kg/ha) Weed free	Weed free	Weedy[a]	Reduction (%)	Reference
3	IA	Soybeans	Row		2384	1645	31	Staniforth (1965)
2	KS	Soybeans	Combination of several		3110	2390	23	Eaton et al. (1976)
2	MS	Seed cotton	Row		2260	0	100	Robinson (1976)
2	MS	Seed cotton	Between rows		2260	895	60	Robinson (1976)
2	MS	Seed cotton	10 cm from rows		2000	1120[b]	44	Chandler (1977)
2	AR	Soybean	Row (30 cm)		2560	1890	26	Oliver (1979)
2	AR	Soybean	Row (61 cm)		2560	2190	14	Oliver (1979)
2	IN	Soybean	Broadcast		3601	2117[b]	41	Hagood et al. (1980)

[a] Full season.
[b] Average of several densities.

Duration of interference of velvetleaf with the crop plants and placement and density had significant effects on yield (Eaton *et al.*, 1976; Robinson, 1976; Chandler, 1977; Oliver, 1979; Hagood *et al.*, 1980). However, these results do not in any way demonstrate that interference of the weed was due to competition, because similar results would be expected if the yield reductions were due to allelopathy. Let us look next, therefore, at the evidence concerning the allelopathic potential of velvetleaf. Gressel and Holm (1964) found that seeds of this species markedly inhibited germination of seeds of several crop plants. Elmore (1980) reported additional evidence of this effect. Colton and Einhellig (1980) reported that several dilutions of aqueous extracts from fresh field-collected leaves depressed germination of radish seedlings and inhibited growth of soybean seedlings. Eluates from chromatograms of the aqueous extract showed that two of three bands containing phenolic compounds were inhibitory to radish seed germination.

The best evidence, however, of allelopathic potential of velvetleaf, in terms of possible crop yield interference, was furnished by Bhowmik and Doll (1979, 1982). They found that water extracts of velvetleaf residues were slightly allelopathic (5–24% inhibition) to radicle and coleoptile growth of corn (*Zea mays*) and to hypocotyl growth of soybeans (Table 2). Residues were highly allelopathic (50% or more inhibition) to height growth and fresh weight increase of shoots of both corn and soybeans in double-pot experiments. Three soil textures (sand; sand plus soil, 50:50; and silt loam soil) and three methods of watering (double-pot, subsurface, and surface) were used. Maximum growth inhibition occurred under the double-pot method of watering followed by the subsurface and surface methods. Growth inhibition was greatest in sand, followed by the sand plus soil mixture, and by silt loam soil.

Obviously, part of the interference of velvetleaf on crop yield could have been

TABLE 2. Allelopathic Effects of Water Extracts of Residues (Radicle and Coleoptile or Hypocotyl Growth) and Residues (1% w/w) of Velvetleaf on Growth of Corn and Soybeans (28 Days)[a,b]

	Crop growth			Height		Shoot (fresh wt)	
				Double pot	Surface	Double pot	Surface
Radicle	Coleoptile	Hypocotyl					
Corn							
S	S	—		H	S	H	S
Soybeans							
—	—	S		H	M	H	H

  [a] Data from Bhowmik and Doll (1979).

  [b] H, highly allelopathic, 50% or more inhibition; M, moderately allelopathic, 24–49% inhibition; S, slightly allelopathic, 5–24%.

TABLE 3.   Interference of Quackgrass on Crop Yields

Years (no.)	State	Crop	Yield (kg/ha) Weed free	Weedy	Reduction (%)	Reference
2	WI	Corn (ear)	6467[a]	647	90	Bandeen and Buchholtz (1967)
2	MI	Potatoes	21,000[b]	14273	32	Islieb 1960

[a] One year for weed free in corn.
[b] Almost free of quackgrass. Controlled by TCA.

due to allelopathy, and this possibility needs to be investigated. Techniques such as those used by Bell and Koeppe (1972) are useful as a partial approach to this investigation, but new techniques need to be developed before such a study could be completed satisfactorily.

Quackgrass (*Agropyron repens*) is an important weed of field crops in many parts of the world and is listed as a serious weed in the United States, as well as in several other countries (Holm *et al.*, 1979). There are many publications on this weed but only a few that clearly indicate its effect on crop yields (Table 3). It is obvious that interference by this species causes serious decreases in corn and potato yields. Buchholtz (1971) showed that a small part of the decrease in corn yield due to quackgrass could be overcome by adding nitrogen and potassium to the infested areas, but not all (Table 4). With heavy applications, only relatively small amounts of the added minerals were taken up by quackgrass (Table 5). He inferred that quackgrass interfered in some way with the uptake of minerals by corn plants, and particularly with nitrogen and potassium. Competition for water was not a factor.

TABLE 4.   Influence of Fertilization on Quackgrass Interference with Corn Yield[a]

Fertilizer (kg/ha) N	K_2O	Grain yield (kg/ha) Without quackgrass	With quackgrass	Reduction (%)
0	0	5900	650	89
220	0	6270	569	91
0	220	6920	894	87
220	220	7570	2520	67

[a] Based on Buchholtz (1971). Reproduced from "Biochemical Interactions among Plants," National Academy Press, Washington, D.C.

TABLE 5.   Uptake of K and N by Quackgrass in Fertilized Plots[a]

Fertilizer application (kg/ha)		Mineral uptake (kg/ha)	
N	K_2O	N	K_2O
120	120	125	65
360	360	148	81
600	600	124	81

[a] From Buchholtz (1971). Reproduced from "Biochemical Interactions among Plants," National Academy Press, Washington, D.C.

Minar (1974) reported that the production of fresh and dry matter of wheat tops was reduced by quackgrass. Addition of fertilizer could not overcome the inhibition completely, even when other possible competitive mechanisms were eliminated. Subsequent work indicated that quackgrass chiefly reduced P-uptake by wheat, even when available P was present in adequate amounts.

It is clear from the work of Minar and Buchholtz that competition for minerals is involved in the interference of quackgrass on crop yield. It appears however, that something else is involved also. As early as 1939, Ahlgren and Aamodt suggested that quackgrass may produce toxins which interfere with growth of crop plants. Le Tourneau and Heggeness (1957) reported that aqueous extracts of quackgrass rhizomes inhibited root growth of pea and wheat seedlings. Kommedahl et al. (1959) reported that stands and dry weights per plant of alfalfa, flax, wheat, oats and barley in field plots were reduced when planted in soil previously infested with quackgrass. Fresh and dry weights of alfalfa were less when ground rhizomes of quackgrass, or water extracts of the rhizomes, were added to soil. When leachate from pots of growing quackgrass was added to soil in which wheat was planted, stands and growth were reduced compared with results when leachate from oat soils was added to the soil.

Ohman and Kommedahl (1960) reported that hot water extracts of roots, stems, rhizomes, and leaves reduced height growth of alfalfa seedlings by 65 to 80%. They reported also that extracts of soil in which quackgrass previously grew were more inhibitory to growth of alfalfa than were extracts of soil in which Canada thistle grew or where no plants had grown. In a subsequent publication (1964), however, they stated that "No evidence was obtained for a phytotoxic substance from living quackgrass roots or rhizomes that could inhibit germination or growth of alfalfa or oats." They pointed out that large amounts of quackgrass residues caused chlorosis and stunting in alfalfa and oats when added to soils, but they obtained similar results with equivalent amounts of cellulose.

Thus, they concluded that the plant responses resulted from the high C:N ratio of the added material which caused a temporary deficiency of available nitrogen. Unfortunately, the only amount of cellulose added was 15 g/400 g of soil, which is a large amount. The likelihood is much greater that significant differences would have resulted if only 1 or 2 g of rhizomes or of cellulose had been added per 400 g of soil.

Harvey and Linscott (1978) found that ethylene was generated in the presence of quackgrass rhizomes, resulting in concentrations as high as 6.4 μl/liter when rhizomes were incubated in sealed vials and 14.3 μl/liter in waterlogged soil. These levels exceeded those previously shown to inhibit growth of crop plants. Ethylene production was reduced following sterilization of rhizomes and soil, indicating that microorganisms were responsible. These authors concluded that ethylene production from quackgrass tissue due to microbial activity in soil may be partially responsible for the "alleged" allelopathic effects of the weed.

Toai and Linscott (1979) found that extracts of soil to which rhizomes and roots, or leaves, of quackgrass were added and allowed to decay greatly reduced growth of alfalfa seedlings. The time of greatest toxicity and duration of the toxicity depended on the temperature of incubation.

Gabor and Veatch (1981) isolated and purified a phytotoxin from a methanol:water extract of dried and ground rhizomes of quackgrass. The substance was golden-brown, crystalline, soluble in water, partly soluble in semipolar solvents, and insoluble in nonpolar solvents. Tests suggested that a single compound was present which contained 42.17% C, 5.06% H, 38.28% O, 4.02% N, 2.03% S, 0.03% Cl, and 0.03% P. Aqueous solutions of the phytotoxin (0.1% w/v) significantly inhibited seedling root growth of corn, oat, cucumber, and alfalfa.

The causes of the interference of quackgrass are clouded even further by the work of Welbank (1961, 1964). In a test of "competition" between this weed and *Impatiens parviflora* in 1961, it was concluded that the "competition" involved both nitrogen and water and that water probably was the more important factor under "normal" conditions. Welbank stated that "There is no need to postulate any toxic root product to account for the experimental results." In a later study of competition for nitrogen and potassium between quackgrass and sugar beet (*Beta vulgaris*), Welbank (1964) concluded that when total dry weights and leaf areas of sugar beets were compared with corresponding nutrient contents, variations in nitrogen content alone could account for most of the effects of competition on growth but that potassium depletion probably contributed a little. Unfortunately, Welbank followed the premise that a decrease in interference due to the addition of a mineral element automatically indicates that the cause of the interference is competition for that element. Actually, as suggested by the work of Buchholtz (1971) and of Minar (1974) discussed previously, such a result could be due to the production of an allelopathic agent by the interfering weed that decreased the uptake of the mineral by associated species, even though the element was present in sufficient amounts in the soil. Danks *et al.* (1975a) found that certain known allelopathic agents significantly decreased uptake of minerals from a nutrient solution.

This discussion of causes of interference by quackgrass illustrates some of the problems that can be encountered in trying to determine the relative effects of competition and allelopathy in the total interference. Such complications should not deter weed scientists, however, from attempting to solve such important problems.

Common milkweed (*Asclepias syriaca*) is a principal weed in Canada and a common weed in the United States and Iraq (Holm *et al.*, 1979). It is increasing in importance in some states, such as Nebraska (Evetts and Burnside, 1973). According to Evetts and Burnside (1972), Evetts (1971) found that this weed caused an average yield loss of 720 kg/ha in sorghum (*Sorghum bicolor*) and that losses were greater with increased densities of the weed. In a 3-year test in Nebraska, Evetts and Burnside (1973) found that milkweed interference decreased grain yield of sorghum by 21%. No tests were made to determine the basis of the interference.

Le Tourneau *et al.* (1956) reported that aqueous extracts of tops of common milkweed (2 g dry wt/100 ml H_2O) markedly inhibited seed germination, coleoptile length, and root length of Mida wheat. Rasmussen and Einhellig (1975) subsequently substantiated the allelopathic effect. They found that aqueous extracts of milkweed leaves significantly inhibited growth of grain sorghum seedlings, and reduced concentrations of the extract resulted in proportional increases in yield. The extracts did not significantly affect growth of corn, however. It appears that at least some of the effect of this weed on crop yield may be due to allelopathy, but more meaningful experiments should be done involving tests other than with plant extracts. Relative effects of competition and allelopathy should be determined, if appropriate techniques can be developed.

Wild oat (*Avena fatua*) is a serious weed in the United States and Canada in addition to six other countries and is a principal or common weed in 23 countries (Holm *et al.*, 1979). In the United States, it is particularly serious in the Northern Plains of the Great Plains Region (Wood, 1953). Interference by wild oats has been shown to reduce yields of several crops markedly (Table 6). Reductions due to high densities of wild oats were much greater than shown in this table because results with all densities were averaged. All the yield reductions were attributed to competition, but no experiments were performed to demonstrate the accuracy of that assumption.

In a 2-year field study in California, Tinnin and Muller (1971) found that wild oat residues prevented germination of certain herbaceous species that normally occur in shrub stands scattered throughout the annual grasslands but not in the grassland proper. They concluded that the inhibition of germination was due to an allelopathic mechanism. Schumacher *et al.* (1982) found that the growth of

TABLE 6.   Effects on Crop Yields of Interference by Wild Oats

Years (no.)	Location	Crop	Yield (kg/ha)		Reduction (%)	Reference
			Weed free	Weedy[a]		
1	Alberta	Wheat	2165	2002	8	Friesen (1961)
3	ND	Flax	1489	616[b]	59	Bell and Nalewaja (1968a)
3	ND	Wheat	1407	1039[b]	26	Bell and Nalewaja (1968b)
3	ND	Barley	1625	1261[b]	22	Bell and Nalewaja (1968b)
3	ND	Flax	1018	336[b]	67	Bell and Nalewaja (1968c)

[a] Full season.
[b] Average of yields with several weed densities.

leaves and roots of spring wheat were significantly reduced by root exudates of wild oat plants at the two through four leaf stages of development of the wild oats. Two allelopathic compounds were isolated from the root exudates, and various tests indicated that they had several characteristics typical of scopoletin and vanillic acid. Here again the relative effects of competition and allelopathy in the interference of wild oats need to be determined.

Downy brome (*Bromus tectorum*) is a serious weed in the dryland winter wheat areas of the Pacific Northwest in the United States (Rydrych, 1974). It is a serious weed of crops in Turkey and a principal weed in the United States and Canada in addition to three other countries (Holm et al., 1979). This weed markedly reduces crop yield, with reductions averaging 40% or more with all densities (Table 7). Rydrych and Muzik (1968) reported reductions in wheat yields up to 92% at some sites with high densities of downy brome. All these results were attributed to competition even though no experiments were run to test this hypothesis.

Rice (1964) found that aqueous extracts (10 g fresh weight/100 ml deionized H_2O) of the entire downy brome plant, as well as similar extracts of individual organs, were inhibitory to growth of nitrogen-fixing and nitrifying bacteria. Thus, downy brome has allelopathic potential, but apparently no one has determined whether it has allelopathic effects on higher plants. Obviously, this should be done and, additionally, the relative contributions of competition and allelopathy to total interference of downy brome on specific crop yields should be determined.

Lambsquarters (*Chenopodium album*) is one of the most important weeds in sugar beet fields in the Pacific Northwest of the United States (Dawson, 1965). It

TABLE 7.   Effects of Interference by Downy Brome on Crop Yields

			Yield (kg/ha)[a]			
Years (no.)	State	Crop	Weed free	Weedy	Reduction (%)	Reference
3	WA	Wheat	3137	1517[b]	52	Rydrych and Muzik (1968)
3	OR	Wheat	3570	2150[b]	40	Rydrych (1974)
1	IL	Alfalfa	2614[c]	839	68	Kapusta and Strieker (1975)

[a] Average of plots at all locations.

[b] Average of results with all weed densities.

[c] Average of all chemically treated plots in which the amount of downy brome was less than 100 kg/ha. Amount in weedy plot was 3984 kg/ha.

is a serious weed in many other parts of the United States, as well as in fourteen other countries (Holm *et al.*, 1979). It is a common or principal weed in 34 additional countries. Dawson (1965) reported that lambsquarters left uncontrolled in sugar beet rows all season in Washington reduced sugar beet yields as much as 94%. Significant reductions in sucrose content also resulted when root yields were reduced by 63% or more. This interference was attributed to competition even though no experiments were performed on effects of this weed on water, light, or minerals. Williams (1964) found that continuous interference by lambsquarters reduced the dry weight yield of kale (*Brassica oleracea*) by 99.8%.

As early as 1956, Le Tourneau *et al.* reported that aqueous extracts of lambsquarters (2 g dry tops/100 ml H_2O) markedly reduced germination, coleoptile length, and root length of Mida wheat. Rice (1964) reported that water extracts of lambsquarters were slightly inhibitory to growth of nitrogen-fixing and nitrifying bacteria. Kossanel *et al.* (1977) found that root exudates of this weed in culture solutions retarded growth of radicles of corn. Water solutions in which lambsquarters previously grew and water extracts of its roots also inihibited growth of corn roots. Additionally, Bhowmik and Doll (1979) found that water extracts of residues of lambsquarters inhibited root and coleoptile growth of corn and hypocotyl growth of soybeans. Moreover, residues incorporated in three different soils were highly allelopathic to height growth and fresh weights of shoots of both corn and soybeans. There is an obvious need for more research on interference effects of this weed on crop production.

Canada thistle (*Cirsium arvense*) is considered a serious weed in the United States and in four additional countries (Holm *et al.*, 1979). It is a common or principal weed in 21 other countries. Hodgson (1958) reported a 40% reduction in wheat yield over a 3-year period in Montana due to interference from Canada thistle. This figure was based on his data for the years 1954–1956 and compared yields in 2,4-D sprayed plots (low weed population) with those in nonsprayed plots (high weed population). Addition of nitrogen fertilizer to weedy plots increased wheat yields only very slightly. No additional data on interference of this weed on crop yield could be found.

Bendall (1975) reported that water extracts of the roots (1 g fresh material: 4 ml H_2O) of Canada thistle reduced germination of subterranean clover (*Trifolium subterraneum*) seed by 87%, and similar extracts of the foliage reduced germination by 14%. Water extracts of the roots reduced radicle growth of seven different plant species by amounts up to 84% in 72 hours, and foliage extracts reduced radicle growth up to 82%. Root residues (3% w/w) reduced the dry-weight increment (4 weeks growth) of seedlings of six species by 20 to 64%, and foliage residues (3% w/w) reduced the weight by 24 to 64%. Root or foliage residues added to soil (2.5% w/w) significantly reduced growth of *Setaria viridis* and *Amaranthus retroflexus* even with added nutrients (Table 8).

Wilson's (1981) results strongly supported those of Bendall. Field examinations showed that as Canada thistle shoots increased in area, the number of kochia (*Kochia scoparia*), *Iva xanthifolia*, and *Hordeum jubatum* plants decreased. In greenhouse studies, roots and shoots of Canada thistle mixed with soil reduced growth of sugar beet (Mono Hy D2), wheat (*Triticum aestivum*— Centurk), alfalfa (*Medicago sativa*—Dawson), and Canada thistle seedlings. Corn (Jacques No. 1004) and dry edible beans (*Phaseolus vulgaris*—Great Northern No. 59) were affected to a lesser extent. Crop growth was inversely proportional to the amount of Canada thistle residue added to the soil. The effects of the residue lasted for about 60 days, and neither autoclaving of the residue and soil nor fertilization of the soil had any effect on residue toxicity. Additionally, the growth of sugar beets watered daily with leaf leachate of Canada thistle was

TABLE 8.   Effects of Canada Thistle Residues on Plant Growth in Greenhouse Soil[a]

| | Fresh weights as percentage of control[b] | | | |
| | Canada thistle roots | | Canada thistle foliage | |
Test species	+ Nutrients	− Nutrients	+ Nutrients	− Nutrients
Setaria viridis	20[c]	96	51[c]	65[c]
Amaranthus retroflexus	9[c]	24[c]	28[c]	39[c]

[a] Data from Stachon and Zimdahl (1980).

[b] Each plant material and soil type had a control with and without added nutrients. Added nutrients consisted of one-half strength Hoagland's solution every 3 days.

[c] Significantly different from respective controls at 0.05 level or better.

decreased. Here again it would be worthwhile to determine the relative effects of allelopathy and competition in the total interference of Canada thistle on crop yield.

Yellow nutsedge (*Cyperus esculentus*) is a serious weed in the United States and Canada along with eight other countries (Holm *et al.*, 1979). It is a common or principal weed in eight additional countries, and according to Stoller *et al.* (1979), this weed seems to be increasing in crops in the United States. Interference by yellow nutsedge markedly reduces crop yield according to results of several scientists (Table 9). These researchers attributed the interference effects to competition without sufficient evidence to support this hypothesis.

In 1973, Tames *et al.* found that the tubers of yellow nutsedge contained compounds that inhibited the growth of oat coleoptiles and the germination of seeds of seven crop species. They identified *p*-hydroxybenzoic, vanillic, syringic, ferulic, and *p*-coumaric acids in the extracts, plus four other active compounds that were not identified. Drost and Doll (1980) found that foliage residues of yellow nutsedge were very inhibitory to root and shoot growth of corn and soybeans (Table 10). Moreover, they reported that water extracts of the tubers reduced growth of soybean roots and shoots. In general, soybeans seemed to be affected considerably more than corn by the allelopathic compounds produced by yellow nutsedge.

Purple nutsedge (*Cyperus rotundus*), listed by Holm (1969) as one of the ten worst weeds in the world, is a serious weed in 52 countries, including the United States and is a principal or common weed in 22 additional countries. Interference by purple nutsedge adversely influences crop yields (Table 11). Reductions in yields in the investigations cited ranged from 23 to 89%, even on the basis of the average effects of various weed densities. Reductions in yield of rice were as high as 38%, for example, in the high purple nutsedge densities. The percentage of reduction in yield of rice due to interference by purple nutsedge was greater with added nitrogen (Okafor and DeDatta, 1976). These researchers found that

TABLE 9.   Effects of Interference of Yellow Nutsedge on Crop Yields

Years (no.)	State	Crop	Weed free	Weedy[a]	Reduction (%)	Reference
			Yield (kg/ha)			
2	CA	Seed cotton	2860	1880	34	Keeley and Thullen (1975)
3	IL	Corn	9130	5360	41	Stoller et al. (1979)
3	AL	Seed cotton	1972	1344	32	Patterson et al. (1980)

[a] Full season.

TABLE 10. Effects of Residues of Yellow Nutsedge on Growth of Corn and Soybeans as Percent of Untreated Control[a]

Nutsedge residue (% w/w)[b]	Soybeans		Corn	
	Shoots	Roots	Shoots	Roots
Foliage residues				
0.125	95	55[c]	93	91
0.250	78[c]	43[c]	93	111
0.500	65[c]	44[c]	81[c]	91
Tuber residues				
0.125	77	61[c]	58[c]	76[c]
0.250	60[c]	46[c]	65[c]	66[c]
0.500	58[c]	45[c]	56[c]	68[c]

[a] Modified from Drost and Doll (1980).
[b] Percentage based on 1500 g silica sand.
[c] Significantly different from control at 95% confidence level.

TABLE 11. Effects of Interference by Purple Nutsedge on Crop Yields

Years (no.)	Place	Crop	Yield (kg/ha)		Reduction (%)	Reference
			Weed free	Weedy		
2	Brazil	Garlic (bulbs)	1060	120	89	William and Warren (1975)
2	Brazil	Okra	25800	9800	62	William and Warren (1975)
2	Brazil	Carrots-Nantes	41400	20600	50	William and Warren (1975)
2	Brazil	Carrots-Kuroda	14400	8800	39	William and Warren (1975)
2	Brazil	Greenbean	6400	3800	41	William and Warren (1975)
2	Brazil	Cucumber	39200	22400	43	William and Warren (1975)
2	Brazil	Cabbage	13000	8400	35	William and Warren (1975)
2	Brazil	Tomato	34400	16200	53	William and Warren (1975)
3[a]	Philippines	Rice	1433[b]	1108[b,d]	23	Okafor and DeDatta (1976)
3[a]	Philippines	Rice	3867[b]	2783[c,d]	28	Okafor and DeDatta (1976)

[a] Three crops in 2 years. [b]No added nitrogen. [c]120 kg/ha of nitrogen added. [d]Average of yields with several weed densities.

this weed competed extensively with rice for moisture. They also found that purple nutsedge reduced the light transmission ratio (ratio of light intensity incident at a height 10 cm above the ground level within the crop row to that incident at 10 cm above the crop) at the panicle initiation stage of rice. The reduction was proportional to the increase in the purple nutsedge density. Competition obviously accounted, therefore, for some of the interference effects.

Singh (1968) tested water extracts of tubers of purple nutsedge on seed germination and seedling growth of 10 species of crop plants. Inhibition of seed germination varied from 0 to about 65% compared with water controls. Only two species showed no effect. Seedling growth (length) was inhibited in all species with a maximum of about 85%. Friedman and Horowitz (1971) reported that soil previously infested with purple nutsedge for 9 to 12 weeks significantly reduced germination of mustard (*Brassica nigra*), barley, and cotton seeds, and soil infested for only 6 weeks significantly reduced germination of mustard and cotton seeds. Ethanol extracts of soil previously infested with purple nutsedge also significantly inhibited radicle growth of barley. Horowitz and Friedman (1971) air-dried tubers of purple nutsedge and mixed them with soil (clay or sand) at rates of 0.1, 0.5, 1.0, and 2.0% (w/w). After 1, 2, or 3 months of incubation at 25°C with the soil kept moist but not water-logged, undecomposed plant material was removed, and the soil was tested directly against growth of barley or was extracted with ethanol. In the latter case the extract was tested against the growth of barley. Both the soil and extract of soil significantly reduced root and top growth of barley in many of the tests. The inhibition was generally proportional to the concentration of plant material originally added to the soil and was greater in sandy soil than in clay soil.

Lucena and Doll (1976) investigated both the competitive and allelopathic effects of purple nutsedge on sorghum and soybeans. They found that addition of tubers to soil in concentrations (w/w) found in field plots in Colombia (South America) caused pronounced growth inhibition of both crop plants. Top residues had very little effect, however. In experiments with living purple nutsedge plants and sorghum, or soybean, the purple nutsedge significantly inhibited the height and fresh weight increment of the crop plants and thus the total amounts of N, P, and K absorbed by the crops. The percentages of N, P, and K in the sorghum and soybean plants were not altered, however, in comparison with control crop plants growing alone.

Meissner *et al.* (1979) grew purple nutsedge (called red nutgrass by them) in sterilized, well-fertilized sandy loam soil for 2 to 3 months at a mean soil temperature of either 23° or 30°C, corresponding to that of the temperate and warm season, respectively, at Pretoria, South Africa. The soil was then sifted to remove plant material, and its effects on crop growth were compared with those of uninfested soil of the same origin. Growth of barley (cv Elses), cucumber (*cucumis sativus* cv Fletcher), and tomato (*Lycopersicum esculentum* cv Rooi

Khaki) was reduced considerably in warm season soil, but no effect was observed when grown in the temperature season soil. Subsequent experiments demonstrated that growth of the same crop plants, plus that of sorghum (cv NK 283) and onion (*Allium cepa* cv De Wildt), was reduced in the previously infested warm season soil and the water economy of all species was also impaired.

Meissner *et al.* (1979) reported also that water extracts of tubers of purple nutsedge markedly reduced the survival of radish (*Raphanus sativus* cv White Icicle), onion, and tomato seedlings but not cucumber seedlings.

Yield of strawberry (*Fragaria* spp.) fruits was reduced considerably when grown together with purple nutsedge, and the strawberries stopped producing fruits 1 month earlier than the controls, in spite of the fact that all tops of nutsedge were cut back at short intervals to prevent shading. Moreover, the plants were regularly watered and fertilized. Production of runners by strawberries was markedly reduced by the nutsedge also. Much work still needs to be done to determine the relative effects of competition and allelopathy on the interference of purple nutsedge on crop yield.

Crabgrass (*Digitaria sanguinalis*) is a serious weed in 23 countries including the United States (Holm *et al.*, 1979). It is a common or principal weed in 16 additional countries. The interference of crabgrass on crop yields can be pronounced as indicated by its effects on cotton and peanuts (Table 12), in which average yield reductions ranged from 42 to 95%. Robinson (1976) reported no yield of seed cotton in 2 out of 3 years when crabgrass was allowed to remain in the rows. The effects were attributed to competition without evidence to substantiate this hypothesis.

Gressel and Holm (1964) noted that water extracts of crabgrass seeds (1 g/16.7 ml H_2O) inhibited seed germination of several crop plants. Rice (1964) found that water extracts of roots or tops (1 g fresh wt:10 ml H_2O) were inhibito-

TABLE 12.   Effects on Crop Yields of Interference by Crabgrass for Full Season

Years (no.)	State	Crop	Yield (kg/ha)		Reduction (%)	Reference
			Weed free	Weedy		
2	OK	Peanuts (nuts)	2447	120[a]	95	Hill and Santelmann (1969)
3	MS	Seed cotton	2447	573[b]	77	Robinson (1976)
3	MS	Seed cotton	2447	1420[c]	42	Robinson (1976)

[a] Mostly crabgrass, some smooth pigweed (*Amaranthus hybridus*).
[b] Crabgrass in the rows.
[c] Crabgrass between the rows.

ry to growth of certain nitrogen-fixing and nitrifying bacteria. Schreiber and Williams (1967) reported that decaying roots of crabgrass inhibited growth of corn roots in a root divider experiment in which one corn plant was employed in each setup, and 12-12-12 fertilizer was applied at the rate of 200 lb/acre to test and control soil. Parenti and Rice (1969) found that whole plant water extracts (1 g fresh wt:10 ml H_2O) of crabgrass markedly reduced seedling growth of six different plant species. Moreover, root exudates of living crabgrass plants significantly reduced growth of seedlings of four of five species. Decaying whole plant material of crabgrass (1 g:454 g of soil) did not significantly affect growth of five test species in comparison with growth in soil to which peat moss was added in equivalent concentration.

Considerable research needs to be done with crabgrass in order to assess its allelopathic potential and to determine the relative importance of allelopathy and competition in determining total interference on crop yield.

Barnyardgrass (*Echinochloa crusgalli*), listed by Holm (1969) as one of the 10 worst weeds in the world, is a serious weed in 32 countries, a principal weed in 10 countries, and a common weed in 4 countries. Its importance in the United States is clear from the number of papers published on its interference with crop yields (Table 13). Yield reduction ranged from 19 to 79% depending on the crop, environmental conditions, and method and effectiveness of weed control. Wellhausen (1962) pointed out that extensive amounts of virgin land was cleared in Venezuela in the late 1950s and early 1960s and planted with rice or corn. After a few years of continuous rice production, barnyardgrass began to dominate the fields so that little or no grain was produced.

TABLE 13.   Effects of Interference by Barnyardgrass on Crop Yields

Years (no.)	State	Crop	Yield (kg/ha) Weed free	Weedy[a]	Reduction (%)	Reference
1	AR	Rice	4152[b]	1861	55	Smith (1960)
2	WA	Field beans	3024	980	68	Dawson (1964)
1	WA	Sugarbeets	74545	38182	49	Dawson (1965)
4	AR	Rice	5488	1131	79	Smith (1968)
2	CA	Rice	6276[c]	2050	67	Oelke and Morse (1968)
3	AR	Rice	7327[d]	3871[d]	47	Smith (1974)
2	WA	Sugarbeets	65455[e]	52727[e]	19	Dawson (1977)

[a] Entire season.
[b] Good control with chemicals but not entirely weed free.
[c] 100% weed free with chemicals.
[d] Average yields of three rice varieties.
[e] Average yields with three sugar beet spacings.

In investigations cited in the previous paragraph, yield reductions were attributed to competition although there were no data to substantiate this. Gressel and Holm (1964) reported that aqueous extracts of seeds of barnyardgrass (1 g/16.7 ml H_2O) significantly reduced germination of seeds of several crop plants. More recently, Bhowmik and Doll (1979) found that water extracts of barnyardgrass residues inhibited hypocotyl growth of soybeans by 25 to 49%. Moreoever, residues (1% w/w), which were incorporated into three different types of soil, reduced height growth of corn and soybeans by 5 to 24% and reduced the increments in shoot fresh weight by 25 to 49%. Thus, the allelopathic potential of this weed needs also to be considered in any experiments concerned with its interference on crop yields.

Kochia (*Kochia scoparia*) was not considered by Holm *et al.* (1979) to be a serious weed in any country. They did list it as a principal weed in Argentina and as a common weed in the United States and Canada. Weatherspoon and Schweizer (1969) stated that kochia is a very troublesome weed in sugarbeets in the central high plains and intermountain western part of the United States. They reported that "competition" from kochia throughout the growing season over a 2-year period reduced the average root yield of sugar beets at Fort Collins, Colorado, by 98%. Weeding by hand for about 4 weeks after emergence of the sugar beets resulted in very little reduction by interference. As indicated above, these scientists attributed the reduction in yield to competition, but they offered no data to support this hypothesis.

Wali and Iverson (1978) reported that decaying leaves of kochia significantly reduced growth of kochia seedlings in soils, even when the soils were fertilized with various amounts of N, P, and K. Subsequently, Lodhi (1979a) identified five phenolic compounds in kochia leaves and found that each of these inhibited radicle growth of radish and kochia seedlings, even in a $1 \times 10^{-4} M$ aqueous solution. Iverson and Wali (1982) and Einhellig and Schon (1982) gave additional evidence concerning the allelopathic effects of this species. It is evident again that both competition and allelopathy need to be considered in any subsequent work on interference by this weed.

Giant foxtail (*Setaria faberii*) is not an important weed in most countries of the world, but it is listed as a common weed in Japan and the United States (Holm *et al.*, 1979). Knake and Slife (1962) stated that it is one of the most prevalent grass weeds in the Midwest in the United States, and it has become a threat to corn and soybean yields.

The interference by giant foxtail on crop yields is appreciable, but it is still rather low as compared with effects of the other weeds previously mentioned in this chapter (Table 14). Yield reductions were greater, of course, when weed densities were higher. For example, Knake and Slife (1962) reported a reduction in soybean yield of 28% at the highest weed density, whereas the average reduction in all weed densities was only 12%. In later studies, Knake and Slife

TABLE 14. Effects on Crop Yields of Interference by Giant Foxtail

Years (no.)	State	Crop	Yield (kg/ha)			Reference
			Weed free	Weedy[a]	Reduction (%)	
3	IL	Corn	5880	5167[b]	12	Knake and Slife (1962)
3	IL	Soybeans	2591	2291	12	Knake and Slife (1962)
3	IL	Corn	8324	7229	13	Knake and Slife (1965)
3	IL	Soybeans	2591	1882	27	Knake and Slife (1965)
3	IA	Soybeans	2618	1636	38	Staniforth (1965)

[a] Interference for full season.
[b] Average yield in all weed densities.

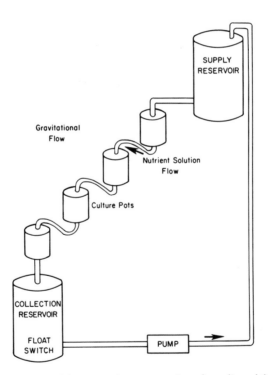

Fig. 1. Stairstep apparatus. Diagrammatic representation of one line of the experimental apparatus used to separate mechanisms of allelopathy from those of competition. These lines contained pots of corn alternating with pots of *Setaria faberii*. Control lines contained only pots of corn. (From Bell and Koeppe, 1972. Reproduced from *Agronomy J.* **64,** 321–325, by permission of the American Society of Agronomy.)

(1965) found that elimination of giant foxtail for 3 weeks after emergence of corn or soybean resulted in yields almost as high as when giant foxtail was eliminated for the entire growing season.

Giant foxtail has been reported to have allelopathic potential, the earliest report being that of Schreiber and Williams (1967). Using the split-root technique previously described for crabgrass, these investigators reported that decaying roots of giant foxtail greatly inhibited root growth of corn. Bell and Koeppe (1972) reported that interference by giant foxtail in greenhouse experiments using mixed culture treatments reduced height growth as well as fresh weight and dry weight increments of corn by as much as 90%. In subsequent experiments, a stair-step apparatus was used so that a culture solution could flow through separate pots of giant foxtail, which were alternated with pots of corn (Fig. 1). Controls consisted of pots containing only corn plants. In some experiments, dead foxtail plants were used and, in others, only giant foxtail residues were incorporated in the sand of the pots alternated with pots of corn. They found that mature living giant foxtail inhibited growth of corn by approximately 35% by an allelopathic mechanism compared with 90% reduction due to both competition and allelopathy. Phytotoxins leached from dead giant foxtail reduced corn growth by as much as 50% (Fig. 2). Thus, it seems that the techniques used by Bell and Koeppe offer a good starting point to separate the allelopathic and competitive components of interference. Additional new techniques need to be developed to effectively separate these components.

Bhowmik and Doll (1979) supported the results of previous investigators concerning the allelopathic potential of giant foxtail. They found that aqueous extracts of its residues reduced radicle and coleoptile growth of corn, and residues incorporated in soil inhibited growth in height and fresh weight of both corn and soybean seedlings.

Holm et al. (1979) listed yellow foxtail (Setaria glauca) as a serious weed in eight countries including the United States, a principal weed in 4 countries, and a common weed in 10 countries. Interference by this weed has been shown to have an appreciable effect on crop yields, particularly of corn, soybeans, and grain sorghum (Table 15). Unlike the results with several other weeds, addition of nitrogen fertilizer caused a lower percentage yield reduction in corn (Staniforth, 1957, 1961). In fact, Staniforth (1957) reported that corn yields were increased 2 to 3 times more than biomass of yellow foxtail by the added nitrogen. According to Staniforth (1958), soybean yield reductions due to yellow foxtail averaged 5% when soil moisture conditions were either adequate for or severely limiting to plant growth. Reductions were also small when moisture supplies were low early in the season and adequate later. Reductions averaged 15%, however, when soil moisture was adequate until late July, but it was severely limiting from that time until the crop matured.

Staniforth's (1957, 1958, 1961) data indicated that competition probably plays a prominent role in the interference of yellow foxtail with crop yields. There is also evidence that this weed has allelopathic potential. Gressel and Holm (1964) reported that water extracts (1 g/16.7 ml H_2O) of seeds of yellow foxtail reduced germination of seeds of several crop plants. Schreiber and Williams (1967) found that decaying roots of this weed significantly reduced root growth of corn in a split-root experiment. Bhowmik and Doll (1979) reported that aqueous extracts of yellow foxtail residues significantly inhibited radicle and coleoptile growth of corn and hypocotyl growth of soybeans. Moreover, decaying residues (1% w/w)

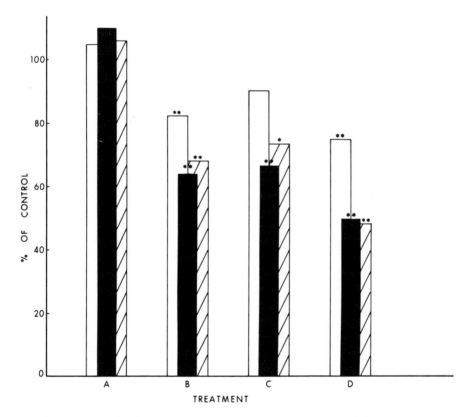

Fig. 2. Allelopathic influence of *Setaria faberii* on *Zea mays*. Height (white bars), fresh weight (black bars), and dry weight (hatched bars) are presented as a percentage of the control after 1 month's association. Treatments included corn seedlings (A) started together with giant foxtail seedlings; (B) growing with mature live giant foxtail; (C) growing with whole dead giant foxtail plants; (D) growing in contact with the material leached from giant foxtail residue that was cut and incorporated into the sand culture pots. (*), Significant difference from control at 0.05 level; (**), significant difference from control at 0.01 level. (From Bell and Koeppe, 1972. Reproduced from *Agronomy J.* **64,** 321–325, by permission of the American Society of Agronomy.)

TABLE 15.   Effects of Full Season Interference by Yellow Foxtail on Crop Yields

Years (no.)	State	Crop	Weed free	Weedy[a]	Reduction (%)	Reference
			Yield (kg/ha)			
2	IA	Corn	4149[b]	3284[b]	21	Staniforth (1957)
2	IA	Corn	6440[c]	5956[c]	7	Staniforth (1957)
2	IA	Soybeans	2073[d]	1936[d]	7	Staniforth (1958)
2	IA	Corn	3309[b]	1425[b]	57	Staniforth (1961)
2	IA	Corn	6847[c]	5091[c]	26	Staniforth (1961)
3	KS	Sorghum grain	4846	3395	30	Feltner et al. (1969)

[a] All season.
[b] No added N fertilizer.
[c] Average yields with two amounts of added N.
[d] Average yields with several moisture conditions.

in soil reduced height growth of corn by 50% or more and fresh weight increase of tops by 25–49%. Such residues reduced height growth of soybeans by 25 to 49%.

Johnsongrass (*Sorghum halepense*), considered to be one of the 10 worst weeds in the world, is a serious weed in 22 countries including the United States, a principal weed in 11 countries, and a common weed in 4 additional countries.

The seriousness of Johnsongrass is indicated by the fact that it caused an average yield reduction of 34 to 87% in four different crops (Table 16). As usual, densities, duration of interference, environmental conditions, and crop varieties determine the actual percentage reduction in yield. McWhorter and Hartwig (1972) reported that interference by Johnsongrass reduced soybean yield by 27 to 42%, depending on the soybean variety tested.

All reductions in crop yields cited in Table 16 were attributed to competition by Johnsongrass without supporting evidence. This is unfortunate because as early as 1967, Abdul-Wahab and Rice reported that aqueous extracts (1 g fresh wt/10 ml H_2O) of Johnsongrass rhizomes, or leaves, markedly inhibited seedling growth of eight species. Moreover, root exudates of Johnsongrass and decaying Johnsongrass residues (1.85 g air-dried leaves and culms, or 1.2 g air-dried rhizomes/454 g soil) significantly reduced growth of several test species.

Bieber and Hoveland (1968) also reported that water extracts of tops, or roots (1 g dry weight/15 ml H_2O), of Johnsongrass reduced seed germination of crownvetch and radicle growth of crownvetch and crimson clover. Horowitz and Friedman (1971) found that decaying underground organs (2% w/w) of Johnsongrass in soil inhibited both root and shoot growth of barley. Ethanolic extracts of soils containing concentrations of decaying underground parts of Johnsongrass

TABLE 16.   Effects of Full Season Interference of Johnsongrass on Crop Yields

Years (no.)	Place	Crop	Yield (kg/ha)		Reduction (%)	Reference
			Weed free	Weedy[a]		
1	Israel	Seed cotton	2972[a]	380	87	Kleifeld (1970)
2	LA	Sugarcane	15455[b]	6818	56	Millhollon (1970)
3	MS	Soybeans	1665[c]	1098[c]	34	McWhorter and Hartwig (1972)
2		Corn	5420[d]	2818[d]	48	Roeth (1973)

[a] Average yield in all plots treated chemically for weed control at all sites.
[b] Average yield in plots that were hand-weeded in 1966 and 1968.
[c] Average yield of six soybean varieties.
[d] Average yield in cultivated and uncultivated plots.

ranging from 0.1 to 2% also retarded radicle growth of barley. Obviously much more work needs to be done to determine the relative contributions of allelopathy and competition to the total interference of Johnsongrass on crop yields.

Other Weeds with Allelopathic Potential

Numerous other weed species that are important in agriculture (Holm *et al.*, 1979), in at least some countries, have allelopathic potential (Table 17). The evidence is strong for some species and weak for others.

Some of the evidence for the allelopathic potential of many of the species of weeds will be presented in this section (see Table 17). For convenience, the species will be discussed in alphabetical order, except where investigations on two or more species were reported in the same paper.

Gajić and her colleagues published a series of papers concerning the allelopathic actions of corn cockle (*Agrostemma githago*) on wheat and of wheat on corn cockle (Gajić, 1966, 1969; Gajić and Vrbaški, 1972; Gajić and Nikočević, 1973; Gajić *et al.*, 1976; Vrbaški *et al.*, 1977). Field tests over a period of years demonstrated that wheat grain yields were increased appreciably when grown in mixed stands with corn cockle as compared to pure stands of wheat. On the other hand, corn cockle yields were decreased in mixed stands with wheat as compared to pure stands of corn cockle. These results occurred on two significantly different soil types—smonitsa and brown soils—and with several wheat varieties. Similar results occurred in fertilized and unfertilized plots. Most subsequent studies addressed the reasons for the stimulating effects of corn cockle on wheat yield and quality. Sterile seedlings of corn cockle stimulated growth of sterile wheat seedlings on agar during the heterotrophic growth phase, thus demonstrating that an excreted metabolite was involved. Several substances

TABLE 17.   Some Additional Weedy Species of Cultivated Fields with Demonstrated Allelopathic Potential

Species	Reference
Agrostemma githago	Gajić (1966)
Amaranthus dubius	Altiera and Doll (1978)
Amaranthus retroflexus	Gressel and Holm (1964), Bhowmik and Doll (1979, 1982)
Ambrosia artemisifolia	Gressel and Holm (1964), Rice (1964, 1965a), Bhowmik and Doll (1979)
Ambrosia cumanensis	Anaya and Del Amo (1978)
Ambrosia psilostachya	Neill and Rice (1971)
Ambrosia trifida	Rasmussen and Einhellig (1979a)
Antennaria microphylla, A. neglecta	Selleck (1972)
Artemisia absinthium	Grümmer (1961)
Berteroa incana	Bhowmik and Doll (1979, 1982)
Boerhavia diffusa	Sen (1976)
Bromus japonicus	Rice (1964)
Calluna vulgaris	Salas and Vieitez (1972)
Camelina alyssum	Grümmer and Beyer (1960)
Camelina sativa	Kranz and Jacob (1977a,b), Lovett and Duffield (1981)
Celosia argentea	Pandya (1975), Ashraf and Sen (1978)
Cenchrus biflorus	Sen (1976)
Cenchrus pauciflorus	Rice (1964)
Centaurea diffusa, C. maculosa, C. repens	Fletcher and Renney (1963)
Cirsium discolor	Le Tourneau *et al.* (1956)
Citrullis colocynthis	Bhandari and Sen (1971)
Citrullis lanatus	Bhandari and Sen (1972)
Cucumis callosus	Sen (1976)
Cynodon dactylon	Horowitz and Friedman (1971)
Daboecia polifolia	Salas and Vieitez (1972)
Digera arvensis	Sarma (1974a)
Eleusine indica	Altiera and Doll (1978)
Erica spp.	Ballester and Vieitez (1971), Salas and Vieitez (1972)
Euphorbia corollata	Rice (1964, 1965a,b)
Euphobia esula	Le Tourneau *et al.* (1956), Le Tourneau and Heggeness (1957), Selleck (1972), Steenhagen and Zimdahl (1979)
Euphorbia supina	Rice (1965b), Brown (1968)
Galium mollugo	Kohlmuenzer (1965a,b)
Helianthus annuus	Rice (1965a), Wilson and Rice (1968)
Imperata cylindrica	Abdul-Wahab and Al-Naib (1972), Eussen (1978), Eussen and Soerjani (1978)

(continued)

TABLE 17. *(Continued)*

Species	Reference
Indigofera cordifolia	Sen (1976)
Iva xanthifolia	Le Tourneau *et al.* (1956)
Lactuca scariola	Rice (1964)
Lepidium virginicum	Bieber and Hoveland (1968)
Leptochloa filiformis	Altiera and doll (1978)
Lolium multiflorum	Naqvi (1972), Naqvi and Muller (1975)
Lychnis alba	Bhowmik and Doll (1979, 1982)
Oenothera biennis	Bieber and Hoveland (1968)
Panicum dichotomiflorum	Bhowmik and Doll (1979, 1982)
Parthenium hysterophorus	Rajan (1973), Kanchan and Jayachandra (1979a,b), Dube *et al.* (1979)
Plantago purshii	Rice (1964)
Polygonum aviculare	AlSaadawi and Rice (1982a,b)
Polygonum orientale	Datta and Chatterjee (1978, 1980a,b)
Polygonum pennsylvanicum	Le Tourneau *et al.* (1956), Gressel and Holm (1964)
Polygonum persicaria	Martin and Rademacher (1960)
Portulaca oleracea	Le Tourneau *et al.* (1956), Gressel and Holm (1964)
Rumex crispus	Einhellig and Rasmussen (1973)
Salsoli kali	Lodhi (1979b)
Salvadora oleoides	Mohnot and Soni (1976)
Schinus molle	Anaya and Gómez-Pompa (1971)
Setaria viridis	Rice (1964), Bhowmik and Doll (1979)
Solanum surattense	Sharma and Sen (1971), Mohnot and Soni (1977)
Tagetes patula	Altiera and Doll (1978)
Trichodesma amplexicaule	Sen (1976)
Xanthium pennsylvanicum	Rice (1964)

that were stimulatory to the growth of wheat seedlings were isolated from corn cockle seeds, including agrostemmin, allantoin, and a gibberellin.

Gajić *et al.* (1976) reported that application of agrostemmin to wheat fields at the rate of 1.2 g/ha increased grain yields of wheat on both fertilized and unfertilized areas. Moreover, the free tryptophan content of the grains was increased, and the quality of the resulting flour and bread were improved. Thus, the evidence is substantial that the stimulating effect of corn cockle on wheat yield in mixed stands is an allelopathic effect. The amount of agrostemmin necessary to increase wheat yield is phenomenally small.

Altiera and Doll (1978) investigated the effects of decomposing residues of *Amaranthus dubius, Eleusine indica, Leptochloa filiformis,* and *Tagetes patula* on seed germination of corn (maize), beans, sorghum, and several weedy species. Chopped fresh shoots (20 g) of one of the species were mixed in the top 5 cm of soil (clayey, pH 6.8; organic matter, 3.5%) in each 1.85 liter pot, and 10 seeds of a given crop plant or 50 weed seeds were planted immediately. Apparently two pots were used for each crop or weed and for each residue. The number of emerged seedlings was counted 20 days later. *A. dubius* and *T. patula* significantly inhibited germination of seeds of one or more of the crop plants, and all four weeds significantly inhibited germination of several weedy species.

Bhowmik and Doll (1979) observed that water extracts of residues of *Amaranthus retroflexus* (redroot pigweed), *Ambrosia artemisiifolia* (common ragweed), *Berteroa incana* (hoary alyssum), *Lychnis alba* (white cockle), *Panicum dichotomiflorum* (fall panicum), and *Setaria viridis* (green foxtail) inhibited growth of one or more of the following: radicle growth of corn, coleoptile growth of corn, or hypocotyl growth of soybean. Results were more striking, however, when residues of the weeds were mixed in sand (1% w/w). Corn and soybean plants were grown for 28 days after which height and shoot fresh weight were determined. The height growth and fresh weight increase of corn and soybean shoots and roots were inhibited by all weed species.

In a separate experiment on effects of the weed residues, Bhowmik and Doll (1979) investigated the effects of three soil textures—sand, sand plus soil (50:50), and silt loam soil—and of three methods of watering—double-pot, subsurface, and surface. Growth inhibition was always greatest with the double-pot method of watering, followed by the subsurface and surface methods. Growth inhibition was greatest in sand, followed by the sand plus soil mixture and the silt loam soil respectively. In general, soybean growth was inhibited more than that of corn.

Soil previously in contact with *Ambrosia psilostachya* (western ragweed) roots, decaying leaves of this weed, rain-drip (leaf leachate), and root exudate reduced the growth of some of the 10 test weedy species, stimulated some, and had no effect on others (Neill and Rice, 1971). Volatiles from the leaves of western ragweed markedly inhibited elongation of the first leaf of *Bromus japonicus* and hypocotyl growth of *Amaranthus retroflexus*.

Giant ragweed (*Ambrosia trifida*), a common invader of waste areas and crop land of much of the United States, often exists in nearly pure stands. Rasmussen and Einhellig (1979a) reported that aqueous extracts of giant ragweed inhibited seed germination and seedling growth of radish and sorghum. The osmotic potentials of the dilutions used were less than −1 bar, an osmotic condition found to have no effect on the test species. Five phytotoxins were isolated and characterized but not identified, and all of them reduced seed germination of radish and sorghum.

Selleck (1972) tested water extracts of *Antennaria microphylla* (small everlasting), *Antennaria neglecta* (field pussytoes), and *Euphorbia esula* (leafy spurge) and water extracts of soil in contact with roots of these species against seed germination and seedling growth of wheat (Thatcher), three weedy species, and one prairiegrass. Most extracts of all species inhibited germination and seedling growth of the test species. Extracts of soil in contact with roots of leafy spurge or field pussytoes reduced root growth of all test species and stem growth of most. Extracts of soil in contact with small everlasting retarded root growth of three of the test species, including wheat, but inhibited stem growth of only one species. Unfortunately, no tests were made to determine if osmotic effects of the extracts might have been involved.

In field experiments, Selleck reported that soil in which small everlasting had previously grown was markedly inhibitory to growth of the usually vigorous, deep-rooted perennial leafy spurge. He found that small everlasting suppressed growth of leafy spurge markedly in pots in the greenhouse, even though small everlasting is a much smaller, shallow-rooted plant. He concluded that small everlasting, field pussytoes, and leafy spurge all have considerable allelopathic potential.

Sen (1976) investigated the effects of aqueous and several organic solvent extracts of five important weedy species from the Jodhpur area of India on seedling growth of two important crop plants—til (*Sesamum indicum*) and bajra (*Pennisetum americanum*). The species tested for allelopathic potential were *Boerhavia diffusa, Cenchrus biflorus, Cumcumis callosus, Indigofera cordifolia,* and *Trichodesma amplexicaule.* At least some extracts of all five weedy species affected growth of seedlings of bajra, til, or both. In most tests, water extracts were the most stringent inhibitors, followed closely by methanol extracts. Unfortunately, no evaluations were made of osmotic effects, and no tests were made to determine if any of the inhibitory materials were released by plants that produced them.

It has been known for many years that yield of flax is greatly reduced when even a small percentage of flax weed, *Camelina alyssum,* is present among the flax plants. Grümmer and Beyer (1960) found no toxic root excretions, but they reported that the leaves are the source of potent plant inhibitors. When using artificial rain, flax plants in close proximity to *Camelina* plants produced 40% less dry matter than the controls, which received the same amount of water added directly to the soil instead of being allowed to drip from the plants. The experimental setup was such that no competition was involved. There is some question as to whether Grümmer and Beyer used *Camelina alyssum,* because Holm *et al.* (1979) listed this as a common weed only in a small part of South America.

Camelina sativa is a common weed of flax in Europe. Kranz and Jacob (1977a,b) investigated the effects of this weed on sulfur (^{35}S), phosphorus (^{32}P), and rubidium (^{86}Rb) uptake by flax (*Linum usitatissimum*). They reported that flax plants took up fewer ^{35}S, ^{32}P, and ^{86}Rb ions in mixed culture with *Camelina* than when growing in monoculture. On the other hand, *Camelina* took up more of all these ions in mixed culture than in monoculture. They stated: "These findings support the opinion that the diminution of the dry weight of *Linum* under the influence of *Camelina* is caused decisively by competition and not by allelopathic factors.'' However, they pointed out (1977a) that participation of allelopathic factors cannot be excluded because a diminished absorption of sulfate also took place in flax plants growing in a nutrient solution in which *Camelina* plants had previously grown but which had been replenished with minerals. This agrees with the observation that allelopathic agents have been repeatedly demonstrated to decrease the uptake of minerals by plants. Hence, absorption of minerals is not a valid measure of ''competitive'' effects, but only of interference.

Lovett and Sagar (1978) reported that aqueous washings of the leaves of *Camelina sativa* consistently stimulated the growth of radicles of flax seedlings. They found, however, that a free-living nitrogen-fixing bacterium had to be present in the *Camelina* phyllosphere for the stimulation to occur. Lovett and Jackson (1980) demonstrated that leaf washings of *C. sativa* were stimulatory to seedling radicles of *C. sativa*, flax, and wheat (cv Songlen). Sterile leaf washings had no effect. Inoculation of sterile leaf washings of *C. sativa* with bacteria cultured from previous leaf washings of this species caused the previously ineffective sterile washings to stimulate growth of flax radicles. However, inoculation of sterile leaf washings of *Helianthus annuus*, *Brassica napus*, or *Salvia reflexa* with the same bacterial cultures from *C. sativa* did not result in stimulation of flax radicles. Thus, an appropriate substrate present on leaves of *C. sativa*, and perhaps close relatives, appeared to be necessary for production of the allelopathic substance.

Lovett and Duffield (1981) identified benzylamine in nonsterile leaf washings of *C. sativa* and demonstrated that concentrations of this compound up to 100 ppm stimulated radicle elongation of flax, whereas concentrations above 200 ppm inhibited growth. They also identified benzyl isothiocyanate in leaves of *C. sativa*, and bacteria similar to those identified in the phyllosphere of *C. sativa* degraded benzyl isothiocyanate to hydrogen sulfide and benzylamine (Tang *et al.*, 1972).

Pandya (1975, 1976) demonstrated that aqueous extracts of fresh leaves, stems, and roots of *Celosia argentea* inhibited shoot and root growth of bajra (*Pennisetum americanum*). Root and leaf extracts were more inhibitory than stem extracts. In 1977, he reported that root exudates of *C. argentea* were very inhibitory to radicle growth of bajra seedlings until the *C. argentea* plants reached an age somewhere between 50 and 75 days after the time of planting. Exudates of the youngest plants (up to 25 days) were most inhibitory. Subsequent experiments indicated that root exudates of 20 day plants were more inhibitory than those of 10 day plants (Pandya and Pota, 1978). The root exudates also reduced growth of *Rhizobium meliloti*, but the inhibition decreased after the *Celosia* seedlings were 40 to 50 days old. Ashraf and Sen (1978) presented supporting evidence for the effects of aqueous extracts of *Celosia argentea* on bajra and demonstrated that the extracts have similar inhibitory effects on til (*Sesamum indicum*).

Two species of *Centaurea, C. diffusa* (diffuse knapweed) and *C. maculosa* (spotted knapweed), were introduced into British Columbia, Canada, during 1890–1900 and by 1970 over 30,000 ha of rangeland were infested with these weeds (Muir and Majak, 1981). *C. repens* (Russian knapweed), introduced from the Caucasus (Fernald, 1950), has also become a serious weed in rangelands (Fletcher and Renney, 1963). Fletcher and Renney found that the growth of tomato and barley was inhibited in soil naturally infested with knapweed previously or in soil with powdered knapweed residues added. Ether extracts (5 g dry weight of residue:100 ml ether) were tested against seed germination and seedling growth of tomato and barley also to determine which of the three species of knapweed listed above was most inhibitory and to determine which organs were most toxic. The leaves contained a higher concentration of the toxic material than did other plant parts, and the order of inhibition of the three species from greatest to least was *C. repens, C. diffusa,* and *C. maculosa,* respectively. An inhibitor was isolated that was soluble in both ether and water, and chromogenic sprays and ultraviolet absorption spectra indicated that the inhibitor was an indole derivative. It was not identified, however.

Bhandari and Sen (1971) made aqueous extracts of fresh material of roots, stems, leaves, and fruit pulp of *Citrullus colocynthis,* a common weed in bajra in western Rajasthan, India. The extracts were tested against seed gemination and seedling growth of bajra over a 48-hr period. Extracts of *C. colocythis* delayed seed germination of bajra but had little effect on final germination. Root and shoot growth were inhibited and root hair formation was suppressed. *Citrullus lanatus* was later demonstrated to also have allelopathic potential (Bhandari, and Sen, 1972).

Horowitz and Friedman (1971) incubated dried subterranean organs of bermudagrass (*Cynodon dactylon*) in light and heavy soil for 1, 2, or 3 months. After removing the decayed plant material, the residual activity of the soil was assayed by barley (*Hordeum vulgare*) sown in the soil and by a barley radicle growth test on an ethanolic extract of the soil. Seed germination and root and top growth of barley were inhibited when growing in soil that had previously contained bermudagrass residues. Extracts of soils that previously contained decaying bermudagrass residue also reduced radicle growth of barley in many of the tests. In general, growth inhibition was proportional to the concentration of plant material in the soil.

Sarma (1974a) reported that *Digera arvensis,* which is a common weed in crops in Rajkot, India, produces toxins that inhibit seed germination and radicle elongation and reduce dry weight increment of bajra.

When areas of Galicia, Spain, which were occupied by various species of heaths (*Ericaceae*), are used for agriculture, serious growth problems result in crop plants, particularly with various grass species (Ballester and Vieitez, 1971; Ballester *et al.*, 1972; Salas and Vieitez, 1972; Vieitez and Ballester, 1972). These investigators found that *Calluna vulgaris, Daboecia polifolia, Erica arborea, E. australis, E. ciliaris, E. cinerea, E. mediterranea, E. scoparia, E. tetralix,* and *E. umbellata* contained substances inhibitory in the *Avena* coleoptile straight-growth test. *E. australis* was most inhibitory and *E. arborea* was least inhibitory. Several phenolic compounds were identified in various parts of heath plants and in the soil under some of the species in field studies. They identified orcinol and orcinolglucoside in plant material and soil, presumably for the first time.

Le Tourneau *et al.* (1956) tested aqueous extracts of oven-dried tops of *Ambrosia trifida, Cirsium discolor, Euphorbia esula, Iva xanthifolia, Lychnis alba, Polygonum pennsylvanicum,* and *Portulaca oleracea* against seed germination and seedling growth of Mida wheat. Usually 2 g dry-weight of tops were extracted in 100 ml of water at 20 lb pressure for 15 min. The liquid was filtered while hot, and the filtrate was autoclaved again to sterilize it. In a few trials, 1–5 g of plant material were extracted in 100 ml water in a similar manner. All extracts inhibited seed germination and coleoptile and root growth of Mida wheat. Measurements of pH and calculations of osmotic pressures of extracts indicated that these factors were not involved in the inhibition.

Le Tourneau and Heggeness (1957) found aqueous extracts of freshly harvested *Eurphorbia esula* (leafy spurge) to be more inhibitory than extracts of oven-dried material to test species. Steenhagen and Zimdahl (1979) observed reductions in frequency and density of quackgrass and common ragweed in field plots in which densities of *Euphorbia esula* were high. Field soil samples taken from areas of moderate and high densities of leafy spurge inhibited growth of tomatoes and crabgrass in the greenhouse. When litter of leafy spurge was incorporated into the soil, growth in height and increase in dry weight of tomatoes and crabgrass was inhibited. This was true for both leaf and root litter. Surface application of the litter, however, did not affect growth of tomatoes or crabgrass.

Kohlmuenzer (1965a,b) reported that water extracts of tops of *Galium mollugo* in dilutions from 1:1 to 1:100 of plant material to water inhibited germination of wheat by 18 to 100% and radish by 19 to 42%. Dilutions of 1:10 to 1:500 retarded growth of seedlings of cultivated sunflower (*Helianthus annuus*) by 10 to 100%. Development of onion bulbs and roots was also markedly inhibited by similar extracts. Dilutions of 1:5 and 1:10 induced necrosis of meristematic tissues of onion.

Artificial raindrip from leaves, exudates of roots, decaying leaves, and soil previously in contact with common sunflower (native *Helianthus annuus*) roots greatly suppressed seed germination and seedling growth of most of nine test species of plants (Wilson and Rice, 1968). No crop plants were included in the tests, so such tests should be made because of the pronounced allelopathic effects of this species.

Cogongrass or alang-alang (*Imperata cylindrica*), one of the ten worst weeds in the world, is widespread in warm regions of the world and forms a dreaded weed in many tropical regions (Holm, 1969). Abdul-Wahab and Al-Naib (1972) identified scopolin, scopoletin, chlorogenic, and ischolorogenic acids in water extracts of leaves and culms of this weed. All of these compounds are known allelopathic agents. Unfortunately, no bioassays were made with the aqueous extracts, or of possible leaf leachates, root exudates, or decaying material.

Eussen and Soerjani (1978) studied the effects of leaves of cogongrass on the growth of cucumber (cv Soloyo) by putting leaves on the soil surface, mixing leaf sections in soil, or mixing dried, ground leaves in the soil. In each test, the dry weight of the added leaf material was 0, 2.5, 5, or 10% of the dry weight of soil in the pots. All treatments inhibited growth of both roots and tops of cucumber plants.

Extracts of fresh leaves (10% dry wt/v) were tested against seed germination of corn (maize), sorghum, rice (*Oryza sativa*), cucumber, tomato, and nine weedy species. The extracts were tested against seedling growth of all crop plants except cucumber and against two weedy species. Germination was not affected appreciably in any species, but root growth was markedly reduced in all test species and top growth was reduced slightly in corn, sorghum, and rice. The effects were not due to osmotic pressures of the extracts because the activity on seed germination and seedling growth of mannitol solutions isotonic to the extracts was consistently lower than the activity of the extracts.

Eussen (1978) reported that aqueous extracts of rhizomes and roots of cogongrass were also inhibitory to root growth of tomato. All extracts stimulated shoot growth at low concentrations but inhibited shoot growth at higher concentrations. Seed germination was delayed by the extracts but the total germination percentage was not affected. The extracts were fractionated using liquid–liquid extraction and paper chromatography and numerous inhibitory compounds were isolated but not identified.

Bieber and Hoveland (1968) tested water extracts of several weed species against seed germination of crownvetch (*Coronilla varia*) and other selected crop species and found that extracts of *Lepidium virginicum*, *Oenothera biennis*, and *Digitaria sanguinalis* were very toxic. Water extract of *Lepidium* also inhibited seed germination of *Festuca arundinacea*, *Trifolium incarnatum*, *Lespedeza cuneata*, and *L. striata* at a concentration of 1:150 (w/v). *Lepidium* residues incorporated into soil for 10 weeks were toxic to germination of crownvetch seed also.

Naqvi (1972) found, in greenhouse and field tests, that Italian ryegrass (*Lolium multiflorum*) suppressed germination and growth of many species in its vicinity. Later, Naqvi and Muller (1975) reported that leachates of living tops (artificial rain), leachates of soil previously occupied by Italian ryegrass, and decomposing residues were toxic to seedling growth of oats, *Bromus* sp., lettuce, and *Trifolium* sp.

Parthenium hysterophorus, a common to serious weed in five countries of the world (Holm *et al.,* 1979), has spread to almost all parts of India within two decades (Kanchan and Jayachandra, 1979a). Its rapid spread and interference with crop growth stimulated extensive research on its possible allelopathic potential. As early as 1973, Rajan investigated the possible inhibitory effects of fruits and receptacles of this weed on germination and seedling growth of wheat (*Triticum*). A mixture of fruits and receptacles added to Petri dishes containing wheat grains and water inhibited coleoptile and radicle growth. Seed germination was slowed, but total germination was not affected. Washing the mixture with water for 72 hr before adding to the Petri dish containing wheat grains eliminated most of the inhibitory action.

Sarma *et al.* (1976) found that 1 and 5% aqueous extracts of thoroughly washed shoot material of *P. hysterophorus* arrested growth of peanut (*Arachis hypogea*), *Crotalaria juncea,* and black gram (*Phaseolus mungo*) seedlings, whereas seed germination was not affected.

Char (1977) observed that tomato, chillies (*Capsicum annuum*), and French bean (*Phaseolus vulgaris*) did not set seed or fruit after pollination with their own pollen mixed with pollen of *Parthenium hysterophorus.* They set seed and fruit well, however, when pollinated with only their own pollen. He confirmed this observation by repeating the process in flowers that had the anthers removed prior to blooming and by using fresh pollen taken from dehiscing anthers. Mixing pollen of the crop plants with an aqueous extract of intact pollen of *P. hysterophorus* before pollinating also inhibited seed and fruit set. This appears to be the first reported case of pollen allelopathy.

Dube *et al.* (1979) reported that aqueous extracts of air-dried leaves, inflorescences, stems, and roots of *P. hysterophorus* decreased seed germination, seedling growth, and chlorophyll production by cotyledons of radish, cabbage (*Brassica oleracea,* Capitata group), and cauliflower (*B. oleracea,* Botrytis group).

The most thorough demonstration of the allelopathic potential of *Parthenium hysterophorus* was made by Kanchan and Jayachandra (1979a,b, 1980). They found that the growth of plants associated with this weed in the field was retarded. Nodulation of Burpees Stringless bean was reduced also, with the inhibition decreasing with increasing distance from the weed. Watering Burpees Stringless bean in soil with leachates from pots containing *P. hysterophorus* markedly inhibited nodulation and root and shoot growth of bean plants. Similar leachates markedly inhibited growth of UP301 wheat plants also. Maximum exudation of inhibitors from the roots of *P. hysterophorus* occurred at the rosette and flowering stages, and the inhibitors remained active for about 30 days.

Roots or leaves were added to sandy loam soil in field plots in amounts equivalent to those actually measured in the field, 333 kg/ha of leaves and 216 kg/ha of roots. Seeds of five crop species were sown in replicated plots. Decaying roots were tested only against bean. Decaying leaves inhibited nodulation and growth of Burpees Stringless bean and cowpea (*Vigna unguiculata*), branching of Pusa Ruby tomato, height increase and tillering in ragi (*Eleusine coracana*, cv Poorna), and in the yield of bean, cowpea, tomato, and ragi. Decaying leaves stimulated growth of bajra, however. Decaying roots significantly inhibited nodulation, growth, and yield of bean. The inhibitors released to the soil from leaves and stems remained active for about 30 days.

Kanchan and Jayachandra (1980) confirmed and extended the results of Char (1977) concerning the allelopathic effects of *P. hysterophorus* pollen. They found that the white dust that settled on the leaves and stigmatic surfaces of plants growing in the midst of *P. hysterophorus* consisted of clusters of *Parthenium* pollen. Pollen grains deposited in large numbers on leaves reduced the chlorophyll content of the leaves of certain test species.

Prostrate knotweed (*Polygonum aviculare*), a bad weed in crops in many parts of the world, is also a pernicious weed in lawns in many countries, including the United States. It rapidly encroaches into bermudagrass lawns, and the bermudagrass dies in patches of prostrate knotweed, whereas bermudagrass at the edges of the knotweed patches turns yellow. Field measurements in Norman, Oklahoma, demonstrated that *P. aviculare* stands increased in diameter at an average annual rate of 1.5 m, even though they were surrounded by a heavy sod of bermudagrass. The rapid invasion of heavy bermudagrass sod by prostrate knotweed and the existence of this weed in pure stands after a few months suggested that allelopathy, as well as competition, might be involved in its interference.

AlSaadawi and Rice (1982a,b) investigated the allelopathic potential of *P. aviculare*. Concentrations of several major and trace elements were determined in knotweed patches and in the surrounding bermudagrass stands, and the only significant difference found was a slightly higher concentration of calcium in the *P. aviculare* stand. Thus, elimination of bermudagrass did not appear to be due to a deficiency of minerals. Moreover, there were no appreciable differences in soil texture or pH in the two stands. Prostrate knotweed is very tiny and does not shade bermudagrass appreciably either. Soil minus litter was collected under a *P. aviculare* stand and under a bermudagrass stand in March and in July, and these soil collections were used to grow five different test species: *Gossypium barbadense* (cotton), sorghum, *Chenopodium album*, *Sporobolus pyramidatus*, and bermudagrass. Soil collected under *Polygonum* in July did not affect seed germination and seedling growth of any of the test species when compared with germination and growth in soil collected under bermudagrass. Soil collected under *Polygonum* in March, markedly inhibited seed germination of all test species except sorghum. Seedling growth of all species, except *S. pyramidatus*, was inhibited by the soil collected under *Polygonum* in March. These results indicated that soil phytotoxicity was closely associated with prostrate knotweed. Experiments were conducted to determine the source of toxins in the soil under *P. aviculare.*

Artificial rain was allowed to fall on fresh mature plants of *P. aviculare.* The leachate, which dripped from the leaves, was collected and used to water the same five test species that were growing in greenhouse potting soil. The leachate reduced germination slightly in all test species, inhibited growth of *C. album,* and stimulated growth of bermudagrass.

When decaying shoots of prostrate knotweed were added to potting soil in the same concentrations as determined in the field, germination of cotton and *C. album* seeds was greatly reduced, and seed germination of the other test species was slightly reduced. Decaying shoots also reduced seedling growth of all test species.

Decaying roots of prostrate knotweed added to potting soil in concentrations equivalent to those measured in field plots, reduced germination of bermudagrass and *C. album* appreciably, but they did not affect germination of other species. Decaying roots also inhibited seedling growth of cotton and bermudagrass significantly, but other test species were not affected.

Effects of root exudates of prostrate knotweed on the same five test species were determined using the methods of Bell and Koeppe (1972) and Tubbs (1973). The root exudate retarded root growth of sorghum, top growth of bermudagrass, and root and top growth of *C. album.* Seedling growth of other test species was not affected.

AlSaadawi and Rice (1982b) isolated four inhibitors from living *P. aviculare* plants, three of which were glucosides. Four different inhibitors were isolated from prostrate knotweed residues and soil under stands of this weed, and none of these occurred in soil from bermudagrass stands. Three of these inhibitors were phenolic glycosides containing both fructose and cellobiose. Color reactions with various phenolic reagents indicated that all the inhibitors isolated were phenolic compounds. All eight inhibitors reduced seed germination and/or seedling growth of *C. album.* Moreover, some of them inhibited growth of different strains of *Rhizobium* and *Azotobacter.*

Polygonum orientale is found throughout India and is often a troublesome weed in rice fields. Datta and Chatterjee (1978, 1980a) tested extracts of various organs of this weed against mustard (*Brassica juncea*), lettuce (*Lactuca sativa*), rice, and pea (*Pisum sativum*). Leaf and stem extracts reduced seed germination in all test species, and flower and root extracts reduced germination in all except pea. Lettuce was not used in growth tests, but growth of all other species was inhibited by certain concentrations of some extracts. Growth was decreased most by leaf extracts, and mature leaves produced more inhibitory material than young leaves or leaves of an intermediate age. Treatment of the aqueous solutions with Norit eliminated much of the inhibitory activity, but autoclaving or inoculation with soil microorganisms did not reduce the activity. Two inhibitory compounds were isolated and tentatively identified as luteolin and apigenin glycosides.

Datta and Chatterjee (1980b) extended their previous work on the allelopathic potential of *Polygonum orientale* by testing aqueous extracts, leaf leachate, decaying leaves, and soil from a patch of the weed against five weedy species. Growth of all five test species was reduced in all tests, indicating that the inhibitors entered the substrate in various ways.

As previously mentioned, aqueous extracts of Pennsylvania smartweed (*Polygonum pennsylvanicum*) were shown by Le Tourneau *et al.* (1956) to inhibit seed germination and coleoptile and root growth of Mida wheat. Later, Gressel and Holm (1964) reported that aqueous extracts of *P. pennsylvanicum* seeds reduced seed germination of alfalfa (*Medicago sativa*), but not of seven other crop species. Coble and Ritter (1978) found that interference from as few as eight Pennsylvania smartweed plants per 10 m of row reduced soybean seed yield by 13%. Higher densities reduced the yield up to 62%. No allelopathic effects of *P. pennsylvanicum* due to root exudates were observed in greenhouse studies using a recirculating nutrient solution and alternate pots of soybeans and Pennsylvania smartweed. More research is needed on possible allelopathic effects of this weed on other crop plants and on pathways of movement of inhibitors out of the weed.

Aqueous extracts of *Portulaca oleracea* seeds arrested seed germination of alfalfa and radish (Gressel and Holm, 1964). Earlier, Le Tourneau *et al.* (1956) showed that aqueous extracts of oven-dried tops of *P. oleracea* inhibited germination and seedling growth of Mida wheat. Obviously, more reserach is needed on allelopathic mechanisms of this weed.

Observations of vegetational patterning suggested to Einhellig and Rasmussen (1973) that curly dock (*Rumex crispus*), a common weed of field margins and wastelands, might be allelopathic. Aqueous fresh-leaf extracts of this species slowed seedling growth of sorghum, corn, and *Amaranthus retroflexus*. Osmotic potentials were too low to account for the growth inhibition. Three phytotoxins were isolated by paper chromatography and were identified as phenolic compounds. All three phytotoxins inhibited seed germination of sorghum and radish.

Russian thistle (*Salsola kali*), a common to serious weed in crops in nine countries of the world (Holm *et al.,* 1979), is a pioneer species in succession on mine-spoils and in revegetating old fields. Lodhi (1979b) observed that Russian thistle disappears rapidly from the pioneer weed stage on mine-spoils in western North Dakota, and he hypothesized that Russian thistle might be allelopathic to itself and other plants of the pioneer weed stage. He found that low concentrations of decaying leaf powder (1 g per 454 g of soil) inhibited shoot and root growth of *Kochia scoparia* and Russian thistle, but not of honey clover (*Melilotus officinalis*). An equivalent weight of peat moss was added to the control soil to keep the organic matter content the same.

Artificial raindrip leachate of fresh tops of Russian thistle retarded shoot growth of honey clover and kochia, but not roots or shoots of Russian thistle. Each of six phenolic compounds identified in leaf extracts of *S. kali* reduced germination and radicle growth of radish. However, only one identified phytotoxin (ferulic acid) inhibited germination and seedling growth of Russian thistle.

Mohnot and Soni (1976) demonstrated that water extracts of air-dried leaves of *Salvadora oleoides* inhibited seed germination and root and shoot growth of seedlings of sorghum.

Anaya and Gómez-Pompa (1971) demonstrated that extracts of leaves and fruits of piru (*Schinus molle*) were strongly inhibitory to seed germination and seedling growth of cucumber (*Cucumis sativus*) and wheat. Piru is a pernicious weed of crop plants in some parts of Mexico, and it is possible, therefore, that allelopathy may play a role in its interference.

Solanum surattense is a prickly undershrub and common weed of Indian desert fields. Mohnot and Soni (1977) found that aqueous extracts of air-dried stems inhibited seed germination and seedling growth of sorghum. Sharma and Sen (1971) demonstrated earlier that aqueous extracts of the fruit pulp from yellow ripe fruits of *S. surattense* reduced seedling growth of bajra and til. Unfortunately, no further studies were undertaken to determine if *S. surattense* was allelopathic to sorghum, bajra, til, or other crop plants.

Very little research was done on allelopathic effects of weeds on crop plants prior to 1970 (Rice, 1974). Although this situation has changed, much more research is needed on this important subject. Especially needed are studies on the quantitative effects of interference on crop yields of most of the weeds discussed in this section. Moreover, virtually nothing has been done toward determining the relative contributions of allelopathy and competition in the total interference of any of the species discussed in Section I. Additionally, only aqueous extracts of many of the species have been demonstrated to inhibit plant growth. Such experiments do not confirm allelopathic potential.

II. ALLELOPATHIC EFFECTS OF CROP PLANTS ON OTHER CROP PLANTS

Theophrastus observed and described inhibitory effects of crop plants on other crop plants over 2000 years ago (see Chapter 1). In spite of this, no scientific research was done to verify such observations until early in the twentieth century.

Schreiner and his associates published a series of papers starting in 1907 in which they presented evidence that exhaustion of soil by single-cropping is due to addition of growth inhibitors to the soil by certain crop plants (Schreiner and

Reed, 1907a,b, 1908; Schreiner and Shorey, 1909; Schreiner and Sullivan, 1909; Schreiner and Lathrop, 1911). Schreiner and Reed, (1907b) demonstrated that roots of seedlings of wheat (*Triticum*), oats (*Avena sativa*), and certain other crop plants exude materials into the growing medium that elicit chemotropic responses by the roots of wheat and oat seedlings. Schreiner and Reed (1908) developed a technique that is still used for determining possible allelopathic effects of compounds obtained from the soil or from plants. They showed that many compounds previously identified from various plants retarded the growth and transpiration of wheat seedlings. Schreiner and Sullivan (1909) extracted an unidentified substance from soil fatigued by the growth of cowpeas and found that the substance strongly inhibited the growth of cowpeas. Moreover, the soil from which the inhibitor was extracted no longer inhibited the growth of cowpeas.

Benedict (1941) studied the reasons for the natural thinning of smooth brome, *Bromus inermis,* and found that oven-dried roots of smooth brome, when placed in soil with seeds of that species, reduced seedling growth. He obtained similar results by adding a leachate from an old culture of smooth brome to seedlings. He thus established that a toxic substance was produced by smooth brome roots.

Bonner and Galston (1944) observed that the edge rows in guayule, *Parthenium argentatum,* plantings in Salinas, California, had much larger plants than the center rows and the differences could not be eliminated by heavy watering and fertilizing. In addition, roots of adjacent plants did not intermingle but grew in entirely separate areas, and seedlings of guayule plants virtually never grew under larger guayule plants. On the other hand, such seedlings were commonly found growing under other kinds of shrubs. Experiments were designed to determine if guayule produces a growth inhibitor. Initial experiments indicated that leachates from pots of 1-year-old guayule plants greatly reduced the growth of guayule but not tomato seedlings. In another experiment, guayule seedlings were planted in sand adjacent to a 1-year-old guayule plant. In addition, other guayule seedlings were planted in fresh sand in a glass jar, and the jar was placed in an excavation in the sand under the older guayule plant so shading effects on all seedlings were similar. The seedlings in the glass jar were, thus, not subjected to any possible inhibiting material which might be present in the sand around the older plant. Seedlings growing under the guayule plant had a high mortality rate and grew slowly if not contained in a glass jar, whereas those grown under the same conditions, but in a separate glass jar, had good growth and lower mortality. These experiments indicate that roots of guayule excreted a toxin.

Bonner and Galston identified the toxin as *trans*-cinnamic acid. This compound is highly toxic to guayule seedlings, with growth reduction resulting from as little as 1 mg/liter of culture solution. Guayule seedlings were at least 100 times as sensitive as tomato seedlings to cinnamic acid, which explained why tomato seedlings were not affected by the leachates of the guayule plants in the initial experiments.

In later work, Bonner (1946) found that cinnamic acid is also toxic to the growth of guayule plants in soil. A concentration of less than 1 part in 100,000 in the soil depressed the growth of the plants over a period of 6 weeks. He found that the toxin is unstable in the soil, however, and decreases with time. It does not disappear in sterilized soil, so obviously it is decomposed by microorganisms. Apparently, it must be added to the soil continuously to be effective as a koline, as has been demonstrated with other kolines.

Evenari (1949) reviewed early research on production of seed germination inhibitors by seed plants. Only a few of his original findings will be mentioned here. He found that bulb juice of *Allium cepa* and *A. sativum* contained inhibitors as does root sap of *Armoracia lapathifolia*, tuber sap of *Brassica caulocarpa*, and fruit juice of *Citrus aurantium, C. limonia,* and *C. maxima.* He stressed the widespread occurrence of potent seed germination inhibitors in various species of the Cruciferae and empahsized that the evidence indicates that mustard oils are the chief inhibiting substances in such cases. Evenari pointed out that when a piece of orange or lemon peel was put in a large Petri dish in which a small Petri dish containing 50 wheat grains on moist filter paper was placed, all of the wheat grains failed to germinate, indicating that a volatile inhibitor was produced by the peels. He gave considerable evidence that the volatile inhibitor is an essential oil.

The majority of the investigations directly concerned with allelopathic effects of crop plants in agriculture have involved effects of decomposing crop residues. This probably resulted from the expanding use of stubble mulch (no till) farming since the "dust bowl" days in the United States to control erosion by wind and water, with resulting decreases in crop yields in numerous instances.

McCalla and Duley (1948) reported that stubble mulch farming reduces the stand and growth of corn under some conditions, and he found that soaking corn seeds in an aqueous extract of sweet clover (*Melilotus alba*) for 24 hours reduced germination and growth of tops and roots on agar in Petri dishes. Alfalfa extracts had less depressing effects and wheat straw extracts either stimulated growth or had no effect. Subsequently, McCalla and Duley (1949) placed soil from the Agronomy Farm at Lincoln, Nebraska, on greenhouse benches and mulched some of it with wheat straw at the rate of 2 to 4 tons/acre and left some unmulched. They planted corn grains in both areas and kept the soil thoroughly wet by watering 2 or 3 times daily. This was done because the adverse effects of stubble mulching on corn were particularly striking during periods of wet, cool weather. In three trials, the average percentage of germination of corn in the mulched plots was 44%, and in the unmulched plots, it was 92%. The authors also pointed out that they had repeated this many other times in pot experiments with similar results.

Another phase of their investigation involved soaking corn grains in aqueous extracts of wheat straw or sweet clover for 24 or 48 hours and then placing them on agar plates. In some instances, ammonium nitrate was added to the water in which the plant material was soaked to give a 0.5% solution. Some corn grains soaked in distilled water were placed on agar plates containing a 1% concentration of several organic nitrogen compounds. The water extract of sweet clover reduced the seed germination and growth of corn roots, but the water extract of wheat had little effect. The extract resulting from the wheat straw and ammonium nitrate reduced germination and growth markedly, and many of the roots grew upward. McCalla and Duley pointed out that there appeared to be increased microbial growth with the added ammonium nitrate.

In the dishes containing organic nitrogen compounds, relatively high percentages of the corn roots grew upward except on portions of the plates where there was limited bacterial growth due to the presence of fungi. Subsequent experiments indicated that the bacteria produced a gas under some conditions that caused the corn roots to grow upward, but the effects were not as striking as they were when the corn grains were in contact with the agar on which the microorganisms were growing. The percentage of germination of corn was not influenced by the microorganisms, but root growth was markedly affected in some cases. These experiments indicated that the inhibiting effects of the mulch resulted from a combination of toxins present in the plant material plus toxins produced by microorganisms whose growth was stimulated by material in the mulch.

Guenzi and McCalla (1962) collected crop residues in September from fields on the Agronomy Farm at Lincoln, Nebraska. The materials, consisting of wheat and oat straw, soybean and sweet clover hay, corn (maize) and sorghum stalks, and bromegrass and sweet clover stems were extracted with hot and cold water. One-half of each water extract was autoclaved for 1 hour at 20-lb steam pressure. Additionally, wheat straw was extracted with ethanol, and the extract was separated into strong and weak acids, and neutral, basic, and water soluble compounds. Electrical conductivity of all water extracts was determined, and the effects of KCl solutions with the same conductivities were determined for all test species so adjustments could be made for the salt content of the extracts. Moreover, the concentrations of reducing compounds were determined so that the effects of glucose solutions with the same osmotic concentrations could be determined for all test species. It was found that the salt content explained only 3 to 8% of the depressive effect of the water extracts, and even the highest osmotic concentration had only a small inhibitory effect on growth of corn, wheat, and sorghum plants.

All residues contained water-soluble substances that depressed the growth of corn, wheat, and sorghum plants. The results of the tests against wheat are shown in Table 18. The general order of increasing toxicity was sweet clover stems, wheat straw, soybean hay, bromegrass, oat straw, corn, and sorghum stalks and sweet clover hay. Nonautoclaved extracts of the residues inhibited

TABLE 18.   Influence of Water-Soluble Substances Extracted from Different Plant Residues on Germination and Growth of Wheat Seeds[a]

		Inhibition (%)					
		Germination		Root growth		Shoot growth	
Crop residues		AC[b]	No AC	AC	No AC	AC	No AC
Sweetclover stems	C[c]	−3[d]	−8	58	7	24	21
	H[e]	−1	−8	51	12	10	10
Wheat straw	C	7	5	36	7	14	21
	H	−5	−5	18	36	7	28
Soybean hay	C	3	−3	80	30	66	45
	H	−1	−5	51	39	48	45
Oat straw	C	3	10	87	64	83	76
	H	−1	−3	84	45	79	62
Bromegrass	C	1	3	71	55	48	59
	H	−1	27	71	62	52	78
Cornstalks	C	−7	89	75	87	62	93
	H	5	38	47	75	62	83
Sorghum stalks	C	9	100	87	100	86	100
	H	3	72	84	82	83	93
Sweetclover hay	C	64	3	95	82	90	83
	H	26	100	95	100	90	100
Mean	C	9.6	24.9	73.6	54.0	59.1	62.3
	H	3.1	27.0	62.6	56.4	53.9	62.4

[a] From Guenzi and McCalla (1962).
[b] AC, autoclaving for 1 hour at 20 pounds steam pressure.
[c] Cold water soluble substances (extracted at 25°C.).
[d] Negative sign indicates stimulation.
[e] Hot water soluble substances (extracted at 100°C.).

seed germination and shoot growth more than the autoclaved extracts in most cases, but the autoclaved extracts reduced root growth more.

All five fractions derived from the ethanol extract of wheat straw contained substances toxic to growth of wheat seedlings. The water-soluble and strong acid fractions had the strongest depressive effects, and the basic fraction had the least effect. In a similar experiment in which corn was used as the test plant, the quantity of material from 2.3 g of wheat straw completely inhibited growth of one corn seed. Guenzi and McCalla estimated that this would be equivalent to about 101 lb of straw per acre, assuming a plant population of 20,000 plants per acre.

Norstadt and McCalla (1963) followed up on the earlier investigation of Mc-Calla and Duley (1949), which suggested that effects of crop residues might be due to a combination of toxins from the residues and from microorganisms caused to grow more profusely by substances in the residues. Norstadt and McCalla used research plots, some of which were subsurface tilled and some were plowed during a 23-year period at Lincoln, Nebraska. A rotation of corn, oats, and wheat was followed in both types, and decreased yields and abnormal appearance of crops occurred in the subsurface-tilled (stubble-mulched) plots in the years with normal to above-normal precipitation. They isolated fungi from the subsurface-tilled soil showing reduced growth and then cultured them in potato dextrose broth. Corn seeds were soaked in the broth for 6 hours and placed in Petri dishes between double layers of filter paper moistened with the broth. The percentage of germination and lengths of roots and shoots were determined after 3 days. In a group of 91 isolates, 14 reduced germination to 50% or less, whereas distilled water controls usually had better than 90 to 95% germination.

A fungus, which produced a particularly potent toxin against growth of corn plants, was identified as *Penicillium urticae*. Norstadt and McCalla (1963) identified the toxin as patulin and compared its inhibitory effect on Cheyenne wheat with 2,4-dichlorophenoxyacetic acid (2,4-D), coumarin, and indole-3-acetic acid (IAA). Germination (as percentage of the control) that occurred in a 50 ppm solution was as follows: 2,4-D, 40%; coumarin, 80; IAA, 85%; and patulin, 85%. The concentration required to reduce root length to 50% of the control was as follows: 2,4-D, 1 ppm; coumarin, 9 ppm; IAA, 25 ppm; and patulin, 20 ppm. Shoot growth was reduced to 50% of the control by 63 ppm of IAA, 20 ppm of coumarin, 40 ppm of patulin, and 7.5 ppm of 2,4-D. Thus, patulin is a potent inhibitor of growth in higher plants in addition to its known inhibitory effect on fungi. Patulin was subsequently found to be inhibitory to Cheyenne wheat seedlings in sand and two soil types.

McCalla and Haskins (1964) reported that in a later test of 318 fungi isolated from soil in stubble-mulched plots at Lincoln, Nebraska, 52 produced toxins which reduced shoot growth of corn by 50% or more and 167 produced toxins, which reduced root growth of corn by a comparable percentage. These fungi were not identified, nor were the toxins. Several fungi produce patulin (Norstadt and McCalla, 1963), and many fungi produce other substances toxic to growth of higher plants (McCalla and Haskins, 1964). According to McCalla and Haskins (1964), Krasilnikov found 5 to 15% of 1500 cultures of actinomycetes tested to inhibit growth of higher plants, about one-third of the 300-plus cultures of non-spore-forming bacteria, and 20 to 30% of the spore-forming bacterial cultures. The possibility is very good, of course, that crop residues stimulate the growth of many of these organisms.

Ellis and McCalla (1973) reported that the *P. urticae* population comprises 90% of the total fungal population in the soil where stubble-mulch wheat farming occurs. They pointed out that the toxic effects of patulin had been demonstrated on young seedlings, germinating seeds, isolated plant tissues, and plants that had continuous applications of patulin until maturity. However, the effect of a single application on wheat plants subsequently grown to maturity had not been tested. Moreover, it was known that *P. urticae* blooms sometimes last for relatively short time periods. Ellis and McCalla found that a single 100 ppm application of patulin to soil in which Lee spring wheat (*Triticum aestivum*) was planted and allowed to grow to maturity, reduced internodal elongation, floret number, seed weight, and seed number. Yields were reduced according to the proximity of application prior to heading. The authors concluded that a single exposure of growing wheat plants to patulin can produce yield reductions similar to those which occur in stubble-mulch farming.

Guenzi and McCalla (1966a) identified and quantified five phenolic acids in mature plant residues of oats, wheat, sorghum, and corn. These five compounds were *p*-coumaric, syringic, vanillic, ferulic and *p*-hydroxybenzoic acids. *p*-Coumaric acid was present in greatest amounts. Both acid and base hydrolyses were used to free the acids from the bound form, and higher concentrations were found in alkaline hydrolyzates, indicating an ester-type linkage. All five acids were inhibitory to growth of wheat seedlings.

Guenzi and McCalla (1966a) estimated on the basis of usual yields of the four crop plants at Lincoln, Nebraska, that the following amounts of *p*-coumaric acid would be added in pounds per acre in the residue: 89 by sorghum, 72 by corn, 8 by wheat, and 23 by oats. They pointed out that even though these acids are mostly bound in the residues, there should be periods during decomposition when rather large amounts could be released in the immediate vicinity of the residue and be sufficiently high to affect plant growth.

Guenzi and McCalla (1966b) next extracted, identified, and quantified phytotoxins in soil from stubble-mulched and plowed plots. They used a number of solvent systems and techniques for extracting and separating the toxins. The extracted material in each case was dried and different concentrations in water were tested against germination and growth of wheat. All fractions from the stubble-mulched soil retarded growth, but the acid fraction was most inhibitory. Five phenolic acids were identified in soil extracts and were identical to those in the residue (Table 19). The concentration of ferulic acid was twice as high in the stubble-mulched plot as in the plowed one. *p*-Coumaric acid was approximately 50% higher in concentration also in the subtilled plot (Table 19).

Guenzi and McCalla (1966b) pointed out that the concentrations of phenolic acids were relatively low compared with those required to inihibit growth of plants. They emphasized, however, that concentrations would unquestionably be higher in localized areas around fragments of residues. Moreover, they stated that there are many phytotoxic substances present in low concentrations in soil, and these would probably have synergistic effects on plant growth, particularly under suboptimal growth conditions.

Guenzi *et al.* (1967) investigated changes in phytotoxic activity of water extracts of residues of corn, wheat, oat, and sorghum during decomposition in

TABLE 19.   Concentration of Phenolic Acids from a 2 N NaOH Extract of a Sharpsburg Plowed and Subtilled Soil[a]

	Concentration			
	Subtilled		Plowed	
Phenolic acids	ppm[b]	% of OM[c]	ppm	% of OM
Ferulic	7.6 ± 4.0	0.021	3.7 ± 1.5	0.010
p-Coumaric	14.4 ± 2.3	0.040	9.4 ± 2.9	0.026
Syringic	T[d]		T	
Vanillic	1.5 ± 0.7	0.004	1.5 ± 0.5	0.004
p-Hydroxybenzoic	1.2 ± 0.3	0.003	1.3 ± 0.1	0.003

[a] Each value is a mean of three replicates. From Guenzi and McCalla (1966b).
[b] ppm calculated on the oven-dry weight of soil.
[c] OM, organic matter.
[d] T represents <1 ppm.

the field during a 41-week period. Wheat was used as the test plant in most assays, but wheat and corn were both used in the initial test of mature residues at harvest time. Nine varieties of wheat straw were tested at harvest time against wheat. The varieties used were Nebred, Warrior, Cheyenne, Ponca, Yogo, Wichita, Pawnee, Bison, and Omaha. The extract of Ponca wheat straw depressed germination of wheat more than any of the extracts of other wheat varieties, but there were no differences among other varieties. The extract of Nebred wheat straw was less toxic to root growth of wheat than were extracts of other varieties, and there were no significant differences among other varieties. The inhibitory effects of extracts of the different wheat varieties on shoot growth of wheat varied greatly, ranging from 11% for Nebred to 36% for Omaha.

Changes in toxicity of extracts of residues during decomposition in the field varied considerably depending on the type of residue (Table 20). The toxicity of extracts of wheat straw remained about the same through the first 4 weeks of decomposition, but virtually all toxicity had disappeared by 8 weeks. The greatest toxicity in the extract of oat straw residue occurred at harvest time, and essentially all inhibitory activity was gone after 8 weeks of decomposition as in wheat. The toxicity of extracts of sorghum residues increased up to 16 weeks during decomposition for the 1963 season but declined slowly during decomposition in the 1964 season. Toxicity of extracts of corn residues remained high during 22 weeks of decomposition but decreased rapidly thereafter.

Martin and Rademacher (1960) incorporated fresh rape (*Brassica napus*) roots in soil and planted wheat seeds in this soil and similar soil without the rape roots. Growth of the wheat seedlings was inhibited at first, but after 4 days, it was stimulated.

TABLE 20. Wheat Seedling Growth in Response to Water-Soluble Phytotoxic Materials from Crop Residues Collected in the Field at Different Periods of Decomposition[a]

Period of decomposition (weeks)	Wheat straw		Oat straw		Sorghum residue		Corn residue	
	Roots	Shoots	Roots	Shoots	Roots	Shoots	Roots	Shoots
0	14 c	11 bcd	49 d	43 c	37 b	59 c	25 c	49 c
1	7 bc	16 cd	33 c	51 c	43 b	57 c	20 b	31 b
2	9 c	12 bcd	34 c	50 c	43 b	58 c	24 c	40 b
4	8 c	18 d	6 ab	18 b	53 c	53 c	43 c	53 c
8	−5 a	9 abc	−4 a	5 a	62 c	54 c	24 c	37 b
12	−7 a	5 a	−1 a	6 a	64 c	67 d	10 ab	40 b
16	−2 ab	8 abc	−4 a	2 a	85 d	77 e	11 ab	35 b
22					40 b	40 b	19 bc	35 b
28					11 a	23 a	6 a	21 a
35	8 bc	5 a	2 a	11 ab				
41	12 c	6 ab	14 b	5 a				

Inhibition of growth (%)[b]

[a] Each value is a mean of three replicates. Modified from Guenzi et al. (1967).

[b] Selections within individual crop residues and in the same column which have a letter in common do not differ in the character measured at the 5% probability level. Negative sign indicates an increase over the untreated.

Patrick and Koch (1958) investigated the effects of decomposing residues of timothy (*Phleum pratense*), corn, rye (*Secale cereale*), and tobacco (*Nicotiana tabacum*) on respiration of tobacco seedlings. Their study also included the effects of different conditions of decomposition on the toxicity of the residues. They obtained soil and plant material for their experiments from the Science Service field plots at Harrow, Ontario, Canada. The four crops mentioned were grown separately in the plots for at least 2 consecutive years.

Plant materials were collected at three different stages of growth: young, intermediate, and nearly mature. The young plants were collected 5 to 6 weeks after planting, the intermediate stage plants 6 to 8 weeks after planting, and the mature plants 10 to 14 weeks after planting. Mature corn and tobacco plants were still green in color, whereas rye and timothy were only partly green. Entire plants were collected, the roots were cleansed of soil, and the plants were cut into pieces about 1 inch long. They were either used immediately or air-dried and stored for future use.

Decomposition experiments involved addition of 250 g fresh weight (or 60 g air-dry weight) of plant material to 1000 g of fresh soil obtained from the field plot in which the particular plant had grown. All controls consisted of similar soil obtained from plots in which no plants were allowed to grow for 2 years and to which no plant material was added. In each case, the soil or soil plus thoroughly mixed plant material was added to 1 gal sterile glazed crocks. In some pots sufficient water was added to saturate the soil, but in others, the excess water was allowed to drain away until field capacity was attained. The pots were covered with aluminum foil and allowed to stand for 0 to 30 days at temperatures of 60° to 70°F. In other cases, similar soil or plant and soil mixtures were autoclaved for 40 minutes at 18 lb pressure in cotton plugged Erlenmeyer flasks, after which they were maintained under the same conditions as previously described.

At various times, soil in some of the flasks and pots was extracted with water for 15 to 20 minutes and centrifuged. The liquid was filtered through Seitz filters to remove soil particles, microorganisms, and plant debris, and the filtrates were tested for phytotoxic effects, usually within 5 days after preparation. The assay involved placing of 6-day-old sterile tobacco seedlings (cv Harrow Velvet) in the filtrates or sterile 0.01 M phosphate buffer for 16 hours at 20°C, after which the seedlings from the filtrates were washed in sterile distilled water and placed in sterile 0.01 M phosphate buffer also. Respiration rates of the various sets of tobacco seedlings were then determined by the Warburg method. Additionally, the effects of the various filtrates on germination and growth of tobacco, barley, and timothy seed, and seedlings were determined in Petri dishes.

Patrick and Koch (1958) found that substances that were inhibitory to respiration in tobacco seedlings formed during decomposition of residues of all four species (Table 21). Moreover, greater inhibition occurred if decomposition took place under saturated soil conditions. They concluded that different toxins were

TABLE 21. Oxygen Uptake of Tobacco Seedlings as Affected by the Products of Decomposition of Plant Residues in Soils Held at Saturation and at Field Capacity[a]

Plant material added to soil	20-Day decomposition period			
	Saturation		Field capacity	
	pH range[b]	O_2 uptake[c]	pH range	O_2 uptake
Soil only	6.4–6.6	128	6.4–6.6	131
Timothy	4.8–5.2	18	5.9–6.4	90
Rye	5.3–5.8	45	6.0–6.6	104
Corn	4.9–5.4	28	6.0–6.5	99
Tobacco	5.5–5.9	68	6.9–7.5	112
Control 125 (±8)[d]				

[a] From Patrick and Koch (1958). Reproduced by permission of the National Research Council of Canada from the *Canadian Journal of Botany* **36,** 621–647.

[b] pH range after the 20-day decomposition period; pH of each extract was adjusted to 5.3 prior to testing.

[c] O_2 uptake, in μl of O_2 after 6 hours at 20°C, of 50 Harrow Velvet tobacco seedlings. In each instance the tobacco seedlings were exposed for 16 hours to the various extracts, then returned to phosphate buffer (pH 5.3) before O_2 uptake was determined. Each figure is based on average of four different decomposition series for each of which three determinations of four replicates each were made.

[d] O_2 uptake, in μl of O_2 (after 6 hours at 20°C) of comparable seedlings preexposed (16 hours) to 0.01 M phosphate buffer solution; based on the average of 12 determinations.

produced or that concentrations were greater under saturated conditions. Another possibility could be that the toxins produced were destroyed faster under aerobic conditions.

Additional experiments were conducted in which soil moisture conditions were varied. In one case, decomposition was carried out for 15 to 20 days in soils at field capacity, followed by saturation. In other experiments, the soil was kept saturated for 15 to 20 days followed by a period at field capacity. Extracts that were made after decomposition for 15 to 20 days at field capacity and followed by flooding for only 3 to 5 days were very toxic. On the other hand, if decomposition was carried out under saturated conditions for the same period and if the soil was then drained to field capacity, toxicity decreased slowly and was reduced by one-half within 10 days after the return to the drier conditions.

Results of experiments in which plant materials of different ages were decomposed under saturated conditions for various periods of time were striking (Table 22). The time required for the formation of toxic substances was markedly affected by the stage of maturity of the plant residues that were added to the soil. When residues from young plants were added, toxic substances were produced relatively early in decomposition, but these substances were also inactivated

TABLE 22.   Effect of Plant Maturity and Period of Decomposition on Relative Toxicity of the Resulting Products[a]

Age of plant material added to soil	Type of plant	Decomposition period (days) at 60°–70°F						
		0	5	10	15	20	25	30
Young (5 to 6 weeks after planting)	Soil only	−[b]	−	−	−	−	−	−
	Timothy	−	+[b]	++	+++	++	−	−
	Rye	−	−	+	++	+	−	−
	Corn	−	−	+	++	+	−	−
	Tobacco	−	−	+	+	−	−	−
Intermediate (6 to 8 weeks after planting)	Soil only	−	−	−	−	−	−	−
	Timothy	−	+	+++	+++	++	+++	++
	Rye	−	+	++	+++	+++	++	++
	Corn	−	−	++	+++	++	+++	++
	Tobacco	−	−	+	+++	++	−	−
Mature (10 to 14 weeks after planting)	Soil only	−	−	−	−	−	−	−
	Timothy	−	−	−	++	++	+++	+++
	Rye	−	−	−	+	++	+++	++
	Corn	−	−	−	+	++	+++	+++
	Tobacco	−	−	−	+	++	++	−

[a] From Patrick and Koch (1958). Reproduced by permission of the National Research Council of Canada from the *Canadian Journal of Botany* **36,** 621–647.

[b] Toxicity rating: −, nontoxic i.e., extracts which inhibit respiration of 6-day-old tobacco seedlings by less than 19%; +, ++, +++, mild, intermediate, and highly toxic, producing inhibition of respiration of 20–40%; 41–60%; and 61–95%, respectively.

relatively early. When residues from mature plants were added, a longer period of decomposition was required for the formation of toxic substances, but toxicity remained high for a longer time. In all cases, the toxic substances from tobacco disappeared much more rapidly than those from timothy, rye, and corn.

Patrick and Koch (1958) investigated the relationship between toxicity of the various extracts and their pH. The pH and toxicity of the soil extracts were determined before addition of the plant residue, immediately afterward, and at subsequent 5-day intervals. They found no appreciable change in pH in any case immediately after addition of the residues and no toxicity. The pH of some extracts shifted from the 6.2 to 6.9 range to the 4.8 to 5.9 range during 5 days of decomposition. Corresponding with this change was an increase in toxicity of the extracts. Overall, they found a high correlation between the relative toxicity of extracts and their pH. These experiments did not enable Patrick and Koch to determine whether the toxicity was due solely to pH or whether a low pH resulting from the residues caused more toxins to be produced. To answer this question, they conducted a series of experiments using different extracts and buffers in which the pH of aliquots of each was adjusted from 4.6 to 7.6. They used an extract of fallow soil containing no decomposing plant residues, a highly toxic extract of soil containing plant residue previously assayed, an extract from decomposition, which was previously found to be nontoxic, and a 0.01 M phosphate buffer. After the pH of each was adjusted in the range indicated, each aliquot was assayed for inhibition of respiration of tobacco seedlings. The data indicated that the toxicity of the solutions was not due to acidity. Thus, they concluded that the toxic substances involved were produced chiefly under acid conditions but, once formed, were inhibitory over a fairly broad range of pH values.

Tests of extracts of soil with plant residues, which was autoclaved before setting for 15 days, indicated that very little production of toxins had occurred. If the autoclaved flasks were kept for as long as 25 days, most of the soil extracts reduced oxygen uptake of tobacco seedlings by as much as 35%. They inferred that decomposition of the plant residues was necessary for the production of toxins and that some of the soil organisms had to become re-established in autoclaved flasks that set longer than 15 days.

Extracts were made directly from the plant residues after grinding them to a fine powder, and these extracts were tested in the tobacco bioassay. No inhibiting effects were found. This was interesting because several investigators demonstrated that extracts of timothy, rye, and corn inhibited seed germination and seedling growth of several species of crop plants (Nielsen *et al.*, 1960; Guenzi and McCalla, 1962, 1966a; Grant and Sallans, 1964; Guenzi *et al.*, 1967). Perhaps the toxins that influence seed germination and seedling growth are different from those that affect respiration of tobacco seedlings. Patrick and Koch (1958) did not think so, however, because they tested extracts of soil containing plant residues against germination and growth of tobacco, barley, and timothy and found that only the extracts that inhibited uptake of oxygen in tobacco seedlings inhibited growth of seedlings. The answer to this dilemma is still not apparent.

Patrick and Koch (1958) noted that when tobacco seedlings were placed in some of the toxic extracts, the apical meristem region soon turned brown, whereas in other toxic extracts the zone of elongation turned brown while the apical meristem remained white. All of the toxic extracts also inhibited root hair formation.

In discussing increased production of toxins in poorly drained and acid soil, Patrick and Koch pointed out that initially all their soils were only weakly acid and some were only temporarily saturated. Thus, they postulated that localized zones probably occur in most soils where ideal conditions exist at least briefly, especially near pieces of plant residue. They felt it was especially significant that the roots of seedlings appeared to be most sensitive to the toxins produced by decomposing residues. They would be likely to contact localized zones of toxin production in the soil. They suggested also that there is a strong likelihood that any conditions that cause a reduction in respiratory activity of seedlings is also likely to cause some retardation of all the physiological functions.

Patrick *et al.* (1964) cited evidence for the fairly rapid breakdown of many phytotoxins in soil and pointed out that this led many investigators to conclude that the phytotoxins have no important effects on plants. They emphasized that the amount of inactivation is often balanced by new production and that the effective quantity is the difference between the amount produced and the amount destroyed.

Patrick *et al.* (1963) extended the investigations of Patrick and Koch (1958) to field studies in the Salinas Valley, California. Plant residue-containing soils were obtained from fields treated in the conventional way by growers. In some fields, cover crops of barley, rye, or wheat with or without vetch (*Vicia*) averaged 10 to 15 tons per acre of green plant material and were disked or plowed under. This was done just before the plants came into full head. In other fields, broadbean (*Vicia faba*), sudangrass (*Sorghum sudanense*), or remnants of a commercial crop of broccoli (*Brassica oleracea*, Botrytis Group) had been plowed or disked under. Composite samples of plant residue with the surrounding soil were taken periodically.

The samples were divided into three fractions: soil and residue in the relative proportions found in the field, soil after all recognizable plant residues were removed, and plant residue free from soil. Each fraction was extracted with water in a fashion similar to the method of Patrick and Koch (1958). When the extraction was completed, the pH and electrical conductivity were determined. The latter was done because the salt content increases in most fields under irrigation and controls were run containing a mixture of salts in water to give the same electrical conductivity. This enabled the effects of the salts to be balanced out in the test samples. Effects of test and control extracts on seed germination and seedling growth of lettuce were quantified. In some cases, broccoli, white beans (*Phaseolus vulgaris*), and tobacco were used to see how results compared.

The extracts of soil and residue in the proportions found in the field and extracts of soil after the residue was removed showed low toxicity, whereas the residue appreciably arrested the root growth of lettuce (Table 23). Soil immediately adjacent to residue and extracts of similar soil were bioassayed and found to be inhibitory to root growth of lettuce. Extracts of decomposing field residues of barley, rye, broccoli, broadbean, wheat, vetch, and sudangrass were found to be toxic to lettuce seedlings. Patrick *et al.* (1963) showed that the salinity of surface layers of soil increased during decomposition of plant residues and that the injurious effects of the decomposition products became greater with increasing salinity.

Careful field observations were made of lettuce and spinach plants to determine whether the stunting, uneven growth, and root injury often observed in the Salinas Valley could be attributed to decomposition products. Roots in contact

TABLE 23.   Effect of Water Extracts of Field Soil and Decomposing Plant Residues on Germination and Growth of Lettuce[a]

Plant residue	Time of decomposition in field (days)	Average length (mm) of radicles	
		Plant residue[c]	Control[d]
Barley	0–3[b]	15.3	21.8
	10–13	10.1	21.7
	15–18	13.6	19.8
	20–23	9.3	20.1
	23–26	14.6	19.6
Rye	10–13	14.1	19.5
	14–17	12.0	20.2
	23–26	13.1	21.0
	29–32	19.5	19.3
Wheat	17–20	14.1	19.6
	22–25	10.1	20.8
Vetch	5–8	17.0	20.0
	23–26	17.6	21.0
Broccoli	0–3	12.2	20.2
	20–23	8.1	19.8
Sudan grass	10–13	17.9	20.1

[a] Modified from Patrick et al. (1963).

[b] Fields (2 to 5) were sampled at each period; for brevity, the decomposition periods were grouped in 3-day ranges.

[c] Average of 20 lettuce seedlings after 100 hr in water extract; tests run at room temperature (68–75°F); tests with each extract replicated 4 times. Electrical conductivity of extracts adjusted to 1.5–1.8 mmhos/cm at 25°C; pH 7.0–7.5.

[d] Water containing equal amounts of NaCl, Na_2SO_4, $MgSO_4$, and $CaCl_2$ salts adjusted to give the above-mentioned conductivity readings and pH.

with or close to fragments of decomposing plant debris often had discolored or sunken lesions where the roots contacted the debris. In addition, there were many instances of browning of the apical meristems and other injuries.

Patrick et al. (1963) demonstrated that toxic decomposition products of plant residues are produced under field conditions. Apparently the toxins do not move far from the loci of production, however, and it appears that the extent of root injury and the total effect on the plant depend on the frequency of encounters of the growing root system with plant residue fragments.

Kimber (1973) performed a very elaborate set of experiments near Adelaide, Australia, to determine why wheat straw residues depressed yield of subsequent wheat crops. He pointed out that it has generally been assumed that the decrease in yield is due to immobilization of nitrogen by the increased population of soil microflora. Several methods of adding the residues were tested in combinations with several levels of nitrogen fertilizers. He concluded that both nitrogen immobilization and toxins affect the yield of wheat when it is grown in the presence of excess straw residues. Seed germination appeared to be depressed most by straw placed on the surface of the soil, whereas nitrogen immobilization affected yield the most when the straw was mixed into the soil. Kimber pointed out that addition of nitrogen did not overcome the effects of the straw. He inferred that this added effect was due to toxins derived from the decaying straw.

Kimber's investigation offers a good model to follow in research on plant interference. There is too much emphasis by some researchers on competition alone, and too much emphasis by others on allelopathy alone. Data are beginning to appear that indicate that certain aspects of competition and allelopathy (stimulation or inhibition at some level) operate in all plant interactions.

Tang and Waiss (1978) reported that the major compounds produced in decomposing wheat straw were salts of acetic, propionic, and butyric acids. Traces of isobutyric, pentanoic, and isopentanoic acids were also identified. Amounts increased gradually up to 12 days, and the toxicity of the straw extracts to wheat seedlings increased accordingly.

Patrick (1971) reported that toxins in decomposing rye residue included acetic, butyric, benzoic, phenylacetic, hydrocinnamic, 4-phenylbutyric, and ferulic acids. Five additional compounds, which were not identified, did not appear to be fatty acids.

Chou and Patrick (1976) subsequently identified vanillic, ferulic, phenylacetic, 4-phenylbutyric, p-coumaric, p-hydroxybenzoic, salicylic, and o-coumaric acids, plus salicylaldehyde in decaying rye residues. Most of these compounds were phytotoxic in the lettuce seed bioassay.

Altiera and Doll (1978) incorporated 20 g of chopped fresh foliage of beans (*P. vulgaris*) corn (maize), or cassava (*Manihot esculenta*) in the top 5 cm of soil in 1.85 liter pots, after which they planted seeds of one of these crop plants or one of nine weedy species. Each type of residue significantly reduced germination of three or more test species.

Chou and Patrick (1976) identified 18 compounds in decomposing corn residues in soil—salicylaldehyde, resorcinol, phloroglucinol, p-hydroxybenzaldehyde, and butyric, phenylacetic, 4-phenylbutyric, benzoic, p-hydroxybenzoic, vanillic, ferulic, o-coumaric, o-hydroxyphenylacetic, salicylic, syringic, p-coumaric, *trans*-cinnamic, and caffeic acids. Most of these compounds were phytotoxic in the lettuce seed bioassay.

Pareek and Gaur (1973) found that concentrations of vanillic, p-hydrobenzoic, p-coumaric, salicylic, and syringic acids were considerably higher in the rhizosphere soils of *Zea mays* than in non-rhizosphere soils. On the other hand, they found pyrocatechol only in non-rhizosphere soil. Concentrations of p-hydroxybenzoic and salicylic acid were also much higher in rhizosphere soils of *Phaseolus aureus* (mung bean) than in non-rhizosphere soils. Gluconic acid was present constantly in both rhizosphere and non-rhizosphere soils, whereas tartaric acid was detected in non-rhizosphere soil of mung beans and in uncultivated soil. Citric acid was found only in the rhizosphere soil of maize.

It has been observed for some time in Senegal, Africa, that growth of sorghum is markedly decreased following sorghum in sandy soils, but not at all in soils high in montmorillonite (Fig. 3) (Burgos-Leon, 1976; Burgos-Leon *et al.*, 1980). Burgos-Leon and his colleagues investigated the reasons for these observations. They noted the same results with sorghum seedlings when roots or tops of sorghum were added to sandy soil in laboratory experiments. No growth inhibition resulted, however, when residues were added to soils high in montmorillonite. The results were similar when sorghum seedlings were grown with the same materials under sterile conditions. Water extracts of roots or tops inhibited growth of sorghum seedlings similarly, and the extracts also reduced growth of *Lolium perenne* seedlings in bioassays.

The principal inhibitors in roots of sorghum were identified as *p*-coumaric, *m*-hydroxybenzoic, and protocatechuic acids. Acid hydrolysis of root extracts liberated large quantities of *o*-hydroxybenzoic acid, which, however, did not occur free in the root extract. Using sterile techniques, he demonstrated that inoculation with *Trichoderma viride* or an unknown species of *Aspergillus* eliminated the inhibitory effects of aqueous extracts of roots of sorghum on sorghum seedling growth in a short time. In laboratory experiments, *Aspergillus* sp. prevented toxic effects of water soluble materials from sorghum. In subsequent experi-

Fig. 3.   Reduction in growth of sorghum following a previous sorghum crop in a sandy soil in west Africa: on the left, growth of sorghum following peanuts; on the right, growth of sorghum following sorghum. (Photograph courtesy of Y. Dommergues.)

ments with uninoculated, nonsterile field soil, several weeks were required to detoxify the soil after addition of root residues of sorghum. It was concluded that the native microflora in the sandy soils of Senegal could not detoxify the soil fast enough to prevent growth inhibition of subsequent crops of sorghum in the same soil.

Stevenson (1967) pointed out that the soils of rice paddies in Japan and India contained high enough concentrations of aliphatic acids to inhibit root growth of rice. He also pointed out that anaerobic conditions were favorable to the microbial synthesis of organic acids.

Chandramohan *et al.* (1973) isolated vanillic, *p*-hydroxygenzoic, *p*-coumaric, and three unidentified phenolic acids from rice field soil at Annamalainagar, South India. They found that cinnamic acid (a related compound) was inhibitory to the growth of rice seedlings even at 0.0001 M concentration. Furthermore, addition of a nitrogen fertilizer increased the productivity of rice cultivar Co. 13, but it also markedly decreased the concentrations of phenolic compounds in the soil. They concluded, therefore, that high nitrogen application to soil played a dual role in increasing productivity: (1) It improved nitrogen nutrition, and (2) it dedreased toxicity of the soil caused by the phenols. This is a very important point to consider for investigators who attribute interference solely to competition. Numerous scientists have concluded that no allelopathic mechanism is operative if fertilizing helps to overcome a growth deficiency. This is questionable in many instances, because many investigators have demonstrated repeatedly that increased levels of nitrogen, phosphorus, and most other major elements decrease concentrations of phenolic compounds in most experimental plants (see Chapter 12).

The unharvested parts of rice plants are customarily mixed with the soil by plowing or other mechanical manipulation, because this has been considered beneficial. It has been commonly observed, however, that yield of the second rice crop in a paddy is less than that of the first crop. Chou and Lin (1976) studied the effects of decomposing rice residues in soil on the growth of rice plants. They found that aqueous extracts of decomposing rice residues in soil inhibited radicle growth of rice (cv Taichung 65) and lettuce (cv Great Lakes 366) seedlings and the growth of rice plants. Maximum toxicity occurred in the first month of decomposition and declined thereafter. Extracts of the soil in rice fields also reduced the growth of rice and lettuce, with the toxicity persistent for 4 months. Root initiation of hypocotyl cuttings of mung beans was suppressed also by extracts of decaying rice residues and of paddy soil. Five phytotoxins— *p*-hydroxybenzoic, *p*-coumaric, vanillic, ferulic, and *o*-hydroxyphenylacetic acids—were identified from decomposing rice residues under waterlogged conditions, and several unknowns were isolated. At 25 ppm, *o*-hydroxyphenylacetic acid caused significant inhibition of the radicle growth of rice and lettuce seedlings and suppressed root initiation of mung bean seedlings. All of the other identified toxins have been demonstrated to inhibit the growth of many species of plants, even when the toxins are present in low concentrations. Chou and Lin concluded that the decline in productivity of rice subsequent to the first crop was due chiefly to the allelopathic effects of decaying rice residues in the paddy soil. This conclusion was subsequently supported by Chou et al. (1977) and Chou and Chiou (1979). Chou et al. (1981) found that the toxicity of decomposing rice straw was highest when incubated at 20° to 25°C. Temperatures above 25°C enhanced decomposition, but they also degraded the phytotoxic substances more rapidly.

Large amounts of chemical fertilizer, including nitrogen, are required to maintain a high rice yield (Huang, 1978). Economically, it would be important to know if a considerable portion of the nitrogen could be furnished by biological nitrogen fixation. This might be accomplished either by rotating legume crops with rice or by inoculating paddies with appropriate blue-green algae, if conditions conducive to nitrogen fixation could be maintained. Chou et al. (1977) reported that the quantities of leachable nitrate (NO_3^-) and ammonium nitrogen (NH_4^+) were lower in paddies in which the stubble was not removed. However, they did not investigate the effects of decaying rice straw on nitrogen fixation.

Filamentous blue-green algae, which possess heterocysts, often have vigorous nitrogen-fixing activity. It has been estimated that free-living blue-green algae add from 13 to 70 lb of nitrogen per acre per year to rice paddy fields (De and Sulaiman, 1950; De and Mandal, 1956; Singh, 1961). Galston (1975) found that blue-green algae associated with *Azolla* in rice paddies gave yields 50 to 100% greater than those obtained in adjoining paddies without *Azolla*. Huang(1978) demonstrated that inoculation of pots of rice with blue-green algae increased the grain production by 34 to 41%, depending on the rice cultivar used.

Based on this, Rice et al. (1980) investigated the effects of the five phenolic inhibitors that are produced by decomposing rice straw on growth and on nitrogen fixation of *Anabaena cylindrica*. Four of the five compounds inhibited the growth of *A. cylindrica,* and four of the five also reduced nitrogen fixation (acetylene reduction). A combination of all five compounds was particularly effective in inhibiting growth and nitrogen fixation, indicating a synergistic effect. This is especially important because the five compounds always occur together in decomposing rice residues. These experiments should be extended to field tests, with and without rice residues, because of the paramount importance of nitrogen fixation by blue-green algae in rice culture.

In southern Taiwan, a crop of rice is often followed immediately by a legume crop, commonly consisting of various types of beans (*Phaseolus* spp., *Vigna* spp. etc.) or soybeans. The AVRDC Soybean Report for 1976 (Asian Vegetable Research and Development Center, 1978) cited numerous data indicating that soybean yields in southern Taiwan were increased by several hundred kilograms per hectare when rice stubble was burned before planting the soybeans, rather than allowing the straw to remain in the field and decompose. No explanation was suggested for this result, but it is possible that phytotoxins produced by the decomposing rice straw reduced nitrogen fixation by *Rhizobium* in the nodules of the soybean plants, inhibited growth and yield of the soybean plants directly, or both. Therefore, Rice *et al.* (1981) investigated the first of these possibilities.

All five phenolic compounds produced by decomposing rice straw retarded the growth of three species of *Rhizobium* on agar plates. One of these was a strain of *R. japonicum* isolated from soybean nodules in Taiwan. Some combinations of the phenolics had marked synergistic effects in addition to additive effects. Sterile aqueous extracts of decomposing rice straw in soil inhibited growth of *R. japonicum* and one other species of *Rhizobium*.

The phenolic compounds also reduced the nodule numbers and the hemoglobin content of the nodules in two bean (*Phaseolus vulgaris*) varieties. Aqueous extracts of decomposing rice straw in soil (same concentration as in the soil) significantly reduced nitrogen fixation (acetylene reduction) in Bush Black Seeded beans. It appears likely, therefore, that the reduction of nitrogen fixation may be responsible, at least in part, for the lowered soybean yields in Taiwan following rice crops when the rice stubble is left in the field.

Successive cropping of eggplant (*Solanum melongena*) or foxtail millet (*Setaria italica*) in Korea results in very poor harvests. Lee *et al.* (1967) suspected that allelopathy might be responsible and found that incorporation of the previous year's roots of eggplant into fresh soil severely reduced growth of eggplant. Moreover, incorporation of the previous year's roots of foxtail millet into fresh soil markedly reduced the growth of foxtail millet. Root exudates of eggplant also inhibited the growth of eggplant, and root exudate of foxtail millet reduced the growth of foxtail millet.

Peters (1968) observed that both thin and dense stands of Kentucky-31 fescue (*Festuca elatior*) are usually relatively free of weeds. He suspected, therefore, that allelopathy might be involved in interference of this species against other plants. He found that water extracts of roots and leaves of fescue reduced the germination and the growth of roots of rape and birdsfoot trefoil (*Lotus corniculatus*). Rape and fescue seeds germinated normally in sand cultures in which fescue was growing, but the growth of rape leaves was significantly reduced. When one-half of the rape and fescue root systems were supplied separately with nutrient solution and one-half were grown together in distilled water, fescue inhibited the growth of rape roots in the distilled water. This experiment eliminated competition for nutrients as an explanation for the inhibitory effects of fescue on the growth of rape.

The findings of Kochhar *et al.* (1980) substantiated the conclusions of Peters (1968) concerning the allelopathic potential of Kentucky-31 fescue. Leaf leachate from fescue inhibited the growth of ladino clover and leaf leachate from fescue plants, which were exposed to O_3 reduced nodulation of ladino clover. Leachate from nonexposed fescue plants, however, did not reduce nodulation. Addition of fescue, clover, or fescue-and-clover debris to soil stimulated the growth of both clover and fescue.

Overland (1966) pointed out that "smother crops" were often planted to suppress the growth of weeds. Some of these included barley, rye, sorghum, buckwheat (*Fagopyrum esculentum*), sudangrass, sweet clover, and sunflower (*Helianthus annuus*). She also stated that, despite the lack of proper evidence, it was assumed that smother crops inhibited weed growth through competition. The high rank of barley as a smother crop was thus attributed to its extensive root growth.

Overland hypothesized that barley might be successful at eliminating weeds because it produces toxins in addition to its competitive effects. Her experiments showed that barley inhibited seed germination and the growth of selected plant species, even in the absence of competition. This occurred both in mixed cultures receiving adequate nutrients and water and in germination tests. Water-soluble root exudates of barley caused similar inhibition of germination and the growth of the same species, indicating the involvement of an inhibitory substance or substances. There was a specificity of action of the inhibitor(s), with the greatest inhibition occurring with *Stellaria media*, less inhibition with *Capsella bursa-pastoris* and tobacco, and no significant effect with wheat. There was a definite concentration effect and a suggested periodic production of the inhibitor(s). Living plants and exudates of living roots were more inhibitory than aqueous leachates of dead roots, which supports the concept of an active metabolic secretion of the allelopathic substance(s). Attempts to identify the inhibitor(s) demonstrated the presence of alkaloids, with a much greater concentration of organic compounds in the exudates of living roots than in the leachates of similar amounts of dead roots. The alkaloid gramine, which is produced by barley, inhibited the growth of *Stellaria*, even when present in relatively low concentrations and when extra nutrients were added to the culture medium. Overland pointed out that wheat, which was not inhibited by barley, is frequently grown with barley in primitive agriculture. Most of the "cover crop" species listed by Overland (1966) have been reported to be allelopathic to certain test species.

Clover soil sickness has been known in Europe since the seventeenth century (Katznelson, 1972). Red clover, *Trifolium pratense*, is allelopathic to adjacent red clover plants, and Tamura *et al.* (1967, 1969) isolated and identified nine inhibitory isoflavonoids or related compounds from the tops of red clover. These were ononin, genistein, biochanin A, biochanin A-7-glucoside, and for-mononetin. Subsequently, Chang *et al.* (1969) investigated the biological activity of inhibitors previously identified by Tamura *et al.*, compared their relative concentrations in red clover, white clover (*T. repens*), and orchardgrass (*Dac-*

tylis glomerata), identified inhibitory compounds in soil in which red clover grew, isolated inhibitors from the culture solution in which red clover grew, and determined the degradation courses of isoflavonoids.

All the identified isoflavonoids at 100 ppm inhibited seed germination of red clover by about 50%. A weak synergistic response was observed when several of the inhibitors were combined. Concentrations of the inhibitors required for 50% inhibition of germination were 300 ppm for white clover and 700 ppm for alsike clover. Germination of common vetch (*Vicia sativa*) seeds was hardly inhibited at all even at 1000 ppm.

All identified isoflavonoids from red clover inhibited red clover seedling growth equally in nutrient solution, with inhibition occurring even at concentrations as low as 10 ppm. Rice seedlings suffered no growth reduction even at 1000 ppm of the toxins. Lettuce hypocotyls were not affected, but lettuce root growth was strongly inhibited at 10 ppm or above. The isoflavonoids arrested seedling growth of red clover in soil over an entire 10-week trial period.

The total amount of daidzein, formononetin, genistein, biochanin A, and trifolirhizin in red clover was more than 1000 times as high as in white clover or orchardgrass.

Careful extraction resulted in the isolation of no isoflavonoids from ''sick'' soil. Relatively high amounts of *p*-methoxybenzoic, salicylic, *p*-hydroxybenzoic, and 2,4-dihydroxybenzoic acids were isolated from this soil, however. These compounds plus other phenolic compounds originated from decomposition of the isoflavonoids present in red clover. Subsequent research demonstrated that isoflavonoids were exuded from roots of red clover into a nutrient solution and that many of the phenolic compounds isolated from soil quickly appeared in the culture solution also. Hence, Chang *et al.* (1969) concluded that ''clover sickness'' results from the exudation by red clover of isoflavonoids that decompose to phenolic compounds, which accumulate in the soil to toxic levels.

Katznelson (1972) investigated the clover-soil sickness problem resulting from continuous growth of berseem (*Trifolium alexandrinum*) or Persian clover (*T. resupinatum*) in Israel. He concluded that nematodes may be the cause of the problem in Persian clover but not in berseem. Leaves of berseem from plots previously cropped to berseem had much lower phosphorus levels than the leaves of plants in control plots, and the nutritional imbalance could not be corrected by fertilizer applications. Thus, he concluded that the berseem soil sickness might be due to allelopathic factors.

Over 2000 years ago, Theophrastus suggested that chickpea inhibits the growth of subsequent crops, and several other legumes, in addition to various species of *Trifolium,* have been reported to have allelopathic potential. Nielsen *et al.* (1960) reported that aqueous extracts of alfalfa hay, timothy hay, corn stover, oat straw, and potato (*Solanum tuberosum*) vines reduced seed germination and seedling growth of several crop plants. Alfalfa had the greatest inhibitory effect on both germination and growth. Unfortunately, no tests were run to determine if the results may have been due to osmotic concentrations of the extracts.

Grant and Salans (1964) tested aqueous extracts of roots and tops of alfalfa, red clover, ladino clover (*T. repens* forma *lodigense*), birdsfoot trefoil, timothy, bromegrass (probably *B. inermis*), orchardgrass, and reed canarygrass (*Phalaris arundinacea*) in the late vegetative or early flowering stage. Each extract was tested against seed germination and seedling growth of all of the same species. Legume extracts lowered the percentage of germination in about one-half of the tests, with the extracts of shoots being generally more toxic than the root extracts. In general, grass extracts had less effect on seed germination than legume extracts. The legume extracts reduced seedling growth in most tests, and the grass extracts reduced growth in several tests. Many of the extracts caused twisted roots, prevented root hair development and, in the cases of top extracts of alfalfa and reed canarygrass, killed the roots of test plants. Grant and Salans rated the species based on decreasing inhibition in the following order: alfalfa, birdsfoot trefoil, ladino clover, red clover, reed canarygrass, bromegrass, timothy and orchardgrass.

Grant and Sallans pointed out the difficulty of extending their results to natural conditions, but they suggested that the suppression of root development of birdsfoot trefoil by all extracts might explain the difficulties experienced in obtaining good stands of birdsfoot trefoil. They suggested also that the pronounced growth inhibition of all species by alfalfa might help explain why alfalfa is so aggressive when grown with other forage crops. They reported that timothy and red clover are frequently planted together and their data indicated that these species are quite compatible.

Kaurov (1970) investigated the growth of birdsfoot trefoil and yellow lupine (*Lupinus luteus*) in pure and mixed cultures and found that production per unit area was greater in mixed cultures than in a pure culture of either. He also found that birdsfoot trefoil was more active in absorption of radioactive phosphorus than lupine and that the rate of exudation of ^{32}P from lupine and absorption by birdsfoot trefoil was higher than the rate of exudation of ^{32}P from birdsfoot trefoil and absorption by other birdsfoot trefoil plants. He concluded, therefore, that lupine in mixed cultures improved the mineral nutrition of birdsfoot trefoil. There was no direct evidence of allelopathic effects, but evidence from other investigations suggests the possibility.

Pronin and Yakovlev (1970) reported that yields of fodder beans (probably *Vicia faba*) and maize increased in mixed cultures under both sterile and nonsterile conditions and that the increase was associated with a favorable influence of root excretions of each plant on the other. They stated that under nonsterile conditions, the microorganisms of the root zone intensified the positive or negative influences of plants. The positive effects of maize on the legume indicated that more was involved in the interaction than improvement of nitrogen nutrition by the legume.

Lykhvar and Nazarova (1970) investigated growth of several species of legumes and maize in pure and mixed cultures. They found that beneficial effects of legumes grown in mixed cultures with maize differed with varieties of the legume species. Many varieties gave detrimental results in mixed cultures indicating truly allelopathic effects. As a consequence of these experiments, new varieties of legumes were developed specifically for use in mixed cultures with maize or other crop or forage plants. Such research appears to hold excellent promise in agriculture.

Gaidamak (1971) grew several crops of tomatoes and cucumbers in "gravel" culture using broken bricks as the substrate. He used fresh culture solution at the start of each new crop but continued to use the same broken brick substrate for each successive crop. He analyzed the nutrient solutions used to grow the crops, and each solution contained amino acids, organic acids of an aliphatic series, phenolics, and several unidentified compounds. Most of the compounds had either a positive or negative effect on growth of certain test species. The solution used to grow the eighth crop of tomatoes contained, in addition to those mentioned, some unidentified toxins and phenolcarbonic acid, all of which had strong inhibitory effects on test species.

Dadykin et al. (1970) found acetaldehyde, propionic aldehyde, acetone, methanol, ethanol, and other unidentified compounds in volatile secretions of beet, tomato, sweet potato (*Ipomoea batatas*) and radish leaves, and carrot (*Daucus carota*) roots. Propionic aldehyde had the greatest reduction in growth activity against test species in closed systems.

Rakhteenko et al. (1973a) reported that substances exuded by roots of pea and vetch (*Vicia villosa*) stimulated photosynthesis and absorption of ^{32}P in barley and oat plants. The exudates also stimulated uptake by the cereals of nitrogen, potassium, and calcium from nutrient solution. In contrast, substances exuded from the roots of the cereals inhibited the same processes in the two legumes.

Rakhteenko et al. (1973b) supplied leaves of several species of plants with $^{14}CO_2$ and studied the exchange of organic root exudates in sand cultures of pure and mixed stands of plants. The exchange of root exudates in mixed cultures of peas and oats was 1.5 to 2 times as great as in pure stands of each of these species. In mixed cultures of buckwheat and white mustard (*Brassica hirta*), the exchange of organic root exudates was 3 to 7 times as great as in pure stands of these species. They concluded that these results helped explain the causes of growth inhibition or stimulation of one species by another in mixed cultures.

Zabyalyendzik (1973) investigated the interactions of buckwheat, lupine, mustard, and oats in field and greenhouse experiments. He reported that yields of tops of buckwheat were 30 to 35% greater and yields of grain were 12 to 35%

greater in mixed crops than in pure stands. Water-soluble root exudates of lupine and mustard stimulated growth and development of buckwheat. In contrast, water soluble root exudates of oats inhibited growth and the yield of buckwheat. Root exudates of buckwheat stimulated the growth of oats and increased productivity of oats by 10 to 20%, whereas exudates of buckwheat inhibited growth and yield of lupine.

Litav and Isti (1974) investigated what they termed *root competition* in two horticultural strains of *Spinacia oleracea*—Early Hybrid and F_1 Selma. Suppression of Early Hybrid by F_1 Selma was not eliminated by increasing nutrient levels up to several times the basic dose. Suppression was not accompanied by reduced nitrogen and phosphorus concentration in the shoots. They inferred that F_1 Selma probably was allelopathic to Early Hybrid because of the above considerations as well as the fact that suppression of Early Hybrid was not accompanied by the expected increase in yield of its adversary.

Yurchak (1974) reported that several bacteria involved in the decomposition of *Lupinus* residues synthesized indole and gibberellin-like growth promoting substances on both organic and synthetic media. These substances were found not only in the culture medium, but also in the biomass of some bacteria (e.g. *Bacterium zopfii*). Nadkernichnyi (1974) discovered that continuous cultivation of *Lupinus* and potatoes increased the toxin-producing forms of *Penicillium* and *Aspergillus* in soil. Some species of *Fusarium* and *Mucor* and some species of the family Dematiaceae, which were present, also produced phytotoxins.

Sajise and Lales (1975) used a root-divider technique to study interactions between cogongrass and *Stylosanthes guyanensis*. They determined that the allelopathic activity of cogongrass accounted for a 38% reduction in the growth of *Stylosanthes*. However, *Stylosanthes,* also inhibited cogongrass growth during the first 8 months after transplanting. This is a good example of dual allelopathy, which probably occurs commonly, but is not generally investigated.

Dzyubenko *et al.* (1977) reported that the amounts of toxins increased in the roots and in the rhizosphere when various crop plants were grown in continuous monoculture. Moreover, the increase in plant and soil toxicity was inversely correlated with a decrease in crop productivity.

Petrova (1977) discovered that volatile compounds from the tops of soybean, chickpea, and bean (presumably *Phaseolus* sp.) reduced uptake of ^{32}P by corn plants. The combined volatile compounds, however, from the tops and roots of the same crop plants stimulated uptake of ^{32}P by corn plants. The overall effect on ^{32}P uptake was in good agreement with results on the yield from mixed sowings of these plants under field conditions.

In soil that has not had recent additions of plant residue or other organic material, microbial respiration proceeds at a low rate (Menzies and Gilbert, 1967). Moreover, fungi apparently exist mostly as spores in a state of fungistasis. This microflora usually responds to the addition of plant residue by spore germination, increased respiration, and growth. Menzies and Gilbert (1967) reported that these responses were induced by a volatile component from alfalfa tops, corn leaves, wheat straw, bluegrass (*Poa pratensis*) clippings, tea (*Thea sinensis*) leaves, and tobacco leaves, even when the residue was separated from the soil by a 5-cm air gap. There was a rapid outgrowth of hyphae from the soil surface toward the residue before any growth of fungi could be seen in the plant material. A dense network of hyphae could be detected within 24 hours, with many of the filaments oriented at right angles to the soil surface and reaching almost across the air gap. This development did not occur where the residue was omitted or replaced by moist filter paper.

Vapors from distillates of water extracts of the various plant residues had similar effects on the growth of fungi and markedly increased the number of bacteria and the respiratory rate of microorganisms in soil samples (Menzies and Gilbert, 1967). This fascinating allelopathic phenomenon may be very important in the initial colonization of residue and, thus, in decomposition.

In Hawaii, greenleaf (*Desmodium intortum*) could not be established in a pasture of bigalta (the tetraploid *Hemarthria altissima*), whereas good survival of greenleaf was obtained in a pasture of the diploid *H. altissima* (greenalta) (Young and Bartholomew, 1981). To determine if the failure of greenleaf to become established in a bigalta pasture was due to allelopathy, root residues of greenleaf, bigalta, or greenalta were added to soil in which none of these plants had grown previously. Greenleaf seeds were then planted alone or in mixtures with cuttings of bigalta or greenalta. All of the pots were limed and fertilized equally, and the plants were allowed to grow for either 77 or 153 days. Fresh weights were determined and the tops were analyzed for P, K, Ca, Mg, and N.

Growth of greenleaf was very poor in soil containing root residue of bigalta, with a fresh weight about one-fourth that of greenleaf growing in soil containing either greenleaf or greenalta root residues. Levels of N, K, Mg, and Ca in the tops of greenleaf were unaffected by the source of residue. The concentration of P in greenleaf tops grown in bigalta residue, however, was 0.15% as compared with 0.20% or greater for plants grown in the other residues. The inhibition of vegetative growth by bigalta residue in the absence of competing species ruled out competition as a factor. Previous work showed that root extracts and root exudates of bigalta inhibited seed germination and radicle growth of greenleaf and lettuce.

Research at the University of Illinois demonstrated that the alfalfa yield declines after a few years in the same soil (Miller, 1982). This occurred even though corrections were made for changes in fertility, and the alfalfa plants were protected from fungal attacks. Field results indicated that the best preceding crop for alfalfa establishment was corn followed by various small grains and soybean, and the worst preceding crop was alfalfa. The results suggested that alfalfa produces allelopathic compounds that are very inhibitory to alfalfa growth.

Putnam *et al.* (1983) reported that amending the soil with living or dead root tissues of asparagus (*Asparagus officinalis*) reduced the growth of tomato and barnyardgrass by as much as 80% as compared with peat moss-amended controls. Rhizome and shoot tissues were somewhat inhibitory but considerably less so than roots. When asparagus root tissues were added to pots containing 1-year-old asparagus crowns, shoot and root growth were inhibited as much as 66% and 51%, respectively. The toxicity could be overcome by adding activated charcoal to the pots. Extracts of the roots also caused severe mortality in populations of isolated asparagus cells. It appears likely that the autotoxicity of asparagus may be responsible, at least in part, for the marked asparagus decline which occurs in asparagus fields after several successive years of production (Takatori and Souther, 1978).

III. ALLELOPATHIC EFFECTS OF CROP PLANTS ON WEEDS

Funke (1941) pointed out that beet seeds produce substances that inhibit the growth of *Agrostemma githago* but not that of white mustard. He investigated numerous other instances of selective toxicity also and concluded that such activity explained the exclusive presence of certain weed species in cultivated fields. The potential economic importance of this phenomenon caused Funke to suggest that much more careful research is needed.

Dzyubenko and Petrenko (1971) investigated the interactions between two weed species, *Chenopodium album* and *Amaranthus retroflexus;* and two crop species, *Lupinus albus* and *Zea mays,* in laboratory and greenhouse experiments with water and sand culture. Root excretions of the crop plants inhibited growth of the weeds and increased their catalase and peroxidase activity. Exudates of the roots of the weed species, however, stimulated growth of the cultivated plants.

Neustruyeva and Dobretsova (1972) reported that wheat, oats, peas, and buckwheat suppressed growth, accumulation of above-ground biomass and leaf surface of lambsquarters. Oats had a marked effect, which peaked at flowering time. Oats did not decrease the uptake of nitrogen, phosphorus, and potassium by *Chenopodium* as buckwheat did in spite of the fact that oats had a much greater interference. They concluded that oats exerts an allelopathic effect in addition to its competitive role. This conclusion was supported by Markova (1972), who found that oats suppressed the growth of *Erysimum cheiranthoides* in both laboratory and field tests due at least in part to an allelopathic mechanism.

Lazauskas and Balinevichiute (1972) tested excretions from seeds of hairy vetch against seed germination and seedling growth of thirteen species of weeds. They found that seed germination and particularly seedling growth were markedly inhibited in many tests.

Prutenskaya (1972) reported that germinating seeds of millet (species not given), wheat, oats, vetch (species not given), maize, and buckwheat stimulated germination of crunchweed (*Brassica kaber* var. *pinnatifida*) seeds, whereas germinating barley seeds inhibited germination of crunchweed seeds. She found that the compounds which stimulated or inhibited germination of crunchweed seeds were produced during germination. Prutenskaya (1974) demonstrated that wheat, rye, and barley strongly inhibited the weedy *Brassica kaber,* whereas millet (*Panicum miliaceum*) stimulated the weed.

Putnam and Duke (1974) screened 526 accessions of cucumber (*Cucumis sativis*) and 12 accessions of eight related *Cucumis* species, representing 41 nations of origin, for allelopathic activity against a forb, *Brassica hirta,* and a grass, *Panicum miliaceum.* One accession inhibited indicator plant growth by 87% and 25 inhibited growth by 50% or more.

Fay and Duke (1977) screened 3000 accessions of the USDA World Collection of *Avena* spp. germ plasm for their capacity to exude scopoletin, a naturally occurring compound shown to have root growth inhibiting properties. Twenty-five accessions exuded more blue-fluorescing materials (characteristic of scopoletin) from their roots than a standard oat cultivar (Garry). Analysis of the exuded materials indicated that four accessions exuded up to 3 times as much scopoletin as Garry oats. When one of these four was grown in sand culture for 16 days with crunchweed, the growth of the crunchweed was significantly less than that obtained when the weed was grown with Garry oats. The plants grown in close association with the toxic accession exhibited severa chlorosis, stunting, and twisting indicative of chemical (allelopathic) effects rather than simple competition.

The work of Overland (1966), Peters (1968), and Altiera and Doll (1978) (see Section II, this chapter) indicates that several other crop plants are allelopathic to various species of weeds. Nevertheless, it is obvious that this very important phase of allelopathy has been neglected, and much more needs to be done.

A. Possible Uses of Allelopathy in Biological Weed Control

The similarity of many allelopathic agents (see Chapter 10) to compounds involved in resistance of plants to many diseases (Schaal and Johnson, 1955; Kuć *et al.,* 1956; Hughes and Swain, 1960; Farkas and Kiraly, 1962; Gardner and Payne, 1964) indicates that it should be possible to breed crop plants for resistance to selected pernicious weeds of a given locality, just as crop plants are bred for resistance to disease.

Based on results of their screening of accessions of cucumber for allelopathic potential, Putnam and Duke (1974) concluded that incorporation of an allelopathic character into a crop cultivar could provide the plant with a means of gaining a competitive advantage over certain important weeds.

Unfortunately, little work has been done on the genetics of allelopathic agents. Panchuk and Prutenskaya (1973) and Grodzinsky and Panchuk (1974) worked, however, on the genetic transfer of toxins in hybrids of wheat grass (*Agropyron glaucum*) and wheat (cv Lutescence 329). They found that water extracts of residues of wheat grass were more toxic than extracts of wheat residues against seed germination of radish and against root growth of *Lepidium sativum*. The first generation hybrids exhibited chiefly wheat grass characteristics and manifested high inhibitory activity. Other hybrids studied had intermediate inhibitory activity. The more wheat characters the hybrid had, the less allelopathic activity was found. The inhibitors were identified in both wheat and wheatgrass, and the first generation hybrids, which exhibited chiefly wheatgrass characters, contained a greater number of inhibitors present in wheatgrass. In one hybrid characterized by a maximum manifestation of wheat characters, only inhibitors generally found in wheat were present. All other hybrids occupied an intermediate position with respect to the types of inhibitors.

Much more research needs to be done on genetics of allelopathic agents, and many more screening programs need to be carried out before large scale breeding programs can get underway to produce commercial cultivars of important crop plants that are allelopathic to selected weeds. Screening programs should include not only commercial accessions of crop plants but also wild-type relatives. If promising types will not hybridize with desirable cultivars, various new techniques may be used such as protoplast fusion to introduce the allelopathic genes into the cultivars. Selections should also be made for accessions or wild types that are resistant to allelopathic agents produced by certain important weed species, and this resistance should be introduced into desirable cultivars.

As commercial fertilizers become more expensive, the use of legumes for biological nitrogen fixation will increase, and many legumes are known to be allelopathic to crop plants. Thus, there will have to be more emphasis on selection and breeding of compatible plants for mixed cropping. Considerable emphasis has been placed on this type of reserach already in the Soviet Union.

Emphasis to this point in this section has been on the breeding of allelopathic crop plants to control weeds. Obviously, allelopathy can be used in weed control in several other ways. One method is to use allelopathic agents from various sources as herbicides. Rhizobitoxine is an allelopathic agent produced by certain strains of *Rhizobium japonicum* in soybean nodules (Owens *et al.,* 1972), which causes the host soybean plant to become chlorotic. Anonymous (1969) pointed out that rhizobitoxine is an effective herbicide in amounts as low as 3 oz/acre. Owens (1973) compared the herbicidal effects of rhizobitoxine with two commercial herbicides—amitrole and metflurazone. In postemergence tests with seedlings of various weed and crop species, rhizobitoxine and amitrole were approximately equal in phytotoxicity on a weight basis, and both were generally more phytotoxic than metflurazone. Phytotoxicity of rhizobitoxine varied markedly among various grass species. Crabgrass was moderately sensitive to it, whereas Kentucky bluegrass was very tolerant. Amitrole, on the other hand, was almost as phytotoxic to bluegrass as to crabgrass and metflurazone had little effect on either. This difference in sensitivity of crabgrass and bluegrass to rhizobitoxine could be important in the postemergence control of crabgrass in bluegrass lawns. Postemergence control of crabgrass in such lawns has been seriously limited by the narrow difference in sensitivity between crabgrass and bluegrass to the few herbicides marketed for this purpose.

Agrostemmin, one of the allelopathic agents produced by corn cockle, decreased numbers of weedy forbs and increased yields of grass and the nitrogen content of the grass when applied to pastures in the Zlatibor (Yugoslavia) at a rate of 1.2 g/ha (Gajić, 1973).

Rizvi et al. (1980) stated that pesticides from plant sources are more systemic and more easily biodegradable than synthetic pesticides. Therefore, they started a survey of effects of ethanolic extracts of leaves and seeds of over 50 species of weeds and crop plants against seed germination of the weed, *Amaranthus spinosus*. The extracts of coffee (*Coffea arabica*) were most inhibitory, but extracts of seeds or leaves of 12 other species caused 30 to 50% inhibition, and eight other species caused 20 to 30% inhibition in the same concentrations.

Because of its pronounced inhibitory effect, the water-soluble residue of the coffee seed extract was fractionated further with several solvents, and the most active fraction was soluble in chloroform. This fraction was subsequently found to be toxic to seven additional species of weeds used in tests, but it had little or no effect on the crop species black gram (*Phaseolus mungo*). Rizvi et al. (1981) identified the most active phytotoxic compound in the chloroform fraction as 1,3,7-trimethylxanthine (caffeine). A concentration of 1200 ppm in water completely inhibited seed germination of *Amaranthus spinosus*, but it did not affect seed germination or subsequent growth of black gram. Thus, this compound appears to hold promise as an effective herbicide in at least some crops.

The term allelopathy is rarely used in plant pathology literature. Nevertheless, pathogenic bacteria and fungi produce phytotoxins that affect growth and even survival of the host plant. This is an allelopathic phenomenon. There is growing use and the promise of applying effective weed pathogens to control weeds (Templeton et al., 1979). Toxins produced by pathogens in large laboratory cultures will probably be widely used eventually for control of some weeds. Toxins produced by fungal pathogens of Canada thistle are being tested for control of this weed, and three toxins have proved promising (Anonymous, 1980). These toxins can work even if the weeds are resistant to infection by the fungi that produce them.

Another way in which allelopathy may be used in weed control is to apply residues of allelopathic weeds or crop plants as mulches or plant an allelopathic crop in a rotational sequence and allow the residues to remain in the field (Altiera and Doll, 1978; Harwood, 1979; Putnam and DeFrank, 1979, 1983; Drost and Doll, 1980). Putnam and DeFrank (1979) tested residues of several fall- and spring-planted cover crops for weed control in Michigan. The plants were desiccated by the herbicides, glyphosate or paraquat, or by freezing. Tecumseh wheat, which was desiccated in spring or fall, reduced weed weights by 76 and 88%, respectively. Fall-killed Balboa rye was similar in effectiveness, but fall-killed Garry oats or spring-killed rye had no toxic action on weeds. Mulches of sorghum (Bird-a-Boo) or sudangrass (Monarch) applied to apple orchards in early spring reduced weed biomass by 90% and 85%, respectively. An equivalent mulch of peat moss had virtually no effect. Tree growth was not adversely affected. In fact, the shoot growth of trees in mulched pots was equal to or better than that of trees in unmulched plots.

This investigation was expanded in a 3-year series of field trials to assess the influence of allelopathic crop residues on emergence and growth of annual weeds and selected vegetable crops (Putnam and DeFrank, 1983). Populations of common purslane (*Portulaca oleracea*) and smooth crabgrass (*Digitaria ischaemum*) were reduced 70 and 98%, respectively, by residues of sorghum. Total weed biomass and weight increases of several indicator species were also consistently reduced with residues of barley, oats, wheat, and rye, as well as sorghum. The use of *Populus* excelsior as a control indicated that the effects of the cover crop residues were chemical and not physical. In general, the larger-seeded vegetables, particularly the legumes, grew normally or were even stimulated at times by the cover crop residues. Several species of smaller seeded vegetables, however, were severely injured. Thus, there appears to be a very promising future for the use of allelopathic cover crops or allelopathic mulches for weed control in agriculture and horticulture.

Putnam and his colleagues found that a major variable in the effectiveness of weed control seems to be the maturity level of the cover (mulch) crop at the time of killing. Early fall planting appears to be crucial if the sorghum is to achieve enough growth before being winter killed to be effective subsequently in suppressing weed germination and growth. He had strong evidence that stressed plants produce a stronger allelopathic effect than unstressed plants. This agrees, of course, with a large body of evidence in the literature (see Chapter 12).

Hilton (1979) reported that G. R. Leather at Frederick, Maryland, discovered that 1 of 13 genotypes of the cultivated sunflower (*H. annuus*) reduced total weed cover by 33%. Leachates from the allelopathic genotype significantly inhibited seedling growth of jimsonweed (*Datura stramonium*), lambsquarters, and morning-glory. Leather (1982, 1983) tested the validity of bioassay results in a 5-year field study with oats and sunflowers grown in rotation. Weed density increased in all plots, but the amount of increase was significantly less in plots in which sunflowers were rotated with oats than in control plots with oats only. He suggested that the use of crops with increased allelopathic potential could restrict the need for conventional herbicides to early season applications, with late season weed control being provided by the crop plants.

Hall *et al.* (1982) grew Russian mammoth sunflowers (*H. annuus*) in pots, both in the field and in the greenhouse. Some were stressed by low nutrient availability, and some by increasing the density of plants in the pots. The total phenolic concentration in the sunflower tissue increased with increasing nutrient stress but not with increasing density stress. Addition of the sunflower residues to the soil reduced seed germination of *Amaranthus retroflexus,* and the percentage of inhibition increased with increasing concentration of total phenolics in the sunflower residues.

Still another way allelopathy can be used in weed control is to utilize a companion plant that is selectively allelopathic against certain weeds and does not interfere appreciably with crop growth (Putnam and Duke, 1978). According to Hunter (1971), the amateur gardener experimenters of the Henry Doubleday Research Association discovered that Mexican marigold (*Tagetes minuta*) destroyed certain starchy rooted weeds associated with it. In some cases, the roots of ground elder were reduced to hollow shells with browning. Some appeared to be eaten through, as if by an acid, and these effects were found even at some distance from the marigold plants. Obviously, considerable research will be necessary to discover appropriate companion plants that will kill or suppress undesired weeds and be compatible with various crop plants in a given area.

B. Biological Control of Aquatic Weeds by Plant Interference

Undesirable weeds often impede the flow of water in irrigation canals and degrade the water quality in reservoirs impounded for irrigation or drinking water (Oborn *et al.*, 1954; Yeo and Fisher, 1970). Oborn *et al.* (1954) reported that pondweed (*Potamogeton* spp.) was eliminated from cultures mixed with needle spikerush (*Eliocharis acicularis*) or with dwarf arrowhead (*Sagittaria subulata*), or both. They stated that either of the last two species "crowds out" the pondweed. This seemed rather surprising because needle spikerush attains a height of only a few centimeters whereas most pondweeds readily reach lengths of 2 to 4 m and have floating leaves.

Yeo and Fisher (1970) reported that *Elodea canadensis* and *Potamogeton crispus* were eliminated from certain drainage canals, irrigation canals, and reservoirs in California after needle spikerush was introduced experimentally or naturally. The elimination of the elodea and pondweed was again attributed to competition from the spikerush.

Frank and Dechoretz (1980) did not believe it was possible for the very tiny spikerush to eliminate the more robust pondweed by competition so they initiated some experiments to determine if allelopathy might be involved. They found that leachates from tubs containing dwarf spikerush (*Eliocharis coloradoensis*) significantly reduce the production of new shoots of both American and sago pondweed. The differences could be attributed neither to altered water quality nor to reduced levels of nutrients. There was also an obvious reduction in biomass of the pondweeds exposed to the leachates from the containers of spikerush. The general appearance of the pondweeds exposed to the spikerush leachate was even more striking than the size differences from the controls. Those plants of both species exposed to the leachate were chlorotic and "unthrifty." It appears, therefore, that allelopathy plays a significant role in the interference of spikerush against the pondweeds and possibly also against elodea and other species.

Although much research needs to be done to bring the potential to full fruition, the use of allelopathy for weed suppression in integrated weed control systems has a bright future.

3

Manipulated Ecosystems: Roles of Allelopathy in Forestry and Horticulture

The discussion of allelopathy in manipulated ecosystems has been arbitrarily divided into its roles in agriculture, forestry, and horticulture. Crop and forage plants are included under agriculture, lumber producers under forestry, and ornamentals and fruit producers under horticulture. Obviously, many other categories could be included, such as floriculture and viticulture. In addition, many other kinds of ecosystems are manipulated, at least to some extent, but I hope my intent is clear.

I. FORESTRY

Prior to about 1970, relatively little research was done on allelopathy in relation to forestry. In the last decade, however, research in this area has greatly accelerated. In the sections that follow, the discussions are organized, where feasible, according to allelopathic genera. The sequence of genera is based chronologically on the earliest reported research.

A. Allelopathic Effects of Woody Seed Plants

Pliny (Plinius Secundus, 1 A.D.) stated that the walnut (*Juglans*) was injurious to anything planted in its vicinity. Much later, Stickney and Hoy (1881) observed that vegetation under black walnut (*Juglans nigra*) was very sparse when compared with that under most other commonly used shade trees. As pointed out in Chapter 1, Hoy claimed that the main reason for the failure of vegetation to grow under walnut is the poisonous nature of the drip.

Cook (1921) described the wilting of potato (*Solanum tuberosum*) and tomato plants that were grown near black walnut, as well as injurius effects of walnut on apple trees. These observations supported those of Stickney and Hoy (1981). Subsequently, Massey (1925) found that alfalfa and tomato plants wilted and died whenever their roots came into close contact with walnut roots. This effect was so dramatic that he could trace the extent of the walnut roots without removing soil just by observing the development of wilt in test plants (Fig. 4). There was no specific relationship between the region of greatest concentration of walnut roots and wilting of tomatoes as would be expected if the plant response were caused by drought. Apparently, there was little or no poisoning of the soil as the roots of the affected plants had to be in close contact with those of the walnut. When several pieces of bark from walnut roots were placed in a water culture of tomato plants, the plants wilted and their roots browned within 48 hours. Addition of bark from walnut roots to soil in which tomato plants were growing inhibited plant growth. Massey suggested that juglone or some similar substance might be the toxic constituent of walnut.

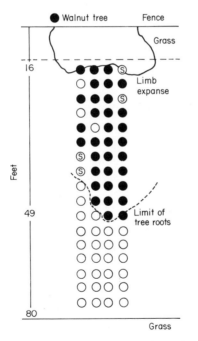

Fig. 4. Diagram showing conditions of tomato plants eight weeks after setting plants in the immediate vicinity of a black walnut tree. Each circle indicates position in which a plant was set. Open circles indicate plants which remained healthy. Circles with S in them represent plants that died soon after transplanting. Black circles indicate plants that wilted and died. (From Massey, 1925.)

Davis (1928) found the toxic substance from the hulls and roots of walnut to be identical with juglone, 5-hydroxy-α-naphthaquinone. The compound was a powerful toxin when injected into the stems of tomato and alfalfa plants.

Perry (1932) reported that black walnut was capable of killing white pine (*Pinus strobus*) and black locust (*Robinia pseudoacacia*) trees growing in its vicinity. A birch arboretum (apparently *Betula papyrifera* according to the photographs) was established in 1962 in the Vermont Agricultural Experiment Station forest near Burlington and, at the end of the first growing season, fifteen trees had died. Virtually all were in the area formerly occupied by a black walnut plantation, an area where some young trees still remained (Gabriel, 1975). The dead birch trees were replaced, and within the next 3 years, an unusually large number of trees died within the boundaries of the old walnut plantation. It was suggested that the death of the birch trees was probably due to allelopathic effects of the dead walnut trees. MacDaniels and Pinnow (1976) summarized the evidence concerning allelopathic effects of black walnut and stated that more research is needed because the evidence conflicts at times.

Fisher (1978) obtained excellent evidence as to one possible cause of the conflicting evidence. He conducted field studies in southwestern Ontario, Canada, in mixed planations of red pine (*Pinus resinosa*), white pine, and black walnut growing on the Brant-Tuscola-Colwood catena of fine sandy loam soils. On excessively drained Brant sites, pine often suppressed the walnut, whereas on imperfectly drained Tuscola and poorly drained Colwood sites, walnut suppressed or even killed the pine. In laboratory studies, the inhibitory activity of juglone readily disappeared from Brant soil under a low moisture regime, but it remained in Brant soil under a high moisture regime (Table 24). Moreover, juglone was still extractable from the soil 90 days after it was added under the wet regime, but it was extractable for only 45 days under the dry regime. Thus, the edaphic conditions played an important role in the interactions between walnut and other species.

Waks (1936) reported that parks of black locust are nearly void of all other

TABLE 24. **Means and Standard Deviations for Radicle Extension of *Pinus resinosa* Seedlings Growing on Soil Treated with 50 ppm of Juglone**[a]

Moisture regime	Days after juglone added and before bioassay						
	0	15	30	45	60	75	90
Dry	16[b] (2)	31[b] (3)	54[b] (4)	73 (4)	87 (5)	98 (5)	99 (5)
Wet	14[b] (3)	18[b] (2)	25[b] (3)	22[b] (3)	28[b] (3)	27[b] (2)	31[b] (3)

[a] Expressed as % of control. From Fisher (1978). Reproduced from *Soil Science Soc. Amer. J.* **42**, 801–803.

[b] Significantly less than control at 5% level.

vegetation, and bark and wood of black locust contain substances that inhibit the growth of barley.

Kuhn *et al.* (1943) reported that mountain ash (*Sorbus aucuparia*) produces parasorbic acid, an unsaturated lactone, that inhibited germination of *Lepidium* seeds in a dilution of 1:1000 and allowed only 10–80% germination at 1:10,000.

Mergen (1959) reported that succession appeared to be remarkably slow in the tree-of-heaven (*Ailanthus altissima*) stands, with virtually pure stands remaining for long periods of time. He found that alcohol extracts of the rachis, leaflets, and stem of the tree-of-heaven caused rapid wilting of other plants of the species when applied to the cut surface of the stems. Similar results occurred when these extracts were applied to 35 species of gymnosperms and 11 species of angiosperms. The only species not adversely affected was *Fraxinus americana*. Approach grafting of *Ailanthus* with several species gave results similar to those with extracts.

Jameson (1961) noted that junipers inhibited plants in the field and demonstrated that water extracts of leaves of three species of *Juniperus* (*J. osteosperma, J. deppeana,* and *J. monosperma*) strongly inhibited the growth of wheat radicles. Extracts of pinyon pine (*Pinus edulis*) were also inhibitory. Later, it was reported that aqueous extracts of *J. osteosperma* and *Pinus edulis* were very inhibitory to root growth of seedlings of several native plants associated with *Pinus* and *Juniperus* (United States Forest Service, 1963). One of these plants was blue grama (*Bouteloua gracilis*), which, according to Arnold (1964), showed a strong zonation response around *Juniperus monosperma*. Jameson (1966) found that tree litter was the major factor associated with reduction of blue grama. Tree cover did not influence blue grama, and in some cases, it appeared to be beneficial. Studies of other investigators indicated that root competition of *Pinus* spp. and *Juniperus* spp. against blue grama is probably slight. Hence, Jameson concluded that allelopathy has a strong influence on the zonation of herbaceous plants under pinyon pine and juniper trees. *Juniperus scopulorum* produces both water-soluble and volatile inhibitors that reduce germination of several herbaceous species, including grasses, that grow at the same altitude as the tree (Peterson, 1972).

Japanese red pine (*Pinus densiflora*) forests are widely distributed in Japan and cover 60–70% of forest land in South Korea. Lee and Monsi (1963) found that vegetation under the trees was sparse and determined that the light intensity was higher in red pine forests than in nine other forest types. Moreover, many of the other forests had dense undergrowths of herbaceous plants.

Lee and Monsi rated understory plants in many red pine stands for vitality using a scale of four classes: class 1 representing the highest vitality and class 4 representing the lowest. They found a very small number of annual and biennial herbs with a rating of 1 or 2 and a relatively small number of perennial herbs with high ratings. A great many herbs that were growing profusely in cultivated fields or other areas adjacent to red pine trees or stands were not found under the red pine trees, or if they were present, they had low vitality and did not reproduce.

Lee and Monsi (1963) suspected that the failure of many species of plants to grow in red pine forests, or to reproduce if present, was due to the allelopathic effects of red pine. They selected 15 test species with 5 representing species characteristic of red pine forests and 10 representing species not characteristic of red pine forests and of low vitality if present. Initially, hot and cold water extracts of fresh and fallen leaves and of roots were tested against germination of seeds of four test species in Petri dishes. Germination of red pine seeds was not affected by any dilutions of any of the extracts, but germination of seeds of three noncharacteristic species was markedly inhibited in all tests.

Numerous experiments were run to determine the effects of soils taken from various red pine forests on germination of seeds and growth of several characteristic and noncharacteristic species. Controls were run with soils taken from experiment stations. There were no marked differences in germination, although red pine seeds regularly germinated somewhat earlier in red pine soils than in control soils. The red pine soils markedly inhibited growth of noncharacteristic species, whereas growth of characteristic species was virtually the same in red pine and control soils (Fig. 5). The results shown in Fig. 5 are representative of their overall results, and measurements of physical and chemical factors of the soils demonstrated no striking or consistent differences between the control and the various red pine test soils used. Lee and Monsi concluded that extracts of various parts of *Pinus densiflora* and soils under this species contain substances toxic to species that rarely occur in red pine forests. They identified a tannin and *p*-coumaric acid in the extracts and in the soil from a red pine forest. Kil (1981) confirmed and extended their results on the allelopathic effects of red pine.

It has been known for some time that hoop pine (*Araucaria cunninghamii*) develops successfully without nitrogen when underplanted to established pine plantations on podzolic soils in eastern Australia (Bevege, 1968). Bevege postulated that *Araucaria* may be stimulated by biochemical factors associated with pine roots. He compared the effects on hoop pine growth of leachates from pots containing either slash pine (*Pinus elliotti*), hoop pine, or crows ash (*Flindersia australis*). Three different substrates were used for the test hoop pine plants, sterilized sand, rain forest soil, and soil from a slash pine plantation. The leachates were prepared by passing complete nutrient solution through sand in which were growing advanced trees of one of the three species listed above. Unmodified nutrient solution was used as a control. The duration of the experiment was 10 months.

All three leachates significantly reduced dry-matter production and caused some mortality. Survivors exhibited varying degrees of necrosis and chlorosis in the sand and stunting and browning in the two soils. Nitrogen and potassium

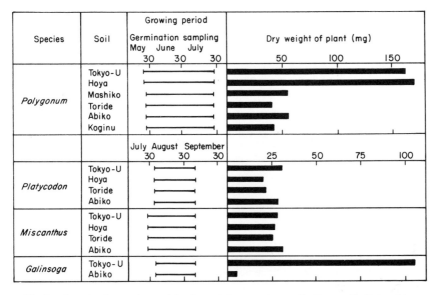

Fig. 5. Germination and growth in dry weight on various soils. Farm soils, Tokyo University campus and Hoya Experimental Field; red-pine-forest soils, Mashiko, Toride, Abiko, and Koginu. *Polygonum blumei,* sown on May 23, 1963; *Platycodon grandiflorum* and *Galinsoga ciliata,* sown on Aug. 5, 1963; *Miscanthus sinensis,* sown on July 20, 1963. (From Lee and Monsi, 1963.)

concentrations of the shoots were unaffected, but phosphorus concentrations and uptake were reduced. Bevege postulated at least two inhibitory systems in the leachates, one inhibiting phosphorus uptake, the other causing symptoms similar to nitrogen and phosphorus deficiency but possibly acting through some other metabolic pathway. The results indicated conclusively that the success of hoop pine, which was underplanted to *Pinus,* could not be attributed to the production of stimulatory root factors by pine. Bevege suggested that the autotoxicity of hoop pine may help to explain its slow regeneration in a hoop pine rainforest.

   Pinus radiata is an important timber tree in New Zealand and Australia. A thick layer of litter accumulates in stands of this species and growth of seedlings of *P. radiata* and other species is sparse or absent (Lill and Waid, 1975). It is also difficult to reestablish a vigorous *P. radiata* stand on the same site after harvesting the lumber (Chu-Chou, 1978). Lill and Waid (1975), Lill and McWha (1976), and Lill *et al.* (1979) found that a volatile substance was produced by *P. radiata,* which inhibited seed germination of perennial ryegrass and *P. radiata* and seedling growth of these species, plus white clover. Germination of white clover was depressed in some tests and stimulated in others (Lill *et al.,* 1979). The volatile *phytotoxin* was identified as ethylene (Lill and McWha, 1976). It is noteworthy that Molisch (1937), who coined the term allelopathy, became interested in the phenomenon primarily because of the growth effects of ethylene produced by ripe fruits of apples and related species.

Chu-Chou (1978) reported that the height growth of *P. radiata* seedlings was reduced by 20–80% when grown on soils incorporating different proportions of roots from old *P. radiata* trees. Aqueous extracts of soil–root mixtures also significantly reduced the growth of *P. radiata* seedlings. The growth-retarding effect was partially overcome by adding nutrients or by soil sterilization. It was postulated, therefore, that growth retardation was caused in part by nutrient deficiency resulting from the addition of organic matter with a high C–N ratio and in part by phytotoxins present in the root tissue. Under aseptic conditions, aqueous extracts from the inner bark of roots of old *P. radiata* trees inhibited growth of a mycorrhizal fungus (*Rhizopogon* sp.) and caused root necrosis and wilting of *P. radiata* seedlings (Fig. 6). Extracts of the root wood and of the outer bark, however, did not inhibit mycorrhizal or root growth.

Peterson (1965) reported that lambkill (*Kalmia angustifolia*) occurs frequently on upland sites in eastern Canada and that there is abnormally poor tree growth on such sites. Bioassays indicated that dry leaves of lambkill contain a substance that hinders primary root development of black spruce (*Picea mariana*) by destroying the epidermal and cortical cells.

Webb *et al.* (1967) found that *Grevillea robusta* does not regenerate in *G. robusta* plantations in Australia, whereas other rain forest species do regenerate. *Grevillea* regenerates freely, however, in adjacent plantations of *Araucaria cunninghamia* and along the edges of rain forests. Blackening and death of the tops of *Grevillea* trees ensued after contact of the seedling roots with the growing roots of older plants of *G. robusta*. Single *G. robusta* seedlings responded similarly in pots with sand and soil watered with nutrient solution plus leachates from pots of older *G. robusta* plants growing in sand. Webb and co-workers concluded that *G. robusta* fails to regenerate in *G. robusta* plantations because of some water-transferable factor associated with the rhizosphere of this species, in which antagonistic microflora may be involved.

Brown (1967) observed that jack pine (*Pinus banksiana*) stands ranging from a few stems to 1000 or more per acre may be found within a few yards of each other. He observed that ground cover plants showed similar extremes in density. Measurements of soil texture, available minerals, depth of water table, water-holding capacity, soil acidity, root competition, slope exposure, microclimate, and seed dispersal did not reveal differences that explained the variations in the jack pine stands. Moreover, there were no differences in logging or fire history, so Brown investigated the possibility of naturally occurring, biologically active materials on jack pine seed germination and growth.

Water extracts of 56 species of plants commonly associated with jack pine

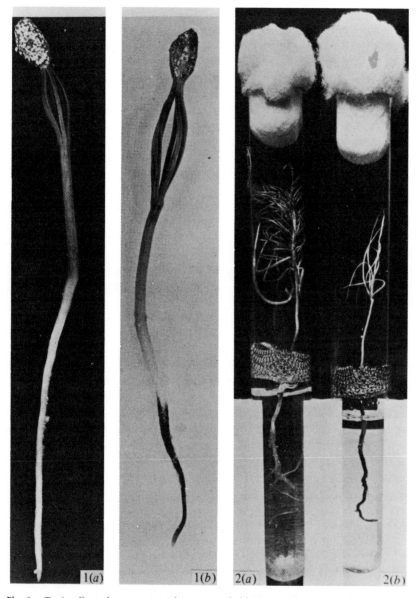

Fig. 6.   Toxic effect of water extract from roots of old *Pinus radiata* on *P. radiata* seedlings growing in White's medium under aseptic conditions. (1) After 48 hr. (2) After 4 wk.   (a) Control without extract, showing healthy root (1a) and shoot (2a).   (b) With root extract, showing blackened and necrotic root (1b,2b) and withered shoot (2b). (From Chu-Chou, 1978.)

were used as the moistening medium for germination of seeds in the laboratory. Nine of the extracts were found to significantly inhibit the germination of jack pine seeds (Table 25). Two species of *Prunus* completely inhibited germination, and two species of *Solidago* and *Salix pellita* almost completely inhibited it. Eleven species were selected in all: three that inhibited jack pine seed germ, six that had no influence on jack pine germ, and two that stimulated jack pine germ. A virtually pure stand of each was selected, and 400 jack pine seeds were planted in each cover type. All species that inhibited germination of jack pine seeds in laboratory tests did so in the field. The two species that were stimulatory in the laboratory allowed high germination rates in the field also. Brown's experiment did not permit any conclusion concerning the role of allelopathy in the subsequent growth of seedlings, however, there were definite interference effects by several of the cover species.

Hook and Stubbs (1967) studied development of vegetation in a cutover bottomland hardwood forest in which an average of seven seed trees were left per acre. They found that tree seedlings, shrubs, and vines grew well under various seed trees but not under three oak species: cherrybark oak (*Quercus falcata* var. *pagodaefolia*), swamp chestnut oak (*Q. michauxii*), and Shumard oak (*Q. shumardii*). Light intensity under the trees was essentially full sunlight except at high noon, and there was no moisture deficiency in the creek bottom area. In fact, the pronounced retardation was more prevalent in the lower wet areas. Hook and Stubbs suggested, therefore, that the three oak species may have had an allelopathic effect on understory plants. In a subsequent study, DeBell (1971) reported that seed germination and growth of cherrybark oak and sweetgum

TABLE 25.   Germination of Jack Pine Seeds in the Presence of Inhibitory Plant Extracts[a]

Species	Plant part extracted	pH of extract	Percentage germination after 14 days[b]
Boletus edulis	Fruiting bodies	—	37
Cladonia cristatella	Plants	5.1	24
Sphagnum capillaceum	Plants	4.6	22
Salix pellita	Leaves	6.2	9
Prunus pumila	Leaves	—	0
Prunus serotina	Leaves	6.2	0
Gaultheria procumbens	Leaves	6.3	21
Solidago juncea	Flowers	—	11
Solidago juncea	Leaves	5.0	9
Solidago uliginosa	Leaves	—	2
Control (average)			82[c]

[a] From Brown (1967).
[b] Significant at 1% level according to Student's *t* test.
[c] Range 62–93%.

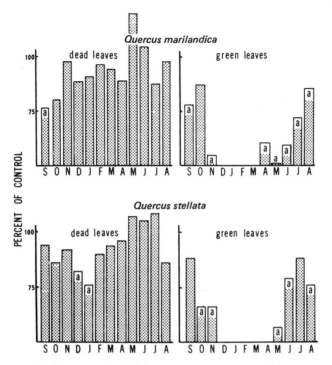

Fig. 7. Monthly bioassays of oak foliage. Values are means of the root growth of 60 seedlings of *Bromus japonicus* treated with leachate and expressed as a percentage of simultaneous distilled water controls. Bars marked "a" are significantly different from control at the 5% level or better. (From McPherson and Thompson, 1972.)

(*Liquidambar styraciflua*) seedlings were reduced in soils collected beneath cherrybark oak. Moreover, cold-water leachates of fresh, whole leaves of cherrybark oak inhibited the growth of sweetgum seedlings. The primary toxin was salicylic acid, and it was suggested that leaching of this compound from the oak crowns by rain probably causes inhibition of the understory plants.

McPherson and Thompson (1972) demonstrated that an upland forest of postoak (*Quercus stellata*) and blackjack oak (*Q. marilandica*) suppresses understory plants. Causative factors were found to be as follows: (1) oak litter inhibits seed germination, largely by physical–mechanical means; (2) oak roots compete strongly with understory plants, slowing growth and indirectly killing seedlings; (3) oak roots have an allelopathic effect that inhibits seedling growth; and (4) green oak foliage produces water-soluble toxins that further restrict the growth of seedlings of understory plants (Fig. 7).

Roshchina (1972a) reported that transpiration water from *Quercus robur* and box elder (*Acer negundo*) contained ethanol, methanol, propanol, acetic aldehyde, propionic aldehyde, and butylene. Sufficient amounts of these compounds could injure plants, but it is not known whether they have any effect under natural conditions.

Roshchina (1974) demonstrated that gases are produced during the decay of leaves of *Quercus robur, Betula verrucosa, Fraxinus pubescens, Rhus typhina, Cotinus coggygria, Tilia cordata, Phellodendron amurense, Sorbus aucuparia, Populus balsamifera, Padus racemosa,* and black locust, which inhibit germination of radish seeds. She stated that the inhibition is due to the presence of volatile organic compounds, as well as methane and CO_2.

Dense forests comprised of several intermixed species of oaks and a great diversity of tropical tree species generally occur on the mesic west slopes of the Sierra de Talamanca in Costa Rica (Gliessman, 1978). The diversity is altered in some areas and practically pure stands of *Quercus eugeniaefolia* occur with very little understory vegetation in spite of abundant light, water, minerals, open spacing, and the presence of bare soil surfaces. Gliessman found that aqueous extracts of green leaves and freshly fallen leaves of *Q. eugeniaefolia* were toxic to cucumber seedlings. The green leaves yielded more toxic materials after a period of a few days without rainfall, indicating that toxin production is an active process in the living leaves. Bioassays of both green and brown leaves, which were collected in the wet season, indicated that they did not inhibit root growth of *Bromus rigidus,* but extracts made from green leaves during the dry season were toxic. This indicates that the toxin, while its input to the substrate in the dry season may be more sporadic due to less frequent rainfall, is definitely effective when the conditions are favorable. High absorbance (265 nm) of water extracts of soil from the top 5 cm indicated a high concentration of phenolics in that soil layer even in the wet season when leaching would be greatest. Gliessman implicated allelopathy as the potential mechanism of dominance of the oak.

Lodhi (1976) observed bare areas under sycamore (*Platanus occidentalis*), rough-leaved hackberry (*Celtis occidentalis*), white oak (*Quercus alba*), and northern red oak (*Q. borealis*) trees in a bottomland forest in Missouri, even though herbaceous species grew well under American elm (*Ulmus americana*), which cast just as dense shade. No significant differences were found in concentrations of most mineral elements and pH under the species with reduced understory growth and under elm. Soil moisture was consistently higher under all test trees than under elm throughout the growing season. Decaying leaves, leaf leachates of all four test species, and soil collected from under test trees significantly reduced seed germination, radicle growth, and seedling growth of seected herbaceous species common to the area. Fourteen phytotoxins were identified from leaf litter of the test species and from soil under the trees (Lodhi, 1976, 1978a). The accumulation of toxins in the soil corresponded with the amount of litter produced and its decay rate. The rate of release of the phytotoxins and their activities depended on whether they were free or bound and on their solubilities in water. Decaying litter of sycamore and red oak showed stronger inhibition of germination and radicle growth in January than in April and August. On the other hand, hackberry and white oak were most inhibitory in April. Further investigation revealed that red oak and sycamore leaves contain mainly inhibitors that are in the free form and that can be extracted in the laboratory without hydrolysis. Therefore, these toxins apparently leached readily into the soil and exhibited peak toxicity in January. White oak and hackberry leaf litter released a large number of toxins only after hydrolysis in the laboratory, and the amounts of phytotoxins were higher in the soil in April than in January in contrast to the situation with northern red oak and sycamore. This answers a very important question often asked, as to whether the isolation and identification of toxins following hydrolysis is of any importance in research in allelopathy. The overall conclusion was that the reduced understory growth under northern red oak, roughleaved hackberry, sycamore, and white oak was due at least in part to allelopathy. This supported the previous conclusion of Al-Naib and Rice (1971) concerning the allelopathic effects of sycamore.

Ellison and Houston (1958) reported that, in experimental plots in central Utah, the total production of four native herbaceous species was greater in the open than under a quaking aspen (*Populus tremuloides*) canopy. Moreover, production was greater in trenched plots than in untrenched plots under the canopy. The reduction in production under the canopy was attributed to competition, primarily among the roots. No mention was made of the possibility of chemical inhibition of growth due to aspen.

Dormaar (1970) reported that *Populus* spp. are encroaching on rough fescue (*Festuca scabrella*) prairie in southwestern Alberta, Canada, and Black Chernozems are being transformed into Dark Gray Lavisols. Aqueous leaf leachates seem to play a role in this transformation, and Dormaar investigated leaf leachates of *Populus* X Northwest, a natural hybrid of *P. deltoides* X *P. balsamifera,* throughout the year. Simple phenolic compounds were present early in the growing season, but complex phenols became more important as the season progressed. No specific compounds were identified, but many phenols are known to have marked allelopathic activity.

Mathes *et al.* (1971) reported that callus tissue isolated from the cambial region of the stem of quaking aspen secreted materials that inhibited growth of *Agrobacterium* spp. The inhibitory materials were not identified, but they were found to be primarily bactericidal.

Younger *et al.* (1980) reported that freshly fallen quaking aspen leaf litter, which was incorporated in red clay soil to achieve 2 and 5% organic carbon, resulted in a substantial decrease in growth of *Festuca elatior, F. rubra,* and Kentucky bluegrass (*Poa pratensis*) as compared with growth in unamended or peat amended soil. The effects were accentuated with added nutrients. Growth of birdsfoot trefoil was not significantly affected, however. Aqueous extracts of the leaf litter (2 and 4%, w/v) did not affect the speed of seed germination or the cumulative germination percentage of these four species plus smooth brome and perennial ryegrass. They suggested that the allelopathic potential demonstrated may indicate a need to revise management plans for aspen forests, including crop rotation.

Olsen *et al.* (1971) pointed out that *Populus tremula* leaves are known to inhibit the growth of mycorrhizal fungi. They isolated and identified benzoic acid and catechol from *P. tremula* leaves and showed that these substances have a strong inhibitory effect on growth of different *Boletus* species. These compounds also had a weaker inhibitory effect on litter-decomposing *Marasmius* species.

Degraded sites in the northern United States and Canada are commonly invaded first by alder (*Alnus*) species, which fix nitrogen symbiotically. These are then replaced by various species of *Populus,* including balsam poplar (*P. balsamifera*) (Jobidon and Thibault, 1981). Moreover, these researchers frequently observed growth depression of alders near a balsam poplar stand that was invading the site. Nevertheless, alders and poplars have been proposed for mixed plantations in short rotation. Therefore, Jobidon and Thibault (1981) decided to

TABLE 26. Morphological Modifications Observed on *Alnus crispa* var. *mollis* Radicle after Treatment with Different Plant Parts of *Populus balsamifera*[a]

Treatments (%, w/v)	Morphological modifications[b]			
	Total inhibition of root hair development	Partial inhibition of root hair development	Total meristematic necrosis	Partial meristematic necrosis
Leaf litter extracts				
2	50	50	62	38
1	14	43	14	40
0.5	0	0	0	0
0.1	0	0	0	0
Bud extracts				
2	80	20	100	0
1	62	38	77	23
0.5	0	73	13	73
0.1	0	0	0	0
Leaf leachates				
2	83	8	8	16
1	62	27	0	18
0.5	0	0	0	0
0.1	0	0	0	0

[a] From Jobidon and Thibault (1981).
[b] Percentage of plants affected per treatment.

TABLE 27. Effects of Different Plant Parts of *Populus balsamifera* on Nitrogenase Activity (mole C_2H_4/3 hour/root dry weight $\times$ 10^{-11}) ($\pm$SE)[a]

Treatments (%, w/v)	Mole C_2H_4/3 hour/root dry weight ($\times$ 10^{-11})[b]	Inhibition (%)
Bud extracts		
2	4.36 ± 0.59[a]	61.72
1	4.12 ± 0.49[a]	63.88
0.5	5.26 ± 0.26[a,b]	53.89
0.1	6.54 ± 0.66[b]	42.67
Control	11.41 ± 0.99[c]	0.00
Leaf litter extracts		
2	4.25 ± 0.96[a]	62.74
1	6.19 ± 0.78[a]	45.74
0.5	4.14 ± 0.54[a]	63.71
0.1	5.20 ± 0.55[a]	54.42
Control	11.41 ± 0.99[b]	0.00

[a] From Jobidon and Thibault (1982).

[b] Within each treatment, means followed by the same letter are not different at 0.05 level, Duncan's new multiple range test.

determine if balsam poplar has an allelopathic effect on green alder (*Alnus crispa* var. *mollis*). They found that water extracts of leaf litter and buds and fresh leaf leachates of balsam poplar inhibited seed germination and radicle and hypocotyl growth of green alder seedlings. Hypocotyl growth was affected least, followed by seed germination and radicle growth, with growth of radicles inhibited by 80%. Growth inhibition was not due to pH changes or osmotic pressure of the extracts. Pronounced morphological modifications occurred, including inhibition of root hair development and necrosis of the radicle meristems (Table 26).

The marked effect of balsam poplar on root hair development of green alder suggested that infection of the root hairs by *Frankia* might be reduced, thus decreasing nodulation and nitrogen fixation (Jobidon and Thibault, 1981, 1982). The average number of nodules per green alder plant, in seedlings treated with any of the three balsam poplar extracts previously described, was only 51% of that of control plants. There was a 62% decrease in acetylene reduction (nitrogen fixation) by green alder seedlings that were treated with the most concentrated bud and leaf litter extracts of balsam poplar (Table 27). Foliar nitrogen content of both nodulated and unnodulated seedlings was significantly lower in extract treated plants than in controls. It is noteworthy that all extracts inhibited height growth, root elongation, and dry weight increment of nodulated and unnodulated green alder seedlings to some degree. However, growth inhibition was 25% less in nodulated than in unnodulated seedlings.

Del Moral and Cates (1971) surveyed the allelopathic potential of 40 species of ferns, conifers, and angiosperms of western Washington. The laboratory phase of the study involved the testing of leaves for volatile inhibitors and the testing of leaves, litter, and bark for water soluble inhibitors of seedling growth of barley, downy brome, and Douglas fir (*Pseudotsuga menziesii*). An index of inhibition was determined for each of the 40 species in the survey. The field phase of the survey involved sampling of vegetation under the species involved in the survey and the separate sampling of vegetation adjacent to the survey species. Cover, frequency, and species composition were determined, and the coefficient of community was calculated for each species. A species was considered to demonstrate interference in the field if it reduced cover under it by 50% or more, the coefficient of community was less than 50%, and the species richness was less than 75% of that of the adjacent area in all samples.

The nine species that reduced growth in the laboratory and caused interference in the field were *Abies amabilis*, *Abies grandis*, *Abies procera*, *Picea engelmannii*, *Taxus brevifolia*, *Thuja plicata*, *Acer circinatum*, *Arbutus menziesii*, and *Rhododendron albiflorum*. Sixteen species produced significant inhibition but no field interference, five species produced no laboratory inhibition but showed interference in the field, and the rest caused no inhibition nor interference.

Del Moral and Cates (1971) concluded that the interference demonstrated by the nine species listed above was primarily due to allelopathy. In the case of the five species that showed interference in the field and no inhibition in the laboratory, they suggested that the interference must be due to competition and not to allelopathy.

Mensah (1972) reported that sycamore maple (*Acer pseudoplatanus*) is allelopathic to yellow birch (*Betula alleghaniensis*). There was a seasonal periodicity in the effect of root exudates of the maple on radicle growth of yellow birch, which was related to the stage of leaf development, rates of photosynthesis and translocation, and the presence of actively growing roots on the maple. Effects were most marked during the middle of the growing season. Leachates of maple seeds strongly inhibited the growth of yellow birch, but leachates of leaves at all stages and leachates of litter had no effect. The chief inhibitory effect on birch was poor root development, which resulted in poor seedling development and death at times. The inhibitory factors were acidic, and those in the seed leachates were phenolics.

Nursery experiments in Michigan showed that the presence of seedlings of sugar maple (*Acer saccharum*) inhibited the growth of yellow birch seedlings, despite the absence of competition (Tubbs, 1973). Root elongation of birch was inhibited by exudates of actively growing root tips of sugar maple. The toxin lost its inhibitory activity after a few days storage. When seedlings of the two species were grown together in an aerated nutrient solution, the number of actively growing root tips of birch formed each day was directly correlated with the activity of the inhibitor.

Kokino *et al.* (1973) isolated water soluble and volatile compounds from decaying leaves of *Acer platanoides, A. pseudoplatanus, A. campestre, A. tataricum, A. laetum, A. turkestanicum, A. ginnala, A. mandschuricum, A. saccharinum, A. negundo, Quercus robur,* and *Fraxinus excelsior.* The compounds were tested against growth of *Lepidium sativum,* and both inhibitory and stimulatory compounds were found. In most species of *Acer,* the decomposing leaves first possessed inhibitory compounds, and by the end of decomposition, they possessed stimulatory compounds.

Baranetsky (1973) found that water extracts of fallen leaves of *Fraxinus excelsior* inhibited several test plants and contained large quantities of a cyanogenic glucoside. Water extracts of *Tilia cordata* contained potent inhibitors also, and extracts of both species contained indole and gibberellin-like growth stimulators.

Thomas (1974) used aqueous extracts of blue spruce (*Picea pungens*) to moisten filter paper on which seeds of grasses and cereals were placed and allowed to germinate. The extracts caused malformed roots and blackened root tips, and they prevented root-hair development.

Matveev *et al.* (1975) reported that plant excretions and substances formed during decomposition of litter inhibited the growth of grass species in black locust and smoketree plantings in the steppe zone of the U.S.S.R. This led to a paucity of species and to a weak development of grass stands.

Al-Mousawi and Al-Naib (1975) observed a pronounced paucity of herbaceous plants in planted *Eucalyptus microtheca* forests in central Iraq, in comparison with the understory in adjacent *Casuarina cunninghamiana* stands. The reduction was not primarily due to soil moisture, nutrient elements, or shading. Leaf extracts, decaying leaves, and soil collected under *Eucalyptus* canopies inhibited seed germination and seedling growth of associated species. Subsequently, Al-Mousawi and Al-Naib (1976) and Al-Naib and Al-Mousawi (1976) found that three volatile inhibitors—α-pinene, camphene, and cineole—and five water-soluble inhibitors—chlorogenic, isochlorogenic, ferulic, *p*-coumaric, and caffeic acids—were produced by the *Eucalyptus* leaves. The volatile inhibitors were those previously identified in *E. globolus* by del Moral and Muller (1969).

Most coastal heath in southern Victoria, Australia, is dotted with stunted individuals or small groves of various *Eucalyptus* species. Some associated heath species are suppressed beneath *E. baxteri* (del Moral *et al.,* 1978). *Leptosperum myrsinoides* and *Casuarina pusilla* are dominant in the surrounding heath but are severely suppressed or absent under *E. baxteri. Leptosperum juniperinum* and *Xanthorrhoea australis* are not suppressed, however. The pattern of suppression is associated with the canopy and not the roots, and soil moisture and nutrient patterns suggest that competition for these resources is not of major significance. Water potential measurements indicated that *E. baxteri* ameliorates drought stress in the suppressed species. Moreover, stem flow and fog drip may also increase moisture beneath the canopy. Competition for light does not appear to be significant either. *E. baxteri* does not cast deep shade at high sun angles because of its sparse foliage and vertical orientation of its leaves. Additionally, the light intensity under *E. baxteri* was higher than beneath *E. nitida*, and the latter did not suppress heath species.

Foliar and litter leachates retarded growth of *Casuarina pusilla, Eucalyptus viminalis,* and wheat. Foliar leachates contained gentisic and ellagic acids; litter leachates contained these plus gallic, sinapic, and caffeic acids. Both leachates also contained several unidentified phenolic aglycones, numerous glycosides, and terpenoids. Topsoil extracts, which contained resins and possibly terpenoids, inhibited growth of the test species also. Del Moral *et al.* (1978) concluded that the suppression zone is associated with the allelopathic activity of *E. baxteri* and is maintained either through the direct transfer of foliar leachates (raindrip) to leaves of understory plants, through root absorption of foliar and litter leachates, or to mycorrhizal inhibition by such leachates.

Zinke (1962) noted a systematic change in pH, nitrogen content, exchangeable bases, exchange capacity, and volume weight with increased distance from the trunks of individual trees. He stated that the pattern was due to the difference between the effect of bark litter, leaf litter, and the adjacent opening or neighboring tree. Smith (1976) demonstrated a pronounced variation in exudation of organic materials from suberized tips of roots of yellow birch, beech (*Fagus grandifolia*), and sugar maple. However, the patterns of inorganic root exudates were similar. Stepanov (1977) reported that the root systems of *Pinus sibirica, Pinus sylvestris, Abies sibirica, Picea odorata,* and *Larix sibirica* produced volatile organic compounds, chiefly monoterpenes. He suggested that some of the peculiarities of the structure of dark coniferous forests might be related, at least in part, to differences in production of volatile root materials.

Water soluble leachates of fallen leaves of catalpa (*Catalpa bignonioides*), mimosa (*Albizzia julibrissin*), horsechestnut (*Aesculus hippocastanum*), Austrian pine (*Pinus nigra*), and coral bean (*Sophora japonica*) were tested against seed germination of various trees and shrubs (Chumakov and Aleikina, 1977). Catalpa had the greatest inhibitory activity, and pine and coral bean the least activity.

June (1976) observed that forests in the northwest Ruahine Range, New Zealand, which are dominated by red beech (*Nothofagus fusca*), are markedly free of understory species. This is particularly true beneath red beech, *Dicksonia lanata, Griselinia littoralis,* and *Pseudowintera colorata.* The last two species listed are shrubs. Because of previous reports of flavonoids, polyphenols, and simple phenols in *Nothofagus* heartwood, June tested the litter of red beech and *P. colorata*, the leachates of twigs of all four species listed above, and the aqueous extracts of soil from red beech forests against growth of red beech seedlings for 58 days. Neither treatment had a significant effect on the growth of red beech, even though all materials except the soil extract had relatively high phenolic contents. June suggested that the phenolics are broken down rapidly in the red beech forest soil. Unfortunately, the potential allelopathic materials were tested against red beech instead of the species that are excluded from the undergrowth.

Petranka and McPherson (1979) investigated the potential role of allelopathy in the intercommunity dynamics that occur between forest and prairie communities in central Oklahoma. Throughout much of the latitudinal extent of this ecotone, a zone of shrubs is present at the forest edge. Under certain conditions, the shrub zone encroaches upon and replaces the tallgrass prairie community. In central and northern Oklahoma, the shrub zone is dominated by *Rhus copallina, R. glabra, Prunus angustifolia,* and *Cornus drummondii.* Of these, *R. copallina* (winged sumac) appears to be the most abundant and aggressive invader of the tallgrass prairie. According to Petranka and McPherson, upland forest tree species are frequently able to invade climax prairie, but these invasions are enhanced by prior invasion of shrubs. In the test plots, the entry and growth of winged sumac clones into a climax prairie resulted in a reduction in the density of prairie vegetation and an increase in tree seedlings, shrubs, and herbs which occurred in the nearby upland forest. Growth ring counts indicated an average rate of encroachment of 1.25 m/year.

Petranka and McPherson found that the rhizomes, flowers, fruits, senescent leaves, and leaf litter of winged sumac contained toxins that significantly inhibited seed germination, seedling growth, or both in twelve species of plants. Most were climax prairie species, but a few were weedy species. Light intensities reached low levels in mature clones of winged sumac, and seedling response in experimental plots indicated that both competition for light and allelopathy reduced the number of prairie seedlings under the clones of *R. copallina.* On the other hand, soil moisture and macronutrient content were generally higher within clones so that competition for these factors was not an important factor in excluding prairie species.

Aqueous extracts of the bark of spruce (*Picea*), pine, birch (*Betula*), and red cedar (*Thuja*) inhibited the growth of rye grass (*Lolium*) (Solbraa, 1979). Bark of all the species is often used as a plant growth medium, so Solbraa decided to find out how long the bark would have to be composted to eliminate the toxicity. He demonstrated that growth inhibition was caused chiefly by various phenolic compounds, and these compounds were consistently reduced below the inhibitory level after 2 weeks of active composting under regulated aerobic conditions. This was true regardless of the type of bark.

Several ground cover plants adversely affect seed germination and the growth of red pine (*Pinus resinosa*) seedlings, and this effect has been attributed to competition (Norby and Kozlowski, 1980). These researchers were aware that several genera of plants common in red pine plantations include species that have been demonstrated to be allelopathic. Therefore, they selected six understory species in a red pine plantation in central Wisconsin for an investigation of their allelopathic effects on red pine. They collected soil from an area where none of the test species were present. The litter and partially decomposed litter were removed, and the soil was autoclaved for 90 minutes at 121°C. This soil was then used for growing the red pine seedlings.

Foliage was collected in September from *Aster macrophyllus, Solidago gigantea, Lonicera tatarica, Solanum dulcamara, Prunus serotina,* and *Rubus idaeus* var. *strigosus.* Water extracts were made from the foliage of each species, and the pH of each extract was adjusted to 5.5. Osmotic potentials of the extracts were measured and found to be too low to cause detrimental effects.

Seed germination of red pine was not affected by the extracts, but radicle elongation was reduced up to 40%. All extracts inhibited height growth, formation of secondary needles, and dry weight increment of red pine seedlings during the 7-week experiment. The total dry weight of seedlings treated with *Lonicera* extract was only 46% of that of control plants. *Lonicera* and *Solidago* extracts also significantly reduced the phosphorus concentration in the red pine needles.

Chou (1980) reported that aqueous extracts of leaves of many dominant woody species in northern Taiwan reduced the growth of test plants by at least 60% in bioassays, even though the extracts were diluted to 10 mosmols osmotic pressure. Species demonstrating allelopathic potential included *Bauhinia purpurea, Bridelia balansae, Ficus gibbosa, F. retusa, F. vasculosa, Glochidion fortunei, Mallotus japonicus, Phyllostachys makinoi, Psidium guajava, Rhododendron* spp., *Sinobamboosa kunishii, Sinocalamus latiflorus,* and *S. oldhami.* Phytotoxicity of the litter of these species was 20 to 40% less than that of fresh leaves, and it was significantly higher in summer than in winter. Chou suggested that these 13 species may be allelopathic to understory plants, but more meaningful experiments need to be conducted, such as the effects of decaying litter, root exudate, rain drip, and possible volatile materials.

Helietta parvifolia is the dominant shrub in the submontane scrub zone of Nuevo Leon, Mexico (Wiechers and Rovalo-Merino, 1982). They found that herbs are very sparse in this ecosystem and postulated that this might result from allelopathic effects of *H. parvifolia.* Crude aqueous extracts of both young and old leaves of this species significantly inhibited the growth of both shoots and roots of the test herbs. Raindrip collected in the field also reduced the growth of the test species, as did some unidentified volatile compounds. Coumarins and furanocoumarins were found in both leaf leachates and leaf extracts. The essential oil from the leaves inhibited stem elongation of test plants, and it was a strong fungicide for some species of *Aspergillus* and some other fungi and a strong bacteriocide for some bacteria isolated from the soil in the *H. parvifolia* ecosystem.

B. Allelopathic Effects of Herbaceous Angiosperms

In northern Wisconsin, there is a prairie community dominated by grasses and bracken fern (*Pteridium aquilinum*); it has been termed a "bracken–grassland" by Curtis (1959). Following glaciation, most of the area was forested and the openings probably followed intensive logging and subsequent fires in the late 1800s (Dawes and Maravolo, 1973). Much of the area has subsequently returned to forest, but the bracken–grassland has remained free of trees. Curtis suggested that several important species in this community might be allelopathic, including orange hawkweed (*Hieracium aurantiacum*). Levy (1970) reported frequencies of 100, 100, and 87% for orange hawkweed in three different stands of bracken–grassland in northern Wisconsin. Dawes and Maravolo (1973) found that orange hawkweed produces phytotoxins that inhibit the growth of both herbaceous and tree species. Phytotoxins, which were extracted from soil taken from orange hawkweed stands, consistently inhibited seed germination and hypocotyl growth of balsam fir (*Abies balsamea*), lettuce, and white pine. Balsam fir and white pine seeds planted and grown in sand, which had previously supported growth of orange hawkweed, had normal germination and shoot growth. However, survival was greatly reduced in both species. Several phenolic and at least two nonphenolic inhibitors were isolated.

Broomsedge (*Andropogon virginicus*), which was reported by Rice (1972) to be allelopathic to several herbaceous species in Oklahoma, is widely distributed throughout the southeastern United States and often in old fields along with loblolly pine (*Pinus taeda*). Priester and Pennington (1978) found that water extracts of live and decaying shoots of broomsedge inhibited height growth of loblolly pine. Root and needle lengths were reduced by extracts of live, dead, and decaying shoots of broomsedge; and crown length was significantly reduced by extracts of live shoots. The extracts also significantly reduced dry weight increase of roots and shoots and decreased the nitrogen concentration in stems but not in roots. Concentrations of potassium and phosphorus were not affected. Thirteen phenolic compounds were identified in the extracts, and all inhibited radish root elongation. Vanillic *m*-coumaric, and *m*-hydroxyphenylpropionic acids were the most inhibitory compounds identified.

Zhamba (1972) mixed pulverized ripe *Heracleum* sp. fruits with soil, or he buried them in soil. After 1 year, he extracted the material with aqueous ethanol, identified the coumarin derivatives, quantified them, and compared them with coumarin derivatives in fresh ripe fruits. Furocoumarins were still present after 1 year in soil, but the amounts were slightly lower. 4-Hydroxycoumarin and four unidentified compounds in fresh ripe fruits were lacking in the compost. Zhamba pointed out that the high resistance of the furocoumarins to decomposition and their strong antibacterial, fungicidal, and insecticidal properties facilitated their accumulation in soil. Plants that are not adapted to the highly toxic furocoumarins die rapidly. Thus, settlements of *Heracleum* are rapidly established in foreign communities.

Junttila (1975, 1976) observed that other plant species rarely grow adjacent to *Heracleum laciniatum* plants in northern Norway. He had evidence that this plant produced fungistatic substances, so he decided to determine if *H. laciniatum* is allelopathic to other plants. He found that both dark and red light germination of lettuce seeds, as well as root and hypocotyl elongation, were inhibited when the seeds were sown in Petri dishes together with seeds of *Heracleum*. *Heracleum* seeds inhibited the germination of *Salix pentandra* and radish seeds, but they had no effect on the germination of spruce (*Picea abies*) seeds. *Heracleum* seeds contained three groups of inhibitors having the following R_f values developed on silica gel with chloroform: (1) 0.60–0.80, (2) 0.20–0.40, and (3) 0.0–0.10. Group (1) inhibited lettuce seed germination and root growth, elongation of *Avena* first internode segments, growth of the fungus, *Cladosporium cucumerinum,* and leakage of β-cyanin from red beet slices. Groups (2) and (3) caused no leakage of β-cyanin, but were inhibitory in other tests. Group (2) contained the furanocoumarins pimpinellin, bergapten, isobergapten, and angelicin, but specific inhibitors were not identified in groups (1) and (3).

The pine–bunchgrass community of northern Arizona consists of ponderosa pine (*Pinus ponderosa*), Arizona fescue (*Festuca arizonica*), and mountain muhly (*Muhlenbergia montana*) (Rietveld, 1975). In park-like stands and openings created by fire or logging, the bunchgrasses capture the sites and develop into dense, exclusive communities where few pine seedlings emerge even though seeds from surrounding trees repeatedly fall into them. Pine seedlings transplanted into such areas grow poorly or do not survive, and even where the grass cover is removed, growth and survival may be depressed. The United States Forest Service (1963) reported that aqueous extracts of foliage of Arizona fescue inhibited radicle growth of selected grass species and ponderosa pine. Rietveld (1975) extended this work and concluded that a growth-inhibitory substance is present predominantly in live foliage, but to a lesser extent in dead residues of Arizona fescue and mountain muhly. This substance could substantially reduce total germination, the germination rate, and radicle development of ponderosa pine. The route of release of the inhibitor was not definitely determined, but it was suggested that leaching from live grass tissues and microbial decomposition of the residues probably were the most significant routes.

Walters and Gilmore (1976) observed that height growth of sweetgum was lower in plots containing tall fescue than in adjacent plots devoid of fescue. Chemical and physical soil factors did not account for the differences. Sweetgum growth was correlated with residual phosphorus and magnesium, but this correlation was achieved across all experimental plots without respect to the presence or absence of fescue. Fescue seeded into pots containing sweetgum seedlings reduced dry weight increment of sweetgum up to 95%. Elimination of compettition through the use of a stairstep apparatus implicated an allelopathic mechanism in the interference of sweetgum growth by fescue. Leachates from the rhizosphere of live fescue, dead fescue roots, and dead fescue leaves reduced dry weight increments of sweetgum up to 60% (Fig. 8). Chemical analysis of sweetgum

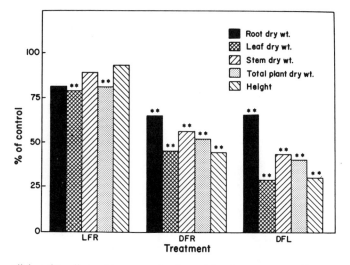

Fig. 8. Allelopathic effects of fescue on the dry weight of sweetgum seedlings grown for 1 month in contact with materials leached from either mature live fescue roots (LFR), dead fescue roots (DFR), or dead fescue leaves (DFL). (**) Indicates significant difference from the control at the 99% confidence level. (From Walters and Gilmore, 1976.)

seedlings from the stairstep experiment suggested an impaired absorption of phosphorus and nitrogen by seedlings that were treated with fescue leachates.

Lakhtanova (1977) reported that water soluble root exudates of *Lupinus poly-phyllus* stimulated growth of *Picea excelsa*. Such stimulatory effects could have important applications in forestry and should be thoroughly investigated.

Horsley (1977a) reported that some previously forested areas in the Allegheny Plateau of northwestern Pennsylvania have failed to return to forest, even though 50 years have passed since clear-cutting. Some of these contain scattered black cherry and red maple (*Acer rubrum*) trees and are called orchard stands (Fig. 9). Small black cherry seedlings grow slowly and soon die in low density orchard stands colonized by a dense cover of bracken fern, wild oat grass (*Danthonia compressa*), goldenrod (*Solidago rugosa*), and flat-topped aster (*Aster um-bellatus*). Careful experimentation eliminated browsing by animals, microclimate, and competition for light, minerals, and water as the primary causes of the retardation in growth of black cherry in the orchard stands. Subsequent studies indicated that foliage extracts of fern, goldenrod, and aster inhibited both shoot and root growth of black cherry seedlings growing on cotyledonary reserves. Foliage extracts of fern, grass, goldenrod, and aster and root washings of goldenrod and aster inhibited shoot growth and dry weight accumulation of seedlings that had exhausted cotyledonary reserves. Soil from the upper horizons of an orchard stand did not moderate the toxicity of the herbaceous foliage extracts or

Fig. 9. A low-density upland cherry–maple orchard stand at Mill Creek, Ridgway District, Allegheny National Forest. Photo taken June 12, 1974 before much height growth had been made by the herbaceous plants. Numerous stumps in orchard stands are evidence of past productivity. (From Horsley, 1977a.) Reproduced by permission of the National Research Council of Canada from the *Can. J. Forest Res. 7.*

root washings, indicating that the resident microorganisms did not rapidly metabolize the inhibitory compounds. In field tests, cherry seedlings did not survive and grow in the orchard stand soil during the first year after removal of the vegetation. They did grow during the second year after removal, however, indicating that an appreciable time-span was required for decomposition or leaching of the inhibitors. This is an important point to keep in mind in any consideration of allelopathy, because phytotoxins have to be produced faster than they are decomposed to be active. If this occurs for even a short time period, however, the effect on susceptible plants can be very pronounced.

Later, Horsley (1977b) compared paired field plots on poorly drained Allegheny hardwood sites and found that the presence of hay-scented fern (*Dennstaedtia punctilobula*), New York fern (*Thelypteris noveboracensis*), short husk grass (*Brachyelytrum erectum*), and club moss (*Lycopodium obscurum*) was correlated with reduced numbers of black cherry seedlings. Numbers of red maple seedlings were fewer where associated with the ferns and grass, but not with club moss; and numbers of sugar maple seedlings were fewer with the ferns but not with grass or club moss. Moreover, the heights of the 10 tallest black cherry seedlings on each plot were less with the ferns and grass but not with club moss. Foliage extracts of all four ground covers inhibited the growth of black cherry seedlings that had exhausted cotyledonary reserves. These tests were made in sand cultures with concentrations of nutrient solutions adjusted so that all solutions contained the same amounts of added nutrients. Horsley concluded that allelopathy appears to play an important role in the reduction in growth of black cherry by woodland ferns and grass but that competition for light is probably very important also.

Sugar maple regenerates well in managed woodlots in Canada's Great Lakes–St. Lawrence forest region, but does not invade abandoned agricultural lands and grows poorly when planted in such lands (Fisher *et al.*, 1978). Interference from old-field plants appears to be a contributing factor in this failure. These investigators found that old-field weed residues inhibited germination and growth of sugar maple even in the absence of competing vegetation. Goldenrod (*Solidago canadensis, S. graminifolia*) and aster (*Aster novae-angliae*) were important producers of water soluble compounds that inhibited germination, nutrient uptake, and growth. These phytotoxins were most readily extracted from decomposing plant residues, but appeared to be natural products rather than microbial breakdown products. The deleterious effects of goldenrod on the nutrition and growth of sugar maple could be overcome by large additions of soluble phosphorus fertilizer, but not by nitrogen. This last point is very important, because many ecologists and others attribute all or most biological effects of added organic matter in soil to the immobilization of nitrogen due to the increased populations of microorganisms. This obviously is important when the added organic material has a high C–N ratio, but is not the overriding effect in many instances. Another important point to remember is that the ability to correct a nutritional deficiency in a test plant by an added nutrient does not necessarily indicate that interference by another species is due to competition. The deficiency could just as well be due to inhibition by an allelopathic agent of uptake of the element in question.

Gilmore (1980) cited numerous references, which stated that the mortality of tree seedlings planted into areas occupied by grass or weeds was due to competition for moisture, nutrients, light, or a combination of these factors. He pointed out, however, that recent evidence showed that more is involved than just competition. This is supported strongly by the numerous examples discussed above.

In 1963, Gilmore and Boggess attributed the low survival and growth of loblolly pine in southern Illinois to competition from ragweeds (*Ambrosia* sp.) and giant foxtail for light and moisture. Gilmore (1980) reinvestigated the reasons for the strong interference of giant foxtail against loblolly pine. When giant foxtail and loblolly pine were grown together in the same pots, height growth of the pines was not affected. By the end of 40 days some pine needles were becoming chlorotic and uptake of P, K, Ca, and Mg was slightly restricted. Shoot weight did not differ from that of the controls during the first 40 days, but after 60 days shoots of treated seedling weighed about 40% less than those of controls. A 12% reduction in increase of root weight of pine occurred after 20 days when grown with giant foxtail, and a 50% reduction occurred after 60 days, compared with controls. After 60 days, Ca in the roots of treated seedlings doubled, whereas K was reduced by 50%, Mg by 30%, and P by 15%.

In a stair-step experiment where no competition occurred, new shoot growth of pine exposed to leachates from the rhizosphere of live or dead giant foxtail became chlorotic after 20 days. After 40 days, needles of some treated seedlings were either dead or dying. Roots of treated seedlings were stunted and appeared to have suffered from dieback. After 60 days, root growth of treated seedlings was reduced by about 30% and shoot growth by 36%. Concentrations of P, K, and Ca were generally lower and Mn higher in shoots of seedlings exposed to leachates from dead foxtail plants than they were in control shoots. Levels of Ca, Fe, Cu, and Mn were generally higher in roots of these treated seedlings than in controls. There was usually more P and less K in roots of pines subjected to leachates from live foxtail plants than in controls. Seedlings exposed to leachates of dead foxtail tops had higher concentrations of N in the shoots and roots than did controls. However, no difference in N occurred in tops or roots of pines exposed to leachates from live foxtail or dead foxtail roots.

Gilmore concluded that allelopathy is very important in the inhibition of loblolly pine growth by giant foxtail in the greenhouse and, probably, in the field also. His data indicated that much of the interference caused by giant foxtail against loblolly pine was due to allelopathy.

Larson and Schwarz (1980) investigated the potential allelopathic inhibition of three N-fixing species by six other species that grow wild or were seeded on strip-mine lands in Ohio. The three N-fixing test species—black locust, black alder (*Alnus glutinosa*), and red clover—have been planted widely in cropland and on disturbed lands in the eastern United States. Aboveground plant parts of tall goldenrod (*Solidago altissima*), broomsedge, crownvetch, wild carrot (*Daucus carota*), tall fescue, and timothy were collected at the time of flowering on old-field or strip-mine areas in eastern and northcentral Ohio. Growth of black locust was inhibited by residues of all six species; goldenrod inhibited growth by 90%; wild carrot inhibited growth by 77%. Black alder growth was not affected by any of the residues except that of crownvetch, which stimulated growth; growth of red clover was inhibited only by goldenrod.

The nodule number on black locust roots was decreased by all residues except broomsedge, and the N-fixation rate was significantly reduced by wild carrot, tall fescue, and timothy. The nodule number on red clover was significantly reduced only by goldenrod residue, but it was significnatly increased by residues of wild carrot, tall fescue, and timothy. The last three residues significantly stimulated N-fixation, but the other residues had no effect on it in red clover. None of the residues affected nodule numbers in black alder, but goldenrod, broomsedge, and crownvetch stimulated the rate of N-fixation.

Many strip mine and old-field areas are seeded with or invaded by herbaceous species that inhibit tree growth and N-fixation in soils (Larson and Schwarz, 1980). They emphasized that black locust appears to be relatively sensitive and that black alder is insensitive to allelopathic inhibition, which suggests that black alder may be better suited to weedy sites. They also pointed out that some legume species may not be compatible with each other, as exemplified by inhibition of black locust by crownvetch. They stressed the need for more research on allelopathic influences in field situations.

C. Allelopathic Effects of Ferns

Research on possible allelopathic effects of ferns has been limited, and most of it has involved bracken fern. Bracken grows under environmental conditions ranging from tropical through Mediterranean to boreal, and much literature attests to its dominating influence over many types of vegetation (Gliessman and Muller, 1972). Most investigators have attributed the dominating influence to competition, but Gliessman and Muller (1972) reported that water extracts of dead bracken fronds collected near Santa Barbara, California, inhibited seed germination and radicle growth of two grass species, while extracts of green fronds were not toxic. Artificial rain leachates inhibited radicle growth of a grass seedling also, as did residues of bracken fronds in soil. Stewart (1975) reported that water extracts of senescent bracken fronds reduced germination of western thimbleberry (*Rubus parviflorus*) and delayed germination of salmonberry (*Rubus spectabilis*) seeds, but not Douglas fir seeds. Unincorporated bracken litter reduced the emergence of all three species in soil, but it did not influence root and shoot growth or dry weight increase. Bracken litter incorporated in the soil, on the other hand, reduced shoot growth and dry weight increase of western thimbleberry seedlings. He concluded that allelopathic interactions may explain the relative absence of woody shrubs from sites dominated by bracken.

The allelopathic effects of bracken on black cherry in Pennsylvania (Horsley, 1977a) were discussed in the previous section in relation to allelopathic effects of herbaceous angiosperms on black cherry. Whitehead (1964) found that soil associated with bracken contained *p*-hydroxybenzoic acid, vanillic acid, *p*-hydroxycinnamic acid, and ferulic acid at concentrations of $3.9 \times 10^{-5} M$, $4.9 \times 10^{-5} M$, $4.2 \times 10^{-5} M$, and $0.4 \times 10^{-5} M$, respectively. Fronds contained these same compounds in high concentrations throughout the growing season, and rain leaching and decomposition of the fronds apparently introduce the compounds into the soil (Glass and Bohm, 1969). Glass (1976) grew six species of grasses and a clover (type not designated) hydroponically in solutions that exactly reproduced the major phenolic acid composition and concentrations of the soil associated with bracken as described by Whitehead (1964). In $CaSO_4$ solution, these compounds severely inhibited root growth as measured by fresh weight, dry weight, and root volume of all species except one grass. Fresh weight increase was inhibited most at 5° and 30°C, and least between 15° and 20°C. The phenolics did not have any effect on barley roots when the phenolics were dissolved in a complete inorganic nutrient solution.

Apparently, bracken has evolved mechanisms of toxin release that allow it to effectively exert its dominance in each particular habitat in which it grows (Gliessman, 1976). In southern California, toxin release is timed for the beginning of the wet season and thus the initiation of germination of associated species. Toxins come primarily from dead standing fronds. In the Pacific Northwest, toxin release is timed to the breaking of dormancy in spring after the snow melts and soil temperatures rise, with toxins coming primarily from litter, roots, and rhizomes. In tropical Costa Rica, toxins are released throughout the year. Gliessman and Muller (1978) demonstrated that competition could not account for the lack of herbs in bracken stands and that uniformity of soil pH, texture, water-holding capacity, and organic matter content eliminated variability in physical factors as a cause. Trapping experiments showed that the higher concentrations of animal activity inside the bracken stands contributed to the pattern of herb suppression of only certain species. In fact, the same pattern was maintained in the animal-free Santa Cruz Island stands. They concluded that phytotoxins leached from the dead standing bracken fronds with the first few rains of the wet season were largely responsible for herb suppression (Table 28). These toxins were isolated in raindrip from bracken foliage in the field and from soil inside the fern stands. Moreover, herbs reinvaded the stands after several seasons when fronds were removed before rains could leach them, and conversely, placing fronds over the herbs in the grassland resulted in inhibition of herb growth.

D. Inhibition of Mycorrhizae and Other Microorganisms

Certain tree species, such as *Betula pendula* and *Picea abies,* frequently fail to develop in association with heather (*Calluna vulgaris*) (Robinson, 1972). Handley (1963) reported that heather inhibits the growth of mycorrhizae on spruce, causing the failure of spruce establishment among the heather. This seemed logical because Melin (1963) pointed out that ectomycorrhizal fungi are very sensitive to substances exuded from plant roots or leached from dead plant material. Robinson (1972) demonstrated that run-off from roots of living *C.*

TABLE 28.   Inhibition of Several Grassland Species by Frond Leachates[a]

Species	Percent of distilled-water controls		
	Unconcentrated extract	Concentrated extract (4×)	
	Growth	Growth	Germination[b]
Bromus rigidus	69.0[c]	45.7[d]	100
Bromus mollis	81.0	46.7[d]	100
Avena fatua	58.9[d]	29.6[d]	60.0[d]
Hypochoeris glabra	57.5[d]	0[d]	0[d]
Festuca megalura	40.0[d]	24.5[d]	57.5[d]
Clarkia purpurea	86.0	17.2[d]	23.0[d]

[a] 30 Seeds per treatment. From Gliessman and Muller (1978).
[b] Germination inhibited only in the extract concentrated 4 times.
[c] Significant at the 5% level, Student's t test.
[d] Significant at the 1% level, Student's t test.

vulgaris and raw humus of this species contained a factor toxic to several mycorrhizal fungi. The data also showed that the inhibitor may have prevented infection of *Calluna* by certain pathogenic fungi. Field observations indicated that the phytotoxin must be continuously produced, because the inhibitory effect disappears soon after the heather is removed.

Trees cannot grow normally on prairie soils unless the seedlings are inoculated with mycorrhizal fungi (Persidsky *et al.*, 1965). These investigators decided, therefore, to try to determine the reasons for the absence of free-living mycorrhizal fungi in prairie soils. They found that extracts of roots of prairie grass did not affect oxygen uptake of mycorrhizal short roots of Monterey pine (*Pinus radiata*) any differently than did the extracts of roots of *P. resinosa*. On the other hand, water extracts of prairie soils inhibited oxygen uptake by the same types of mycorrhizal roots, as compared with the effects of extracts of forest soils. Theodorou and Bowen (1971) reported that decomposition products of grass roots reduced the numbers of an ectotrophic mycorrhizal fungus of Monterey pine in the soil and the frequency of mycorrhizal infections of Monterey pine roots. Apparently, sufficient toxins are present, at least in some prairie soils, to preclude the presence of free-living mycorrhizal fungi.

Olsen *et al.* (1971) found that aqueous extracts of *Populus tremula* leaves strongly inhibited the growth of several species of *Boletus,* a mycorrhizal fungus. They had a weaker inhibitory effect on litter-decomposing species of *Marasmius.* The inhibitors, identified as catechol and benzoic acid, had a strong inhibitory effect on *Boletus.*

Brown and Mikola (1974) studied the influence of fruticose soil lichens on mycorrhizal and seedling growth of forest trees. They demonstrated first that extracts of four species of *Cladonia, Cetraria islandica,* and *Stereocaulon paschale* inhibited numerous species of fungi on agar, most of which were mycorrhizal fungi. Next, they found that extracts of *Cladonia alpestris, Cladonia rangiferina,* and *Cetraria islandica* markedly lowered ^{32}P uptake in axenic, nonmycorrhizal *Pinus sylvestris* seedlings. The *Cladonia alpestris* extract also markedly inhibited ^{32}P uptake by mycorrhizal seedlings of *P. sylvestris. Cladonia alpestris* inhibited the growth and survival of pine and *Picea abies* seedlings in paired nursery plots but did not affect *Betula verrucosa* seedlings. *Cladonia alpestris* also severely limited the number of mycorrhizal root tips of *P. sylvestris.* Removal of the lichen in field tests resulted in an accelerated growth of pine and spruce. Thus, it appears that certain lichens affect survival and growth of some forest trees through production of phytotoxins that inhibit growth of mycorrhizae on those trees.

It is noteworthy that lichens were used for medicinal purposes in Egypt in the seventeenth and eighteenth centuries B.C. and are still brought to Egypt from Europe (Vartia, 1973). Vartia concluded that over one-half the lichen species have antibiotic properties due to lichen substances, the most effective of which are usnic acid, the lichesterinic acid group, and the orcinol-type depsides and depsidones.

Stillwell (1966) isolated an imperfect fungus, *Cryptosporiopsis* sp., from yellow birch and found in plate culture that this fungus inhibited the growth of 31 Basidiomycetes that were isolated from decaying material in coniferous and deciduous forests, an Ascomycete, the dutch elm disease fungus (*Ceratocystis ulmi*), and a Phycomycete (*Phytophthora infestans*). *Cryptosporiopsis,* in liquid culture, produced a substance highly inhibitory to *Fomes fomentarius,* the fungus most commonly associated with decay in living branches of yellow birch. He found also that growth of *Fomes* was inhibited markedly by *Cryptosporiopsis* in sterile yellow birch wood. The amount of decay was reduced in peeled balsam fir (*Abies balsamea*) logs that were treated with a water suspension of mycelial fragments of *Cryptosporiopsis.* If such an antibiotic effect is widespread, then it could be of tremendous significance in preventing the decay of living trees and in regulating the rate of decay of dead plant parts.

Krogstad and Solbraa (1975) tested water extracts of fresh and composted spruce bark against several microbial proteinases and DNases; hemolysins of *Bacillus cereus, Pseudomonas aeruginosa,* and *Escherichia coli;* α- and β-toxins of *Staphylococcus aureus;* lecithinases of *B. cereus* and *Clostridium perfringens;* colicins; different antibiotics; and the bactericidal effects of lysolecithin and peptides. Fresh extracts inhibited a number of the activities tested and extracts of composted bark lost inhibitory activities to varying degrees, depending on the biological system, composting time, and concentration of the bark extract.

Shukla *et al.* (1977) found that culture filtrates of *Mortierella subtilissima, Aspergillus candidus, A. flavus, A. niger, Penicillium rubrum, Papulaspora* sp., *Staphylococcus aureus,* and *Bacillus subtilis* contained antibiotics that inhibited the growth of various fungi isolated from leaf litter of *Shorea robusta.* Such antibiotics may be important in slowing the rate of decomposition of some kinds of litter, thus slowing mineral cycling.

Much more information on inhibition of bacteria, fungi and algae is included in Chapters 1, 4, 6, 7, and 9.

II. HORTICULTURE

A. Introduction

In the early part of this century, Pickering (1917, 1919) observed that apple (*Malus sylvestris*) trees were injured by interference from grass. He thought at first that the effects were due to competition for minerals or exclusion of oxygen from the tree roots by the grass. His early experiments ruled out these factors as causes. He next investigated the possibility that the grass was producing a toxin. He planted grass in a tray completely separated from the container in which the apple seedlings were growing and allowed drainage from the soil in which the grass was growing to flow into the container of apple seedlings. Despite the fact that there was no competition, seedling growth was inhibited, indicating that allelopathy accounted for the interference.

Walnut trees have been known for many years to produce a toxin injurious to apple trees (Massey, 1925, Schneiderhan, 1927). In fact, Schneiderhan (1927) reported that the toxin sometimes kills neighboring apple trees.

Molisch (1937) performed many experiments that demonstrated that fruits of apple, pear (*Pyrus communis*), and related plants produce ethylene, which diffuses away from the fruits and affects other plants in striking ways. He pointed out that ethylene influences the growth in length and thickness of seedlings, hastens the ripening of fruit in amazing ways, promotes proliferation of lenticels, hastens callus formation, hastens leaf fall, prevents the negative geotropic curvature of hypocotyls, and cancels epinastic curvatures.

Tukey (1969) discussed several important allelopathic responses in horticulture resulting from grafting or budding. Among these were dwarfing, resistance to diseases, change in time of maturation of fruits, change in size, and color and quality of fruits.

Krylov (1970) reported that cultivation of potatoes in the space between rows of young apple trees resulted in accumulation of toxins that inhibit tree growth. These toxins caused the total nitrogen content in the roots and tops to decrease, and they also caused a change in the composition of proteins in the bark. The amount of soluble albumins increased and the quantity of residual proteins decreased. The toxins also depressed photosynthesis.

B. The Peach Replant Problem

The difficulty of replanting fruit trees following the removal of old orchards has been recognized for many years in the United States and Europe (Klaus, 1939; Proebsting and Gilmore, 1941; Koch, 1955). The problem is usually species specific, and other plants grow well in the same soil. This replant problem has been reported for apples, grapes, cherries, plums, peaches, apricots, and citrus (Proebsting and Gilmore, 1941; Patrick et al., 1964). Comprehensive investigations covering possible competitive and allelopathic interference have been carried out apparently only for peaches, apples, and citrus.

According to Koch (1955), aboveground symptoms of affected peach (*Prunus persica*) replants are usually retarded growth, eventual stunting, and different degrees of intercostal chlorosis. The roots display different degrees of discoloration and necrosis with brown lesions invariably occurring on otherwise white lateral roots. Occasionally, the effect is so severe that the peach replants die. The effects on the roots show up rapidly, often within 24 hours after emerging from the parent root tissue.

According to Proebsting and Gilmore (1941), a survey of peach growing districts in California indicated varying success of replanted orchards, from almost complete failure to complete success. The differences in response could not be correlated with climate, soil texture, or obvious cultural practices. Consequently, they investigated the following possible causes for the failure of replants: (1) depletion of common nutrients either by the preceding orchard or by soil organisms using root residues as a source of energy, (2) depletion of minor elements, (3) diseases carried over from the previous orchard, and (4) direct toxicity of the roots or their decomposition products.

They laid out several fertilizer plots in districts where the peach replant problem was most serious. These involved addition of various amounts of ammonium sulfate, potassium sulfate, superphosphate, manganese, zinc, copper, vitamin B_1, indolebutyric acid, indolepropionic acid, and peat moss. None of these eliminated the basic problem, although minor responses did occur in some cases.

Proebsting and Gilmore (1941) pointed out that certain diseases, such as crown gall caused by *Pseudomonas tumefaciens,* could be important factors in some failures, but they found many cases of subnormal development where the peach trees were completely free of pathogens. Thus, they concluded that nutrition and disease must play only minor roles in the peach replant problem.

Subsequent experiments were designed by Proebsting and Gilmore to determine if an allelopathic mechanism might be involved in the peach decline.

Initially, peach seedlings were grown in 5-gal cans in the greenhouse either in screened soil from a peach orchard exhibiting the trouble or in screened soil from adjacent land not previously seeded with peach trees. No significant differences in growth of the seedlings were obtained. However, when 500 g of peach roots were added to virgin soil in the same containers, growth inhibition followed. The total average shoot length per control tree was 481 cm versus 326 cm for the test, and the average weight per control tree was 102.5 g versus 53.7 g for each test tree.

In another experiment, almond, apricot, and myrobalan seedlings were planted in a field where peaches had been removed, with peach seedlings planted as checks. The myrobalan seedlings displayed strong growth; the growth of almond and apricot was satisfactory, but that of peach seedlings was very poor.

An apricot orchard, made up of alternating rows of apricot and peach roots, was planted following the removal of a 40-year-old peach orchard. The apricot trees on apricot roots showed strikingly better growth than the apricot trees on peach roots, whereas normally apricot trees on peach roots grow as well as apricot trees on apricot roots. The effect was still apparent when the trees were 9 years old.

In another experiment, bark and wood of peach roots were separated, dried and ground. The ground material was added in various amounts to peach seedlings in 3 gallon sand cultures. Usually, 10 g of bark added around the root area and next to the stem killed the seedlings in 4 to 5 days.

In one experiment, seven types of tests were examined in sand culture, plus the controls: (1) ground root bark, (2) ground root bark plus an equivalent amount of ground peat moss, (3) ground root wood, (4) material extracted from root bark with alcohol, (5) root bark residue after alcohol extraction, (6) water extract of root bark, (7) root bark residue after water extraction, and (8) sand as a control. Seedlings were planted in all and a complete nutrient solution was supplied. Except the root bark residue after alcohol extraction, all test material caused injury to the tops and roots of the peach seedlings.

Amygdalin occurs in the root bark of peach, so an experiment was run in which 1 g of amygdalin was added to 250 ml of nutrient solution and applied once or twice a week to the trees grown in sand. No injury occurred. When the same amount of amygdalin was applied along with a trace of emulsin, however, injury was severe. Two such treatments 3 days apart killed the trees in 30 to 40 days after the first treatment. Emulsin digests the amygdalin to benzaldehyde and hydrogen cyanide, so Proebsting and Gilmore tested the effects of benzaldehyde and potassium cyanide in amounts ranging up to 1 g of benzaldehyde per week in two treatments and 0.1 g of potassium cyanide once a week. The benzaldehyde had little effect, but the 0.1 g dosage of potassium cyanide caused severe injury to the peach seedlings.

Proebsting (1950) reported a carefully documented case history of the peach replant problem based on field observations at Davis, California. Two rows of apple trees and two rows of Lovell peach trees planted in 1922 were removed in 1942 and all four rows (twenty-two trees per row) were planted to Faye Elberta peaches. It was apparent by the end of the first year that the trees succeeding the apple trees were growing better than those succeeding peach trees. At the end of the sixth growing season, the average circumference of the trees following apple trees was 58.2 cm, whereas it was only 42.4 cm for peach trees following peach trees. At that time the average yield was 216.4 lb per tree after apple trees and 118.8 lb per tree after peach trees.

Havis and Gilkeson (1947) did an experiment similar to the sand culture experiment of Proebsting and Gilmore (1941). The materials added to the sand medium were (1) oven-dried (80°F) ground root bark of peach, (2) root bark frozen in carbon dioxide and then ground, (3) whole small fresh roots frozen in carbon dioxide and then ground, and (4) whole fresh roots of various sizes cut into pieces not over 0.75 inches long. The plants used were 1-year Lovell seedlings budded to Elberta in the first experiment; in the second two types were used: (1) Elberta budded on Lovell stocks and (2) Elberta seedlings. The duration of each experiment was 7 months. They reported that there was no evidence at any time in either experiment of toxic effects due to any of the materials added.

Koch (1955) reviewed the evidence for the peach replant problem and the efforts to determine the cause. He reported that the cause had been attributed to insects, nutritional disturbances, spray residues, phytotoxins, and nematodes. He suggested that a coordinated research effort covering several possibilities should be carried out. Such an effort was initiated by the Canada Department of Agriculture and carried on over a period of several years (Patrick, 1955; Ward and Durkee, 1956; Mountain and Boyce, 1958, Mountain and Patrick, 1959; and others).

Patrick (1955) investigated the role of toxins in the peach replant problem. He was chiefly interested in how toxins arise, whether or not microorganisms are necessary for their production, and whether or not they are due entirely to amygdalin. In his first series of experiments, he used soil obtained from site (under peach trees) and intersite areas in an orchard in which the replant problem was evident. The soil was passed through a screen to remove all pieces of roots, stones, and other foreign material. The peach root residue used was obtained from roots of 2-year-old Lovell peach seedlings grown in soil in which no peach trees previously grew. The smaller roots and the bark from the larger roots were ground to a coarse powder.

The residues were added to site and intersite soil at the rate of 1 to 5% (w/w); some of the samples were autoclaved. The flasks containing the soil were incubated for several days at 25°C. Water extracts of the soil were then tested against respiration of Lovell Peach seedling root tips using the Warburg technique.

The rate of breakdown of amygdalin in the various substrates was determined. The results indicated that microorganisms are necessary for decomposition of the amygdalin because no HCN was produced when the soil was autoclaved. It appeared that the soil from the site areas contained a higher proportion of microorganisms capable of hydrolyzing amygdalin than soil from the intersite area because breakdown of the amygdalin began more rapidly in the site soil. Similar results were obtained when pure amygdalin was added to the soils at the rate of 0.25%. The addition of pure amygdalin eliminated, of course, the possibility of introducing into the soil samples specific microorganisms occurring with the peach root residues.

The water extracts of soil samples incubated with peach root residues or amygdalin greatly reduced respiration in peach roots (Table 29). The inhibition increased with increasing amounts of residue and with increasing time of incubation of the soil and residue up to 3 or 4 days from the start. Again, it was evident that the production of toxic substances from amygdalin or root residue depended on the presence of microorganisms that are capable of utilizing the amygdalin. Sour cherry root residue either stimulated respiration or had virtually no effect depending on the time of incubation. Later experiments in which 10% root residues of tobacco or pepper were added to site soil indicated that these residues stimulated respiration of peach roots.

Experiments were conducted to determine the effects on respiration of benzaldehyde and cyanide, which, along with glucose, are the breakdown products of amygdalin. Amygdalin, emulsin, and a combination of these were also tested. Amygdalin or emulsin alone had no effect on respiration, but they markedly inhibited respiration of peach roots when combined. When calcium cyanide was added to site soil the extracts were also very inhibitory to respiration. Benzaldehyde in water greatly reduced respiration in peach roots, with a dilution of 1:5000 (w/v) causing some inhibition.

Patrick (1955) cut down some 14-year-old peach trees and established plots that included the former location of a tree plus the surrounding 20 ft. Some plots were fumigated with ethylene dibromide, others with methyl bromide and some were not fumigated. Lovell peach seedlings about 5 months old were planted 1 ft apart in all plots. Populations of nematodes and microorganisms were determined throughout the following growing season. Peach seedlings in the site areas showed poor growth in both fumigated and unfumigated plots, whereas seedlings 4 ft beyond the site in the intersite grew well. There was slight improvement in growth in the fumigated site areas but the contrasts with intersite areas were still great. It was concluded, therefore, that nematodes or other organisms that would have been greatly reduced by the soil fumigants were not responsible for all the stunting in the site areas. As much as 40% inhibition of respiration in peach roots resulted from water leachates of soil in the fumigated site areas.

Roots of 4-week-old intact peach seedlings were placed in water extracts of soil from site areas, and after 3 to 7 days the seedlings showed wilting symptoms and drying of leaves starting at the base of the plant. The meristematic regions of

TABLE 29. Influence of Incubation Time, Soil Type, and Other Factors on the Relative Toxicity of the Resulting Soil Leachates as Determined by the Intensity of Their Inhibiting Effects on the Respiration of Excised Peach Root Tips[a]

					Total inhibition (or stimulation) of respiration in 5 hours at 20°C (%) [c]			
					Substances added to old peach orchard soil (site area)			
Incubation time (hours) (at 25°C before extracts made)	Peach root residue		0.25% Amygdalin		Peach soil autoclaved			
					Peach root residue	Amygdalin	Sour cherry root residue
	2%[b]	5%	Site area soil	Intersite area soil	2%	0.25%	2%
60	48	—	51	—	—	—	—
70	60	—	55	—	8	10 stimulation[d]	8 stimulation
100	40	78	72	—	6	15 stimulation	—
150	—	68	82	7 stimulation	3 stimulation	7 stimulation	15 stimulation
200	—	59	73	19 stimulation	—	—	2
300	—	—	74	14	—	12	—

a Modified from Patrick (1955). Reproduced by permission of the National Research Council of Canada from the Canad. J. Bot. 33, 461–486.

b Amount of peach root residues, amygdalin, or other root residues, calculated as % by weight of the total soil contents added to each flask.

c Percent total inhibition of respiration in 5 hours of 15 excised peach root tips placed in the various soil–water leachates (containing various amounts of the decomposition products of the substances indicated) was calculated in the usual manner using as check (or normal respiration rate) the respiration of identical root tip samples placed in soil–water leachates containing none of these products.

d If total O_2 uptake was greater than that of the check, it was calculated as percent total stimulation of respiration.

the roots turned brown also. All these symptoms appeared also in seedlings placed in a 1:2000 dilution of benzaldehyde in water, whereas there were no injurious effects even after 2 weeks in extracts of soil from intersite areas.

A large number of microorganisms capable of utilizing amygdalin were isolated from peach root agar inoculated with soil from the site areas in old peach orchards. Both fungi and bacteria were involved, but few actinomycetes grew on the media containing only peach root residue or amygdalin as a carbon source. Many more organisms capable of utilizing amygdalin were isolated from site soil than from intersite soil.

Patrick (1955) concluded that whenever amygdalin and the microorganisms capable of utilizing it are present in the soil, soluble toxic substances highly detrimental to living peach roots are likely to be produced in that soil. The amounts of these toxins and, thus the degree of toxicity produced, would depend on the amount of old peach roots remaining in the soil after the old trees have been removed as well as on their amygdalin content. The toxic effects would be greatest for the first year or two after the trees are removed and should gradually diminish. Patrick pointed out that this diminishing toxicity would explain, at least in part, the observation that the replant problem diminishes 2 or 3 years after removal of the old trees.

Because Patrick's (1955) research demonstrated that amygdalin is the source of phytotoxins produced from peach root residues, Ward and Durkee (1956) determined amounts of this glycoside in various parts of 2- and 3-year-old Lovell peach seedlings and in roots of several varieties of peach trees of different ages. They found that concentrations varied somewhat with different trees of the same variety and age and that amygdalin occurs in leaves, stems, and roots (Table 30). The stems have the lowest concentration followed in increasing amounts by leaves and roots. The roots had their highest concentration in the spring, whereas the tops had their highest concentration in the fall. The concentration of amygdalin was highest in the root bark.

TABLE 30. Amygadalin Content of 2-Year-Old Lovell Peach Seedlings[a]

		Amygdalin (mg/g dry weight)			
Sampling date	Tree number	Leaves	Stem, branches, twigs	Large roots	Small roots
June 25, 1954	A	7.4	4.9	41.5	53.8
June 25, 1954	B	7.1	4.7	39.8	36.6
October 22, 1954	C	13.0	9.3	15.5	22.8
October 22, 1954	D	18.3	9.3	32.8	37.8
October 22, 1954	E	16.9	4.8	21.4	26.1

[a] From Ward and Durkee (1956). Reproduced by permission of the National Research Council of Canada from the *Can. J. Bot.* **34,** 419–422.

Sampling was not adequate to draw definite conclusions, but the concentrations of amygdalin in the root bark of the single trees tested of the varieties Yunnan and Shalil were considerably lower than in any of the other varieties. These two varieties are somewhat resistant to the replant problem.

Mountain and Boyce (1958) investigated the relation of nematodes, and particularly of *Pratylenchus penetrans*, to the growth of young peach replants in old peach orchards. They found that the first roots produced by the young trees were attacked within a relatively short time by *P. penetrans*, which is an endoparasite. It propagates rapidly in the newly formed succulent tissues, and during the early spring period, degeneration of the root system sometimes becomes evident. The populations drop rapidly, however, as the soil temperatures increase. As populations of *P. penetrans* decrease, ectoparasitic nematodes with stylets capable of penetrating into the sieve tubes of the roots appear. During the second year, the endoparasites continue to decrease, and the ectoparasites increase.

Mountain and Boyce (1958) concluded that the precise role of *P. penetrans* in the peach replant problem was not clear, even though relatively large populations appear to be connected with the failure of replanted trees. They suggested that nematodes may be related to the problem in one or both of the following ways: (1) Nematodes may create infection courts for certain bacteria and fungi of peach soils, thus affording an opportunity for the production of toxins through the breakdown of amygdalin, and (2) nematodes may be pathogens whose parasitic activities profoundly affect plant growth directly.

Mountain and Patrick (1959) extended the work of Mountain and Boyce (1958) and reported that *Pratylenchus penetrans* hydrolyzes amygdalin directly by its own enzymes and indirectly through mechanical damage of the host's root cells. This allows the host emulsin and substrate to interact and to produce phytotoxins.

Patrick *et al.* (1964) stated that, although many questions still remained, the evidence suggested that the production of toxic substances through the hydrolysis of amygdalin appeared to be the main mechanism involved in peach root degeneration. They pointed out that any lesion-producing agency that can rupture or penetrate root cells can cause the release of phytotoxins and produce root damage. These agencies include nematodes, fungi, insects, and physical factors. In addition to producing toxins from living roots, microorganisms also release them from peach root residues. Thus, it appears that allelopathy is the primary cause of the peach-decline problem.

C. The Apple Replant Problem

Börner (1959) reported that nurserymen in Germany encounter the replant problem with apple trees after cultivation of apples for even 1 or 2 years, and they find it essential to plant apple seedlings in soils that have never before been used to grow apples. The symptoms of affected apple trees are retarded growth and shortened internodes resulting in a rosette-like appearance. In addition, the roots are often discolored and growth of the tap root is reduced. Börner pointed out that many causes have been suggested, including nutrition, nematodes, and toxins. He stated that Fastabend had found that chopped apple roots, or water leached through affected soils, produced the toxicity symptoms when added to virgin soil in which apple seedlings were growing.

Börner (1959) decided to study further the effects of apple root residues on the growth of apple seedlings and to identify the toxins. Siberian crabapple (*Malus baccata*) seedlings were grown in solution culture with root bark from a 16-year-old apple tree (var. Stayman Winesap) added in amounts of 0.2, 1, and 10 g per 500 ml of solution. After 33 days, the plants were harvested and various kinds of measurements were made (Table 31). The tops and roots were markedly inhibited but especially the roots, even at a concentration of only 0.2 g per 500 ml of solution. Preliminary tests indicated that one or more phenolic compounds were present in the flasks with bark. The phenolics were identified as phlorizin, phloretin, *p*-hydroxyhydrocinnamic acid, phloroglucinol, and *p*-hydroxybenzoic acid.

Börner decided next to determine if the same phenolics could be detected in soil to which he added apple roots. When the soil was leached immediately after adding the apple roots, only phlorizin was found in the leachate. Soil leached 1 to 2 days later yielded phlorizin, ploretin, *p*-hydroxyhydrocinnamic acid, and phloroglucinol. After 8 to 13 days, only the last two compounds were present in the soil. The only compound detected in the earlier water culture experiment and not in soil was *p*-hydroxybenzoic acid.

Roots of two apple species were extracted with ethanol, and the extracts were

TABLE 31.   Influence of Root Bark of a 16-Year-Old Apple Tree on the Growth of Apple Seedlings in Water Culture[a,b]

| Growth response | Growth with standard error in flask (500 ml) containing different amounts of bark | | | |
	0 g	0.2 g	1.0 g	10.0 g
Number of leaves	13.0 ± 0.9	9.8 ± 0.8	7.8 ± 0.9	No growth
Length of stem, cm	16.1 ± 3.4	10.7 ± 2.8	4.1 ± 0.3	No growth
Length of root, cm	19.4 ± 0.6	9.9 ± 0.7	5.7 ± 0.3	No growth
Dry weight of leaves, mg	715.0 ± 65.9	337.5 ± 69.5	141.3 ± 29.9	No growth
Dry weight of stem, mg	220.0 ± 57.2	110.0 ± 35.8	35.0 ± 6.4	No growth
Dry weight of roots, mg	155.0 ± 9.6	96.3 ± 7.1	70.0 ± 9.3	No growth

[a] From Börner (1959).
[a] Duration of experiment was 33 days.

chromatographed. Phlorizin was the only phenolic found that had been previously identified, indicating that it was the only one of the five that is a natural constituent of apple roots. Börner concluded, therefore, that the other phenolics are microbial decomposition products of phlorizin. He did find a flavonoid glycoside, quercitrin, in the root extract of a 16-year-old tree in addition to phlorizin.

In another experiment, Börner (1959) added pure phlorizin to soil. Some soil was sterilized in an autoclave and some was not, and all of the soil was subsequently allowed to stand at 22°C. The same five phenolic compounds were found after a period of time as in the original experiment in which apple root bark was added to nutrient solution, if the soil was not sterilized. Sterilized soil continued to contain only phlorizin. From the sequence of events, Börner concluded that phloretin, p-hydroxyhydrocinnamic acid, phloroglucinol, and p-hydroxybenzoic acid are decomposition products of phlorizin and that the decomposition is accomplished by microorganisms. He inferred also that the sequence of decomposition is as follows:

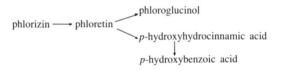

He demonstrated also that species of *Aspergillus* and *Penicillium* isolated from apple root residues could decompose phlorizin in the same way.

Phlorizin, phloretin, and the three phenolics resulting from decomposition of phloretin were tested in solution culture against the growth of apple seedlings. All five compounds inhibited growth. Root growth and dry weight increment of leaves were most strongly affected, and phlorizin and phloretin appeared to be the most inhibitory. However, after more time other decomposition products were present in these flasks. Börner concluded that the importance of the identified phytotoxins in the apple replant problem depends on (1) the physiological effectiveness of the toxins, (2) their concentrations in the soil of orchards, and (3) the ability of microorganisms to destroy them.

Holowczak *et al.* (1960) reported that numerous isolates of *Venturia inaequalis* were able to decompose phlorizin to glucose and phloretin and phloretin to phloroglucinol and p-hydroxyhydrocinnamic acid.

Börner (1963a,b) later found that *Penicillium expansum* isolated from soil in an apple orchard produced patulin and an unidentified phenol when growing in soil containing leaf and root residues of apple. Patulin is a potent phytotoxin as indicated previously (Norstadt and McCalla, 1963).

Berestetsky (1970, 1972) investigated the apple replant problem in the U.S.S.R. and concluded that the inhibition of apple seedlings in old apple orchards was due at least in part to decomposition products of root residues. When he inoculated these root residues with *Penicillium claviforme* and *P. martensii,* phytotoxic substances were produced.

Williams (1960) listed numerous phenolic acids and flavonoid glycosides produced by the leaves, bark, and fruit of apple trees, and most are known phytotoxins under experimental conditions. Nobody has investigated these in relation to the apple replant problem, and they may be important.

The apple replant problem is certainly not completely solved, but present evidence indicates that allelopathic mechanisms are important and may be the primary cause of the problem.

D. The Citrus Decline and Replant Problem

After citrus trees have been grown on some soils for several years, the yields begin to diminish, the trees become less thrifty, abnormal dieback occurs, and new growth is slow in spite of standard fertilizer, pest-control practices, and generally good management (Martin, 1948). When young citrus trees are re-planted in such groves, growth is very slow compared with that of similar young trees in non-citrus soil.

Preliminary soil-fumigation experiments conducted by Martin (1948) suggested that some kind of soil biological factor was at least partly responsible for the replant problem. He pointed out that the chief microbiological agents in the soil that might affect citrus growth are nematodes (chiefly *Tylenchulus semipenetrans*), pathogenic microorganisms, and saprophytic organisms. Martin decided therefore to determine if the fungus flora is different in citrus soils than in non-citrus soils. He obtained soil samples at various depths from many old citrus groves and similar numbers and kinds of samples from noncitrus areas at intervals of 4 to 6 months over a 2-year period. These were plated out and the numbers and types of saprophytic fungi were determined.

Sixty-three species of fungi were found in old citrus soil and 52 in non-citrus soil. More than one-half of these were encountered only occasionally. Usually, the commonly encountered species were found in both types of soil. *Torula* sp. 1 was found only in non-citrus soil, however, and *Pyrenochaeta* sp. and an unidentified fungus (D_1) were found only in old citrus soil. Although *Fusarium* spp. were found in both types of soil, they occurred in much greater numbers in old-citrus soil.

On the basis of additional work, Martin (1950) listed the most abundant species of fungi in old citrus soil in order of decreasing number as *Fusarium solani, Pyrenochaeta* sp., fungus D_1, blue-green penicillia, *Aspergillus versicolor,* and *Penicillium nigricans.* He tested many fungi singly and in combinations against germination of sweet orange seeds and seedling development by using ground citrus roots mixed in quartz sand as the medium for seed germination and for growth of the inoculated fungi. Most of the fungi tested had very little effect on seed germination or seedling development.

Cylindrocarpon radicicola cuased decay of 80% of the orange seeds and infected the roots of all the seedlings that developed. *Fusarium oxysporum* exerted a similar effect except that more seeds germinated. In the *Fusarium solani* cultures, 67% of the seeds germinated, and the root tips of all seedlings except one became infected. In the presence of *Phyrenochaeta* sp., 73% of the seeds germinated. Fungus D_1 alone had little effect, but this species in combination with *Pyrenochaeta* sp. and *Fusarium solani* inhibited germination and caused decay of all but one of the 60 seeds tested. This also happened when *Fusarium solani* and *Pyrenochaeta* sp. were combined. Martin concluded, therefore, that the type of fungus population developing as a result of prolonged growth of citrus on the same soil might be partly responsible for the decay of citrus feeder roots and for the reduced growth of citrus in second or third plantings in the same soil.

Martin *et al.* (1953) fumigated old citrus and non-citrus soils in the greenhouse and the field using five different fumigants. In the greenhouse studies, orange seedlings were planted 6 weeks after fumigation. The plants were harvested and weighed after 9 months, and the leaves and feeder roots were ground and analyzed for 18 mineral elements. Leaves collected from orange and lemon trees in the field plots were analyzed for eight elements. The growth of sweet orange seedlings was markedly improved in old citrus soil that was fumigated in greenhouse studies, and marked improvement of growth of both lemon and orange trees in old groves occurred in fumigated field plots. In some cases, the growth was nearly doubled.

Citrus seedlings grown in old citrus and non-citrus soils were not significantly different in chemical composition. Fumigation did not change the level of any nutrient from a deficiency level to a sufficiency level with the possible exception of manganese in a few tests. Moreover, fumigation of non-citrus soils produced similar chemical changes in the citrus plants, but it did not increase growth, and growth of citrus seedlings in nonfumigated noncitrus soils was better than that in fumigated old citrus soils. Martin *et al.* (1953) concluded that the increased growth in old citrus soils following fumigation must have been caused by destruction of detrimental soil organisms and not to changes in the nutrient status of the soils.

Martin *et al.* (1956) fumigated two kinds of old citrus and non-citrus soils in pots in the greenhouse and then inoculated the soil in the pots with one fungal species or a combination of species. Sweet orange seedlings were then planted in the pots and allowed to grow for 6 to 9 months. The soil was also sampled periodically for kinds of fungi present. They found that the destruction of the existing population in old citrus soil resulted in about as good a growth of orange seedling as in non-citrus soil. Reinoculation of the fumigated soils with selected fungi did not have much influence on seedling growth in most cases. The one exception was inoculation with *Thielaviopsis basicola,* which greatly reduced the growth of orange seedlings in two types of soils. The reduction in growth due to this fungus was less when it was combined with some other fungi such as *Penicillium nigricans* or *Stachybotrys atra.*

Martin and Ervin (1958) grew 12 different legume companion crops along with sweet orange seedlings in old citrus soil. All but one slightly decreased the growth of the orange seedlings. A 1-year rotation to a different crop increased growth considerably when sweet orange seedlings were subsequently planted in the soil. Grasses appeared to be the most effective, but they did not cause the growth of orange seedlings to even approach that of seedlings in non-citrus soil. Of 22 kinds of organic materials added to old citrus soil, only pine needles increased the growth of sweet orange seedlings. Martin and Ervin did not comment on the possible reason for the stimulation by pine, but it may have been due to inhibition of microorganisms by the condensed tannins in pine needles (Rice and Pancholy, 1973). This would agree with the previous conclusions of Martin and his colleagues that certain microorganisms in old citrus soils were responsible for the citrus replant problem.

The design of all experiments conducted by Martin and his co-workers was such that they could conclude only that certain microorganisms (particularly fungi) apparently were responsible for the poor growth of citrus in old-citrus soils. Unfortunately, nothing definite could be concluded as to how the detrimental microorganisms reduced the growth of citrus seedlings. There were some implications, but no clear evidence, that the effects might be due to infection of the roots by certain fungi. No suggestion was made regarding the possibility that the fungi might be producing phytotoxins either directly or by decomposition of some specific compound in citrus roots. The basic question, therefore, as to the primary cause of the citrus decline and replant problem was left open.

In South Africa, the citrus replant problem has reached such serious proportions in the Sundays River Valley that the economic viability of large production areas is threatened (Burger, 1981). Therefore, Burger carried out investigations to determine the causes of the decline. These included studies of the influences of rootstock species, root parasites, soil aeration, phytotoxins in the soil and in citrus residues, salinity, and pH.

Burger found that soil in which Valencia trees on rough lemon (*C. jambiri*) rootstock had grown previously caused the greatest reduction in the growth of a second planting of Valencias budded on rough lemon. Trees on Troyer citrange rootstock (*C. sinensis* × *Poncirus trifoliata*) were the least affected by the old-citrus soil. Rough lemon was used as the rootstock for more than 90% of the citrus planted before 1960 in the Sundays River Valley and is still used in over 60% of all new citrus plantings.

Subsequent tests indicated that old citrus soils contained acid–ether extractible substances that caused a severe reduction in the radicle growth of rough lemon. These substances were accumulated in soil below 500-mm depth, apparently due to downward leaching by irrigation water. Laboratory experiments showed that anaerobic soil conditions were a prerequisite for the formation of these decomposition products.

Homovanillic acid was tentatively identified as the major growth inhibitor occurring in partly decomposed rough lemon roots. It stopped root cell elongation and stimulated radial cell expansion, resulting in the arrest of root elongation and a swelling of the root tip. Four lipid-soluble phenolic compounds, which were found in fresh and partly decomposed citrus roots, appeared to be involved in the growth regulation of citrus roots. Two of the compounds were identified positively as seselin and xanthyletin, pyranocoumarins. The growth inhibitory action of the four compounds was concentration dependent, and they occurred in different concentrations in rough lemon, Empress mandarin, and *Poncirus trifoliata* rootstocks. An increase in salinity of the growth medium caused seselin to be toxic at lower concentrations.

Fumigated old citrus soil increased the total phenolic content of rootstocks of Troyer, Empress, rough lemon, and Swingle trees by 31, 30, 17, and 9%, respectively, in comparison with virgin soil. This effect could be observed in the root extracts used for analyses, the extracts from roots in old citrus soil being much darker colored than extracts of roots from virgin soil. This could, of course, increase the toxicity of the root residues. Rough lemon root residues added to virgin soil inhibited the growth of rough lemon seedlings, whereas root residue of *Poncirus trifoliata* had no effect.

Elimination of root parasites by fumigation caused considerable improvement in tree growth in old citrus soils, but the effect of the root parasites in the fine textured alkaline soils of the Sundays River Valley was not as great as the effect of chemical (and perhaps physical) soil factors.

Burger pointed out that a marked decline in the numbers of beneficial fungi, such as *Trichoderma viride,* occurred when soils in the Sundays River Valley were cropped to citrus. These fungi are capable of breaking down phenolic compounds in the soil, and their absence probably leads to the accumulation of these compounds in the soil. In sufficient quantities, they could directly reduce the growth of citrus. Indirectly, these compounds could cause the decline of citrus trees by rendering the root system more susceptible to a variety of soil organisms, such as the citrus nematode, *Phytophthora* spp., and low grade pathogens, such as *Fusarium* spp. (see Chapter 4).

E. Allelopathic Potential of the Coffee Tree

Coffee is one of the most universal beverages used by humans, and the coffee tree is also used for the production of caffeine, chiefly from the fruit (Chou and Waller, 1980a,b).

Evenari (1949) reported that coffee seeds inhibited the germination of selected test seeds. Moreover, he pointed out that coffee seeds contain 0.3 to 2.36% caffeine and that caffeine is a strong inhibitor of wheat seed germination. Coffee plantations in southeast Mexico exhibited what appeared to be significant allelopathic effects, and Anaya et al. (1978) reported that aqueous extracts of fresh leaves, dried leaves, and roots of coffee plants inhibited growth of weeds associated with a coffee plantation. Anaya et al. (1982) tested four varieties of coffee for their allelopathic effects on common weeds in a coffee plantation and found that all were inhibitory to growth of most test weeds, but there were varietal differences. They also tested leaves and roots from the various species of the tree stratum and leaves from the herb stratum and found that most of them inhibited the majority of the test weeds.

Chou and Waller (1980a) reported that 5% aqueous extracts of dried fallen leaves and roots of coffee plants markedly inhibited radicle growth of lettuce, rye grass (*Lolium multiflorum*), and fescue and the germination of lettuce and fescue. Even a 1% extract of these organs reduced radicle growth of lettuce.

The phytotoxins present in coffee leaves included the purine alkaloids (caffeine, theobromine, theophylline, and paraxanthine) and the phenolics (*p*-hydroxybenzoic, vanillic, *p*-coumaric, ferulic, chlorogenic and caffeic acids, and scopoletin) (see Chou and Waller, 1980a,b). All these compounds, except caffeic acid, significantly inhibited lettuce growth at a concentration of 100 ppm.

F. Allelopathic Effects among Ornamentals and between Ornamentals and Turf Grasses

There has been surprisingly little research on allelopathic phenomena in ornamental plants. This is puzzling because many gardeners and greenhouse workers are very much aware of allelopathic problems. Greenhouse supervisors have told me that they have to discard soil in which they have grown stock (*Malcomia maritima*). Some have told me that all sorts of crucifers poison soil so that plants do not grow satisfactorily afterward, including crucifers. The prevalence of such ideas makes it desirable for considerable research to be done in this area.

Kozel and Tukey (1968) found that *Chrysanthemum morifolium* produces a potent phytotoxin that leaches from the leaves and is very inhibitory to the growth and development of chrysanthemum. Tukey (1969) stated that chrysanthemum cannot be grown continuously in the same soil for several years because of accumulation of inhibitors in the soil.

Oleksevich (1970) reported that barberry (*Berberis*), horse chestnut (*Aesculus*), rose (*Rosa*), lilac (*Syringa*), viburnum, fir (*Abies*), and mockorange (*Philadelphus*) have considerable allelopathic activity. He found that they inhibit neighboring plants and cause soil toxicity. He stated that barberry produces much berberine, an alkaloid, which is a strong inhibitor of plant growth and development.

Fales and Wakefield (1981) studied the effects of turf grasses on the establishment and growth of flowering dogwood (*Cornus florida*) and forsythia (*Forsythia intermedia*). One phase involved a field study over two growing seasons in which the shrubs were planted in an established sod of Kentucky bluegrass and red fescue (*F. rubra*) on an Enfield silt loam soil. Treatments included different sized areas of turf-free space, surface, and subsurface placement of fertilizer and irrigation and two mowing heights. A second phase involved the application of aqueous leachates from pots of either Kentucky bluegrass, red fescue, or perennial ryegrass to pots containing forsythia.

In the field studies, turfgrass significantly reduced the growth of both shrubs. Supplementary fertilizer applied as a top dressing failed to benefit the growth of the shrubs, but subsurface treatments resulted in considerable increases in growth. Competition for moisture did not appear to be responsible for the observed differences in growth, because maintaining a high level of soil moisture did not overcome inhibitory effects of the turfgrass. Competition for N was indicated, however, by the results of leaf analyses.

The aqueous root leachates of perennial ryegrass, Kentucky bluegrass, and red fescue reduced top growth (dry weight increment) and increase in the length of branches of forsythia. Root leachates of red fescue and perennial ryegrass significantly reduced root growth of forsythia. None of the leachates affected the N, P, and K concentrations in the forsythia leaves, however. Fales and Wakefield concluded, therefore, that suppression of the growth of forsythia and flowering dogwood by the three turf grasses may involve chemical inhibition as well as direct competition for available N.

4

Roles of Allelopathy in Plant Pathology

The term allelopathy is rarely used in plant pathology literature. Nevertheless, development and morphogenesis of pathogens, antagonism of pathogens by non-host organisms, development of disease symptoms, and host plant resistance to pathogens appear to involve allelopathic agents (Bell, 1977). One other item should be added to this list: the promotion of infection due to allelopathic compounds in the environment of the potential host (Patrick and Koch, 1963). Some researchers of allelopathy do not include these topics because they believe the chemical compounds involved do not escape into the environment. However, phytoalexins and other compounds produced by the hosts do enter the environment of the pathogen, and phytotoxins produced by the pathogens escape from the pathogens and affect the hosts in many ways. This subject is, of course, a very involved one and would require several books to review it. There has been extensive research activity in this field for many years, and the pace of research on phytotoxins, phytoalexins, elicitors of phytoalexin production, and mechanisms of action of these various allelopathic agents has accelerated. Thus, this research field will be alluded to, and relatively recent reviews will be cited for further information.

I. ALLELOPATHY IN DEVELOPMENT AND MORPHOGENESIS OF PATHOGENS

Spores of most parasitic fungi are formed in dense populations in or on infected host tissue, and the spores usually remain ungerminated while located at their site of production (Bell, 1977). This can be due to several factors, one of which is the production by the spores of fungistatic agents that are excreted into the water around the spores. These self-inhibitors generally assure dispersal of viable ungerminated spores. Endogenous germination stimulators that counteract inhibition of germination are produced by some spores (Bell, 1977).

The results of Menzies and Gilbert (1967) indicated that decomposing plant residues produce volatile components that stimulate spore germination and fungal growth in soil (see Chapter 2). King and Coley-Smith (1968) reported that a volatile principle evolved from onion and leek seedlings caused germination of dormant sclerotia of *Sclerotium cepivorum*.

Schenck and Stotzky (1975) found that volatile compounds released from germinating seeds have marked effects on several microorganisms, and they later surveyed several herbaceous and tree species for production of volatile compounds from germinating seeds and seedlings (Stotzky and Schenck, 1976). These included pea (var. Pluperfect), bean (var. Kinghorn Wax), cumumber (var. Marketer), loblolly pine, and slash pine. Aldehydes were evolved by seeds of all species tested in amounts significantly greater than those found in ambient air. The amounts evolved were greatest during the first 3 days after imbibition and then decreased rapidly to undetectable levels. The faster germinating seeds gave sharper evolution peaks, but the more slowly germinating seeds evolved volatiles over longer periods of time. These researchers suggested that the volatiles probably stimulated the development of microorganisms in the vicinity of the seeds.

Allen and Newhook (1974) demonstrated that ethanol suppresses the chemotactic response of zoospores of *Phytophthora cinnamomi*. Later, ethanol was found in the rhizosphere of radicles of *Lupinus angustifolius* (Young *et al.*, 1977). Concentrations were commonly in the range of 1 to 5 mM, which were previously shown to exert a positively chemotactic influence on zoospores of *P. cinnamomi*.

Gerrettson-Cornell and Humphreys (1978) found that 1.5% aqueous extracts (w/w) of raw bark of *Pinus radiata* almost completely eliminated sporangium formation by *Phytophthora cinnamomi* in Petri plates. Later, it was reported that aqueous leachates of jarrah (*Eucalyptus marginata*) forest leaf litter and of soils from Mt. Tambourine, Queensland, and Tallaganda State Forest, New South Wales, markedly reduced the number of zoospores produced by *P. cinnamomi* inoculated into the leachates. The two soils were previously reported to suppress root-rot disease caused by *P. cinnamomi*.

Root exudates from cowpea (cv. Mala) and sorghum (Guinea race) seedlings enhanced germination of conidia of four species of *Fusarium* previously isolated from the rhizosphere and rhizoplane of cowpea and sorghum plants (Odunfa, 1978). The germ tubes and some conidia were lysed by cowpea exudate in 48 hours, however. Mycelial growth in liquid culture was enhanced in cowpea root exudate medium, but it was retarded in sorghum exudate medium. It was suggested, therefore, that the sorghum root exudate contained an antifungal substance.

The survival of most parasites depends on the formation of resting propagules, because they must spend prolonged periods of time apart from the host plants. There is much evidence that formation of sclerotia and chlamydospores may be conditioned by allelopathic agents (Bell, 1977).

II. ALLELOPATHY IN ANTAGONISM OF PATHOGENS BY NONHOST ORGANISMS

Most propagules of pathogens do not survive to infect suitable host organisms because of many adversities, one of which is antagonism by other organisms (Bell, 1977). One aspect of antagonism is allelopathy, and the evidence for this is voluminous. This phenomenon merges into that of host–plant resistance to pathogens, as I will demonstrate shortly. Most research in this area has concerned the production of antibiotics, both by other pathogens and by saprophytes (Bell, 1977). Many references (including reviews) were cited by Bell (1977) concerning the role of antibiotics in this antagonism. Recent research on production of allelopathic compounds by higher plants that prevent propagules from surviving to infect suitable hosts will be the area of concentration. Also, the sources of allelopathic agents are often not known, and this is certainly true in the case of fungistasis in soil. This phenomenon is widespread in soil (Bell, 1977) and has been attributed to many factors. Lockwood (1959) suggested that diffusible antibiotics produced by *Streptomyces* spp. might be an important cause. Mishra and Pandey (1974, 1975) investigated numerous factors affecting soil fungistasis and concluded that their data supported ''a biological origin of the inhibition of spore germination.'' Many compounds, known to inhibit spore germination and fungal growth, are produced by higher plants and enter the soil in various ways. Thus, it is possible that some of these compounds may be, at least, partially responsible for fungistasis.

Agnihothrudu (1955) investigated varieties of pigeon-pea, *Cajanus cajan,* that are resistant and susceptible to wilt caused by *Fusarium udum.* In the rhizosphere microorganisms isolated from the resistant variety, 13 to 33% strongly inhibited *Fusarium udum,* whereas most such isolates from the susceptible variety were not inhibitory to the fungus. All the active organisms were species of *Streptomyces.* Jackson (1965) gave much evidence for the ecological importance of antibiosis between soil microorganisms, and thus this appears to be an extremely fruitful area for future research.

Santoro and Casida (1962) described antibiotics that occur in the mycelium of various laboratory cultures of *Boletus* and *Amanita,* which are mycorrhizal fungi. Subsequently, Krywolap and Casida (1964) assayed seven strains of *Cenococcum graniforme,* a mycorrhizal fungus, for the production of antibiotics. All strains produced an antibiotic that strongly inhibited gram-positive bacteria,

Rhizobium meliloti, and *Saccharomyces cerevisiae.* It inhibited selected test actinomycetes, but it did not inhibit any of the test fungi, other than yeast.

Krywolap *et al.* (1964) isolated an antibiotic from *C. graniforme* mycorrhizae and from roots and needles of red pine (*P. resinosa*), white pine, and *Picea abies.* Chromatographically, the antibiotic resembled that obtained *in vitro* from *C. graniforme.* In trembling aspen, antibiotic activity was observed only in *C. graniforme* mycorrhizal roots. The antibiotic was found also in soil that contained *C. graniforme* sclerotia. These results were strongly supported by the experiments of Grand and Ward (1969) involving the same tree species in pure plantations on two soil types in Pennsylvania.

Marx (1969a) tested seven mycorrhizal species known to occur in loblolly pine and shortleaf pine (*Pinus echinata*), against the growth of 48 different fungal root pathogens in agar plates. The root pathogens were species of *Phytophthora, Fomes, Polyporus, Sclerotium, Armillaria, Cylindrocladium, Fusarium, Poria, Thanatephorus, Rhizoctonia,* and *Pythium.*

Laccaria laccata, Lactarius deliciosus, Leucopaxillus cerealis var. *piceina, Pisolithus tinctorius,* and *Suillus luteus* inhibited growth of nearly one-half of the root pathogens. *Leucopaxillus cerealis* var. *piceina* inhibited 92% of the pathogens. Culture filtrates of *L. cerealis* var. *piceina* inhibited soil bacteria and zoospore germination of *Phytophthora cinnamomi. P. cinnamomi* is the primary cause of littleleaf disease of shortleaf and loblolly pine (Marx, 1969a). Subsequently, Marx (1969b) identified an antifungal and antibacterial antibiotic produced by *Leucopaxillus cerealis* var. *piceina* as diatretyne nitrile. Concentrations causing minimum inhibition (20%) to germination of zoospores of *Phytophthora cinnamomi* were 50 to 70 ppb, with total inhibition at 2 ppm.

Diatretyne nitrile did not inhibit germination of aseptic shortleaf pine seeds exposed to concentrations up to 40 ppm for 1 or 2 hours. Concentrations up to 6 ppm did not adversely affect height growth, needle color, or radicle development of shortleaf pine seedlings, whereas seedlings developed poorly in the 8 and 10 ppm concentrations. Concentrations of 20 ppm and higher inhibited seedling growth completely.

Marx and Davey (1969a) reported that *Leucopaxillus cerealis* var. *piceina* produced diatretyne nitrile and a related antibiotic, diatretyne 3, in mycorrhizae and rhizospheres of shortleaf pine seedlings in aseptic culture. Only 25% of nonmycorrhizal short roots adjacent to mycorrhizae formed by *L. cerealis* var. *piceina* were infected by zoospores of *Phytophthora cinnamomi,* whereas 100% of the short roots of control seedlings (no mycorrhizae) and the short roots adjacent to mycorrhizae formed by *Pisolithus tinctorius* and by *Laccaria laccata* were infected. When adjacent to mycorrhizae formed by *Suillus luteus,* however, only 77% of the short roots on shortleaf pine and 85% of the short roots on loblolly pine were infected.

Fully developed mycorrhizae formed by *Laccaria laccata, Leucopaxillus ce-*

realis var. *piceina,* and *Suillus luteus* on shortleaf pine were resistant to *P. cinnamomi.* Mature mycorrhizae formed on loblolly pine by *L. laccata* were resistant also. Marx and Davey also found that short root initials covered by fungal mantles from adjacent mycorrhizae resisted infection. They concluded that at least some ectotrophic mycorrhizae function as biological controls against pathogenic root infections.

Resistance of naturally occurring ectotrophic mycorrhizae of shortleaf pine to infection by *Phytophthora cinnamomi* was demonstrated by Marx and Davey (1969b). Five morphological forms of mycorrhizae, one of which was formed by *Cenococcum graniforme,* were inoculated with zoospores of *P. cinnamomi.* The fully formed mycorrhizae were resistant to infection, but some mycorrhizae with incomplete fungal mantles at the root meristem were infected (Table 32). The pathogen did not penetrate the Hartig net region, however, in those that were infected. Mycorrhizae with artificially exposed cortex tissue also were resistant to infection. Zoospores were attracted to excised root tips with exposed stelar tissue but not to intact roots.

The antagonistic activity of certain microorganisms can be used in biological control of plant diseases and several good reviews of this subject have appeared (Baker and Snyder, 1965; Baker and Cook, 1974; Bruehl, 1975). A few examples will be cited. *Fomes annosus* is a root pathogen of loblolly and slash pine, and its basidiospores germinate on cut stumps, grow down through the stump, and infect living roots through root grafts. McGrath (1972) suggested the biological control of this disease by inoculating cut stumps with basidiospores of

TABLE 32. *Phytophthora cinnamomi* **Infection of Roots of Shortleaf Pine Seedlings Grown in Nonsterilized Humus in Greenhouse Pot Culture**[a]

| | Infection by *Phytophthora cinnamomi*[b] | | | |
| | 3 days after inoculation | | 10 days after inoculation | |
Root type	Inoculated (No.)	Infected (%)	Inoculated (No.)	Infected (%)
Mycorrhizal form 1	12	25[c]	14	43[c]
Mycorrhizal form 2	17	0	7	0
Mycorrhizal form 3	23	0	11	0
Nonmycorrhizal short root	16	100	11	100
Nonmycorrhizal lateral root tip	6	100	6	100

[a] From Marx and Davey (1969b).
[b] Approximately the same number of each respective root type were incubated free from zoospores of *P. cinnamomi* in the cylinders and accordingly remained free from pathogenic infections.
[c] Mycorrhizae with incomplete fungal mantle and Hartig net development.

Peniophora gigantea. When this is done, *P. gigantea* prevents entry of *F. annosus.* Elliston *et al.* (1976) reported that bean plants (*Phaseolus vulgaris*) were protected locally and systemically by *Collectotrichum* spp., nonpathogenic to bean, against anthracnose caused by *Colletotrichum lindemuthianum.* The protection in this case was attributed to induced resistance, but the boundary between antagonist activity and induced resistance is very obscure in many instances.

Crown gall caused by *Agrobacterium tumefaciens* is of great economic concern to nurseries growing rosaceous plants, *Rubus* species, grapevines, and various nut-bearing trees (Moore and Warren, 1979). Epidemics of 80 to 100% galled nursery stock have occurred in the Pacific Northwest, but fortunately they are of infrequent occurrence. Economic losses to nurseries occur primarily because most states prohibit interstate shipment or receipt of galled plants, which then have to be destroyed.

A new approach to crown gall control was suggested in 1972 by Kerr and a colleague (Kerr, 1972; New and Kerr, 1972). A nonpathogenic *Agrobacterium* strain, *A. radiobacter* (strain 84), was inoculated onto peach seeds, and the seeds were planted in soil infested with *A. tumefaciens.* After 3 months, only 31% of the plants from seeds inoculated with strain 84 were galled compared with 79% for the plants from uninoculated seeds. Since 1972, strain 84 has been used worldwide on thousands of plants, including species of *Prunus, Rubus, Malus, Salix, Vitis, Libocedrus, Chrysanthemum, Crataegus, Carya, Rosa, Pyrus,* and *Humulus* (Moore and Warren, 1979). It is remarkable that a single strain of bacteria has been disseminated so widely and used successfully by so many investigators in such a short period of time.

Pathogenic strains sensitive to strain 84 are prevented from transferring their T_i plasmid to the wounded host, apparently because of a bacteriocin (agrocin 84) produced by the antagonist that either kills or prevents attachment of the pathogen to the host receptor site. Strain 84 has not prevented crown gall in some instances because the pathogens are insensitive to agrocin 84, produce a bacteriocin against strain 84, or produce an inhibitor against agrocin 84 (Moore and Warren, 1979). Other antagonists inhibit some of the insensitive pathogenic strains *in vitro* and prevent infection in the greenhouse, but they have not been as effective in field tests.

According to Walton (1980), inoculation of the soil in lettuce fields at Beltsville with the fungus *Sporidesmium sclerotivorum* was effective in reducing lettuce drop disease caused by *Sclerotinia minor.* One treatment was effective for two lettuce crops, and the disease was controlled far more effectively than with chemicals.

Li and his colleagues (Li *et al.,* 1969a,b, 1970, 1972, 1973; Trappe *et al.,* 1973; Li, 1974, 1977) presented much evidence concerning production by higher plants of compounds antagonistic to a root-rot pathogen, *Poria weirii.* Li *et al.*

(1969a) described an excellent quantitative method for assaying soil for inhibitory fungi. They found that many phenolics and other compounds, which have been repeatedly isolated from soil, inhibited the growth of *Poria weirii in vitro*. They found also that the combination of *p*-coumaric, syringic, and ferulic acids, as found in the roots of the *Poria*-resistant red alder (*Alnus rubra*), also inhibited the growth of *P. weirii*. A highly susceptible species, Douglas-fir, contained *p*-coumaric acid but not ferulic or syringic acids, and *p*-coumaric acid is not inhibitory to *P. weirii* alone. Subsequently, Li (1977) found that red alder contains an ezyme that hydrolyzes *p*-coumaric acid to caffeic acid and *p*-hydroxyphenylacetic acid to 3,4-dihydroxyphenylacetic acid, but Douglas-fir does not contain such an enzyme. Caffeic and 3,4-dihydroxyphenylacetic acids are both inhibitory to *P. weirii*. The *p*-hydroxyphenylacetic acid is found in soil under red alder, but is not inhibitory until hydrolyzed as indicated above. Many species of plants that form an understory in red alder stands produced several of the same phenolic compounds that inhibited *P. weirii*.

Yakhontov (1973) suggested the possibility of using the antagonistic activity of nonhost plants in controlling plant diseases also. He found that water-soluble excretions of numerous weeds and crop plants inhibited *Dactilospheara viticola*, which causes the disease phylloxera in grape plants. He observed also that there were different degrees of infection of grape plants by phylloxera in areas with various grasses growing alongside the grape plants in vineyards.

Much more information on the role of antagonistic organisms in the control of plant diseases can be found in Bruehl (1975).

III. ALLELOCHEMICS AND THE PROMOTION OF INFECTIONS BY PATHOGENS

Ludwig (1957) reported that *Helminthosporium sativum* produces a toxin in a culture medium that inhibits its own growth and also conditions susceptible seedlings for infection by the fungus.

Patrick *et al.* (1964) pointed out that many plant pathologists believe that toxins play causal roles in some, if not most, plant diseases. The toxins involved are generally thought to be produced by the pathogenic microorganisms involved in the diseases. Cochrane (1948) suggested, however, that some root rots are initiated by direct toxic action of plant residues. Considerable evidence was obtained by Patrick and his colleagues to support this hypothesis (Patrick and Koch, 1958; Patrick *et al.*, 1963, 1964; Patrick, 1971). In field studies in Salinas Valley, California, Patrick *et al.* (1963) observed that discolored or sunken lesions were often present on roots of lettuce or spinach plants where the roots grew in contact with or in close proximity to fragments of plant residues. When isolations were made from the lesions no known primary pathogen was consistently obtained, and the microorganisms most frequently found were common soil saprophytes. They concluded that the toxins produced by the decomposing residues conditioned roots to invasion by various low-grade pathogens.

Toussoun and Patrick (1963) found that the incidence of root rot of bean caused by *Fusarium solani* f. *phaseoli* was greatly increased if the bean roots were exposed to toxic extracts from decomposing plant residues before inoculation with the pathogen. Patrick and Koch (1963) reported that the extent and severity of black root rot of tobacco caused by *Thielaviopsis basicola* were much greater when tobacco roots were exposed to toxic extracts of decomposing plant residues prior to inoculation. Moreover, they found that the pathogen was equally destructive to susceptible and resistant varieties of tobacco after treatment of roots with the toxic extracts. Thus, another dimension is added to the harmful effects of phytotoxins on plants.

IV. ALLELOPATHY IN DEVELOPMENT OF DISEASE SYMPTOMS

More research has been done on phytotoxins produced by plant pathogens than on any other allelopathic aspect of plant pathology (Wood *et al.*, 1972; Strobel, 1974; Bell, 1977). The reader is referred to these references for a comprehensive analysis of this subject.

V. ALLELOPATHY IN HOST PLANT RESISTANCE TO DISEASE

In the terminology of Grümmer (1955), most of the compounds involved in the resistance of hosts to pathogens fall in the category of phytoncides. These are generally divided into two categories: (1) secondary compounds generally present in the host, but which may increase subsequent to infection and(2) phytoalexins, new compounds formed only after infection.

Evenari (1949) pointed out that the seeds of *Brassica oleracea* contain a microbial inhibitor belonging to the mustard oils and that the resistance of crucifers to clubroot disease caused by *Plasmodiophora brassicae* is attributed to such oils. According to Evenari, Focke reported in 1881 that plants producing essential oils are protected against attacks by parasitic fungi.

Many plants produce compounds, either prior to or after infection by certain pathogens, which render the plants resistant to diseases caused by the pathogens (Schaal and Johnson, 1955; Farkas and Kiraly, 1962; Ingham, 1972; Levin, 1971, 1972; Kuć, 1972; Bell, 1974, 1977; Wood and Graniti, 1976; Swain, 1977). Many of the same compounds that have been implicated in other aspects of allelopathy have been reported to be important in many instances in the resistance of plants to diseases.

As early as 1911, Cook and Taubenhaus found that tannins are very inhibitory to many parasitic fungi and suggested that they may be important in the resistance of some plants to fungal infection. According to Cruickshank and Perrin (1964), Maranon reported that strains of *Oenothera* sp. resistant to *Erysiphe polygoni* had higher tannin contents than susceptible strains. Cadman (1959) found that raspberry leaves contain a substance that prevents the infection of plants by viruses when it is mixed with the inoculum. He identified the substance as a tannin. Somers and Harrison (1967) reported that certain wood tannins are very inhibitory to spore germination and hyphal growth of *Verticillium alboatrum*, especially the condensed tannins. They suggested that the tannins may be important, therefore, in host resistance to *Verticillium* wilt disease.

In 1943, Grosjean found that the bark of *Populus candicans* contains substances with fungistatic activity and that these substances can be isolated from the bark by extraction with boiling water (Grosjean, 1950). He later found that *Populus trichocarpa* has as strong fungicidal activity as *P. candicans* and that eight other species of *Populus* had weak fungicidal activity in the bark. Klöpping and van der Kerk (1951) identified several fungistatic substances from the bark of *P. candicans,* including pyrocatechol, salicin, saligenin, salicylic acid, probably benzyl gentisate, and an unknown. Hubbes (1962) added an eleventh species to the number of *Populus* species known to contain fungistatic agents in the bark. He isolated two main fungistatic agents from the bark of quaking aspen, one of which he identifed as pyrocatechol.

Johnson and Schaal (1952) reported a good correlation between scab (caused by *Streptomyces scabies*) resistance of several potato varieties and the phenolic content of the peels of their tubers. They suggested that the chief compound involved was chlorogenic acid. Other workers have been unable to repeat their results, however (Farkas and Kiraly, 1962). Kuć et al. (1956) found that slices of potato tubers (var. Netted-Gem) inoculated with *Helminthosporium carbonum* produced appreciable amounts of chlorogenic and caffeic acids subsequent to the infection. Both compounds inhibited the growth of *H. carbonum*, with caffeic acid being most inhibitory. The combined activity, however, could not account for the total activity of the crude potato extract. Cysteine, which was present in the extract, gave a pronounced synergistic effect when added to caffeic or chlorogenic acid. Later, Clark et al. (1959) found an amino acid addition product of chlorogenic acid in the peel of two varieties of potatoes—Russet and Netted-Gem. This compound is highly toxic to Race 1 of *H. carbonum.*

One of the best known examples of the protective role of phenolics that are formed prior to infection is that of onion in relation to infection by *Colletotrichum circinans* (Farkas and Kiraly, 1962). The resistance of onion varieties is correlated with the red or yellow pigmentation of the bulb scales. The pigments are flavones and anthocyanins, which are not inhibitory to the pathogen, but protocatechuic acid and catechol occur along with them. These phenols are water soluble, and they diffuse from the dead cell layers of the scales and inhibit spore germination and hyphal penetration of the pathogen. Apparently, cases of preformed resistance factors are rare (Farkas and Kiraly, 1962; Cruickshank and Perrin, 1964).

There are many examples of increased production of protective compounds following infection that are similar to that reported by Kuć et al. (1956). Only a few will be discussed, however. Hughes and Swain (1960) reported a 10- to 20-fold increase in concentration of scopolin and a 2- to 3-fold increase in chlorogenic acid in potato tuber slices infected with Phytophthora infestans. The tremendous increase in scopolin caused a blue fluorescent zone around the infected area. Cruickshank and Perrin (1964) stated that increases in concentrations of phenolic substances have been reported around lesions resulting from the infection of rice leaves by Piricularia oryzae and Helminthosporium sp., of the leaves of Paulownia tomentosa by Gloeosporium kawakamii, of the leaves and peels of apple by Venturia inaequalis and Podosphaera leucotricha, and of the sweet potato roots by Ceratocystis fimbriata. Minamikawa et al. (1963) found that two coumarins, umbelliferone and scopoletin, increase markedly due to infection of sweet potato roots by C. fimbriata. Farkas and Kiraly (1962) found that resistant wheatrust combinations also accumulate phenolics more rapidly than susceptible combinations. After a thorough literature review, Farkas and Kiraly (1962) concluded that a pronounced, postinfectional rise in concentration of phenolics is one of the best documented characteristics of resistance to fungus diseases. They pointed out that a postinfectional increase in concentration of phenolics has also been demonstrated for some bacterial and virus diseases. Cruickshank and Perrin (1964) reported that significant increases in concentrations of scopoletin, kaempferol, quercetin, caffeic acid, chlorogenic acid, and isochlorogenic acid have been associated with virus activity in various hosts. They stated also that coumarin has been shown to inhibit reproduction of tobacco mosaic virus.

According to Farkas and Kiraly (1962), hypersensitivity is one of the most important types of defense reactions of plant tissues to diseases caused by rusts, mildews, Phytophthora, and some viruses. This reaction consists of the rapid death of a few cells in highly resistant varieties, thus confining the pathogen to a restricted area. The breakdown of cellular structure apparently results from an excessive oxidation of polyphenols.

Cobb et al. (1968) suggested that several monoterpenes produced by Pinus ponderosa may play a part in its resistance to infection by Fomes annonus and Ceratocystis spp., especially in conjunction with accumulation of phenolics.

Black root rot of tobacco caused by the soil-borne fungus Thielaviopsis basicola is one of the common diseases of tobacco in Canada. Gayed and Rosa (1975) reported that, in 3- and 4-week old tobacco seedlings inoculated with T. basicola, higher levels of chlorogenic acid were recovered from the immune cultivar Burley 49 than from the susceptible White Mammoth. At the 5-, 6-, and 7-week stages, chlorogenic acid concentrations were higher in both roots and leaves of Burley 49 than in White Mammoth. At these stages, the chlorogenic acid concentration in the roots of both cultivars was 20 to 25 times higher than that in the leaves.

According to Cruickshank and Perrin (1964), Muller and Borger proposed the "Phytoalexin Theory" of disease resistance in 1939, that is, that a compound designated as a phytoalexin, which inhibits the development of the pathogen, is formed or activated only when the parasite comes in contact with the host cells. This theory differs from the previously discussed concept in that the phytoalexins are thought to be unusual metabolites not found at all in uninfected tissues. By 1964, five phenolics and one non-phenol were identified as apparent phytoalexins (Cruickshank and Perrin, 1964): (1) ipomeamarone, which was isolated from sweet potato roots infected with *Ceratocystis fimbriata* and identified as a furanoterpenoid; (2) orchinol, which was isolated from orchid tubers infected with *Rhizoctonia repens* and identified as 2,4-dimethoxy-7-hydroxy-9,10-dihydrophenanthrene; (3) 3-methyl-6-methoxy-8-hydroxy-3,4-dihydroxyisocoumarin, which was isolated from carrot root tissue inoculated with *Ceratocystis fimbriata;* (4) pisatin, isolated from pea pods inoculated with *Monilinia fructicola* and identified as 3-hydroxy-7-methoxy-4',5'-methylenedioxychromanocoumarin; (5) phaseollin, which was isolated from French bean pods infected with *Monilinia fructicola* and identified as 7-hydroxy-3',4'-dimethylchromenochromanocoumarin; and (6) trifolirhizin, which was isolated from roots of red clover (not inoculated, but not under aseptic conditions) and identified as a glucoside of an isoflavonoid. Cruickshank and Perrin (1964) illustrated the structures of all these proposed phytoalexins.

The Phytoalexin Theory has failed to gain wide acceptance. Farkas and Kiraly (1962) did not even mention the theory or the identified phenolic phytoalexins in their rather comprehensive review of the role of phenolic compounds in disease resistance. Nevertheless, the rate of research on phytoalexins has accelerated during the 1970s and several reviews have appeared on this subject (Ingham, 1972; Kuć, 1972; Bell, 1974; Wood and Graniti, 1976; Bell, 1977). Much research has been done also on elicitors of phytoalexin production. Albersheim and his colleagues have been extremely active in this area (Anderson-Prouty and Albersheim, 1975; Ayers *et al.* 1976a,b,c; Ebel *et al.,* 1976).

The importance of disease resistance to the survival of plants is certainly unquestioned. The evidence is clear that many of the same toxins that have been implicated in other phases of allelopathy are important in protecting plants against diseases. In fact, it appears that some of the same compounds may be involved in various allelopathic effects against microorganisms from the period before seed germination, through seed germination, plant growth, reproduction, and seed maturation.

5

Natural Ecosystems: Allelopathy and Patterning of Vegetation

I. CONCEPTS OF PATTERNING

The term *patterning* is used in a restricted sense by many biologists to refer to a mathematical expression of the spatial distribution of organisms within a community (random dispersion, hypodispersion, hyperdispersion). The term is also widely used to refer to spatial arrangements of individuals that are visually apparent in the field, such as bare areas under tree species. Appropriate sampling would obviously enable one to place a visual type in one of the mathematical groups.

Most ecologists have attempted to explain the patterning of vegetation and the general distribution of plants largely on the basis of competition. There is little doubt that competition always plays a role in spatial distribution, but there is growing evidence that allelopathy probably plays a role also in most, if not all, spatial distributions of plants. It is possibly unwise to try to assign a primary role in patterning to either of these phenomena alone. It is very important, however, to try to determine if allelopathy plays a role in such distributions along with competition.

II. ALLELOPATHIC EFFECTS OF HERBACEOUS SPECIES ON PATTERNING

A. Forbs

Cooper and Stoesz (1931) observed the fairy ring pattern of the prairie sunflower (*Helianthus rigidus*), which is due to a pronounced reduction in plant numbers, size, and inflorescences in the center of the clone. Curtis and Cottam

(1950) observed the same phenomenon and became curious as to its cause. They located a group of plots within several of the fairy rings and manipulated them in different ways. Some had the soil removed and replaced without any other treatment; the soil was removed from some, and all roots and rhizomes were removed before the soil was replaced; some plots were fertilized; and the soil was removed from some plots and replaced with soil taken from a spot in the prairie without the prairie sunflower. They found that fertilizing or removing the soil and replacing it directly did not alter the growth of the sunflower. On the other hand, removal of roots and rhizomes of the prairie sunflower from the soil in the fairy rings or replacement with soil from outside the rings resulted in normal growth and flowering of the sunflower.

Wilson and Rice (1968) observed striking patterns of distribution of herbaceous species around individuals of the common sunflower, *Helianthus annuus,* in the pioneer weed stage of old-field succession in Oklahoma (Fig. 10). Field sampling through two growing seasons indicated that *Erigeron canadensis* and *Rudbeckia serotina* were significantly inhibited near sunflowers; *Haplopappus ciliatus* and *Bromus japonicus* were slightly inhibited, but the effect was small and variable; and *Croton glandulosus* growth was stimulated near sunflower

Fig. 10. Zonation of species around *Helianthus annuus* in field plots near Norman, Oklahoma. S, *H. annuus; Bromus japonicus* near *H. annuus; Erigeron canadensis, Rudbeckia serotina,* and *Haplopappus ciliatus* in zone away from *H. annuus.* Photographed by Dr. Roger Wilson.

TABLE 33. Effects of Field Soils (Previously in Contact with Sunflower Roots) on Germination and Growth[a]

Test species	Date soil taken	Mean dry weights of seedlings (mg)		Germination[c]
		Control[b]	Test	
Helianthus annuus	July	24	15[d]	63
	October	28	18[d]	48
Erigeron canadensis	July	19	2[d]	29
	October	24	3[d]	41
Rudbeckia serotina	July	14	6[d]	78
	October	9	5[d]	62
Digitaria sanguinalis	July	28	16[d]	94
	October	24	12[d]	86
Amaranthus retroflexus	July	50	12[d]	83
	October	50	8[d]	64
Haplopappus ciliatus	July	12	9	64
	October	16	11[d]	72
Bromus japonicus	July	11	10	79
	October	34	16[d]	66
Croton glandulosus	July	28	27	89
	October	26	24	76
Aristida oligantha	July	9	10	91
	October	9	12[d]	96

[a] Modified from Wilson and Rice (1968).
[b] Control soils were from same field as test soils but not from around sunflower plants.
[c] Expressed as percent of the control.
[d] Weight significantly different from that of the control.

plants, but the variability was such that the effect over the 2-year period was not statistically significant.

The patterns around annual sunflower plants could not have been due to competition for light, because they were the same on all sides of the sunflower plants, and these plants do not cast a very pronounced shadow when growing alone. Analyses of pH and mineral elements, which were reported to be most likely deficient in the soils involved, suggested that the zonation was not likely due to competition for minerals or pH.

Nevertheless, soil obtained near sunflower plants in the field inhibited seed germination and the growth of *Erigeron canadensis* and *Rudbeckia serotina* when compared with results in soils taken at least 1 m from sunflower plants, whether the soil was obtained in July or October from the field (Table 33). The soil collected in October after leaf fall inhibited growth of *Haplopappus ciliatus* and *Bromus japonicus* but not *Croton glandulosus*. Results with soil taken near sunflower plants correlated well with the patterns around sunflower plants even though there was no competition with the sunflower. This suggested that some compounds must have been added to the soil by the sunflower while it was present, namely, that the patterning might be due to allelopathy. Subsequent experiments indicated that small amounts of decaying sunflower leaves in the soil inhibited the germination and growth of *Erigeron canadeniss* and *Haplopappus ciliatus* and inhibited the growth of *Rudbeckia serotina* and *Bromus japonicus* (Table 34). Root exudate of sunflower was inhibitory to growth of *Erigeron canadensis* and *Rudbeckia serotina* but not to the other three species under consideration here (Table 35). Leachate of sunflower leaves inhibited the growth in soil of *Erigeron canadensis, Rudbeckia serotina,* and *Haplopappus ciliatus* but not the growth of *Bromus japonicus* and *Croton glandulosus* (Table 36).

Croton glandulosus was not inhibited by any of the three sources of toxins from sunflower, and it grew better in the field close to the sunflower than it did a meter or so away. *Erigeron canadensis* and *Rudbeckia serotina* were inhibited in all tests and were found to be significantly inhibited near sunflower plants in the field. Growth of *Bromus japonicus* was inhibited by only one source of toxin and

TABLE 34.   Effects of Decaying Sunflower Leaves on Growth of Seedlings and Germination[a]

Test species	Experiment number	Mean dry weight of seedlings (mg)		Germination[b]
		Control	Test	
Helianthus annuus	1	44	22[c]	52
	2	36	21[c]	40
Erigeron canadensis	1	54	19[c]	87
	2	32	16[c]	71
Rudbeckia serotina	1	17	3[c]	95
	2	12	2[c]	81
Digitaria sanguinalis	1	126	16[c]	106
	2	97	11[c]	97
Amaranthus retroflexus	1	78	12[c]	56
	2	91	16[c]	32
Haplopappus ciliatus	1	13	8[c]	71
	2	26	10[c]	64
Bromus japonicus	1	47	17[c]	97
	2	39	15[c]	94
Aristida oligantha	1	15	21	97
	2	19	23	102

[a] Modified from Wilson and Rice (1968).
[b] Expressed as percent of the control.
[c] Dry weight significantly different from control.

Haplopappus ciliatus was inhibited by only two sources, and these two species were inhibited somewhat in the field but not significantly so over the 2-year sampling period. Thus, field results highly correlated with results of laboratory tests of allelopathic effects of the sunflower plant. *Rudbeckia serotina, Erigeron canadensis,* and *Bromus japonicus* occur as winter annuals in revegetating old fields, and thus the patterns are obvious around dead sunflower stalks during the late winter months also. The chief phytotoxins identified were chlorogenic and isochlorogenic acids in aqueous extracts of all organs of the sunflower plant, scopolin, and a suspected α-naphthol derivative in leaf leachate (Wilson and Rice, 1968).

Helianthus mollis, a species of scattered occurrence throughout the tall grass prairie region, forms dense colonies from creeping rhizomes and usually excludes other species from those portions of the clones where it occurs with greatest densities (Anderson *et al.,* 1978). These researchers devised a bioassay method for rapidly screening species suspected of producing allelochemics and used it to determine the potential for allelopathic action of *H. mollis.* Extracts of whole sunflower plants significantly inhibited both radicle and shoot development of radish and wheat, but inhibited only the radicle of little bluestem (*Schiz-*

TABLE 35. Effects of Sunflower Root Exudate on Seedling Growth[a]

| | | Mean dry weight (mg) | |
Test Species	Experiment number	Control	Test
Helianthus annuus	1	44	32[b]
	2	58	40[b]
Erigeron canadensis	1	53	36[b]
	2	59	45[b]
Rudbeckia serotina	1	36	18[b]
	2	33	14[b]
Digitaria sanguinalis	1	53	39[b]
	2	43	25[b]
Amaranthus retroflexus	1	71	23[b]
	2	64	15[b]
Haplopappus ciliatus	1	63	53
	2	48	42
Bromus japonicus	1	63	61
	2	54	52
Croton glandulosus	1	34	32
	2	47	46
Aristida oligantha	1	99	95
	2	96	91

[a] Modified from Wilson and Rice (1968).
[b] Dry weight significantly different from control.

TABLE 36.   Effects of Sunflower Leaf Leachate on Germination and Seedling Growth[a]

Test species	Experiment number	Mean dry weight of seedlings (mg)		Germination[b]
		Control	Test	
Helianthus annuus	1	47	41	78
	2	48	37	63
Erigeron canadensis	1	137	75[c]	103
	2	147	50[c]	89
Rudbeckia serotina	1	73	23[c]	106
	2	60	30[c]	91
Digitaria sanguinalis	1	37	8[c]	95
	2	26	15[c]	91
Amaranthus retroflexus	1	228	80[c]	75
	2	206	72[c]	68
Haplopappus ciliatus	1	16	6[c]	127
	2	29	17[c]	92
Bromus japonicus	1	73	72	102
	2	60	51	97
Croton glandulosus	1	73	67	92
	2	82	80	81
Aristida oligantha	1	15	12	104
	2	22	24	101

[a] Modified from Wilson and Rice (1968).
[b] Expressed as percent of the control.
[c] Dry weight significantly different from the control.

achyrium scoparium). Inhibition of radish shoots and wheat radicles was greater at high concentrations of the extracts. Extracts from plants collected from the center of the clone inhibited radicle growth of radish more than extracts from plants growing at the edge of the clone. Anderson *et al.* did not separate the influence of competition from that of allelopathy in the success of *H. mollis,* but suggested that allelopathy played a part.

Becker and Guyot (1951) tested root extracts, root leachates, and extracts of the rhizosphere soil of 15 species of forbs against seedling growth of wheat, flax, vetch, and radish. The forbs are common in revegetating old fields in France, and at least some extracts of each species inhibited growth of some or all test species. Allelopathic plants tested included one or more species of each of the following genera: *Achillea, Aphyllanthes, Asperula, Barkhausia, Helianthemum, Hieraceum, Origanum, Papaver, Picris, Pterotheca, Solidago, Teucrium,* and *Thymus.* The most inhibitory species overall were *Hieracium umbellatum, H. vulgatum,* and *Solidago virgaurea.* Several other inhibitory species were added to this list in subsequent tests (Becker *et al.,* 1951). Guyot (1957) reported that a bunch-grass stage appears in about 10 years after abandonment from cultivation in dry calcareous soils in northern France. This stage has mosaic-like dominance patches of forbs, which Guyot attributed to allelopathic influences of the dominant species in each patch.

One of the species found by Becker and Guyot (1951) to be allelopathic was *Hieracium pilosella*, and Widera (1978) reported that it caused a marked decline in growth of red fescue when the two species grew together on the Sudetic Plateau in Poland. *Hieracium pilosella* had a pronounced allelopathic effect against red fescue, and Widera concluded that the allelopathic activity "is the mechanism of winning the competition of *H. pilosella* over *F. rubra.*"

Extensive areas of potentially productive but depleted rangeland in the western United States are occupied by cluster tarweed (*Madia glomerata*) (Carnahan and Hull, 1962). Seeding of forage species on tarweed-infested sites fails consistently unless most of the tarweed is eliminated first. Even then there is considerable loss of both seedlings and established plants on sites previously infested with tarweed. Laboratory tests by Carnahan and Hull (1962) indicated that both ground air-dried tarweed and leachate of tarweed inhibited germination of intermediate wheatgrass (*Agropyron intermedium*). As the concentration of the leachate increased, the inhibitory effects increased. The effect of leachate from 10 g of dried tarweed per square foot, which was approximately equal to the average amount in the field, was significantly different from the control. Even one-sixteenth of that concentration caused inhibition of germination.

There was no difference in emergence of treated wheatgrass and control plants in greenhouse cultures, but at the end of 3 months the air-dry weights of plants from all tarweed treatments were significantly lower than those of the controls. In field tests with different densities of tarweed, grass emergence was the same in all plots. However, as few as four tarweed plants per square foot increased mortality of wheatgrass seedings, and with 16 to 1,024 tarweed plants per square foot, the mortality of grass plants was 90 to 100%. The field effects resulted from a combination of competition and allelopathy.

Olney (1968) observed massive monospecific stands of *Veratrum* in the Colorado Rockies and also observed that subalpine fir (*Abies lasiocarpa*), Engelmann spruce (*Picea engelmannii*), and species of *Delphinium* and *Aconitum* at the periphery of the stands had a spindly and pale appearance. Based on these observations, plus the knowledge that the massive stands of *Veratrum* had developed in 25 years, he suspected that at least some species of this genus might have pronounced allelopathic potential. Olney collected *V. tenuipetalum* from the Black Mesa Experimental Range in Colorado over a 4-year period and did both acid- and alkaline-ether extractions of various parts of the plants. Clinton oats or Balboa winter rye seedlings were used for bioassays, and growth inhibition was judged with respect to controls by measuring the lengths of coleoptiles and roots. Young leaves and buds of veratrum yielded non-indolic growth accelerators and inhibitors in the acidic ether fraction. The accelerators decreased in concentration as leaves matured, whereas the inhibitors increased.

The alkaline ether fraction of leaf base meristems, buds, roots, and rhizomes was rich in crystalline and amorphous alkaloids and phenolic acids. The unchromatographed mixture and certain of the purified major and minor alkaloids strongly inhibited growth of oat and rye seedlings. In addition, profound changes in morphology and cytology of the seminal roots resulted. DNA disappeared partially to completely in seeds germinated for 2 days in solutions of some of the alkaloids. More than 1 mg per gram fresh weight of both the principal alkaloid and two phenolic acids were obtained from rhizomes, leaf base meristems, and leaf buds. Roots yielded about one-half these concentrations, and above-ground parts yielded very small amounts. No field tests were made, and this obviously needs to be done before inferences can be made concerning the allelopathic potential of *V. tenuipetalum.*

Muller (1969) reported that mustard (*Brassica nigra*) forms pure stands that have invaded slopes in the annual grasslands of coastal southern California. All the species involved are annual plants that pass the dry summer and early fall as seeds, and the supply of grass seed is plentiful both inside and outside the mustard stand. Seed germination, which occurs at the time of the first significant winter rain, is usually completed within a 2 to 3-day period while the rain is still falling and the ground is saturated. There is no competition during this period in the grass or mustard area because only the dead parts of the plants of the previous growing season are present. Nevertheless, Muller repeatedly observed over many years that the grass seeds germinated in gerat density in the grassed areas but not at all in immediately adjacent areas within the mustard stand, despite plentiful supplies of grass seeds and moisture. Mustard seeds germinated well in the mustard stand.

Bell and Muller (1973) investigated the causes for the failure of grasses to invade the mustard stands, and initial tests indicated that soil factors were not responsible. There were no significant differences in soil texture, pH, temperature, minerals or moisture, and, as Muller previously pointed out, the pattern is established at the time of germination when there is no competition. Foraging studies indicated that seeds of *Avena fatua* could be markedly reduced in the mustard stands by animals, but none of the other species was appreciably affected. It was concluded, therefore that the almost total lack of grass seedlings in the mustard zone could not be due solely to foraging activities. Light measurements and the use of artificial shading screens demonstrated that light was not a factor in the patterning. Thus, they concluded that some sort of allelopathic mechanism is involved and proceeded to test this hypothesis.

Their early tests for inhibitors were concerned with the possible role of volatile toxins because several persons had previously reported that certain volatile compounds resulting from the hydrolysis of mustard oil glycosides are toxic to some microorganisms. Using gas chromatography, they found that large quantities of allyl isothiocyanate are produced when living vegetative parts of *B. nigra* are crushed. Tests of the volatile materials from macerated tissue and of the pure compound using the sponge bioassay method of Muller *et al.* (1964) showed that both were very inhibitory to radicle growth of *Bromus rigidus*. The compound also was adsorbed onto soil to the extent that the soil was very inhibitory to radicle growth of *B. rigidus*. Tests made periodically after the soil was allowed to stand indicated that all inhibitory activity was gone after 9 weeks. They found that the germination of seeds exposed to the vapors was retarded, as was root growth initially, but after open-air storage for 6 weeks, very little inhibitory activity remained. They concluded that the volatile inhibitors were not responsible for the patterning.

Experiments were performed next to determine if germinating mustard seeds were inhibitory to seed germination of some of the grasses involved. No effects were found.

Bell and Muller observed that the invasion of grassland by mustard was more marked on the downhill side, which made them suspect that water-soluble toxins were involved. Water extracts of the tops of living seedlings, dead roots, dead stems, and dead leaves were tested against root growth of three grass seedlings in the sponge bioassay, and the extracts of the dead stems and leaves were very inhibitory. Extracts of living tops and of dead roots had no effects. Leachates of dead stems collected in the field during rains also inhibited the root growth of grass seedlings. Subsequent studies indicated that all the toxins were washed out by the first rain of the season. Thus introduction of toxic leachates into the soil coincided with the time during which the vegetational pattern was established. The incorporation of macerated dead leaves into the soil markedly inhibited root growth of test seedlings when added in amounts even lower than calculated amounts in the field.

Field plots were established in mustard stands in which some were cleared of all debris. Some had the dead stalks cut, broken into pieces, and replaced; others were left as controls. Other plots were established in the neighboring grassland. The density of seedlings of each species in each plot was determined following the first rainfall. The resulting seedling density of the various grass species was virtually the same in the plots cleared of mustard stalks as in the undisturbed grassland. However, plots with broken up pieces of mustard stalks had significant reductions in numbers of seedlings of both grasses and mustard.

Bell and Muller (1973) concluded that the establishment and maintenance of *Brassica nigra* in virtually pure stands results from toxins in the rain water leachates from dead stalks and leaves of the *Brassica* crop of the previous year.

Neill and Rice (1971) observed in an old field near Norman, Oklahoma, abandoned for 25 years, that the composition of vegetation adjacent to western ragweed, *Ambrosia psilostachya,* was different from that apart from it. These observations were quantified using 0.25 m² quadrats. Clippings were made in quadrats with western ragweed, in quadrats 1 m removed from the first quadrat, and in an area that included no *A. psilostachya.* The second quadrat was always located in the same direction from the first unless the above conditions were not met in that location. The clippings were separated by species, and dry weights were taken. Five sets of quadrats were clipped every 2 weeks from June through October.

The mean oven-dry weights of the forbs were not statistically different in the quadrats with *A. psilostachya,* as compared with those 1 m away. On the other hand, *Andropogon ternarius* had a significantly lower mean oven-dry weight near *A. psilostachya,* and *Leptoloma cognatum* had a significantly greater mean dry weight near western ragweed than 1 m away. *Tridens flavus* also had a higher mean dry weight near western ragweed, but not at the 0.05 level of significance.

Several chemical and physical soil factors were measured to a depth of 30 cm; the factors were no different near the western ragweed plants than at least 1 m away. Soil moisture content measured in July after a lengthy period without rain indicated no significance difference near the ragweed plants and 1 m away.

It was decided, therefore, to determine whether the differential growth patterns observed and confirmed by sampling would occur in soil obtained near ragweed plants without the ragweed competition. In July and again in January, a series of eight soil samples, minus litter, was taken within 0.25 m of several *A. psilostachya* plants, and another series of eight was taken over 1 m away from the same plants to be used for controls. The samples were collected with a sharp-nosed shovel and placed in plastic pots. The soil collections from the two dates were treated as separate experiments. Seeds of test species collected from abandoned fields near Norman, Oklahoma, were planted in their respective pots and grown in the greenhouse immediately after the soil was collected each time. Germination was recorded at the end of 2 weeks after which the plants were thinned to the five largest plants per pot. They were grown for 3 more weeks, and dry weights were then determined. Field soil taken in July from near the *A. psilostachya* plants proved to be significantly stimulatory to *Andropogon ternarius, Bromus japonicus, Leptoloma cognatum, Rudbeckia serotina,* and *Tridens flavus* (Table 37). However, soil that was collected in January prior to the onset of germination in the field and after accumulation of *A. psilostachya* leaves stimulated only *Leptoloma cognatum* and *Tridens flavus,* but not at the 0.05 level of significance. The January soil significantly inhibited growth of *Rudbeckia serotina.*

There were some interesting correlations with field patterns. *Leptoloma cognatum* and *Tridens flavus,* which grew better near *A. psilostachya* in the field, were stimulated by soil collected near *A. psilostachya* during both sampling periods. The results to this point suggested that the *A. psilostachya* plants were

TABLE 37.   Effects of Field Soils Previously in Contact with *Ambrosia psilostachya* Roots on Germination and Growth of Test Species[a]

Species	Date soil taken	Mean dry weight (mg) with standard error		Germination[b]
		Control	Test	
Andropogon ternarius	July	25 ± 1.4	33 ± 2.1[c]	80
	January	10 ± 0.6	10 ± 0.5	66
Aristida oligantha	July	31 ± 1.8	26 ± 1.7	100
	January	11 ± 0.5	11 ± 0.5	105
Bromus japonicus	July	22 ± 1.1	46 ± 3.3[c]	112
	January	17 ± 1.0	14 ± 0.6	97
Erigeron canadensis	July	13 ± 1.0	13 ± 0.8	92
	January	8 ± 0.6	7 ± 0.6	85
Haplopappus ciliatus	July	18 ± 1.2	20 ± 1.9	77
	January	10 ± 0.5	7 ± 0.5[c]	83
Leptoloma cognatum	July	20 ± 1.9	36 ± 1.8[c]	90
	January	13 ± 1.0	15 ± 1.1	114
Rudbeckia serotina	July	16 ± 0.8	25 ± 1.6[c]	83
	January	8 ± 0.7	6 ± 0.4[c]	130
Tridens flavus	July	27 ± 1.6	43 ± 3.2[c]	102
	January	6 ± 0.8	8 ± 1.4	100

[a] Modified from Neill and Rice (1971).
[b] Expressed as percent of control.
[c] Dry weight significantly different from control at 0.05 level or better.

producing organic compounds that stimulated some plants and inhibited others. Experiments were designated to investigate this possibility.

Initial experiments were concerned with the effects of decaying air-dried fresh leaves of western ragweed. Seeds of test species often associated with *A. psilostachya* were germinated in pots containing soil mixed with 1 g air-dried powdered leaf material per 454 g of soil or with 1 g of air-dried peat moss in control pots. Rice (1968) determined that mature stands of *A. psilostachya* produced more than 1 g of air-dry weight of leaves per 454 g of soil to the depth of plowing, the top 17 cm. Percentage germination of the greenhouse-grown plants was determined after 14 days, the plants were thinned to the five largest ones per pot, and the oven-dry weights of the seedlings were determined after 3 more weeks.

The air-dried leaves inhibited the growth of *Andropogon ternarius* seedlings but stimulated *Aristida oligantha, Bromus japonicus, Haplopappus ciliatus,* and *Rudbeckia serotina.* Growth of other species tested was stimulated to some degree. Growth of *Leptoloma* and *Tridens* was slightly stimulated also, but the effect was not statistically significant. This experiment was repeated using old leaves that had aged on the *A. psilostachya* plants and were starting to drop. These leaves produced contrasting results, with significant inhibition in dry weights occurring in four of the test species. The percentage germination of *Haplopappus ciliatus* was reduced appreciably.

Leachates of the tops of western ragweed obtained by artificial rain were used to water the same test species growing in soil. The leachates stimulated the growth of *Aristida oligantha, Leptoloma cognatum,* and *Tridens flavus;* they inhibited the growth of *Bromus japonicus* and *Haplopappus ciliatus.*

Root exudates inhibited the growth of *Andropogon ternarius, Bromus japonicus, Erigeron canadensis,* and *Rudbeckia serotina,* whereas they stimulated the growth of *Aristida oligantha* and *Tridens flavus.*

Correlations between the field patterns of associated species and the results of the various experiments were generally good. Growth of *Andropogon ternarius,* which was inhibited near ragweed in the field, was inhibited by root exudate and decaying material of western ragweed. Growth of *Leptoloma cognatum* was stimulated by soil obtained near western ragweed and by leaf leachate of ragweed, and it grew slightly better near ragweed in the field. *Tridens flavus* grew better near ragweed in the field, and it was stimulated by field soil, leaf leachate, and root exudate. *Bromus japonicus* was inhibited by decaying overwintered leaves, leachate, and root exudate, and it did not grow quite as well near western ragweed in the field as it did 1 m or more away. *Aristida oligantha* growth was inhibited in some tests and stimulated in an equal number of tests, and it grew about as well in the field near ragweed as it did 1 m or more away.

Neill and Rice (1971) concluded that the vegetational patterns around *Ambrosia psilostachya* in revegetating old fields are probably due primarily to organic compounds produced by western ragweed, which stimulate growth of some associated species and inhibit others.

Quarterman (1973) pointed out that the relative simplicity of cedar glade communities in Tennessee and the sharp delimitation of zones within them provided a good opportunity for investigating allelopathy on a community level and for considering its interactions with other environmental factors. She reported that areas of open glades with soil from zero to 5-cm deep (zone I) are dominated in winter and spring by three annual forbs—*Arenaria patula, Leavenworthia* spp., and *Sedum pulchellum.* In summer, there are two dominants—*Cyperus inflexus* and *Talinum calcaricum*—occurring as sporadically distributed populations. Areas of open glades with soil 5 to about 20 cm deep (zone II) are dominated by two grasses, *Sporobolus vaginiflorus* and *Aristida longespica;* a legume, *Petalostemon gattingeri;* and a moss, *Pleurochaete squarrosa.*

Previous research had demonstrated that some species of cedar glade communities produced compounds inhibitory to germination of other cedar glade species (Quarterman, 1973). Therefore, she decided to test aqueous extracts of all dominant species of zones I and II for the reciprocal effects on germination. She found a widespread occurrence of leachable germination inhibitors among these species. She concluded that the allelochemics appeared capable of affecting local species distribution and zonation in open cedar glade communities. Subsequently, Turner and Quarterman (1975) reported that *Petalostemon gattingeri* influences the distribution pattern of *Arenaria patula* by means of inhibitors present at all seasons of the year in some or all parts of the *Petalostemon* plant. Inhibitors are leached from the living plant and from litter shed in September and October and are retained by the soil. They are neither degraded nor leached away. Correlation of life cycles of the two species and interaction with seasonal precipitation ensures that inhibitors are present at the crucial time of germination and seedling establishment of *A. patula*. In deeper soils where the cover of *P. gattingeri,* and therefore the production of inhibitors, is greater per unit area, the silt and clay fractions are sufficient to retain the inhibitors in quantities that prevent germination and growth of *Arenaria*. In more shallow, coarse-textured soils with a lower cover of *Petalostemon,* only sufficient quantities of inhibitors are retained by the soil to stunt the growth of *A. patula*. In a previous study, Harris *et al*. (1973) isolated and identified 2-(4-hydroxybenzyl)malic acid from *P. gattingeri* and found a 5 mM solution of the compound inhibited germination of *A. patula* seeds by 70%.

Groner (1974) observed that plantlets of *Kalanchoe daigremontiana,* which grew in the same substrate with one or more established plants of the same species, developed slowly. Application of fertilizers caused little or no increase in the growth rate of the plantlets, whereas the older plants showed a positive response. These reactions suggested a possible allelopathic effect of the older plants on the plantlets. Subsequently, Groner found that the height of plantlets, which were detached from their parent leaves and planted in fresh substrate, increased at about 2 times the rate of sister plantlets grown in substrate previously used to grow *K. daigremontiana*. Addition of water extracts of finely cut stems and leaves from older plants caused a retardation of growth similar to that observed in a substrate containing *K. daigremontiana* roots, even though nutrient levels were adequate in all cases. It was concluded that inhibition was caused by one or more phytotoxins that were exuded from roots of established plants and that were present also in stems and leaves of those plants. Groner (1975) also investigated the influence of *K. daigremontiana* on six grass species, six forbs, a cactus, and a fern. Inhibition of seed germination and seedling development occurred in all species except three grass species when planted in a substrate in which *Kalanchoe* either had grown or was still growing. Extracts of *Kalanchoe* shoots induced similar responses. Mineral nutrients were again supplied in adequate amounts in all tests.

Croton bonplandianum is prevalent in various types of habitats in West Bengal, India, and often forms very "aggressive" stands (Datta and Sinha-Roy, 1975). Datta and Sinha-Roy decided to determine if some of the aggressiveness might be due to allelopathic effects. Treatment of seeds of fifteen commonly associated species with extracts of *Croton* leaves prevented germination of those of eleven species at the highest concentration (100 g leaves in 250 ml water). There was still remarkable retardation of five species even when the extract was diluted 1:4 with distilled water. Leachates of croton leaves reduced seed germination of most of the 15 test species also. Decaying leaves in soil inhibited seed germination and root and hypocotyl growth in 13 of the 15 test species. Soil was collected to a 10 cm depth in an area infested with croton, in October (senescent period for croton) and in July (peak of growth season for croton). Soil collected at least 1 m away from croton plants was used as control soil. Germination and root and hypocotyl growth were lowered in some species in the July test soil as compared with controls. There were no effects due to the October soil. Apparently, the phytotoxins were exuded from roots or leached from leaves of croton primarily during its period of most active growth. Two phytotoxins were isolated but not identified. Absorption spectra in ultraviolet light suggested a relationship to phaseic or abscisic acid.

Lodhi (1979a) found that *Kochia scoparia* produces phytotoxins that cause autotoxicity in this species. The various phytotoxins isolated from *Kochia* leaves did not affect seed germination or early radicle growth of *K. scoparia,* but subsequent radicle growth was reduced. Overall seedling growth was markedly reduced by the phytotoxins, which were identified as phenolics and flavonoids.

In some parts of southern California, *Pholistoma auritum* dominates understory vegetation beneath isolated *Quercus agrifolia* trees that occur in annual grasslands (Parker and Muller, 1979). Pure stands of *P. auritum* were maintained in successive years, although adjacent to grasses with easily dispersed propagules. The fresh litter of *P. auritum* completely inhibited germination of seeds of *Avena fatua* and *Bromus diandrus,* the two most important grasses in the surrounding grasslands. In the control, using leached litter as the germination bed, over 92% of the seeds of both species germinated. Thus, *P. auritum* possesses a dominance mechanism that can severely limit the presence and growth of other understory species, even though it is dependent on the oak tree for its own establishment.

Only minimal amounts of research have been done on the allelopathic effects of forbs on the rhizosphere mycoflora. Mishra and Kanaujia (1972) demonstrated that the diluted juice of *Calotropis procera* or *Datura metel* leaves cause marked changed in the rhizosphere mycoflora of *Pennisetum typhoides* when sprayed on the grass. Singh (1977) reported that both root exudates and root extracts of *Solanum nigrum* and *Argemone mexicana* had varying effects on the growth of fungi in culture. Root extracts of *S. nigrum* retarded the growth of *Trichoderma lignorum,* but root exudate had no marked effect. An alkaloid, solanine, extracted from the roots was responsible for reduction in the growth of the fungus. Root extracts and exudates of *S. nigrum* stimulated the growth of *Mucor luteus, Aspergillus flavus, Fusarium solani,* and *F. chlamydosporum.* Root extracts of

Argemone mexicana accelerated the growth of *Cladosporium herbarium* and *Curvularia lunata,* but did not affect other fungi tested, and root extracts of *A. mexicana* did not affect the growth of any test fungi.

B. Grasses

A few years back, I observed that *Sporobolus pyramidatus* often expanded the size of its stands in the University of Oklahoma Golf Course from a few plants to rather large areas in a short time in spite of the heavy stand of bermudagrass on the course. As *Sporobolus* spread, it almost completely eliminated the bermudagrass and occurred in virtually pure stands. Later some of the *Sporobolus* plants died in the center of the stands, and the remaining ones were less vigorous in vegetative growth and inflorescence production than around the margin (Fig. 11). Thus, very striking patterns were created, and similar patterns were found in the Wichita Mountains Wildlife Refuge in southwestern Oklahoma in natural buffalograss (*Buchloe dactyloides*) areas (Rasmussen and Rice, 1971).

Sporobolus pyramidatus is a small grass, usually attaining a height in our area

Fig. 11. *Sporobolus pyramidatus* stand associated with *Cynodon dactylon* showing edge effect. *S. pyramidatus* in foreground. (From Rasmussen and Rice, 1971.)

of Oklahoma of no more than 1 dm or so, even in flower. Thus, it casts very little shade on other plants. Its small size, the pure stands in which it occurs, its edge effect, and its rapid invasion of other vegetation caused Rasmussen and Rice (1971) to investigate the possibility of allelopathic effects. Sampling of the *Sporobolus* stands and the areas surrounding them in the golf course site demonstrated that moderate amounts of three species, other than *Sporobolus,* occurred in the stands in the early part of the growing season. Some *Cynodon* plants were present and two winter annuals—*Bromus catharticus* and *Hordeum pusillum.* These last two were not present in the August sampling. Outside the *Sporobolus* stand, bermudagrass was the only species of consequence. In the Wichita Mountains Wildlife Refuge, the only species that occurred in the *Sporobolus* stands was buffalograss, and it was present only in minute amounts. Outside the *Sporobolus* stands, buffalograss was the only species of any consequence present.

The species chosen to be tested for possible allelopathic reaction to *Sporobolus pyramidatus* were *Cynodon dactylon, Buchloe dactyloides, S. pyramidatus, Bromus japonicus, Amaranthus retroflexus, Hordeum pusillum,* and *Aristida purpurea. Cynodon dactylon, Buchloe dactyloides, S. pyramidatus,* and *Hordeum pusillum* were obvious choices. *Bromus japonicus* and *Amaranthus retroflexus* were chosen because they are commonly used indicator species of allelopathy in our area. Also, good seed sources of both species are available. Although *Aristida purpurea* was not sampled, its seeds were found in great quantity in the *Sporobolus* stands at the Wichita Mountains Wildlife Refuge site, and plants grew profusely in the areas surrounding buffalograss.

Initial experiments were designated to determine if *Sporobolus* was causing marked changes in selected physical and mineral factors in the soil to a depth of 30 cm, which would certainly include most of the roots of the species involved. No differences were found in soil pH, organic matter, total nitrogen, total phosphorus, or soil moisture. Soil moisture was also measured in August when it is most likely to be deficient in central Oklahoma. Obviously, the analyses were not inclusive enough to exclude all soil factors as possible causes of the virtual exclusion of *Cynodon* from the *Sporobolus* stand, but they did indicate no obvious changes.

In order to eliminate competition from *Sporobolus* and to determine if inhibitors were present in soil in the *Sporobolus* stands, soil minus litter was taken from within such stands and outside them in the golf course site in July and in January, and each collection was used as a separate experiment. All test species listed above were grown in these soils.

The July soils collected within *Sporobolus* stands significantly reduced the oven-dry weights of *Hordeum pusillum, Bromus japonicus,* and *Aristida purpurea,* but stimulated the growth of bermudagrass. *Sporobolus pyramidatus* and buffalograss were also stimulated, but not to a statistically significant level. Germination was appreciably reduced in all species at the second day, but the germination process recovered considerably in some after 2 weeks. The stimulation of bermudagrass and buffalograss growth demonstrated there was no deficienty of minerals in the soil in *Sporobolus* stands.

Soils collected in January from *Sporobolus* stands significantly inhibited all test species except *S. pyramidatus* itself, and germination of all tested species was markedly affected. Dry weight of *Sporobolus* was reduced in the test, but not significantly. These results indicated a phytotoxic effect of the soils closely associated with *Sporobolus* and eliminated any competitive mechanism associated with the presence of the *Sporobolus* plant.

Studies were undertaken next to determine the sources of the toxins in the soil of the *Sporobolus* plots. In one type of experiment, leachate collected from shoots of *Sporobolus* by use of artificial rain was used to water the test species in soil. Control pots were watered with water that had not passed over shoots of *Sporobolus*.

The leachate was not generally inhibitory to the growth of most test species. Only the growth of *Bromus japonicus* was reduced in one experiment; the growth of *Bromus, Amaranthus retroflexus,* and *Sporobolus* itself was reduced in the second experiment.

Field observations indicated that even in early June, *Sporobolus* leaves began to die and were associated in large quantities with the living plant. As much as 8–10 g of dead tissue could be found on a single plant. To determine the effect of decaying *Sporobolus* shoots on the test species, 1 g of shoots (air-dried for 3 weeks) was added to each 454 g of soil. A control series was run with 1 g of peat moss added per 454 g of soil. The test and control series were watered equally with distilled water.

Dry weight increments of all test species except *Aristida purpurea* were decreased by decaying *Sporobolus* shoot material. The growth of *Cynodon dactylon* and *Amaranthus retroflexus* was greatly reduced in the second test, and the latter never developed beyond the cotyledon stage in the month that it grew. Seed germination was appreciably reduced in all species at both check periods.

A similar experiment was performed using decaying root material of *Sporobolus* rather than shoot material. *Cynodon dactylon* was significantly inhibited in both germination and growth, but no other species was inhibited.

Next, an experiment was designed to determine if exudates from *Sporobolus* roots were inhibitory to the test species. The staircase structure was used as a method for collecting exudate each day of the experiment, and the water that passed over *Sporobolus* roots was used to water a set of pots containing soil. A control series of pots was watered with distilled water that had not passed over *Sporobolus* roots.

Root exudates significantly reduced the dry weight increments of bermudagrass in both tests and markedly reduced its seed germination. The dry-weight increases of *Hordeum pusillum, Bromus japonicus, Sporobolus pyra-*

midatus, and *Aristida purpurea* were significantly reduced in one test. *Buchloe dactyloides* was not inhibited in either experiment but its seed germination was reduced in one test.

Five *Sporobolus* stands on the University of Oklahoma Golf Course were measured, and stakes were placed around the perimeter of each stand in July of 1969 to determine the rate of spread. By June, 1970, there was a rapid spread of *Sporobolus.* The stands increased in diameter by 1–4 m, and this rate of spread occurred even though a heavy bermudagrass sod completely surrounded each stand of *Sporobolus.*

The evidence seems clear that *Sporobolus pyramidatus* is able to spread rapidly into heavy sods of *Cynodon dactylon* or *Buchloe dactyloides* because it produces toxins that exude from living roots or diffuse from decaying roots or shoots and inhibit seed germination and growth. Bermudagrass and buffalograss, both perennials, generally spread rapidly vegetatively, bermudagrass by rhizomes, and buffalograss by runners. It might appear, therefore, that inhibition of seed germination of these species is not important in the patterning. As previously pointed out, however, the *Sporobolus* stands are very open, especially at the center, and they could be reseeded even though encroachment by rhizomes or runners is eliminated.

Pioneer weed species are apparently prevented from invading the bare centers of the *Sporobolus* stands because of the toxic effects of leachates, root exudates, and decaying materials of *Sporobolus* on seed germination and growth of representative species. *Hordeum pusillum,* which was found in small amounts in the *Sprorbolus* stands in late winter and spring, was adversely affected by root exudate and decaying material of *Sporobolus.* It grows to maturity and flowers, however, at a time when inhibitors produced by *Sporobolus* are at their lowest concentration in the soil.

To survive the allelopathic effects on itself, *Sporobolus* must move constantly into new areas. To do this it must be less inhibitory to itself than to the surrounding vegetation, and this is apparently true because field soils did not affect *Sporobolus* at either collection date. This probably means that the concentrations that are inhibitory to *Sporobolus* can be maintained only by continued association with its own decay material.

Aristida purpurea probably is prevented from germinating and growing in the *Sporobolus* stands in the buffalograss areas of southwestern Oklahoma by a combination of unfavorable soil factors and the allelopathic effects of *Sporobolus.*

The initial entry of *Sporobolus pyramidatus* into dense bermudagrass or buffalograss sod apparently requires some disturbance. *Sporobolus* seeds require light for germination, and the *Sporobolus* stands start, therefore, in disturbed areas such as divots in the golf course, or cattle tracks in the buffalograss areas.

p-Coumaric acid and ferulic acid were extracted from shoot residue in large

Fig. 12. The three herb zones, as viewed from grass zone, showing *Adenostoma fasciculatum* in background (shrub zone), the adjacent conspicuous *Dodecatheon clevelandii* in the border zone, and primarily *Avena fatua* in the foreground (grass zone). (From Tinnin and Muller, 1971.)

quantities, and both compounds were very inhibitory to germination of *Amaranthus palmeri* seeds.

Tinnin and Muller (1971) reported that the annual grasslands of California are best developed on deep clay soils, whereas the hard chaparral dominated by deep-rooted evergreen shrubs, such as *Adenostoma fasciculatum* (chamise), occurs generally on better drained and shallower soils. Contacts between the hard chaparral and grassland are generally occupied by soft chaparral dominated by such species as *Salvia californica* and *Artemisia californica*. Tinnin and Muller found one area in the foothills of the San Rafael Mountains near Santa Barbara, California, where direct contact occurred between hard chaparral dominated by chamise and the annual grassland. In that area, clay soils were present over Careaga Sandstone at an elevation of 560 m. Herbaceous species in the contact area were separated into three zones with one group occupying primarily the shrub zone, another group chiefly the normal grassland, and a third group mainly the border zone between the other two zones (Fig. 12). This third zone was about 1 m wide and extended completely around the margin of the shrub stand. The grassland zone adjacent to the border zone consisted almost exclusively of *Avena fatua*. This fact together with the fact that *A. fatua* dominates vast areas of annual grassland, sometimes representing virtually the entire complement of vascular plants in sizeable areas, caused Tinnin and Muller to suspect that interference from A. *fatua* might be involved in the pattern of zonation around chamise.

The three herb zones were sampled separately by permanently located quadrats, and of a total of 28 annual species, eight occurred only in the grassland zone, three only in the border zone, and three only in the shrub zone. In addition, three other annual herbs occurred only in the grassland and border zones, and one species only in the border and shrub zones. This demonstrated that there was a clearcut pattern of species distribution, not just of growth.

After the first rainstorm of the growing season, seedlings were identified and counted in all the marked quadrats and were followed throughout the growing season. Seed germination occurred within a 3-day period after the first rain, and no differential mortality among species occurred during the growing season. The pattern of distribution of seedlings was similar to that identified by previous sampling. They concluded, therefore, that the pattern was established at the time of seed germination.

Investigations were carried out to determine the reasons for the pattern of germination, particularly for the failure of annual herbs from the shrub and border zone to occur in the grassland. It was previously shown by McPherson and Muller (1969) that chamise produces toxins that inhibit germination and growth of annual herbs from the grassland. Initially, a study was made of the possible role of small animals because there was evidence of light grazing in all zones. Of the seedlings that were marked, however, one occasionally had a leaf clipped, but seedling density was never influenced greatly. There was additional evidence that animals did not significantly influence the density of seeds.

Soil analyses were made of texture, moisture, pH, electrical conductivity, cation exchange capacity, available phosphorus, nitrate, and total nitrogen in the top 6 cm. No differences were found that could account for the pattern of herbs. Soil moisture was consistently higher in the grassland, although the difference from the other zones was small.

Light intensities at the soil surface averaged approximately 20,200 lux in the grass zone, 12,160 lux in the shaded portions of the border zone, and 8500 lux in the shrub zone. Openings were present in all zones, which allowed 100% of full sunlight to reach the soil surface. Competition for light was obviously not responsible for the failure of shrub and border zone seeds to germinate in the grass zone, even if they required light for germination.

Soil temperatures in the top 2 cm were similar in the shrub and grass zones even on sunny days, and this was true in all zones on the cloudy, rainy days when germination occurred and the herb pattern was established.

Tinnin and Muller (1971) decided, therefore, that an allelopathic mechanism must be involved in the failure of shrub and border zone herbs to grow in the grass zone. They designed experiments to test this hypothesis. Field plots were established in the grass zone; and part of them were clipped to remove dead growth from the previous growing season. The clipped material consisted primarily of *Avena fatua*. The clipping was done in early summer in each year of the study. Abundant supplies of one of the shrub-border zone species, *Centaurea melitensis*, were introduced into clipped and unclipped grass zone plots by hand. Seed germination and survival of herb species were observed throughout the winter growing season, and it was found that *C. melitensis* germinated in clipped and unclipped plots. A density of only 200 individuals per m² occurred in the unclipped plots, whereas 1200 individuals per m² occurred in the clipped plots. These investigators concluded, therefore, that the results supported the hypothesis of an allelopathic control of germination because germination success was directly correlated with the absence of dead grass straw, primarily of *Avena fatua*. Late in the growing season, the *Centaurea* plants in the border zone were 0.5-m tall and had many floral heads per plant; those in the unclipped grass zone that survived were less than 1-dm tall bearing 0 to 3 poorly developed floral heads; and in the clipped grass zone plots, the plants were about 2 dm in height with 1 to 10 poorly developed heads. These were obviously growing better than in the unclipped area, but their growth was definitely inhibited in comparison with those in the border zone. Those in the clipped plots were growing with living *Avena fatua* plants, of course, which indicated interference from the living *Avena* plants.

In other experiments, Tinnin and Muller (1971) found that the dry straw of *Avena fatua* from the previous growing season contained water soluble toxins that were leached into the soil by the first rains of the growing season and selectively inhibited germination.

Results of previous studies by McPherson and Muller (1969) plus results just described caused Tinnin and Muller (1971) to conclude the following concerning the three herb zones at the study site: (1) Several grass zone plants were excluded from the shrub and border zones by toxins produced by chamise; (2) the herbs that occurred in the shrub and border zones did so only because of their tolerance for the toxins produced by chamise; (3) some of the shrub and border zone herbs were excluded from the grass zone by toxins present primarily in dry straw of *Avena fatua;* (4) after the toxic materials were washed out of the dead straw, the living grass zone seedlings helped maintain the herb pattern through mechanisms of allelopathy and competition; (5) because of the differential susceptibility of the herbs to the inhibitors, the herbs were sorted initially into the three zones; (6) seeds of each species were deposited mainly near the parent plants and fell mainly in the same herb zones as the plants that produced them; (7) some seeds did reach other zones but germination was partially inhibited; (8) of those that did germinate, only a small percentage reached maturity due to competition and inhibition by chemicals; (9) the pattern was thus maintained.

Imperata cylindrica is a perennial grass, which is a pernicious weed in many parts of the world. Abdul-Wahab and Al-Naib (1972) found that water extracts of the leaves and culms contained scopolin, scopoletin, chlorogenic acid, and iso-chlorogenic acid, all known phytotoxins. Unfortunately, they did not extend their investigation to determine if these phytotoxins enter the environment where true allelopathic effects could result. Eussen (1978) and Eussen and Soerjani (1978) demonstrated a definite allelopathic effect of this species because leaves placed on the surface of soil or incorporated into soil inhibited growth of test species.

Zaikova (1973) analyzed roots and soils under various meadow communities in the Karelian, A.S.S.R., for quantities of phytotoxins. He found the highest levels in true mesophilic meadow communities and the lowest levels in the grass-sedge and horsetail-sedge communities in peat soils. The levels of phytotoxins changed when the meadows were fertilized, and a correlation was found between the phytotoxin content, yield of grass, biological activity of the soil, and the mass of fungi present.

Miscanthus floridulus is a dominant grass, distributed widely in mountainous country throughout Taiwan (Chou and Chung, 1974). There is a unique pattern of herb exclusion by the *Miscanthus* stands, which caused Chou and Chung to investigate the possible allelopathic activity of this species. They found that leaf leachate resulting from artificial rain on foliar crowns of *Miscanthus* significantly inhibited seed germination and radicle growth of lettuce. Aqueous extracts of leaves and of soils collected from a *Miscanthus* stand were also inhibitory to lettuce. Osmotic effects were eliminated in all cases. Eight phytotoxins were isolated from leaf extracts and seven were identified, all of which were phenolic acids. It was concluded, therefore, that allelopathy is important in the exclusion of herbs from *Miscanthus* stands.

Field studies of the germination of teasel (*Dipsacus sylvestris*) seeds revealed that litter (mainly quackgrass, *Agropyron repens*) inhibits germination (Werner, 1975). When litter was removed just prior to sowing teasel seeds, 32% of the seeds germinated compared with 0.8% when the litter remained. Moreover, when litter was removed from the ground surface up to 1 year after seeds were planted, there was an immediate increase in the number of seeds germinated. Greenhouse studies with quackgrass litter, forb litter (mainly *Aster pilosus, Daucus carota,* and *Melilotus officinalis*), vermiculite litter, and no litter demonstrated that there was no difference in teasel seed germination under no litter and vermiculite litter, but there was a highly significant reduction under *A. repens* litter and a significant but smaller reduction under forb litter. In fact, the reduction under *A. repens* was significantly greater than that under forb litter. Werner pointed out that *A. repens* is known to be allelopathic and that results of her studies do ''not preclude a notion that quack grass was allelochemic to teasel germination.'' She stated further, however, that other microhabitat changes due to the presence of litter were undefined and might be involved. It is notable that an equivalent amount of a chemically inert litter, which would also affect numerous habitat variables, had the same effect on germination as no litter did.

Chou and Young (1975) surveyed 12 species of subtropical grasses for the presence of phytotoxins: *Acroceras macrum, Andropogon nodosus, Brachiaria mutica, Chloris gayana, Cortaderia selloana, Cynodon dactylon, Digitaria decumbens, Eragrostis curvula, Panicum maximum, Paspalum plicatulum, Setaria sphacelata,* and *Tripsacum laxum.* Aqueous extracts of leaves of all species inhibited lettuce seed germination and radicle growth, even though osmotic concentrations were much below those required to cause inhibition. *A. macrum, C. gayana, D. decumbens,* and *P. maximum* extracts were most inhibitory, and *C. selloana* extract was least. Six phenolic acids were identified in ether extracts of the twelve species, and they were differentially distributed in the grasses. Most of the compounds were also found in soil collected from under the various species in the field, and control soil with no grasses or other herbs had significantly lower concentrations of the phytotoxins than did the grass soils. The authors suggested, therefore, that the phytotoxins in soils under the test grasses probably originated from the grasses due to leaching from the plants during rain, exudation from the roots, decomposition of grass residues, or all of these. In subsequent research, Chou (1977) reported that numerous phytotoxins remained in the water fraction of extracts of the twelve grasses after ether fractionation. None was identified, however.

Newman and Rovira (1975) selected eight species from a permanent neutral British grassland, which gave no particular indication from field observations that they were involved in allelopathic interactions. Four were grasses, *Anthoxanthum odoratum, Cynosurus cristatus, Holcus lanatus,* and *Lolium perenne,* and four were forbs, *Hypochoeris radicata, Plantago lanceolata, Rumex acetosa,* and *Trifolium repens.* Leachates of donor pots of each species were tested against each of the eight species in receiver pots with extra nutrients being added on a regular schedule. Leachates of all donor species were significantly inhibitory as compared with controls having no plants in the donor pots. Analysis of some plants for N, P, and K showed that the growth reductions were not due to nutrient deficiency. Growth of four of the species, *L. perenne, H. radicata, P. lanceolata,* and *T. repens,* were inhibited more by pot leachates of their own species than by leachates of other species. All other species, except *R. acetosa,* showed the opposite response. *R. acetosa* showed an intermediate response to its own leachate. Subsequent field observations indicated that the most autoinhibited species were normally found as isolated individuals or as a few individuals in a group, not as pure stands. The three species that were alloinhibited were all capable of dominating a permanent grassland. The authors concluded that the operation of autoinhibitory exudate effects may turn out to be a key process in controlling species diversity in grasslands.

Newman and Miller (1977) investigated the effects of root exudates from *Anthoxanthum odoratum, Lolium perenne, Plantago lanceolata,* and *Trifolium repens* on uptake of N, P, and K by the same species as receiver species. With every receiver species, *Plantago* exudate caused a lower ^{32}P uptake than in the controls (by 18–29%). Uptake from the four root exudates was always in the order *Trifolium > Lolium > Plantago* but with the position of *Anthoxanthum* variable. In some cases, P uptake was promoted by the root exudate relative to the control; in other cases, it was inhibited.

Bokhari (1978) investigated the allelopathic effects of four kinds of short grass prairie litter and three kinds of extracts from living plants of blue gramma and western wheatgrass (*Agropyron smithii*) on seed germination of blue gramma, western wheatgrass, and buffalograss. The extracts were made from plants in three distinct phenological stages of growth. The fresh extracts were more toxic to germination than litter extracts, and of the fresh extracts, those made at the earlier phenological stages were more inhibitory than those made at the advanced stage. Both blue grama and western wheatgrass exhibited autotoxicity, and Bokhari reported that both appeared to grow better in a mixture of other species. These observations support the conclusions of Newman and Rovira (1975).

In California grasslands in which purple needlegrass (*Stipa pulchra*) has regained dominance, wild oats (*Avena fatua*) is markedly reduced in importance as compared with surrounding annual grasslands without purple needlegrass (Hull and Muller, 1977). These investigators decided to examine the ability of needlegrass to reduce the importance of wild oats. They determined that moisture, animal activity, and reduced light intensity were not responsible for the reduction in wild oats in the perennial grassland. Analyses of nutrients and fertility experiments indicated that available nitrogen occurred at concentrations limiting to growth of *A. fatua* in both grasslands, but the nitrogen concentration was lowest in the annual grassland where the *Avena* grew well. Extracts of needlegrass straw were slightly toxic to wild oats, as were root exudates of needlegrass. There was no evidence for the concentration of toxins in the soil, however. The authors suggested that a soil-borne phytotoxin from *Stipa* may reduce the root volume of wild oats and thus reduce its capacity to take up the limited supply of available nitrogen.

Panova (1977) found that the 0 to 10 cm soil horizon, which was most saturated with roots, was the horizon with the greatest inhibitory activity regardless of the floristic composition of the particular steppe community under study. Seasonal dynamics of toxin concentration were most clearly manifested in *Festuca sulcata*. The greatest quantity of water soluble inhibitors occurred in the roots in the bud formation period, and there was a sharp decrease in concentration during the fruit ripening period. This agreed with the results of Bokhari (1978) in his studies of allelopathic activity of blue grama and western wheatgrass.

Survival of foxtail grass in undisturbed sites was tested in a 4-year study of giant green foxtail (*Setaria viridis* var. *major*), robust purple foxtail (*S. viridis* var. *robusta-purpurea*), robust white foxtail (*S. viridis* var. *robusta-alba*), yellow foxtail (*S. glauca*), and giant foxtail (*S. faberii*) (Schreiber, 1977). At the end of the experiment, giant foxtail dominated the stand along with small amounts of yellow foxtail, regardless of the types seeded. Because robust white foxtail and robust purple foxtail have greater seedling vigor, exhibit more rapid growth rates, and produce more seed than giant foxtail, Schreiber concluded that the rapid dominance of giant foxtail must be due to more than competition for light, water, and nutrients and that a logical basis for its dominance would be its allelopathic potential (see Chapter 2, Section I).

Vogl (1974) and Peet *et al.* (1975) suggested that the decline in productivity of an undisturbed prairie might be due to toxins that leach out of litter and inhibit growth of prairie plants. Rice and Parenti (1978) found, however, that prairie litter incorporated in prairie soil significantly stimulated growth of seedlings of the four dominant prairie grasses. The litter did inhibit growth of *Amaranthus palmeri,* an important species in the pioneer weed stage of old-field succession.

C. Emergent Aquatic Plants

Szczepańska (1971) pointed out that beds of emergent aquatics are usually monospecific, and most ecologists who have investigated the ecology of such plants have been impressed with this same fact. This sort of situation is always at least suggestive of an allelopathic mechanism. Nevertheless, very little research has been done concerning the allelopathic effects of emergent aquatics, and these minimal results are not conclusive. The following discussion is meant to accentuate the need and potentialities for further research.

McNaughton (1968) studied the autotoxic effects of cattail, *Typha latifolia,* and found that aqueous extracts of its leaves completely inhibited germination of cattail seeds. In addition, water from a cattail marsh slightly inhibited seedling growth of this species, and water squeezed from soil in which cattail was growing was highly inhibitory to its seedling growth. Seed germination was only partially inhibited when the water extract of the cattail leaves was treated with Polyclar AT (a polyamide) to remove phenolic compounds.

McNaughton did not investigate the allelopathic effects of cattail on other neighboring species. Thus, the reason for the common occurrence of monospecific stands of cattail was not clarified. In most instances where an allelopathic mechanism has been found to be operative, however, toxins produced by a given species are generally more toxic to other species than to the one that produces them. This is perfectly logical, of course, from an evolutionary standpoint. It is certainly possible, therefore, that *T. latifolia* is allelopathic to other emergent aquatics and that this is primarily responsible for its virtually pure stands, accentuated of course by competition.

Szczepańska (1971) investigated the growth of *Phragmites communis, Typha latifolia, Equisetum limosum,* and *Schoenoplectus lacustris* in pure cultures and in various two-species combinations in several types of natural substrates. He found pronounced interference on the part of some species and pronounced stimulation on the part of others. These experiments did not elucidate the mechanism of interference.

Additional experiments were performed, however, on effects of decaying aerial parts of *Typha latifolia, T. angustifolia, Eleocharis palustris, Glyceria aquatica, Schoenoplectus lacustris,* and *Acorus calamus* on growth of *Phragmites communis* seedlings in lake mud. Ten grams of plant material per 300 ml of mud killed the seedlings regardless of which species was used for the plant material. Even 3 g of plant material of each test species per 500 ml of mud killed most of the *Phragmites* seedlings, and those which survived were very stunted and had necrotic areas. Szczepańska concluded that the interference against *Phragmites* when growing with other emergent aquatic species was due primarily to allelopathy on the part of the interfering species.

Szczepański (1971), a colleague of Szczepańska, stated that *Glyceria aquatica* seems to have the strongest allelopathic effects of any of the species studied in his laboratory. Substances that leach out of dead leaves of this species are very inhibitory to seed germination of other species. These substances become most toxic on the ninth day after the leaching begins, which suggests that they may be a product of microbial decomposition. One part of the leachate per 10,000 parts water causes inhibition of seed germination, so Szczepański estimated that inhibition starts at 1 ppm or lower due to the fact much of the leachate is material other than the toxin. He pointed out that this makes the activity at least as great as that of some well-known herbicides, such as trichloroacetate (TCA).

In a study of seed banks of shallow, glacial prairie marshes in northcentral Iowa, Van der Valk and Davis (1976) observed that certain substrate samples with high organic matter content usually produced fewer seedlings and often had significant seedling mortality. Many of the seedlings were deformed, also, and it was suggested that these results could be a result of allelopathic effects of the organic matter in the substrate. Germination of *Typha* seeds was inhibited in substrates with a high *Typha* litter content as compared with their germination in seed banks in open water without *Typha* litter. Van der Valk and Davis pointed out that their results supported the hypothesis of McNaughton (1968) that *Typha* produces phytotoxins that inhibit germination of *Typha* seeds.

Bonasera *et al.* (1979) tested extracts of vegetative organs of four freshwater tidal marsh species against germination and seedling growth of lettuce, radish, tomato, and cucumber. Two of the marsh species, *Peltandra virginica* and *Typha latifolia,* were emergent aquatics. *P. virginica* leaf and petiole extracts inhibited germination of all seeds except cucumber and inhibited growth of all test species. Extracts of leaves, roots, and rhizomes of *Typha* significantly inhibited germination of lettuce seeds only. Leaf extracts inhibited growth of all test seedlings, but rhizome and root extracts inhibited growth of lettuce seedlings only. Unfortunately, no tests were made of osmotic pressures of the extracts, and no native species were used as test species. This is particularly unfortunate in the case of *P. virginica* because there is no other research on its allelopathic potential.

III. ALLELOPATHIC EFFECTS OF WOODY SPECIES ON PATTERNING

Vegetational patterns associated with trees and shrubs generally are more striking that those associated with herbaceous species, regardless of the type of interference responsible. It was probably for this reason that more accounts occurred in the early literature on allelopathy suggesting that several observed patterns were due to chemical inhibition by certain woody species. Unfortunately, some of the earliest suggestions were not supported with experimental evidence, and most investigations prior to the present decade failed to eliminate adequately possible mechanisms other than allelopathy as primary causes of the observed results.

A. Shrubs

Bode (1940) obtained evidence that many herbaceous species fail to grow near hedges of *Artemisia absinthium* because of a toxin produced by the shrub. Bode's results were supported by the investigations of Funke (1943).

Went (1942) did a very careful study of the association between annual plants and shrubs in southern California deserts. He found that annual plants were never associated with some shrubs unless the shrubs were dead. One of the shrubs that fit this category was *Encelia farinosa,* and Went suggested that this shrub might produce toxins that inhibit growth of annuals. He did not have any experimental evidence to support his suggestion, but Gray and Bonner (1948a,b) obtained such evidence, and they isolated and identified a toxin from the leaves of *Encelia.*

Deleuil (1950, 1951a,b) observed that annual plants rarely occurred in the Rosmarino–Ericion association in France even where light was obviously adequate. He subsequently found that field soil from this shrub association contained potent toxins that could be leached out with water and inhibit the growth of numerous annuals when added to soil in which the annuals were growing.

Muller *et al.* (1964) became intrigued with the striking patterns of vegetation in and around patches of *Salvia leucophylla* and *Artemisia californica* in the California annual grasslands (Fig. 13). Virtually no herbaceous species occur within the shrub stands, a bare zone approximately 1 to 2 m wide occurs around the stands, a zone 3 to 8 m wide containing stunted plants of *Bromus mollis,*

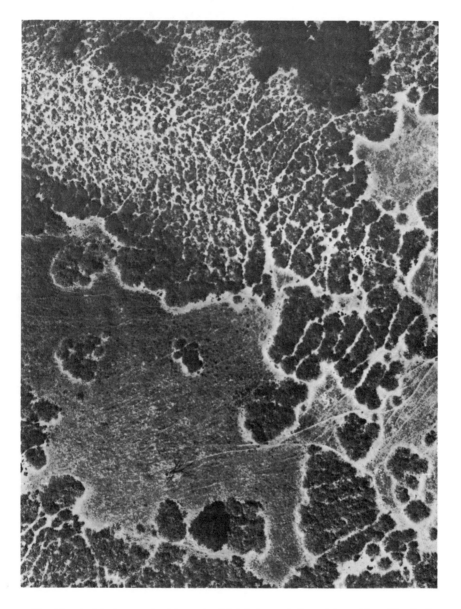

Fig. 13. Aerial photograph of intermixed *Salvia leucophylla* and *Artemisia californica* invading annual grassland in the Santa Ynez Valley, California. This pattern is widespread on Zaca clay soils. (From Muller, 1966.)

Fig. 14. *Salvia leucophylla* producing differential composition in annual grassland: (1) to left of A, *Salvia* shrubs 1–2 m tall; (2) between A and B, zone 2 m wide bare of all herbs except a few tiny seedlings of same age as the large herbs to the right; (3) between B and C, zone of inhibited grassland consisting of several grass species but lacking *Bromus rigidus* and *Avena fatua*; (4) to right of C, uninhibited grassland with large plants of numerous grass species including *Bromus rigidus* and *Avena fatua*. (From Muller, 1966.)

Erodium cicutarium, and *Festuca megalura* occurs around the bare zone, and this is surrounded by normal grassland (Fig. 14). The grassland consists chiefly of *Avena fatua, Bromus mollis, B. rigidus, B. rubens, Erodium circutarium*, and *Festuca megalura* (Muller, 1966).

The zonation did not appear to be correlated in any way with edaphic factors because the shrub thickets spread onto the deeper soils of grasslands even though they often centered on areas of shallow, rocky soil (Muller, 1966). Trenches dug across contact zones revealed no recognizable soil differences between adjacent shrub and grass areas. The inhibition extended far beyond the range of the shrub roots also. Soil analyses indicated that no consistent differences in mineral content or physical factors occurred in the different zones. There was no evidence of salinity stress in any of the zones. Results from addition of cattle droppings reinforced the conclusion that mineral deficiencies were not involved. There was no response to the manure in the bare zone, a slight increase in vigor resulted in the inhibited zone and a great increase in growth and vigor resulted in the uninhibited zone. Observations and measurements during periods of very favorable moisture indicated that the differential growth was maintained, and thus moisture was eliminated as the determining factor in initiating and maintaining the observed zones.

The possibility of predation damage was investigated, and it was found that there was some damage to seedlings during the first few days of growth, especially by the golden-crowned sparrow, *Zonotrichia atricapilla;* the damage was greater near the shrubs than away from them (Muller, 1966). When individual seedlings were marked along permanent transects, it was found that few seedlings were totally lost to grazing, and that many seedlings remained stunted even if they were protected from grazing damage; some inhibition zones showed no grazing damage at all. It was concluded that grazing by small animals may augment the effect of inhibition but cannot initiate or maintain the zones of inhibition.

The evidence suggested that chemical inhibitors might be responsible for the zones, so investigations were made to check this possibility. It was noted that inhibition zones on the uphill side of *Salvia* thickets were about equal to those on the downhill side. This indicated that water-soluble inhibitors could not account for the bare or retarded zones and suggested the possibility that volatile inhibitors might be involved.

An assay technique was developed in which assay seeds were placed between sheets of filter paper on a moist sponge on the floor of a storage dish of 500 ml capacity. Two beakers, each containing 1 g of the plant material to be assayed, were placed in the storage dish beside the sponge. The dish was covered with parafilm and incubated at 25° to 28°C for 48 to 96 hours depending on the assay seeds. Results with cucumber, *Cucumis sativus,* were always recorded as the average length of radicles produced by the germinating seeds in 48 hours (Muller, 1966). Initial experiments indicated that leaves of *Salvia leucophylla, S. apiana,* and *Artemisia californica* produced volatile inhibitors of root growth of cucumber and *Avena fatua* seedlings (Muller *et al.,* 1964). Roots of *S. leucophylla* failed to inhibit the growth of cucumber seedlings, even when the filter paper on which the cucumber seedlings were growing was in contact with the roots. Muller and Muller (1964) identified six terpenes in ether extracts of leaves of each of three species of *Salvia, S. leucophylla, S. apiana,* and *S. mellifera,* using gas chromatography. The identified terpenes were α-pinene, β-pinene, camphene, camphor, cineole, and dipentene. The same compounds were identified also from the atmosphere in dishes in which leaves of the species were sealed.

The identified terpenes were tested for inhibitory activity against growth of roots of seedlings of cucumber. A measured amount of each was placed separately on filter paper disks, and each disk containing a terpene was placed in a beaker inside a container with a moist sponge and filter paper on which were placed soaked cucumber seeds as previously described. The containers were sealed and after 48 hours at 28°C, the lengths of the radicles were measured. Muller and Muller (1964) found that all the terpenes were toxic, with camphor being most toxic and the two pinenes the least toxic. When equal amounts of leaves of the three *Salvia* species were placed separately in sealed containers of the same size, the order based on decreasing inhibitory activity was *S. mellifera, S. leucophylla,* and *S. apiana.* This was the same order as that based on decreasing amounts of terpenes emanating from the leaves. They concluded, therefore, that inhibition of growth of annual grassland species in and around *Salvia* thickets is due to the production of volatile terpenes.

C. H. Muller (1965) collected air samples from among leafy branches of *Salvia leucophylla* and *S. mellifera* in the field and identified cineole and camphor from these samples. The same two compounds were identified consistently from air around *S. leucophylla* in the greenhouse and from the liquid collected in a dry ice-cold trap in the field when air was forced through the trap near *Salvia* patches.

Muller et al. (1964) suggested that the terpenes on inhibited plants might be deposited in dew, and they collected artificial dew from cooling coils near a group of *Salvia* plants in the greenhouse and found that the artificial dew was slightly toxic in some tests to the growth of roots of cucumber seedlings. Later, C.H. Muller (1965) decided that perhaps the terpenes might accumulate in the seedling cuticle because oils and waxes are efficient solvents for volatile terpenes. He tested this possibility by injecting air from a flask containing *S. leucophylla* leaves into sealed flasks containing unbroken ampoules of granulated paraffin. Immediately after this air was injected, a 1 ml sample was withdrawn to determine the original concentration, and the jars were next shaken to break the ampoules to release the paraffin. Aliquots of 1 ml were withdrawn subsequently at 5 and 45 minutes after the original injection of terpene containing air into the flasks. The data indicated that there was a rapid uptake of the terpenes by the paraffin (Fig. 15).

C. H. Muller (1965) suggested from the evidence just presented that the volatile terpenes may dissolve in the cuticular layer of the epidermis or mesophyll cells and then pass through the plasmodesmata into the cells. He suggested further that in young leaves in which formation of the cuticle is still in progress, the terpenes may dissolve in fatty acids and lipids present on the exposed cell wall surfaces, in the adjacent plasma membranes, and in the cytoplasm.

Muller and del Moral (1966) suggested a third possible method by which the volatile terpenes might be absorbed or adsorbed and brought in contact with the affected plants. Muller noted from the beginning of his work on the zonation around certain shrubs in the chaparral that the patterns were most pronounced in fine textured soil and particularly on heavy Zaca clay soils, which are widely distributed in the Santa Ynez Valley, California. He and de Moral hypothesized, therefore, that perhaps the volatile terpenes may be adsorbed on colloidal material in soil and come into contact in this way with roots of affected plants. They obtained Zaca clay from gopher mounds in grassland portions of the patterned area and placed aliquots of this soil in vacuum flasks. The side arms of the flasks were stoppered with rubber serum caps and the mouths were sealed with rubber

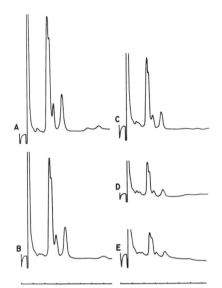

Fig. 15. Extraction of vapors of *Salvia leucophylla* terpenes from atmosphere by paraffin: (A) Control at beginning of experiment; (B) control at end of experiment; (C) treatment at beginning of experiment; (D) treatment after 5 min exposure to paraffin; (E) treatment after 45 min exposure to paraffin. (From C. H. Muller, 1965.)

stoppers wrapped in cellophane to reduce solution of atmospheric terpenes in the rubber. The soil was kept air dried in some flasks, moist to field capacity in some and still others had layers of 5-mm glass beads to serve as controls. A measured amount of air obtained from a flask containing fresh *S. leucophylla* leaves was injected into each of the test flasks. A 1-ml sample was taken from each flask immediately to determine the initial concentrations, and other 1 ml samples were analyzed at 10 minutes and 1 hour after the original injection of air containing terpenes. The results clearly indicated that the terpenes were adsorbed by the soil and a greater amount was adsorbed by dry soil than by moist soil (Fig. 16).

Subsequently, the Zaca clay soil was dried; some was exposed to *Salvia* volatiles for 18 hours (heavily charged), and some for 1 hour (lightly charged). The soil was moistened to field capacity, and *Bromus rigidus* seeds were planted in aliquots of the charged soils and sealed in containers. Controls consisted of flasks of similar soil not exposed to *Salvia* terpenes. All the flasks were kept in darkness for 48 hours at 25°C, after which the germination percentage and root growth were determined. No germination occurred in the heavily charged soil when assayed immediately. Therefore, open containers of the heavily charged soil were kept in a growth chamber for 2 months and assayed again. Germination reached 43% this time, and the growth of roots averaged only 7% of that of the controls. The lightly charged soil was very inhibitory to radicle growth of

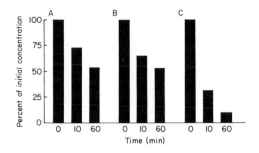

Fig. 16.   Rate of terpene adsorption by soils; (A) control, 5-mm glass beads; (B) moist soil; (C) dry soil. Each sampled initially and at 10 and 60 min. Each quantity is expressed as percentage of initial concentration. (From Muller and del Moral, 1966.)

Bromus rigidus seedlings when assayed immediately and was still inhibitory 4 months later when root growth of seedlings in test soil was only 60% of control growth (Fig. 17). It was clear from these experiments that the volatile terpenes from *Salvia*, which were adsorbed on soil, remained in an active state and that they migrated from the surfaces of soil particles to the sites of inhibition within the plants.

Based on all investigations to this point in time, Muller and del Moral summarized as follows their inferences as to the steps by which *Salvia* and other volatile terpene producing plants initiate and perpetuate the patterns of zonation: (1) Terpenes are evolved into the air at a maximum rate during periods of high temperatures and are adsorbed in greatest amounts at this time by the soil because the period of highest temperatures corresponds to the period of driest soils. (2) The terpenes are held on the soil at least until the early part of the following growing season. (3) Seeds and seedlings that are in contact with the terpene-containing soils extract some of the terpenes by solution in cutin, which is in direct contact with the soil particles. (4) The terpenes are transported into the cells by means of the phospholipids in the plasmodesmata.

Tyson *et al.* (1974) measured the amounts of camphor and 1,8-cineole volatilized from *Salvia mellifera* and also the CO_2 fixation rate. They concluded that the loss of carbon in the two main terpenes volatilized represents a substantial metabolic cost, which implies that it confers some benefit on the plant. They suggested this benefit might be the allelopathic effects of camphor and 1,8-cineole on other species as reported by Muller and del Moral (1966).

Muller (1966) analyzed numerous species of plants from the California chaparral for production of volatile terpenes and found that, in addition to the species of *Salvia* and *Artemisia* previously discussed, the following species produced high amounts of terpenes: *Lepechinia calycina, Heteromeles arbutifolia, Prunus ilicifolia, Prunus lyonii, Umbellularia californica,* and *Artemisia tridentata.* All of these species were toxic in the usual assay tests, and many are common species in the chaparral.

Halligan (1973, 1975, 1976) investigated the causes of bare areas associated with stands of *Artemisia californica* in grasslands. He pointed out that several investigators had claimed different causes of the bare areas around shrub zones in grassland areas of southern California, including cattle, other animals, moisture, and phytotoxins produced by the shrubs. He concluded that allelopathy, animal activity, and an irregular pattern of throughfall and stemflow of precipitation in the shrub stand affected the pattern of herb distribution in and around *A. californica* stands. He considered activity by small animals, however, to be the most important factor and believed that Muller (1966) overrated the role of allelopathy in the zonation. Halligan (1975) reported that the essential oil of *A. californica* contains five major terpenoids, the most toxic ones being camphor and 1,8-cineole. Many minor terpenoids occur also. The soil and litter in stands of *A. californica* contain many of the same volatile substances that occur in *Artemisia* foliage.

The most common dominant in the chaparral of southern California, chamise (*Adenostoma fasciculatum*), does not produce volatile terpenes and yet virtually no herbs are found in mature chamise stands (Fig. 18). McPherson and Muller (1969) investigated the reasons for the lack of herbaceous species in the chamise stands. Initial experiments were designed to determine if the lack of herbs might be due to a deficiency of minerals. Excavations showed that the root system of chamise is directed downward and that there are very few roots in the upper part of the soil. Moreover, herbs grew readily on disturbed areas where the organic horizons were removed. Test plots were fertilized also at the rate of 33 lb of N, 55 lb of phosphoric acid, and 22 lb of K per acre plus traces of Ca, Fe, Mn, S,

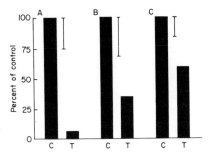

Fig. 17. Bioassay of lightly charged soils: (A) and (B), results of two typical preparations, assayed immediately; (C), assay of soil from (B) after 4 months storage. The quantity of growth under each treatment is expressed as percent of a simultaneous control. Least significant difference at 5% confidence interval is shown by vertical lines. (From Muller and del Moral, 1966.)

Fig. 18.   Overview of relatively young (12 years since burning) open stand of *Adenostoma fasciculatum* showing bare ground between shrubs. (From McPherson and Muller, 1969.)

and Zn. Subsequent sampling in mid-January when seeds of most herbs had germinated in the study area showed no differences between fertilizer plots and control areas. A second sampling about 1 month later indicated a decline in density of herbs in both areas, and there was no increase in size of herbs in treated or control areas since the January sampling.

Four clearings (6 × 6 m) were made in chamise stands by cutting off the tops 5 to 10 cm above the soil in August. The tops were removed and the clearings were fenced to prevent small animals from entering the clearings. Comparable areas of undisturbed brush were fenced in the same way to serve as control areas. Sprouts formed from some chamise stumps in the cleared areas soon after clearing. No herb growth occurred, however, until the first rain in early November. By the end of January large numbers of herbs were found in the cleared areas but very few in the control areas. Twenty-nine herbaceous species were found in the cleared areas, and only a few in the control areas. An average of 40 seedlings per m² occurred under the shrubs, and they were stunted and visible only with careful searching. Herb populations in the clearings averaged over 1000 per m², and these developed normally, reaching several times the sizes of the same species under the shrubs. The herb populations in the control plots reached a peak in mid-December, remained constant until late January, and then declined through the rest of the growing season. Herb numbers in the clearings continued to increase until late January and then remained almost constant (Fig. 19). There was no alteration in mineral supply in this experiment furnishing more evidence that the lack of herbs in mature chamise stands was not due to mineral depletion by chamise.

Light intensity did not appear to McPherson and Muller to be a critical factor because of the open nature of many chamise stands, but they investigated this factor anyway. They found a luxuriant ground cover of various herbaceous species in a grove of *Quercus agrifolia* and *Arbutus menziesii* near their study sites. They measured light intensity at 1 dm above ground along a transect in the

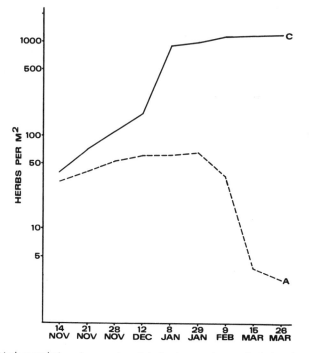

Fig. 19.   Herb populations in experimental clearings and controls during the 1966–1967 growing season. (C) In clearings; (A) in adjacent, undisturbed *Adenostoma fasciculatum* stands. (From McPherson and Muller, 1969.)

chamise chaparral and along a similar transect in the *Quercus-Arbutus* stand except measurements were taken at 4 dm to prevent shading by the herbs. Light values were uniformly low along the transect in the *Quercus-Arbutus* stand with most values being less than 30% of full sunlight. Nevertheless, thriving plants of several herbaceous species were present along the entire transect. On the other hand, light intensity was relatively high on the average in the chamise stand, with a mean approaching 65% of full sunlight. Actually, values equal to full sunlight alternated with lower values. Virtually no herbaceous plants were found along the entire transect in spite of the relatively favorable light conditions.

Light requirements of herbaceous species were checked further by placing artificial shading devices over parts of freshly cleared areas. Sampling of the shaded and unshaded cleared areas during the following growing season indicated no differences in size and numbers of herbs in the two areas. They concluded that competition for light could not account for the lack of herbs in chamise stands.

McPherson and Muller concluded that competition for soil moisture was not an important factor in the exclusion of herbaceous species from the chamise stands because of the deep root systems of chamise and the fact that the rainfall pattern is such that annual plants were not subjected to drought stress. Rainfall is generally confined to the winter months and germination of most annuals outside the influence of shrubs occurs simultaneously within a few days after the season's first significant rain. They pointed out that during the germination period, soils are near field capacity throughout the region and thus differences in soil moisture could not account for the differential germination. Droughts of sufficient magnitude to kill annuals are uncommon. In spite of logical evidence against soil moisture as a determining factor in the absence of herbs in chamise stands, McPherson and Muller investigated the possibility further. They measured soil moisture in chamise stands and in adjacent stands of herbaceous plants along roadsides throughout a growing season and found no appreciable differences. At no time did the soil moisture values in either type of stand reach the wilting range of the soils involved.

Counts of herbaceous seedlings in fenced and unfenced areas within the chamise chaparral indicated that small animals had a slight impact on numbers of herbs in such areas. However, when the number of herbs per m² in the cleared area (over 1000) was compared with the number per .n² in the fenced, uncleared area (70), it was obvious that the effect of the animals was slight when compared with that of the shrub.

All the evidence to this point suggested an allelopathic mechanism of herb suppression. Tests were initiated, therefore, to determine if tops of chamise produce water-soluble toxic materials. Aqueous extracts of the leaves of *Adenostoma fasciculatum* inhibited the growth of roots of seedlings of *Bromus rigidus* in the usual moist sponge bioassay procedure, but similar extracts of roots had no effect. Aqueous extracts of stems of chamise, with and without a common lichen encrustation, did not affect root growth of Bromus rigidus, but fresh leaves of chamise placed in direct contact with the seedbed had an inhibitory effect.

A leachate of leafy branches of chamise obtained by means of artificial rain in the laboratory was very inhibitory to root growth of Bromus rigidus, after the leachate was evaporated to about one-ninth of its original volume. Leachates of chamise collected in the field from intact plants, both from artificial and natural rainfall, were inhibitory to Bromus in the assay. A leachate, which was applied to surface soil collected in one of the field sites, also inhibited root growth of Bromus.

Seed germination and root growth of Calandrinia ciliata, Helianthemum scoparium, and Silene multinervia were inhibited by the the natural-rainfall leachate of chamise tops. This was particularly striking because germination of seedlings of Bromus rigidus was not affected in any of the tests. Moreover, root growth of seedlings Helianthemum and Silene was reduced much more by the leachate than were the seedlings of Bromus rigidus. The effect of raindrip on Calandrinia, Helianthemum, and Silene was particularly significant because these species are prominent soon after the tops of chamise are killed by fire.

McPherson and Muller concluded (1) that leaves of mature Adenostoma fasciculatum shrubs accumulate toxins on their surfaces during the dry summer season, (2) that these toxins are washed off the leaves by the rains that initiate a new growing season, (3) that the toxins are retained in the soil and possibly in the upper 1 to 3 cm, (4) that additional increments of toxins are added with each rain, (5) that most seeds in the soil of mature chamise stands are prevented from germinating by the toxins, and (6) that seeds of a few resistant species do germinate but the seedlings usually fail to mature and reproduce.

Species of Arctostaphylos (manzanita) are other important shrubs in the California chaparral that do not produce volatile inhibitors (Muller et al., 1968). These investigators found, however, that Arctostaphylos glauca and A. glandulosa inhibit herb growth for a distance of 1 to 2 m from the edge of the drip lines of their canopies. This results in a virtually bare zone between the shrubs and adjacent grassland, which is occupied by only a few species of perennial herbs. Sampling of transects at right angles to lines of contact demonstrated an inverse correlation between litter and density of seedlings of herbs (Table 38). Pure stands of either of the species of Arctostaphylos as well as mixed chaparral stands having significant amounts of Arctostaphylos have the same dearth of herbs as chamise stands. Leaf litter of both species of Arctostaphylos contained water soluble inhibitors of seed germination and seedling growth. Leachate from uninjured foliar branches was also found to be inhibitory.

Based on results of the early experiments on species of Arctostaphylos, Chou and Muller (1972) investigated more thoroughly the allelopathic effects of A.

TABLE 38.   Relationship of Herb Seedling Populations to Litter Cover about the Margins of *Arctostaphylos glauca* Stands[a]

Distance in m from canopy edge:	0.0	0.5	1.0	2.0	4.0	8.0
Percent cover by litter:	55	22	18	3	8	1
Seedlings per m²:	30	60	60	620	1210	2650

[a] From Muller *et al.* (1968).

glandulosa var. *zacaensis*. This shrub occurs in extensive pure stands on Zaca Ridge in the San Rafael Mountains of California. The same type of zonation occurs around these stands as described above and initial experiments were designed to determine if the zones could be due to differences in mineral levels or pH. The amount of sodium, copper, zinc, manganese, iron, nitrogen, phosphorus, and potassium in the soil was determined; in general, the mineral content in the shrub soil was higher than in herb zone soil, except for sodium and manganese. There was no significant difference, however, in total exchangeable bases, and the pH was virtually the same in both areas. There was a considerably higher organic matter content in the shrub soil (29%) than in the bare zone soil (9 to 18%) or the herb zone soil (5%).

Competition for light was evaluated next by Chou and Muller. They found no herbs under or between *Arctostaphylos* shrubs, even when the canopies were very open (as little as 50% covered). Moreover, herbs were absent in artificially cleared quadrats during the first growing season. Additionally, zones around *Arctostaphylos* often lacked any herbs, even though some of these bare zones were observed several years earlier to have a dense cover of annual herbs. They concluded that light was not the limiting factor in herb growth.

They found that *Arctostaphylos* roots were distributed principally at soil depths between 20 and 70 cm. Only a few roots were found in the top 10 cm of soil, where most of the herb roots occur. *Adenostoma* has a similar root distribution, and McPherson and Muller (1969) demonstrated that soil moisture was not important in the elimination of herbs from the chamise stands. Chou and Muller (1972) concluded that soil moisture is not the deciding factor in the elimination of herbs from the manzanita stands or the bare zones around them and that the overall similarity with the situation in chamise indicates a similar allelopathic role for manzanita.

Several plots, 10 × 15 m each, were marked in manzanita stands, and the downhill one-half of each was cleared of shrubs in the summer, with the upper one-half of each left as a control. The shrubs were cut off about 5 cm above the soil surface, and the tops were removed from the plots. Each plot was fenced to keep small animals out of both the control one-half and the cleared one-half. Stump sprouts were removed periodically from the cleared areas over the 3-year duration of the experiment.

Only four seedlings occurred in the total cleared area during the first year. There was one seedling of each of the following species *Emmenanthe penduliflora*, *Avena fatua*, *Rhamnus californica*, and *Ceanothus oliganthus*. In the second year, the density of seedlings in the cleared area rose to 36 per m², and it fell again in the third growing season to about 15 per m². These results strongly contrasted with the results from the clearing of *Adenostoma* because McPherson and Muller (1969) found a luxuriant growth of seedlings during the first growing season after clearing the chamise, 1000 seedlings per m². These differences suggested the presence of persistent phytotoxins in manzanita.

A zone 1.5 to 2 m broad in which no seed germination occurred developed adjacent to the control in each clearing during the third growing season. This zone contained many manzanita seedlings from the previous growing season, but no new seedlings appeared in spite of the fact that they appeared in abundance elsewhere in the cleared areas during the third growing season. Clearing had thus caused favorable conditions for the germination of *Arctostaphylos* seeds for only 1 year in the zone next to mature *Arctostaphylos* plants. In the following year, the toxin content of the soil had apparently already become too great for seeds of this species, or any other, to germinate. Chou and Muller concluded that water-borne toxins from the manzanita were probably responsible for the observed results.

They used seeds of *Bromus rigidus* and *Avena fatua* in the standard sponge or sand assay to determine whether aqueous extracts of various parts of *Arctostaphylos glandulosa* plants produce phytotoxins. They tested extracts of fresh fallen leaves, bark, roots, and pericarp, and they determined the osmotic concentration of each extract. All extracts were inhibitory to some extent. Those of leaves and bark were most inhibitory, and the effects were not due to the osmotic concentrations. The osmotic concentrations of the pericarp extracts were so high, however, that Chou and Muller concluded that the growth inhibition of *Bromus* and *Avena* in this case was probably due to the osmotic effect of the extracts.

Leachates of leafy branches of manzanita obtained by use of artificial rain were also inhibitory and not due to an osmotic effect.

Living leaves and leaf litter of different ages were collected in the field. The leaf litter was divided into three stages: The oldest was 2 years old and partially rotted, the youngest was less than 1 year old and partially intact, and the last was intermediate in age. Extracts of living leaves and the youngest-leaf litter inhibited root growth of test seedlings by 40% or more, extracts of leaf litter of intermediate age inhibited growth by 20%, and extracts of the oldest litter were only slightly inhibitory. Extracts of fresh leaf litter were tested against several important herbaceous species by the usual assay procedures, and all were greatly inhibited.

Soil was collected from the shrub, bare, and herb zones in field plots and wetted to field capacity with distilled water; these soil samples were used in assays with *Bromus rigidus*. The *Arctostphylos* soil arrested root growth by 15 to

20%, and the bare zone soil inhibited root growth by about 40%. Thus, Chou and Muller concluded that water-soluble toxins are leached out of the manzanita stands and accumulate in the bare zone soil downhill.

Numerous toxins were identified in leaves and leaf litter of manzanita, in the soil under manzanita, and in the bare zone. One potent toxin identified in the leachate of leafy branches of *Arctostaphylos glandulosa* was hydroquinone. This was tested in the sand assay against *Avena fatua* and *Bromus rigidus,* and it inhibited root growth of both species in a concentration at least as low as 50 ppm, which was the lowest tested. Other identified inhibitors were tested against lettuce seedlings, and all the inhibitors were found to inhibit root growth.

Chou and Muller tested the effects of heat on the toxins in fresh leaf litter of *Arctostaphylos* by grinding the litter and exposing the ground material to various temperatures up to 240°C for 2 hours. The samples were subsequently extracted with distilled water, and the extracts were tested against *Bromus rigidus* in the sponge assay. Toxicity increased with heating up to 160°C, but it rapidly declined after that and disappeared at 200°C. They concluded that the phytotoxins produced by *Arctostaphylos* can apparently be denatured under conditions of a severe fire, thus permitting the flush of herb growth that follows a fire in the chaparral.

Chou and Muller concluded from their many experiments (1) that the exclusion of herbaceous species from areas strongly influenced by *Arctostaphylos glandulosa* var. *zacaensis* results from the allelopathic effects of this shrub, (2) that the similarity of this mechanism to the mode of dominance exerted by *Adenostoma fasciculatum* is very close, and (3) that the toxic effects of *Arctostaphylos* are somewhat greater and persist longer than those of *Adenostoma.*

As was pointed out, mature California chaparral usually has few or no herbaceous species. Fires are very common, however, and in the southern California chaparral, fire usually consumes the crowns of the vegetation in addition to the litter on the ground (Horton and Kraebel, 1955). Seeds of herbaceous species germinate in large numbers following the start of the rainy season after a fire. Herbaceous species are the most prominent plants in the first growing season, and these consist of annual herbs, biennials, bulb-forming perennials, and short-lived subshrubs. In the second growing season following a fire, weedy introduced herbaceous species appear, and a rapid decrease in numbers and vigor of herbs begins after the second growing season (McPherson and Muller, 1969).

C. H. Muller and his colleagues suggested on the basis of their early research that fires release the allelopathic effects of the shrubs by removing the sources of toxins and by destroying some of the toxins present in the soil and litter (Muller et al., 1968; McPherson and Muller, 1969; Muller, 1969; Chou and Muller, 1972). Thus, a flush of herbs appears from seeds that have been dormant in the soil since the previous fire and that are no longer inhibited by toxins. As the shrubs again develop in prominence after the fire, interference increases due to the increased addition of toxins and to competition.

Christensen and Muller (1975) re-examined the causes of the herb flush following burning of chamise and concluded (1) that the increased germination of seeds resident in the chamise soil following fire results from the breaking of dormancy by the heat or some unknown agent in those seeds possessing endogenous dormancy mechanisms and the removal of the source of phytotoxins that cause dormancy of otherwise nondormant seeds and (2) that the growth of both herb and shrub seedlings is enhanced in the burned chaparral as a result of enriched soil–nutrient conditions, removal of the source of phytotoxins, and lower mammal densities and reduced grazing.

Research on effects of shrubs on patterning has been rather slow recently compared with earlier activity. Numerous investigators have presented evidence that several species of *Artemisia* have allelopathic effects against neighboring species, and most recent research on shrubs has concerned this genus. Much of this research has been concerned with the identification of the phytotoxins produced and the mechanisms of action. Shafizadeh and his colleagues identified several coumarins and sesquiterpene lactones from several species of *Artemisia* (Shafizadeh and Melnikoff, 1970; Shafizadeh *et al.*, 1971; Shafizadeh and Bhadane, 1972a,b).

Hoffman and Hazlett (1977) tested aqueous extracts of *Artemisia tridentata* litter and foliage extracts of this and several other species of *Artemisia* against seed germination of numerous species that are associated with the various *Artemisia* species under field conditions. Volatile compounds from the foliage were also tested against seed germination. Numerous instances of inhibition or stimulation of germination were reported. The authors concluded that their results suggest possible allelopathic influences of *Artemisia* on species distribution patterns in *Artemisia*-dominated vegetation. Unfortunately, no tests of possible osmotic effects were made in the experiments with extracts, and none of the experiments was designed to clearly demonstrate true allelopathic potentials of the species of *Artemisia*.

Friedman and Orshan (1975) reported that 85% of the achenes of *Artemisia herba-alba* fell under the canopy of the parent plant and were attached to the soil by their mucilaginous surface. Nevertheless, only 3 to 13% of the seedlings of this species were found under the canopy, 40% emerged within a radius of 10 cm just outside the canopy, and the remaining ones were widely scattered at greater distances. They suggested that an antagonistic influence was operating either through inhibition of germination or by killing of seedlings at a very early stage of emergence. They further suggested allelochemics or pests as the antagonistic influences.

In the northern Negev Desert (Israel), annual plants are more numerous on south-facing slopes that are dominated by *Zygophyllum dumosum* than on north-facing slopes that are dominated by *Artemisia herba-alba*, in spite of the fact that moisture conditions are more favorable on the north-facing slope (Friedman *et al.*, 1977). Moreover, annuals are fewer in the vicinity of *Artemisia* than elsewhere. In the year following removal of *Artemisia,* the number of annuals increased but remained smaller than on the south-facing slope, suggesting a residual inhibitory effect. Tests demonstrated that shoots of *A. herba-alba* released both volatile and water soluble substances that strongly inhibited germination of *Helianthemum ledifolium* and *Stipa capensis* seeds. Both of these species are rare annuals in the populations of *Artemisia* but common in adjacent associations. They were not affected by volatiles or water soluble materials from *Zygophyllum dumosum.* It was concluded, therefore, that allelopathy may be mainly responsible for the absence or rarity of sensitive annual species in the neighborhood of *Artemisia.*

The results of Friedman *et al.* (1977) supported the suggestion of Friedman and Orshan (1975) that allelochemics might be responsible for the rare occurrence of seedlings of *A. herba-alba* under parent plants.

The residual effects of various species of heath when heathlands in Galicia in the northwest corner of Spain are used for agriculture have already been noted (see Chapter 2, Section I). Galicia is mostly covered with unproductive heathlands since removal of the oak forests (Ballester *et al.,* 1977). The natural occurrence of annual grasses and other herbaceous species in this humid region appears to be limited by several shrubs, particularly species of the Ericaceae.

Minute amounts of leaf material of *Erica scoparia* (2 g fresh wt/100 ml H_2O) inhibited radicle growth of red clover by 25 to 30%. At higher concentrations of the extracts (over 10 g fresh wt/100 ml H_2O), strong root curvatures resulted, and no root hairs were formed. It was demonstrated that osmotic pressure was not responsible for the effects. Ten phenolic compounds were identified in the water extracts: *p*-hydroxybenzoic, vanillic, syringic, caffeic, ferulic, *p*-coumaric, protocatechuic, gentisic, and 2-hydroxyphenylacetic acids, and 2,4-dihydroxyphenyl acetonitrile.

Aqueous extracts of aerial parts of *Erica australis* and *E. arborea* (20 g fresh wt/100 ml H_2O) significantly inhibited root and hypocotyl growth of red clover, and the effect was not due to osmotic pressure (Ballester *et al.,* 1979). A thin layer of dead leaves or flowers of either of these species significantly inhibited the growth of red clover when applied to soil in the greenhouse.

Carballeira (1980) identified fifteen phenolic compounds in methanolic extracts of flowers, leaves, stems, and roots of *E. australis* and associated soil. These included *p*-hydroxybenzoic, protocatechuic, vanillic, syringic, gentisic, caffeic, *p*-coumaric, ferulic, and sinapic acids along with esculetin, scopoletin, orcinol, quercetin, myricetin, and kaempferol. All were found in the roots, 8 in leaves, 7 in stems, 10 in flowers, and 4 in the soil. It appears, therefore, that the suppression of herbs in heathlands may be explained, at least in part, by the allelopathic effects of associated shrubs.

Clerodendrum viscosum is a robust undershrub that grows to a maximum height of 1.8 m and gives off a fetid odor (Datta and Chakrabarti, 1978). It is common throughout India, Burma, and Ceylon, forming rather dense thickets. It spreads rapidly to the exclusion of other species in diverse habitats in the vicinity of Calcutta. In stands of *C. viscosum*, the population of other species is very low, and those plants that emerge exhibit poor growth.

Aqueous extracts of roots, stems, leaves, and inflorescenses of *C. viscosum* inhibited the germination and seedling growth of mustard and lettuce, and some of the extracts inhibited the germination and growth of pea and rice. Soil collected to a depth of 10 cm near *C. viscosum* in the field arrested germination and growth of the four crop species and four common weed species in the area, in comparison with the effects of soil collected 1 m away from *C. viscosum* plants. Thus, it seems likely that allelopathic effects of *C. viscosum* may be at least partially responsible for the lack of herbs in field stands of this species.

B. Trees

Stickney and Hoy (1881) suggested that the failure of most herbs to grow under black walnut is due to toxins produced by the walnut tree. These observations were supported by Cook (1921), Massey (1925), and Davis (1928), and Massey and Davis presented evidence that the failure of herbs to grow under walnut trees is due to a toxin produced by the walnut trees.

Waks (1936) observed that parks of black locust are nearly void of all other vegetation, and he found that the bark and wood of black locust contained toxins that inhibit the growth of barley (*Hordeum*). His results suggested, therefore, that the failure of herbs to grow under the black locust trees was due to allelopathy.

Mergen (1959) reported that *Ailanthus altissima* remains in virtually pure stands for long periods of time and found that extracts of the rachis, leaflets, and stem of *Ailanthus* caused rapid wilting of 45 of 46 species of seed plants when applied to the cut surface of the stems of the test species. The virtually pure stands may be due to allelopathic effects of *Ailanthus*, therefore, but numerous additional types of experiments should be performed before such an inference is drawn.

Arnold (1964) demonstrated definite zones of herb growth around one-seeded juniper (*Juniperus monosperma*) in Arizona. He suggested that the zonation was probably due to competition for light, moisture, and minerals. He also suggested that the zonation warranted more detailed ecological studies. Subsequently, Jameson (1961) found that water extracts of leaves of three species of *Juniperus*, including *J. monosperma*, strongly inhibited growth of wheat radicles. Later, Jameson (1966) found that tree litter was the major factor associated with the reduction of blue grama under and near *Juniperus* spp. and *Pinus* spp. (see Chapter 3, Section I).

Baker (1966) reported that *Eucalyptus globulus* produces volatile materials that inhibit root growth of cucumber seedlings and also the growth of hypocotyls but not the roots of *Eucalyptus* seedlings.

Later, del Moral and Muller (1969) reported the abundance of this species in parts of California, both as a planted and naturalized species. Moreover, they described the bare areas that regularly occur under trees of this species and pointed out that most mature, undisturbed stands are almost devoid of herbaceous annuals. This species often occurs in annual grasslands around Santa Barbara, California, and only stunted grasses are found under the outer part of the canopy. Fully developed grassland begins at the edge of the canopy, and this situation prevails in a variety of edaphic and microclimatic conditions. A similar lack of herbaceous species under *E. globulus* also occurs on the campus of the University of California at Santa Barbara, despite the fact the trees are kept trimmed and litter is not allowed to accumulate. Del Moral and Muller reported an incidence on the campus at Santa Barbara when heaps of fresh top soil were placed under two *E. globulus* trees. In the subsequent growing season, a heavy growth of annual plants occurred on the fresh soil. During the 2 years following the first growing season, however, no herbs grew on the piles of soil even though no litter accumulated. Thus, they concluded that the failure of annual herbs to grow under *E. globulus* is due primarily to something other than the litter.

Del Moral and Muller (1969) demonstrated in several ways that light, nutrients, and moisture are adequate for herb growth under *E. globulus* and that small animals do not inhabit or visit *Eucalyptus* stands frequently enough to influence the herbaceous vegetation. At one time after a period of drought, the average soil moisture content in one stand was 12.5%, whereas it was only 8.4% at a distance of 6 m from the canopy in the grassland. One very large tree of *E. globulus* on the Santa Barbara campus had its first branches 4 to 5 m above the ground so that full sunlight penetrated diagonally to the trunk. A new lawn was seeded, fertilized, and regularly watered under and to the southwest of the tree. The lawn developed rapidly except under the canopy where no germination occurred in spite of adequate light, water, and minerals.

Fog is very common in the area studied by del Moral and Muller, and large amounts of fog drip are common under trees. They hypothesized, therefore, that the fog drip might be primarily responsible for the lack of herb growth under the canopy of *E. globulus*. They collected fog drip under trees of this species on the Santa Barbara campus, filtered it to remove debris, and tested it against *Bromus rigidus* in the sponge assay procedure. Distilled water was used for controls. The fog drip was found to be very inhibitory to the growth of radicles of *Bromus* seedlings in six separate trials over a 6-month period.

Subsequently, drip was obtained by spraying a fine mist of distilled water over freshly collected leafy branches of *E. globulus* and the drip was tested against six species found in annual grasslands adjacent to *Eucalyptus* groves, plus *E. globulus* itself. Significant reductions in radicle growth resulted in all test species due to the action of the artificial drip.

Natural and artificial fog drip from *E. globulus* was tested against *Bromus rigidus* using the soil bioassay technique previously described. Milpitas loam soil was substituted for sponges as the bed for germination, and both types of fog drip arrested root growth of *Bromus* in the assays. This demonstrated that the toxicity of fog drip is not lost in soil and supported the hypothesis that fog drip is responsible for the lack of annual herbs under *E. globulus*.

No terpenes were found in the fog drip from *E. globulus*, but numerous phenols were found including ellagic, chlorogenic, *p*-coumarylquinic, gentisic and gallic acids, and tannins. Del Moral and Muller stated that ellagic acid is a nontoxin, but work in my laboratory indicates it is toxic to nitrifying bacteria (Rice and Pancholy, 1973).

Del Moral and Muller (1970) observed characteristic zonations of herbaceous plants around groves of *Eucalyptus camaldulensis* trees and a virtual absence of herbs in the groves in Santa Barbara County, California. These groves occur in annual grasslands in which the most important species are *Avena fatua*, *Bromus mollis*, *B. rigidus*, *Erodium cicutarium*, *Lolium multiflorum*, *Medicago hispida*, and *Trifolium hirtum*. *Festuca megalura* and *Hemizonia ramosissima* are sometimes important also.

The pattern associated with *E. camaldulensis* groves consists of (1) a litter zone covered by a moderately thick layer of leaves, bark, branches, and capsules, which extends throughout the stand and to about the edge of the canopy, (2) a "bare" zone 2 to 5 m broad, devoid of litter and located between the edge of the litter zone and the edge of the grassland, and (3) the normal grassland (Fig. 20). Del Moral and Muller (1970) pointed out that virtually no annual herbs occur in the litter zone, but of those sporadic ones that occur, *Bromus mollis* and *Festuca megalura* are the most common. Occasionally a few seedlings of *Bromus rigidus*, *Trifolium hirtum* and *Avena fatua* may appear. Seedlings of *Eucalyptus* itself are common but do not survive beyond the first season.

The "bare" zone appears bare but actually has herb densities higher than the litter zone. The densities are still low, and the plants are very stunted. *Bromus mollis* and *Festuca megalura* are also most important here, just as in the litter zone. *Trifolium* seedlings are common but usually do not survive to maturity. *Avena fatua* and *Bromus rigidus* are uncommon in this zone. The grassland adjacent to the bare zone is dominated by *Bromus mollis*, *Festuca megalura*, *Trifolium hirtum*, and *Erodium*, and this dominance slowly shifts to *Avena fatua* and *Bromus rigidus*. This pattern is re-established each year with the advent of winter rains, and the germination of seeds of all the annual species within a period of a few days.

Del Moral and Muller decided to determine what phases of interference are responsible for the pattern of herbs associated with *E. camaldulensis*. Their initial thrust was aimed at the role of competition for minerals. They determined the levels of potassium, calcium, manganese, sodium, iron, magnesium, zinc, copper, nitrogen, phosphorus, and sulfur in each of the three zones described.

Fig. 20. Pattern of inhibition surrounding stand of *Eucalyptus camaldulensis*. (From del Moral and Muller, 1970.)

No statistically significant differences in levels of any of the elements were found between zones. To test the possible role of competition for minerals still further, fertilizer containing most of the elements for which analyses were made was applied in bands 1.2-m broad across all zones. Herb growth was increased in all zones by the fertilizer, but the difference in growth between the litter zone and the grassland was even greater than in the untreated bands. Moreover, growth in the fertilized litter zone still did not approach even that in the unfertilized grassland. They concluded, therefore, that nutrient limitations were not responsible for the pattern of herb growth inhibition.

Next, light intensity in several stands of *E. camaldulensis* was compared with that under *Quercus agrifolia*, where herbaceous growth was good. The mean intensity under the *Quercus* trees was 45% of full sunlight, whereas it was 64% of full sunlight in *Eucalyptus* stands. It was thus concluded that competition for light is not likely to be an important mechanism of inhibition of herb growth.

Soil-moisture measurements through two growing seasons indicated that the moisture content in the bare zone was consistently lower than in the grassland, whereas it was similar in the litter and grassland zones. Despite these facts, the density and the vigor of the litter zone plants is less than for the bare zone plants. Thus, moisture differences could not explain the lack of herbs in the litter. Growth measurements were made of herbs in the litter zone and of similar species in the grassland zone early in the growing season when rains were common and soil moisture conditions were optimum in all zones. Litter-zone plants were found to be markedly inhibited in growth, even under the favorable moisture conditions existing at the time.

In another investigation of the soil moisture factor, del Moral and Muller dug a trench around a plot in the litter zone to sever all roots of *Eucalyptus* to a depth of 0.5 m, which was the depth of the hard pan. Heavy plastic was placed in the trench before refilling to prevent subsequent growth of roots into the plot. The results during a growing season with normal rainfall indicated virtually no difference in herb density and growth in the plot without root competition from *Eucalyptus* and the area in the litter zone with root competition. However, there was a marked increase in density and growth of species in the plot without competition from *Eucalyptus* in an unusually wet year (160% of normal rainfall). Del Moral and Muller concluded that competition for water alone does not account for the establishment of the inhibition pattern, but it could be important in the intensification and extension of the pattern.

Elimination of competition for light, water, and minerals as the basic cause of the pattern of inhibition in and around *E. camaldulensis* groves left only one aspect of interference to be investigated—allelopathy.

Much previous work indicated that various species of *Eucalyptus* produce several terpenes. Therefore, del Moral and Muller (1970) tested the leaves for the presence of volatile inhibitors by use of the sponge bioassay procedure, with *Bromus rigidus* as the test species. They found that the leaves of *E. camaldulensis* produce strong volatile toxins, and they subsequently identified the major volatiles as α-pinene, β-pinene, α-phellandrene, and cineole. Milpitas loam soil taken from the field site adsorbed the volatile terpenes and became toxic to radicle growth of *Bromus rigidus*.

Del Moral and Muller obtained soil from the three zones in a *Eucalyptus* stand and heated it to drive off terpenes, if present, into sealed flasks. The atmospheres of the flasks were subsequently analyzed by liquid–gas chromatography for terpenes, and large quantities of α-pinene and cineole were obtained from litter-zone soil. The same terpenes were found in bare-zone soil but in lower concentrations, and no terpenes were detected in the grassland soil. The concentrations of terpenes in litter-zone soil were great enough to completely stop seed germination of the herb species involved, based on previous tests. Subsequent tests confirmed a trend of decreasing terpene concentration from the litter zone to the grassland zone, and a decrease in concentrations in all zones as the growing season progressed. They concluded that leaves of *E. camaldulensis* produce and volatilize large quantities of terpenes during the hot, dry summer and that these are adsorbed on dry soil in large amounts. Thus, the concentrations of terpenes in the soil are at a peak at the beginning of the rainy season when seed germination occurs.

Experiments were carried out next to determine if various plant parts of *Eu-*

calyptus and litter in the groves contain water soluble toxins. Aqueous extracts of green leaves, red leaf litter, brown leaf litter, bark, fresh roots, and partially decomposed roots were tested against *Bromus rigidus* in the sponge assay. The brown-leaf litter was older than the red-leaf litter. Both types of litter and bark were toxic to *Bromus* prior to the rainy season, but the toxicity of the leaf litter decreased markedly subsequent to the advent of the rains. They concluded that leaching by rain is an important mechanism for transfer of toxins from the litter into the soil.

Another test of toxicity of the litter was made by placing four or five layers of leaves from the litter in a *Eucalyptus* grove on the surface of trays of soil in which four test species were planted. One-half the trays of each species were irrigated by spraying from above so that the water would have to pass over the litter. The other trays were subirrigated so that the litter remained dry and was not leached (control). The growth of all test species was significantly inhibited by leachate from the litter.

The effect of litter in the field was investigated by establishing two adjacent plots in a grove of *E. camaldulensis*. The litter was removed from one plot and allowed to remain on the other. One hundred seeds each of *Avena fatua* and *Festuca megalura* were carefully sown in each plot to insure that all seeds fell in favorable spots for germination. Growth and survival of both species were significantly reduced in the plot with litter. This was especially significant because the soil of the plot without litter was drier.

Ten toxins were found in the aqueous extracts of leaf litter, five of which were identified: gallic, ferulic, *p*-coumaric, chlorogenic, and caffeic acids.

Extracts of soil from the litter zone and the grassland zone were tested against *Bromus rigidus* in the sponge assay. The extracts were made from soil taken at 2-cm increment levels to a depth of 6 cm. Toxins were present in all three levels of soil from the litter zone and increased in amount with increasing depth. No toxins were found in the grassland soil. Attempts to identify the toxins in the soil from the litter plot failed, but it was suggested that they might be tannins.

Del Moral and Muller found that herbs abounded throughout groves of *E. camaldulensis* that were growing in sandy soils, and subsequent tests demonstrated that soils from such areas did not adsorb terpenes or phenolic inhibitors to the same degree as fine-textured soils. Moreover, they found that the toxins involved were degraded in soil more rapidly under aerobic than anaerobic conditions. Coarse-textured soils also generally have better aeration, so perhaps the inhibitors that were adsorbed in a coarse soil were broken down faster.

Del Moral and Muller concluded that the zonation of herbs in and near *Eucalyptus camaldulensis* is primarily due to allelopathy that is mitigated to some extent by other factors. They concluded further that both terpenes and phenolics are important phytotoxins produced by this species of *Eucalyptus* and that the pattern of shifting of dominance with distance from the canopy indicates that terpenes are capable of selectively inhibiting plants. Phenolic acids act strongly where litter accumulates but they cannot materially influence the establishment of the bare zones or the inhibited grassland, according to these investigators.

The landscape director at the University of Oklahoma was not able to grow *Cynodon dactylon, Lolium multiflorum,* or *Poa pratensis* under sycamore (*Platanus occidentalis*) trees, even with repeated attempts and adequate irrigation and fertilization. Al-Naib (1968) investigated the problem and found that extracts of decaying materials of leaves, fruits, and buds of sycamore were inhibitory to the growth of the above mentioned grasses. This suggested that the bare areas under the sycamore trees might be due primarily to an allelopathic mechanism.

Subsequently, Al-Naib and Rice (1971) observed that virtually no herbaceous plants grow under sycamore trees in natural areas, except in the case of isolated trees in exposed areas where the fallen leaves are blown completely away from the vicinity of the tree. A project was undertaken, therefore, to determine whether the bare areas under the sycamore in natural habitats were due primarily to competition for light, water, or minerals or to chemical inhibitors produced by the sycamore.

Stands of sycamore were selected in Pottawatomie County, Oklahoma, for intensive studies. The area in which the stands were located had not been grazed for a considerable time period prior to the study. Very few species were found within the stands and individuals were scattered and had poor vitality (Fig. 21). Species in the stands included *Ambrosia trifida, Elephantopus carolinianus, Elymus virginicus, Spartina pectinata,* and *Symphoricarpos orbiculatus.* The following species were found away from the edge of the canopy and area of leaf fall and accumulation: *Andropogon glomeratus, Andropogon virginicus, Ambrosia psilostachya, Cynodon dactylon, Elymus virginicus, Panicum anceps, Panicum scribnerianum, Panicum virgatum, Setaria viridis, Symphoricarpos orbiculatus,* and *Tridens flavus.*

Sampling was accomplished by locating thirty quadrats at the edge of the canopy, 30 quadrats 0.5 m outward from the first series, and 30 additional quadrats 0.5 m outward from the second series. All plants in the quadrats were clipped and separated by species; oven-dry weights were determined. The weights of all species were significantly lower in the quadrats at the edge of the canopy than in those farther away. The area immediately below the sycamore canopy was not sampled because virtually no herbaceous plants grew there.

Initial experiments were designed to determine if there were differences in selected minerals, pH, or soil moisture under the sycamore canopy and away from the canopy in the dense herb zone. Analyses of total nitrogen, total phosphorus, and exchangeable and easily soluble copper, iron, and zinc indicated no differences in the two zones. The pH was virtually the same in the two zones also. Soil moisture was determined in the two zones in June, July, and August

Fig. 21.   View of sycamore stand showing bare area beneath canopy. (From Al-Naib and Rice, 1971.)

when it was most likely to be limiting to plant growth. It was measured at two soil depths, 0 to 15 cm and 15 to 30 cm and was greater under the canopy at both depths at all sampling times.

Most of the species that did not grow under sycamore grew well under other nearby tree species where the light intensity was just as low as under the sycamore. Nevertheless, numerous shading devices designed to cast the same shade as the sycamore trees were placed in the herb area in the field in the spring of 1968. These were checked every week or two, but by the time chosen for sampling in October, they had been removed and destroyed by vandals. The exact location could not be located because there were no noticeable differences in growth. The experiment was repeated in 1969, and determination of oven-dry weights of species per unit area under the shading devices and in the open grassland indicated no significant differences in growth.

All evidence to this point indicated that competition for light, minerals, and water probably was not responsible for the failure of herbaceous plants to grow under sycamore trees. Moreover, growth of herbs did not occur or was inhibited any place sycamore leaves were blown by the wind and accumulated in appreciable numbers away from the canopies of the stands. Experiments were thus undertaken to determine if allelopathy might be an important influence in the failure of herbs to grow well under the sycamore canopy and near it. The tests were concerned with studying the effects of decaying sycamore leaves, leachates from living sycamore leafy branches, and soil collected under sycamore trees.

Field sampling of sycamore leaves on the ground after leaf fall indicated that there were 6.16 g air-dried weight of sycamore leaves per 454 g of soil to a depth of approximately 17 cm. In order to determine the effects of decaying sycamore leaves on selected test species, seeds of these species were planted in pots containing soil mixed with 1 g air-dried ground leaf material per 454 g of soil. Control pots contained 1 g of air-dried peat moss per 454 g of soil to keep the organic matter content the same. Fifty test seeds were planted in each pot with the exception of *Ambrosia psilostachya,* seedlings of which were transplanted from the field, because the seeds germinate very poorly even after cold treatment. All plants were kept under greenhouse conditions, and after germination was completed, the plants were thinned to the five largest plants per pot. The plants were grown for an additional 3 weeks, at which time they were harvested and oven-dry weights were determined. Decaying sycamore leaves inhibited seed germination and seedling growth of all test species (Table 39). The data indicated that leaves of sycamore contain toxins that are released into the soil during the decay process causing inhibition of herbaceous species that are present.

Leachate was obtained by spraying a fine mist of water over fresh, mature leafy sycamore branches, and this leachate was used to water pots of soil containing the same test species used in the previous experiment. Controls were irrigated with water that did not pass over sycamore leaves. Germination and growth of the same species were meausred as before. The leachate reduced the percentage of germination appreciably in all species except *Panicum virgatum* and *Elymus virginicus.* The leachate also significantly reduced the oven-dry weight increments of all species except *Elymus virginicus* and *Panicum virgatum.* The failure of the leachate to inhibit germination and growth of *E. virginicus* seems significant in view of the fact that this species was one of few which grew at least sparsely under the canopy.

To determine effects of soil under the sycamore canopy on germination and growth of test species, soil (minus litter) was taken by means of a posthole digger to prevent disturbing the stratification of the soil and placed directly in 4-inch plastic pots. Control soil was obtained in a similar way outside the canopy and away from the area of leaf fall. These soil collections were made in July when the sycamores were in full leaf and in November after the leaves and some fruits had fallen. The two soil collections were treated as separate experiments with the same test species.

Soil from under the sycamore canopy significantly reduced both seed germina-

TABLE 39.   Effects of Decaying Sycamore Leaves on Seed Germination and Seedling Growth[a]

| | | Mean oven-dry weight of seedlings (mg) | | |
Species	Experiment number	Control	Test[b]	Germination[c]
Ambrosia psilostachya	1	937.0	236.7	—
	2	823.0	220.6	—
Andropogon glomeratus	1	620.7	64.9	37
	2	624.8	63.1	21
Andropogon virginicus	1	589.8	15.8	33
	2	566.4	15.9	25
Elymus virginicus	1	123.6	23.8	51
	2	128.3	27.2	62
Panicum virgatum	1	376.7	100.9	77
	2	355.5	30.0	88
Setaria viridis	1	296.7	17.5	60
	2	390.0	18.2	75
Tridens flavus	1	944.6	67.0	56
	2	905.5	54.5	52

[a] Modified from Al-Naib and Rice (1971).
[b] Dry weight significantly different from control at 0.05 level or better in all cases.
[c] Expressed as percent of control.

tion and oven-dry weights of all test species during both sampling periods. The soil collected under the canopy in November caused a greater degree of inhibition of both germination and seedling growth than that collected in July. This was probably due to a greater accumulation of sycamore debris on the soil surface in November with the resulting leaching of substances from the debris into the soil.

Toxins produced by the leaves were identified as chlorogenic acid, scopoletin, and scopolin; those produced by the fruits as the same three plus isochlorogenic acid, band 510, neochlorogenic acid, and o-coumaric acid; and those present in leachate of leaves as chlorogenic acid and scopolin. All of them inhibited germination of seeds of a plant commonly used as a test species, *Amaranthus retroflexus*.

On the basis of the evidence obtained, Al-Naib and Rice (1971) concluded that the production of chemical inhibitors by sycamore is the basic cause of the failure of most herbaceous species to grow under sycamore trees or in areas where sycamore leaves accumulate. They concluded further that once the herbaceous species are slowed in germination and growth, even moderate competitive effects serve as a feedback mechanism that accentuates the retardation in growth.

I observed over several years on the University of Oklahoma campus at Norman that grasses and other plants did not grow under hackberry (*Celtus leavigata*), even though the lower branches were pruned to a height of 5 to 6 m, which allowed full sunlight to penetrate entirely to the trunk during part of the day. This condition persisted even with addition of fertilizer and thorough watering. This condition was present in natural areas also, both in bottomland and in upland. Consequently, the causes of interference on the part of the hackberry were investigated (Lodhi and Rice, 1971).

A plot containing scattered hackberry and plum (*Prunus mexicana*) trees in a tall grass prairie area was established in the University of Oklahoma Grasslands Research Plots near Norman, Oklahoma. The grasslands plots were on a gently rolling upland with moderately deep sandy loam soil over a soft red sandstone bedrock. Vegetation of the area consisted of tall-grass prairie, which had been invaded by woody species since the elimination of burning and grazing in 1949. Dominant species in the plots were little bluestem, big bluestem (*Andropogon gerardii*), switchgrass (*Panicum virgatum*), and Indiangrass (*Sorghastrum nutans*). Growth of herbaceous species was considerably better under plum than under hackberry.

The herbaceous vegetation under several hackberry and several plum trees was sampled by means of clip quadrats. The plants were separated by species, oven-dried, and weighed. All four important prairie grasses had significantly higher yields under plum trees than under hackberry trees. No differences were found however, in yields around the two species away from their canopies.

Light intensities were measured several times during June and July at numerous positions under several plum and hackberry trees, and the range of intensities under both species was similar: 28,000 to 35,500 lux. Thus, differences in competition for light could not account for the variations in herb growth under the test (hackberry) and control (plum) trees.

Determinations of soil moisture, pH, texture, and several selected mineral concentrations were made to see if the differences in vegetation under the plum and hackberry trees were due primarily to the physical and chemical properties of the soil. Soil moisture was determined during three summer months by taking soil samples at each of two depths (0 to 15 cm and 15 to 30 cm). For physical and chemical soil analyses, soil samples minus litter were collected at the 0 to 30 cm depth under hackberry and plum trees. Analyses were made of total nitrogen, total carbon, total phosphorus, and exchangeable and readily available iron, zinc, copper, and manganese.

There were no significant differences in pH, sand, silt, clay, organic carbon, or concentrations of mineral elements under plum and hackberry trees. Obviously, all essential elements were not quantified, but enough were measured to indicate that the reduction in herb growth under hackberry probably was not due to competition for minerals, and it was not due to textural or pH differences in the soil.

Soil moisture was significantly higher under hackberry than under plum at both depths and at all sampling times, so the reduction in herb growth under hackberry was not due to competition for water.

Allelopathy was the only remaining mechanism of interference, and we investigated the effects of decaying leaves, leachate of fresh leaves, and soil under hackberry versus that under plum on seed germination and the growth of the four major prairie grasses.

To determine the effects of decaying hackberry leaves on the test species, 30 seeds each of big bluestem, little bluestem, and switchgrass and large numbers of seeds of Indiangrass (to obviate poor germination) were planted in each test pot containing 1 g of air-dried fresh hackberry leaf powder per 454 g of soil. A similar number of seeds was planted in each control pot containing 1 g of peat moss per 454 g of soil in order to keep the organic matter content the same as in the test pots. Germination was determined after 2 weeks, and the plants were thinned to the four largest seedlings per pot. Seedlings were allowed to grow for 2 additional weeks and then harvested, oven-dried, and weighed. Decaying leaves significantly inhibited germination and growth of all test species.

Leaf leachate was obtained by use of artificial rain on freshly collected leafy hackberry branches, and the leachate was used to water test species in the soil. Controls were irrigated with water that had not passed over hackberry leaves. The leachate inhibited all test species, as did the decaying leaves. Apparently, toxins produced by hackberry leaves can leach out of living leaves into the soil or they can be released by decay of the leaves. To determine the biological activity and stability of toxic compounds in the soil, soil collections were made in July and January, under hackberry (test) and plum (control). Collections were made with a sharp-nose shovel, and the soil was transferred directly into the pots in order to disturb the profile as little as possible. Seeds of test species were placed in appropriate pots with 30 seeds per pot. Germination was determined at 2 weeks, and dry weight was determined at 4 weeks after planting.

Soil collected in July under hackberry did not significantly reduce seed germination or seedling growth. However, the January sample significantly reduced seed germination and seedling growth of all test species. Apparently the toxic compounds are more active in soil in late fall and winter after the accumulation of hackberry leaves and other plant parts on the soil surface. Without doubt, the levels of toxins in the soil under hackberry would still be high in March when the grass species involved are just starting to renew growth, because low soil temperatures during the winter months would slow down oxidation and microbial degradation. Retardation of growth of herbs in the early part of the growing season probably causes increased inhibition due to competition. Of course, there would be an additional inhibitory effect due to the leachate from the leaves during rains, and thus toxins would continue to be renewed in the soil throughout the growing season. The dominant prairie grasses are perennials, which reproduce chiefly vegetatively, so inhibition of seed germination is probably chiefly important in preventing the invasion of the bare area under hackberry trees by annuals. Lodhi and Rice concluded that the paucity of herb growth under *Celtis laevigata* is probably due primarily to allelopathic effects of hackberry accentuated by competition.

The chief phytotoxins produced by *Celtis laevigata* leaves were identified as sopolin, scopoletin, ferulic acid, caffeic acid, *p*-coumaric acid, and gentisic acid (Lodhi and Rice, 1971). All these inhibited the germination of *Amaranthus palmeri* seeds.

Lodhi (1975a,b) extended the research on *Celtis laevegata* to patterning in bottomland forests in central Oklahoma. Investigations indicated that the relatively bare areas under hackberry were probably not due to competition for minerals, light, or water or to differences in texture or pH. Decaying hackberry leaves, leaf leachate, and soil collected from under hackberry trees significantly reduced seed germination and seedling growth of test species. Ferulic, caffeic, gentisic, and *p*-coumaric acids and scopolin and scopoletin were identified as phenolic phytotoxins produced by hackberry leaves. Ferulic, caffeic, and *p*-coumaric acids were isolated and quantified from the soils under hackberry trees in January, April, and September from 0 to 15 and 15 to 30 cm depths. Seed germination bioassays indicated much higher phytotoxicity levels of individual toxins in January and April, and the toxicity was greater when combinations of inhibitors were tested. Individual phytotoxins extracted from soil in September were not very inhibitory to seed germination of test species, but cumulative effects were still allelopathic. This indicated a synergistic action of the toxins, which had been previously postulated but rarely demonstrated. It was inferred that the reduced plant growth under hackberry in bottomland forests is due primarily to allelopathy, with the initial inhibition accentuated by competition.

Grodzinsky and Gaidamak (1971) observed that concentric zones of specific herbs form around scattered trees of *Pinus sylvestris*, *P. strobus*, *Picea exclsa*, *Larix decidua*, *Thuja occidentalis*, *T. plicata*, and *Abies concolor* in parks in the Kiev and Chernigov regions of the U.S.S.R. They suspected that allelopathy was responsible for the patterns, and tests of aqueous extracts of the soil confirmed their suspicion. They subsequently quantified the phytotoxins in the upper layers of soil in the various herb zones. The concentration of phytotoxins was highest near the tree trunks, and a secondary peak occurred some distance beyond the crown of each tree. They consistently found the lowest concentration just outside the crown of each tree, and they termed this the *neutral zone*.

The neutral zone was occupied by characteristic herbaceous species such as *Urtica dioica*, *Chelidonium majus*, *Taraxacum officinale*, *Galium ruthenicum*, *G. intermedium*, *Veronica chamaedrys*, and *Leonurus quinquelobatus*.

Grodzinsky and Gaidamak (1971) found a similar neutral zone with a low phytotoxin content between beech–hornbeam (*Fagus-Carpinus*) forests and neighboring fescue–meadow grass associations in the Carpathian Mountains.

Parpiev (1971) investigated the relationships between shrubs and small trees in the middle Asian deserts and the herbaceous vegetation under the shrubs and trees. In one phase of his study, he tested extracts of seeds, fallen leaves, and leafy shoots of such woody species as *Haloxylon aphyllum, Salsola richteri, Populus pruinosa, Tamarix hispida,* and *Calligonum* spp. against seed germination of herbaceous species such as *Artemisia ferganensis* and *Chenopodium album,* which are common understory plants. He found that extracts of genetically and ecologically similar species of shrubs and small trees favored the germination of seeds, but extracts of genetically and ecologically distinct species inhibited germination. Thus, patterning was strongly influenced by allelopathy.

Parpiev's results emphasized a very important point, namely, that toxic interactions are generally much more likely to occur and to be more striking if species have only recently been brought in contact with each other. This is probably the chief reason that introduced weeds and trees exhibit such striking allelopathic effects at times (e.g., *Eucalyptus* in California, pioneer weeds in old-field succession). Species that have evolved together for thousands of years in natural climax ecosystems are unlikely to retain strong detrimental allelopathic interactions.

Frei and Dodson (1972) studied the epiphytic orchid population of five species of oak in a cloud forest in Oaxaca, Mexico. They found that orchids were abundant on the bark of *Quercus vicentensis* and *Q. castanea* but were fewer in number on *Q. peduncularis* and on *Q. scytophylla.* Orchids did not grow on *Q. magnoliaefolia.* Subsequent laboratory studies indicated that the bark of the last three species was either toxic to orchid seeds and protocorms or inhibitory to their development. The bark of *Q. magnoliaefolia* was rich in ellagic acid and possibly other gallic acid derivatives, whereas the bark of the other four oak species had no gallic acid derivatives. The bark of the other species did have condensed tannins in different amounts. In combination, gallic and tannic acids and ellagic and tannic acids were toxic to test orchids in all concentrations employed. It was concluded that the distribution of epiphytic orchids on certain of the Mexican oaks was correlated with the absence of inhibitors in the bark of these species.

Chou and Hou (1981) observed a unique pattern of weed exclusion by bamboo on mountain sides in Taiwan. Density of understory herbs in a pure stand of the bamboo, *Phyllostachys edulis,* was significantly lower than that in a pure stand of the conifer *Cryptomeria japonica.* Light intensity, soil moisture, and soil nutrient content were more favorable in the former than the latter, indicating that the low density in the bamboo stand was not likely due to competition. Aqueous extracts of leaves and associated soils of 14 bamboo species were tested against germination and radicle growth of lettuce, rye, and rice. Extracts of six species distributed among three genera had significant allelopathic effects on the test species. The aqueous extracts of some of the associated soils also inhibited germination and growth of test species, and these generally correlated well with the leaves that were inhibitory. The inhibitory effects were not due to osmotic pressures of the extracts. Six phenolic acids and some unidentified flavonoids were isolated from the leaf and soil extracts. It was concluded that the exclusion of herbs from the bamboo stands was probably due to the allelopathic effects of the various species of bamboo.

IV. PATTERNING DUE TO ALLELOPATHIC EFFECTS OF MICROORGANISMS

Perhaps this topic may seem out of place here, but there is no question that chemical activities of microorganisms can be very important in determining the distribution and growth of higher plants. Obviously, the topics of phytotoxin production by pathogens and allelopathic effects of microorganisms on nitrogen fixers and nitrifiers belong here, but these topics are discussed elsewhere. Hence, only the appropriate studies that do not fall under other categories will be reviewed here.

Turner (1971), in a monograph on fungal metabolites, reviewed the types of compounds produced by fungi, many of which are important phytotoxins that can inhibit growth of other microorganisms and of higher plants. Many are important antibiotics, which are also used in the treatment of human ailments. A great many of the metabolites are growth stimulators of various types of organisms also, and probably many of these may be more important than inhibitors in ecology.

Olsen (1973a,b) discussed the production of triterpene glycosides by fungi and the inhibition of fungi by these compounds. His experimental work primarily concerned the inhibitory effect of aescin on fungi with reduced sterol contents. Kogan *et al.* (1973) reported that *Actinomyces olivaceus, A. coelicolor,* and *A. chromogenes* reproduced well in sterile soil, actively added B vitamins to the soil (especially nicotinic acid and vitamin B_{12}), and stimulated root and stem growth of corn seedlings.

Kushnir (1973a) found large numbers of bacteria, actinomycetes, and fungi in the rhizosphere of *Crambe cordifolia* and *Heracleum sosnowski,* even though these two plants are strongly allelopathic. Several of the bacteria and fungi also synthesized allelopathic compounds. Kushnir (1973b) reported that bacterial cultures, which were isolated in the spring from plant rhizospheres in steppe communities, had weak inhibiting effects in bioassays, whereas cultures isolated in the middle of the growing season had considerably stronger inhibiting effects. Kushnir and Shrol (1974) isolated rhizosphere fungi from *Agropyron repens, Bromus inermis,* and *Poa angustifolia* in steppe communities, and culture fluids of each were tested in selected bioassays. Those with strong inhibitory action were analyzed to determine what compounds they contained. Culture fluids of species of *Penicillium, Mucor, Fusarium,* and *Trichothecium* were very toxic and contained amino acids, organic acids, and phenolic compounds. It was concluded that an accumulation of phytotoxins in the soil around the three plant species studied could be due to metabolites of the rhizosphere fungi as well as to root exudations. Kushnir (1977) isolated numerous actinomycetes from the rhizosphere of Leto and Zhemchuzhnaja crysanthemums in the Central Republic Botanical Garden in Kiev, U.S.S.R. Many of the actinomycetes isolated inhibited growth of *Botrytis cinerea, Fusarium oxysporum,* and *Cylindrocarpon madhesic.*

There is an urgent need for an extension of this type of research in natural ecosystems.

⑥

Natural Ecosystems: Ecological Effects of Algal Allelopathy

I. EFFECTS ON ALGAL SUCCESSION

The tremendous fluctuations in abundance of phytoplankton in all kinds of bodies of water have intrigued phycologists, limnologists, and oceanographers for many years (Rice, 1954). Various terms such as blooms and pulses have been used to refer to the rapid increases in numbers of phytoplankton above the numbers that previously existed in a given area. Sometimes the blooms are primarily due to the increase in numbers of one species of algae and sometimes to increases in numbers of several species. Another striking fact about the blooms is that they often disappear as rapidly as they appear, or even faster.

Many different suggestions have been advanced to explain the abundance of each species, the size of the total population, and the succession of important (dominant) species during the growing season (Rice, 1954). Most workers have felt that these conditions and changes are due to variations in physical factors, to a lack of necessary nutrients, or to a combination of these. Some persons have even suggested that the variations in populations result from the action of filter-feeding animals (Rice, 1954). A further suggestion was made that the increased population of one species might affect the growth of another species or several species by the production of toxins, thus influencing seasonal succession.

Algal allelopathy was apparently first noted by Harder (1917) who recorded autoinhibition in old cultures of *Nostoc punctiforme*. Akehurst (1931) postulated allelopathy as a factor in algal succession. He investigated phytoplankton in several freshwater ponds over a period of several years and attempted to correlate the fluctuations in populations with chemical and physical factors in the generally accepted fashion of that time. He failed completely in that effort and in-

189

ferred, therefore, that other kinds of factors must be involved. He suggested that phytoplankton might produce substances that inhibit the growth of some species of phytoplankton and stimulate the growth of others. Unfortunately, there was little available evidence in 1931 to support Akehurst's suggestion, but his hypothesis did stimulate research along these lines.

Pratt (1940) investigated the growth and reproduction of the unicellular green alga, *Chlorella vulgaris,* in great detail. He found that the maximum density of population attained was independent of the size of the inoculum, that the rate of multiplication throughout the growth period varied inversely with the initial density of the population, and that the rate of multiplication as measured by the increase in number of cells per hour per cell decreased during nearly the entire period of growth. He pointed out that these data could be interpreted as evidence for production of a growth inhibitor by the cells.

Pratt and Fong (1940) tested Pratt's hypothesis that *Chlorella vulgaris* produces an autotoxin that limits size of the population it can produce in a given culture. They studied the increases in the populations of cultures in media prepared by adding varying proportions of a medium in which *Chlorella* had previously grown to a new medium. This results in variations in the total amounts, and possibly the relative proportions, of the various elements in the media. They performed preliminary experiments, therefore, to determine the effects of various concentrations and relative proportions of the elements in the medium. They could dilute the medium to twice its original volume with distilled water or double the concentration of the standard solution without seriously impairing growth. In fact under the conditions employed, growth was unaffected by changes in the total salt concentration of the nutrient solution from 0.01 mole to 0.1 mole per liter. Moreover, growth of *Chlorella* was not affected by changes in the salt proportions within relatively broad limits.

The test media were made by adding uniform quantities of salts to each solution consisting of distilled water and sufficient amounts of filtrate of an old culture of *Chlorella* to make concentrations of filtrates varying from 0 to 90%. The minimum concentration of salts, therefore, was 0.063 M in the medium prepared with distilled water, and the possible maximum was 0.12 M in the medium prepared with 90% filtrate and 10% distilled water. It was probably never that high, and thus the range of concentrations was within the optimum range of 0.01 to 0.1 M previously mentioned. The pH of each medium was set at 4.45. Three sets of cultures were prepared for each test medium with initial populations of 1, 100, and 1000 cells per mm^3.

Growth of *Chlorella* was inhibited by the filtrate from old *Chlorella* cultures, the depression of growth increased as the percentage of filtrate in the medium increased, and growth varied inversely with the initial population of the inoculum, for a given concentration of filtrate in the medium.

Similar experiments were subsequently performed, except that the filtrates of previous cultures came from cultures of different ages. Results indicated that the depression of growth increased with the physiological age of the filtrate.

In other experiments, the age of the parent cultures from which inocula for new cultures were withdrawn influenced the early rate of growth of the daughter colonies but had very little, if any, effect on the final populations attained. The early growth rate was somewhat retarded in cultures inoculated with cells taken from relatively old colonies.

Pratt and Fong presented excellent evidence against the possibility that their results might be due to pH changes or reduction in trace elements. Therefore, they concluded that *Chlorella* cells produced and liberated into the growth medium a substance that retarded their growth. According to Rice (1954), Rodhe (1948) found the planktonic alga *Asterionella formosa* had a lower rate of division when cultured in the presence of *Chlorella* than when grown alone.

Additional research on possible allelopathic effects of *Chlorella* was conducted by Rice (1954). Two species of freshwater algae were used in these experiments, *Chlorella vulgaris* and *Nitzschia frustulum* (Bacillarieae), to ascertain whether a species could influence its own growth as well as the growth of another species. He first devised a medium in which each of these two rather different species could grow well. He determined the growth curves, daily division rates, and the effects on pH of the culture medium each day for a 7-day period with pure cultures of each of the species. Next, he determined the growth curves and daily division rates for each species and the pH of the medium each day when the species were grown in mixed culture. *Chlorella* grown in mixed culture attained a population size only 60% of that attained when grown alone under the same conditions and in the same medium (Fig. 22), and the division rate was also significantly less than when grown alone. *Nitzschia*, however, reached the same population size in mixed culture as when grown alone. Rice devised repeat experiments in which the medium was enriched daily to replace any depleted minerals and used buffers to prevent marked pH changes. His results were not significantly different in pure or mixed cultures from those previously obtained. He concluded, therefore, that the reduced population size and division rate in *Chlorella* grown in mixed culture was definitely not due to nutrient deficiencies or pH effects. In subsequent experiments in which the cultures were aerated, *Nitzchia* attained a significantly smaller population size in mixed culture than when grown alone, indicating that *Chlorella* was inhibitory to *Nitzschia*. Later, Rice found that *Nitzschia*, when placed on an agar slant with *Chlorella*, never grew until it contacted *Chlorella*, indicating that *Chlorella* produced some diffusible inhibitor of the other alga.

Conditioning the medium by allowing one species to grow in it for 2 days prior to adding the second species resulted in a much more pronounced inhibition of the second species, regardless of which was added first. In another type of experiment, the medium was conditioned by growth of one of the species for

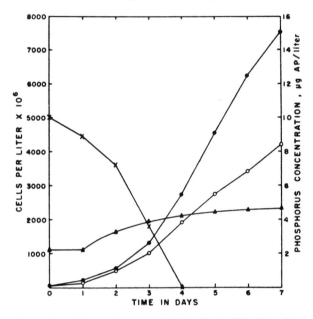

Fig. 22. Comparison of growth curves of *Chlorella* in *Chlorella* culture and in mixed culture with *Nitzschia* prepared with standard culture medium. Closed circles, growth curve in *Chlorella* culture; open circles, growth curve in mixed culture; X, phosphorus concentration; triangles, pH. (From T. R. Rice, 1954.)

several days. The algal cells were then removed by centrifugation, and the pH of the filtrate was adjusted to 7.0. The conditioned medium was filtered through a Berkefeld filter and one-half of this filtrate was used directly as a growing medium, after the same concentration of nutrients was added as used in the standard culture medium. The other one-half of the filtrate was washed with 1% Norit A (carbon) and filtered to remove the carbon. The same concentration of nutrients used in the standard culture medium was added, and the filtrate was autoclaved. This same procedure was followed with *Nitzschia*-conditioned and *Chlorella*- conditioned media. Each organism was subsequently inoculated separately into all four resulting types of conditioned media and the control medium. The size of the *Chlorella* population reached in 7 days in the Norit-washed and autoclaved medium, whether conditioned with *Nitzschia* or *Chlorella,* was not significantly different from the control population. The size of the *Chlorella* population reached in the medium that was conditioned by either *Chlorella* or *Nitzschia* and not Norit-washed and autoclaved was greatly reduced. The population reached in the *Chlorella*-conditioned medium was larger than when grown in the *Nitzschia*-conditioned medium.

The results were similar with *Nitzschia* cultures. In this case, *Nitzschia* was inhibited more by the *Chlorella*-conditioned medium than by the *Nitzschia*-conditioned medium. It is clear, therefore, that both *Nitzschia* and *Chlorella* produce substances that reduce their own growth and that of the other species. These inhibitors are either destroyed by autoclaving, absorbed by Norit A, or both.

While experiments with conditioned media were underway, a bloom of *Pandorina* appeared on a local pond that had been fertilized. Early in the bloom, Rice (1954) determined a population of 73 million cells per liter. About 2 weeks later, the population had declined to 46 million cells per liter. A water sample was collected from the pond at that time, centrifuged, and Berkefeld filtered. This was then used to make two types of *Pandorina*-conditioned media just as previously described for other conditioned media. *Chlorella* and *Nitzschia* were then grown separately in each of the *Pandorina*-conditioned media, and populations were determined each day. The population size reached by *Chlorella* after 7 days in *Pandorina*-conditioned water, which had been Berkefeld filtered only, was 81% of that obtained in similar medium that had also been washed with Norit A and autoclaved. The population size of *Nitzschia* after 5 days in the pond water, which was Berkefeld filtered, was only 70% of that obtained in the conditioned pond water medium that had been washed with Norit A and autoclaved. Rice inferred that a substance was present in the pond water that inhibited the growth of both *Chlorella* and *Nitzschia*. According to Rice, Lefevre and some colleagues grew several species of algae in media prepared with filtered medium in which *Pandorina* had previously grown and found that the majority divided poorly, and some shrank in size and died.

Pratt (1942, 1944, 1948) named the inhibitory substance from *Chlorella* chlorellin and determined a number of its properties. Chlorellin is soluble in 95% ethanol, ether, and petroleum ether and is more readily extractable from aqueous alkaline solution than from acid solution. It diffuses through a collodion membrane indicating that its molecules are probably less than 15 Å units in diameter. It is destroyed by heat. Pratt also studied a number of factors that affected the production and accumulation of chlorellin.

Lefevre and Nisbet (1948) reported that a culture of *Phormidium uncinatum* increased the concentration of dissolved organic matter in the medium by almost eightfold in 40 days. Twenty-four hours after *Scenedesmus quadricauda* was placed in the medium, it became chlorotic, and the cells had a granular appearance. Absolutely no cell division occurred in 40 days. The experiment was repeated using a medium in which *P. uncinatum* had grown for only 12 days, and the results were identical. Their evidence indicated that the required ingredients in the medium were sufficient for good growth of *Scenedesmus*. When the medium was considerably diluted with distilled water, *Scenedesmus* started dividing immediately.

Subsequently, Lefevre and Nisbet found that the medium in which *S. quadricauda* grew for 30 days completely prevented development of *Pediastrum boryanum.* Moreover, media in which either *S. quadricauda* or *P. uncinatum* had grown were found to inhibit growth of *Cosmarium botrytis.*

In October 1949, a canal in a park in Rambouillet, France, had a bloom of *Aphanizomenon gracile,* and the filtered water from the canal was found to completely arrest multiplication and to kill *Pediastrum boryanum, P. clathratum* var. *punctulatum, Cosmarium lundellii,* and *Phormidium uncinatum* (Lefevre *et al.,* 1950). The water also inhibited the growth of *Micrasterias papillifera* but not so severely as the others.

In April 1950, a bloom of *Oscillatoria planctonica* occurred in the same canal, and the filtered canal water completely inhibited multiplication and killed *Chlorella pyrenoidosa, Cosmarium lundellii, C. obtusatum, Pediastrum boryanum, Phormidium uncinatum, P. autumnale,* and *Scenedesmus quadricauda.* There was a less marked activity against *Micrasterias papillifera.*

Lefevre *et al.* (1950) found that all required nutrients were present in sufficient quantities in the canal water to give excellent growth of the test algae. They noted that by June 1950, the *Oscillatoria* had completely disappeared, and algae, which were previously inhibited, were reproducing rapidly.

Lefevre and his colleagues (Lefevre *et al.,* 1949, 1952; Lefevre, 1950; Jakob, 1954) performed many other experiments in both field and laboratory relating to the effect of certain species of algae on other species of algae. They reported that algae produce substances in culture medium that may be algastatic, algacidal, or algadynamic, and they suggested that it is the relative proportion of these substances that determines their final effect. They concluded that a species of alga that multiplies abundantly in natural waters secretes substances in sufficient amounts to inhibit the development of other algae present, algae which eventually die except for a few individuals in each species that are resistant to the inhibitors. As soon as the dominant alga disappears, perhaps due to autotoxins, the antibiotic action ceases and the resistant individuals of other species multiply rapidly. Jakob's (1954) research indicated that similar kinds of chemical interactions also occur between soil algae. Some of the genera of algae that were reported by these researchers to have species inhibitory to other algae (other than those discussed above) were *Pandorina, Nitzschia, Nostoc, Cylindrospermum, Mesotaenium, Ankistrodesmus, Anabaena, Microcystis,* and *Ceratium.*

Denffer (1948) reported that *Nitzschia palea* produces a toxin, both in agar and in water culture, that inhibits its own growth. He found that assimilation goes on unimpeded in spite of the inhibited growth, but the stationary final state of the cultures is characterized by the occurrence of many blocked mitoses. He concluded, therefore, that the allelopathic agent is probably a specific mitosis toxin.

Fogg and Westlake (1955) reported that in pure cultures of *Anabaena cylindrica,* 5 to over 50% of the total nitrogen assimilated dissolves in the medium as peptide nitrogen. They demonstrated that species of algae from other phyla liberate extracellular peptides also and that peptides occur commonly in natural waters. They suggested that peptides may exert important effects on the growth of organisms in water by forming complexes with other dissolved substances. Jørgensen (1956) investigated the chemical interactions in all two species combinations of *Nitzschia palea, Asterionella formosa, Scenedesmus quadricauda,* and *Chlorella pyrenoidosa.* He found that the filtrate of *Chlorella* inhibited the growth of *Nitzschia* but accelerated the growth of *Scenedesmus.* Filtrate of *Scenedesmus* retarded growth of *Nitzschia* and *Chlorella* and was also autotoxic. Filtrate of *Nitzschia* was autotoxic and also reduced the growth of *Asterionella;* however, it accelerated the growth of *Chlorella.* Filtrate of *Asterionella* inhibited the growth of *Nitzschia* but accelerated its own growth.

Haematococcus pluvialis, a motile green alga, is a common occupant of ephemeral rainwater pools but does not occur in permanent ponds (Proctor, 1957a). In a study of 10 permanent ponds in central Missouri, Proctor found that the unautoclaved water from all ponds was inhibitory to the growth of *H. pluvialis* at some or all sampling times over a period of several months. The water from the ponds was particularly toxic during times of heavy algal blooms. In laboratory tests, *H. pluvialis* was found to be killed when grown with *Chlamydomonas reinhardii* and severely inhibited in growth when grown with *Scenedesmus quadricauda.* Proctor concluded that the failure of *H. pluvialis* to grow in permanent ponds is due to the presence of toxins in the ponds and that these toxins are associated with algal blooms in the ponds.

Proctor (1957b) did an excellent series of experiments that substantiated his previous conclusions concerning inhibition of *H. pluvialis* and strengthened the evidence for the widespread inhibition of phytoplankton by other phytoplankton. He tested seven species of phytoplankton for production of antibiotics effective against algae. In initial experiments, he tested the relative growth of five algal species grown in the 10 possible two-membered culture combinations and found that one species failed to grow in 3 of the 10 combinations, and very little growth of one species occurred in three other combinations (Table 40). One or both species were inhibited in all combinations. It was particularly striking that *Anacystis nidulans* almost completely prevented the growth of all four other test species. *Chlamydomonas reinhardii* completely prevented growth of *Haematococcus pluvialis,* causing the latter to form resting spores or generally to die. Proctor decided to do detailed studies with these two species because of the pronounced inhibition and because the two species could be distinguished readily even under low magnification.

Proctor experimented with cell-free media conditioned by the growth of *Chlamydomonas,* using procedures very similar to those used by Rice (1954). *Haematococcus* actually grew better in a medium conditioned for only 1 to 2 days by the growth of *Chlamydomonas,* but growth of *Haematococcus* was markedly inhibited by a medium conditioned for 4 to 6 days by growth of *Chlamydomonas* at pH 8.5. The same medium was considerably less inhibitory

TABLE 40.   Relative Growth of Five Different Algae Grown
in 10 Possible Two-Membered Cultures Shown[a,b]

Combinations	Relative growth
Scenedesmus quadricauda	25
Chlorella vulgaris	100
Chlamydomonas reinhardii	25
Chlorella vulgaris	75
Chlamydomonas reinhardii	87
Scenedesmus quadricauda	12
Haematococcus pluvialis	20
Chlorella vulgaris	100
Haematococcus pluvialis	29
Scenedesmus quadricauda	38
Haematococcus pluvialis	0
Chlamydomonas reinhardii	100
Anacystis nidulans	100
Chlorella vulgaris	5
Anacystis nidulans	100
Scenedesmus quadricauda	10
Anacystis nidulans	100
Chlamydomonas reinhardii	0
Anacystis nidulans	100
Haematococcus pluvialis	0

[a] From Proctor (1957b).
[b] Growth of the algae in single-membered cultures under
the same conditions is taken as 100%.

when the pH was 7.5. The toxin from *Chlamydomonas* was relatively heat stable;
the conditioned medium could be autoclaved for 10 to 15 min without any
appreciable loss of inhibitory activity. Autoclaving for even 2 hr lowered the
inhibitory activity by only approximately 50%.

In subsequent experiments, Proctor (1957b) boiled the inhibitory cell-free
conditioned medium and allowed the steam to pass through a water-cooled
condensor column packed with glass wool or beads. A substance collected on the
column and packing as a yellowish-white film or as droplets. This material was
insoluble in water but dissolved slowly in a basic aqueous solution, and it was

very toxic to *Haematococcus* cells, completely killing them within 24 hr after inoculation in a medium to which the material from the condensor was added. Much larger amounts of the toxin were obtained when conditioned medium containing *Chlamydomonas* cells was boiled in a similar apparatus.

Proctor inferred that the inhibitor, or inhibitors, produced by *Chlamydomonas* was probably a long-chain fatty acid or several such fatty acids. He did not attempt to identify any of the compounds from *Chlamydomonas* cells or from the conditioned medium, but he tested the effects of six saturated and two unsaturated fatty acids, under various pH conditions, on the growth of six species of algae, including *Haematococcus*. All fatty acids tested inhibited *Haematococcus* in dilute concentrations of 5 mg per liter or less at a pH of 8.2. Growth of all test species was reduced by all fatty acids, but some required concentrations were relatively high. Of the compounds tested, oleic, palmitic, and linoleic acids were most inhibitory to all species. Lowering the pH decreased inhibitory effects of the fatty acids, and this finding coincided with the effects of inhibitory material from *Chlamydomonas*.

McCracken *et al.* (1980) purified the inhibitory extract from the same strain of *Chlamydomonas reinhardii* used by Proctor (1957b) and concluded that the inhibitory fraction consisted of at least 15 free fatty acids. Assays with *H. pluvialis* and other algae indicated that the unsaturated fatty acids were most inhibitory and the toxicity increased with an increase in double bonds.

Hartman (1960) reviewed research done through 1957 on allelopathic effects of algae. He summarized in tables the research done under laboratory and field conditions and listed species found to be allelopathic and test algae found to be affected.

Fogg *et al.* (1965) reported that radioactivity was found in dissolved organic matter in the water as well as in the cells of phytoplankton following exposure *in situ* for 3 to 24 hr of samples of lake or sea water to which [^{14}C]bicarbonate had been added. The amount in the water was between 7 and 50% of the total carbon fixed in the photic zone of the water column. The presence of extracellular [^{14}C]labeled organic matter was found under a wide variety of conditions and with many types of phytoplankton communities. The extensive liberation of extracellular products by phytoplankton suggests a strong potential for growth effects on other microorganisms.

Weekly observations over a 7-year period indicated that the phytoplankton of Narragansett Bay, Rhode Island, was alternately dominated from May through October by brief blooms of the diatom *Skeletonema costatum* and the flagellate *Olithodiscus luteus* (Pratt, 1966). The two species were almost never abundant simultaneously. Culture experiments were conduced in which each species was grown in a medium conditioned by growth of the same or the other species, with nutrients restored. Each species inhibited its own growth; the growth of *Skeletonema* was reduced by high concentrations of *Olithodiscus*-conditioned medium, but stimulated by low concentrations. The *Skeletonema* medium had no effect on growth of *Olithodiscus*.

Pratt hypothesized that *Olisthodiscus* achieves dominance by the production of large amounts of an ectocrine that inhibits *Skeletonema* and that *Skeletonema* achieves dominance primarily due to its superior reproductive rate. Moreover, in small competing populations, growth of *Skeletonema* is accelerated by the ectocrine from *Olithodiscus*.

The dinoflagellate *Peridinium polonicum* produces large amounts of a substance, glenodinin, which is very toxic to freshwater and marine fish (and to mice) at a pH above 8.0 (Nozawa, 1968). Culture experiments with a freshwater alga *Scenedesmus obliquus* and a marine alga *Dunaliella tertiorecta* demonstrated that glenodinin was toxic to both algae above a pH of 7.4. The toxin markedly inhibited photosynthesis in the algae, but it was not clear whether this was the basic cause of death of the cells.

By using a multiple diffusion chamber, Martin *et al.* (1974) found that a blue-green alga (*Gomphosphaeria aponina*) was allelopathic to the red tide alga, *Gymnodinium breve*, but *G. breve* had no effect on *G. aponina*. An unspecified ciliate was found to stimulate the growth of *G. aponina*, however.

According to Berglund (1969), various investigators on the Swedish west coast demonstrated that certain green algae grew better when cultivated in seawater taken from the *Fucus-Ascophyllum* zone than when cultivated in surface water collected 100 m from shore. Growth was poor also in seawater taken from a depth of 30 m, even after enrichment with nutrients. Growth was stimulated in free surface water and in deep water if certain living algae (*Ulva, Chorda, Ceramium, Ascophyllum, Chondrus, Fucus*) were previously placed in the water for 24 hr. Thus, Berglund (1969) decided to determine if a particular green alga, *Enteromorpha linza*, produces substances stimulatory to the same alga and to *Enteromorpha* sp. Axenic *E. linza* was grown in an artificial medium for an unspecified period of time, the alga was removed, and one part of this medium was added to four parts of a fresh nutrient medium. Growth of each species in the amended medium was compared with that in a fresh nutrient medium and was found to be markedly stimulated. Water-soluble and water-insoluble fractions were isolated from the medium in which *E. linza* had grown, and both fractions stimulated the growth of both species of *Enteromorpha*. The fractions were not identified. Such growth stimulants could certainly affect algal succession by causing stimulated species to have a selective advantage in competition.

During the autotrophic growth of various strains of *Chlorella pyrenoidosa* and *C. vulgaris* formic, acetic, glycolic, lactic, pyruvic, α-ketoglutaric, and acetoacetic acids were found in the medium (Maksimova and Pimenova, 1969). Glyoxylic acid was present also in filtrates of *C. vulgaris*. The total quantity of organic acids varied considerably, depending on the phase of growth and general state of the culture. In some cases, it amounted to about 25% of the total quantity of extracellular compounds. It should be recalled here that acetic and pyruvic acids were two of the allelochemics found to be produced in decaying wheat straw. In spite of a rather high concentration of organic acids produced in the medium by *Chlorella,* the growth of the alga was not affected.

Fitzgerald (1969) reported that laboratory cultures of the aquatic weeds *Myriophyllum* sp., *Ceratophyllum* sp., and *Lemna minor;* liquid cultures of barley, and cultures of the filamentous green algae, *Cladophora* sp., and *Pithophora oedogonium* remained relatively free of epiphytic algae or competing phytoplankton if the cultures were nitrogen-limited. He suggested two possible reasons for this antagonistic activity: (1) Aquatic weeds or filamentous algae may use most of the limited supply of available nitrogen, or (2) certain bacteria may be present under the low N conditions, which may produce toxins selective against certain algae. Nitrogen stress has been shown to increase the amounts of phenolic allelochemics manyfold in plants (Chapter 12). Therefore, I suggest this as another possible explanation of Fitzgerald's observations.

Filtrate of the green alga *Hormotila blennista* was found to be autostimulatory (Monahan and Trainor, 1970). It was also stimulatory to one strain of *Scenedesmus* and to a soil bacterium (*Bacillus* sp.) at pH 6.3. The filtrate was inhibitory to the soil bacterium and to a different strain of *Scenedesmus* at pH 7.7. Monahan and Trainor (1971) reported that acid, basic, and volatile acid filtrate extracts of *H. blennista* reduced the lag time of this species at low concentrations. Whole filtrate did not affect the lag time, however. Stimulatory properties of the filtrate were dialyzable and heat labile. They suggested that heat labile, low molecular weight organic extracellular products were responsible for the stimulatory property of the filtrate.

Harris (1970) found that axenic cultures of *Platydorina caudata* (Volvocaceae) achieved quite high peaks of growth, followed by a rapid decline in colony number, and then culture death. Careful testing indicated that the decline and death were not due to a limitation of nutrients. Subsequently, a substance was isolated that was autoinhibitory, had several characteristics of a protein, but was not identified. Harris thought it was just autoinhibitory from initial tests, but in a later study of 11 species of the Volvocaceae for self- and crossinhibition, the substance inhibited the growth of species in at least three other genera (Harris, 1971a). All 11 species were found to be allelopathic to at least some of the other species, and some were autoallelopathic. All inhibitors were destroyed by autoclaving. Subsequently, Harris (1971b) isolated a phytotoxin from culture filtrates of aseptically grown *Pandorina morum,* which was inhibitory to most members of the Volvocaceae. The substance could be diluted 15 to 20 times with the retention of at least partial activity, was relatively stable to high temperatures, moved slowly through a dialysis membrane, and possessed antibacterial activity also.

Harris (1971c) reported that the filtrate of *P. morum* reduced the photosynthetic rate of *Volvox globator* by 65% in a 1-hr exposure and by 91% after 12 hr. There was no effect on the respiration rate as measured by oxygen consumption. The mode of action of the photosynthetic inhibitor was examined by exposing *Volvox globator* and isolated spinach chloroplasts to a partially purified inhibitor preparation (Harris and Caldwell, 1974). Oxygen evolution of *Volvox*, whole chloroplasts, and broken chloroplasts were reduced indicating that the substances inhibit the light reaction of photosynthesis. The inhibitor markedly reduced the rates of O_2 evolution by *Eudorina cylindrica* and *Gonium pectorale* but had little effect on *Pandorina morum* itself.

The inhibitor from *P. morum* was stable when exposed to acid, although exposure to a pH of 2.0 for 30 min destroyed most of the activity (Harris and Parekh, 1974). It was partially soluble in benzene and chloroform. All attempts to degrade or destroy it with several proteolytic enzymes were unsuccessful, indicating that it was not a protein. Use of Sephadex G-10 indicated that the inhibitor probably has a molecular weight below 100 (Harris and Caldwell, 1974).

Khailov (1971, 1974) gave extensive data on the kinetics of release of dissolved organic matter (DOM) by unicellular and multicellular marine algae, the kinetics of uptake of the DOM, the chemical nature of the compounds, and the mechanism of action resulting in growth stimulation or inhibition.

Analysis of two consecutive *Anabaena flos-aquae* waterblooms in Lake Nelson, New Jersey, in July and August of 1970 showed strong declines for major phytoplankters as the blue-green approached maximum population density (Williams, 1971). Representative genera responded similarly in laboratory studies with medium conditioned by log-phase *Anabaena*. Degenerative changes noted in inhibited algae included increases in vacuole formation and cytoplasmic density, loss of flagellae, altered morphology, and death. Lipid extracts of lake water during early bloom production were stimulatory to *Anabaena* while having either no effect or an inhibitory effect upon other algal genera present in the lake.

Murphy *et al.* (1976) reported that excretion of hydroxamate chelators by certain blue-green algae ties up iron and makes it unavailable to other algae, thus inhibiting their growth. This is an allelopathic action because the iron deficiency resulted from the addition of an organic compound to the environment. In a survey of seven blue-green algae and 10 green algae, only *Anabaena flos-aquae*, *Microcystis aeruginosa*, and *Phormidium autumnale* produced hydroxamate chelators in iron-deficient media.

Conditioned seawater, which was removed by filtration from an exponentially growing mixed phytoplankton population, inhibited growth of a small inoculum of cells from the same source without reducing the lag period (Huntsman and Barber, 1975). When the two populations, one in exponential growth and the other freshly inoculated, were separated by a filter membrane allowing passage of excreted compounds, the growth rate of the freshly inoculated cells was depressed, but the lag phase was reduced also. Thus, both inhibitory and stimulatory compounds appeared to be excreted by the cells during exponential growth, but the stimulatory group, apparently involved with trace metal metabolism, was more easily degraded.

Keating (1977) reported convincing evidence of the direct role of algal allelopathy in algal succession in a eutrophic lake, Linsley Pond in Connecticut. She studied the bloom sequence over a period of 3 years and correlated this with the effects of cell-free filtrates of dominant blue-green algae on both their successors and predecessors. She found an "unbroken correspondence between the effects of heat-labile probiotic and antibiotic filtrates and the rise and fall of bloom populations in situ." Keating used only axenic or unialgal (bacterized) isolates from Linsley Pond in all tests. She also collected water samples from the pond before, during, and after bloom maxima and found that heat-labile effects in these samples correlated well with data obtained from filtrate studies and with the natural sequence in the pond. Keating (1978) pointed out that one of the most significant long-range, successional changes in community dominance as lakes age is the replacement of diatom blooms of mesotropic lakes by the blue-green algal blooms of eutrophic lakes. She decided, therefore, to examine the relationships between blue-green algal populations in Linsley Pond and diatom populations. Diatom bloom populations vaired inversely with the levels of the preceding blue-green algal populations over a 5-year period. Moreover, cell-free filtrates of axenic or bacterized cultures of the dominant blue-gree algae from the pond inhibited the growth of diatoms isolated from the same pond. Lake waters, collected during the blue-green algal blooms, also inhibited diatom growth. Keating concluded, therefore, that blue-green algal dominance of eutrophic lakes is probably due to allelopathy.

Wolfe and Rice (1979) investigated the possible allelopathic interactions of five planktonic green algae isolated from the Cleveland County, Oklahoma, area and one yellow-green alga obtained commercially. Sterile filtrates of axenic cultures were tested in all possible combinations. Numerous instances of significant stimulation or inhibition were observed. Filtrates of *Cosmarium vexatum* significantly inhibited growth of the other five test species in most experiments but stimulated growth of *C. vexatum*. This combination is probably important in the common production of waterblooms by *C. vexatum* in ponds and swamps. *Botrydium becherianum* was significantly inhibited by filtrates of all species including that of itself. Moreover, the *Botrydium* filtrate did not inhibit growth of any other species. *B. vexatum* is largely a terrestrial species in nature even though it grows luxuriantly in a liquid culture medium. The sensitivity of *B. vexatum* to allelopathic substances produced by phytoplankton species, and the stimulatory effect of *B. vexatum* on those species may be important factors restricting it to mainly terrestrial situations.

Two marine diatoms, *Thalassiosira pseudonana* and *Phaeodactylum tricor-*

nutum, were studied singly and in mixed culture (Sharp *et al.*, 1979). *Thalassiosira pseudonana* was capable of a higher growth rate than *P. tricornutum,* but in two-species cultures, *P. tricornutum* dominated in the latter portion of the exponential phase, indicating a strong interference against *T. pseudonana.* The filtrate of *P. tricornutum* caused an initial lag phase and a reduced terminal population density for *T. pseudonana.* Results from two-species continuous cultures caused these researchers to infer that *P. tricornutum* was allelopathic to *T. pseudonana.*

Mason and Gleason (1981) reported that cell-free extract of a freshwater cyanophyte *Scytonema hofmanni* consistently inhibited the growth of other blue-green algae, selected green algae, and a bacterium (*Bacillus brevis*). Growth of two actinomycetes was stimulated.

In an excellent review, Hellebust (1974) stated that the production of extracellular growth-inhibiting and -promoting substances by many algae in culture indicates that such substances may play important roles in the succession of species commonly observed in aquatic ecosystems. He pointed out further that the considerable differences in abilities of samples taken from different water bodies to support growth of selected algal assay organisms support this hypothesis.

Despite the lack of completely desirable confirmatory data from natural bodies of water, the evidence indicates that fluctuations in phytoplankton populations and succession of species with time are controlled, at least in part, by allelopathic interactions.

One of the important milestones in allelopathic research occurred when Gupta and Houdeshell (1976) included a category of allelopathic effects of algae on other algae and allelochemic effects of algae on zooplankton in a differential-difference equations model of a dynamic aquatic ecosystem. There is no doubt in my mind that all meaningful, functional ecological models will eventually have to include a category on allelopathic and other allelochemic effects.

II. ALLELOPATHIC EFFECTS OF ALGAE NOT RELATED DIRECTLY TO ALGAL SUCCESSION

The choice of words in this heading are emphasized because *all* allelopathic effects of algae probably play at least some minor indirect role in their succession.

Mautner *et al.* (1953) reported that a marine red alga, *Rhodomela larix,* was allelopathic to several gram-positive and gram-negative pathogenic bacteria. He suggested that the active compound might be a brominated phenol. Von Glombitza and his colleagues identified many brominated phenolic antibiotics from several red algae (Rhodophyta) (von Glombitza and Stoffelen, 1972; von Glombitza *et al.*, 1974). Fenical (1975) reviewed the halogenated allelopathic compounds produced by various red algae. These include halogenated phenols, halogenated monoterpenes, halogenated sesquiterpenes, and halogenated diterpenes. He also discussed the various kinds of organisms whose growth is known to be affected by the compounds.

Fogg and Boalch (1958) reported that four ether-soluble yellow compounds were found in the culture medium of the brown alga *Ectocarpus confervoides,* which was bacteria-free. The compounds resembled carotenoids but were not identified. Later, Craigie and McLachlan (1964) found that another brown alga (*Fucus vesiculosus*) also released yellow substances into the medium and suggested that the compounds were probably flavonols or catechin-type tannins. These compounds were very inhibitory to the growth of several marine algae.

Conover and Sieburth (1964) observed that branch tips of vigorously growing plants of *Sargassum natans* and *S. fluitans* appeared to be almost devoid of periphytes, epiphytes, and parasites. Subsequent counts substantiated this observation, and extracts of branch tips markedly inhibited a *Vibrio* species isolated from both species of *Sargassum* washed ashore near Miami, Florida.

Ragan and Craigie (1978) identified numerous phenolic compounds in *Fucus vesiculosus* and *Polysiphonia lanosa.* They pointed out that algae from the classes Phaeophyceae, Rhodophyceae, Cyanophyceae, Chlorophyceae, Bacillariophyceae, Xanthophyceae, and Chrysophyceae have yielded some 50 phenolic compounds. Many of those identified are important allelopathic agents.

Bhakuni and Silva (1974) reviewed the chemistry and biological activity of the brominated compounds, the nitrogen heterocyclics, the nitrogen–sulfur heterocyclics, the sterols, the terpenes, the dibutenolides, the sulfated polysaccharides, and the proteins and peptides isolated from marine flora.

Water and phytoplankton samples from the Weddell Sea were assayed for their antibacterial activity, and four of seven seawater samples contained sufficient phytoplankton in volumes of 1 to 500 ml to inhibit bacterial growth (Sieburth, 1959). The antibacterial substance was active in fresh preparations, stable at 60°C for 30 minutes labile when heat dried at 85°C for 40 minutes, water soluble, and filterable.

Burkholder *et al.* (1960) tested extracts of 150 species of marine algae against three known bacterial species, a fungus, and several isolates of marine bacteria. Sixty-six of the algal species showed some degree of activity on the growth of one or more of the test microorganisms. The more active species included *Amansia multifida, Chondria littoralis, Dictyopteris justii, Falkenbergia hillebrandii, Goniaulax tamarensis, Laurencia obtusa, Murrayella periclados,* and several species of *Wrangelia.* The dinoflagellate *G. tamarensis* showed remarkable activity against *Staphylococcus aureus* and *Candida albicans.*

An antibacterial substance was isolated from the mucilaginous colonial chrysophyte *Phaeocystis* (probably *P. poucheti*) and identified as acrylic acid

(Sieburth, 1960). It was tested against nine species of bacteria and three fungal species and inhibited growth of all of them. Guillard and Hellebust (1971) reported that *Phaeocystis poucheti* released up to 7 μg/liter of acrylic acid during a bloom. The chrysophyte, *Ochromonas malhamensis,* produces two compounds that are antibiotic to several bacterial species (Hansen, 1973). One of the compounds increased in activity on exposure to visible light and after being boiled. The other was not affected by light but had enhanced activity after being boiled for 5 minutes. Filtrates from cultures of *Anabaena cylindrica, Chlorella pyrenoidosa, C. vulgaris, Nitzschia palea,* and *Scenedesmus quadricauda* stimulated or inhibited the growth of *Staphylococcus aureus* in many tests (Jørgensen and Steemann-Nielsen, 1961). Both growth-accelerating and growth-inhibiting substances were present simultaneously in the culture solution.

Chlorophyllides separated by paper chromatography from ether and ethanol extracts of either *Chlorella vulgaris, Chlamydomonas reinhardii,* or *Scenedesmus quadricauda* inhibited the growth of *Bacillus subtilis* after the chlorophyllides were transformed by light (Jørgensen, 1962). Another unidentified derivative of chlorophylls also inhibited *B. subtilis* only after illumination. A third substance, unrelated to plant pigments, was inhibitory only after illumination in extracts of *Chlamydomonas,* whereas, in extracts from *Chlorella* and *Scenedesmus,* it was inhibitory in both light and darkness.

Katayama (1962) summarized the volatile constituents identified in several green, brown, and red marine algae. He separated these into sulfur compounds, acids, aldehydes, alcohols, terpenes, phenols, and hydrocarbons. The antibacterial activity of the fatty acids, aldehydes, and terpenes was determined against *Staphylococcus aureus* and *Escherichia coli.* All fractions inhibited both bacteria in at least some concentrations.

Aubert *et al.* (1970) demonstrated that a diatom, *Asterionella japonica,* produced antibiotics effective against bacteria.

In short term experiments using cultures of *Chlorella pyrenoidosa, Anabaena flos-aquae,* and *Asterionella formosa,* Nalewajko and Lean (1972) found that the percentage of photosynthetic products released from the cells increased with increasing growth rates. In similar tests with *Navicula pelliculosa,* however, the reverse was true. Larger molecular weight compounds were released from cells as the cultures aged. On the other hand, in several lakes a preponderance of large molecular weight compounds was apparent in filtrates even in short-term experiments. Filtrates of cultures of planktonic bacteria growing on [^{14}C]glycolate were found to contain large molecular weight organic compounds. Subsequently, it was demonstrated that in both nonaxenic cultures of algae and in lake water, bacteria utilize low molecular weight extracellular metabolites of algal origin and form larger molecular weight compounds. Such chemical interactions can result in compounds inhibitory or stimulatory to both algae and bacteria.

The antibacterial activity of sea water has been known for about 100 years (Moebus, 1972a). It has been shown that bacteria carried to the sea by rivers become rapidly eliminated, but there has been no commonly accepted reason for this decline. The antibacterial activity of North Sea water was investigated by Moebus over a 2-year period, using *Escherichia coli, Serratia marinorubra,* and *Staphylococcus aureus* as test organisms. Seasonal changes in inactivation of test cells and multiplication of marine bacteria were correlated with the life cycles of several diatom species. A breakdown of phytoplankton blooms produced the most pronounced influence on antibacterial activity of sea water. Sometimes, however, enhancement of kill rate could be established during growth periods of various algal species. Moebus concluded that both production of harmful compounds by phytoplankton and food consumption by marine bacteria cause the antibacterial activity of raw sea water.

Subsequent work by Moebus (1972b) in which low amounts of organic matter were added to natural and synthetic seawater, resulting in increased antibacterial activity, caused him to lean toward a nutrient-dependent hypothesis as the chief cause of the antibacterial activity of seawater.

Bioko (1973) reported that water extracts of *Protococcus viridis* and *Trentepohlia umbrina* inhibited radish seed germination, whereas a water extract of *Hormidium nitens* stimulated germination. He found that water extracts and volatile excretions of 5 of 10 species of lichens tested inhibited radish seed germination.

Two strains of the marine bacterial species *Vibrio anguillarum* enhanced growth of all, or most, of 10 species of phytoplankton algae on an enriched agar medium (Ukeles and Bishop, 1975). *Escherichia coli* had little effect on growth of the algae, however. *V. anguillarum* did not stimulate growth of the phytoplankters in an artificial seawater medium, and subsequent studies suggested that the stimulating effect on agar was due to the release of growth substances through bacterial hydrolysis of the agar. Ukeles and Bishop pointed out that there have been many reports of stimulating effects of marine bacteria on algal growth. One role could be the release of required nutrients or growth factors from bound forms by the bacteria. However, as these researchers pointed out, under natural conditions an intricate relationship must exist between bacteria and algae, which is poorly understood at present and needs considerable study.

7

Natural Ecosystems: Allelopathy and Old-Field or Urban Succession

I. OLD-FIELD SUCCESSION IN OKLAHOMA

Plant succession and its causes have been topics of interest to ecologists for many years. Most investigators have attributed the causes of succession to changes in physical factors in the habitat, to availability of essential minerals, to differences in seed dispersal and seed production, or to competition or combinations of these. Cowles (1911) was very energetic in his attempts to explain the causes of plant succession, and he emphasized the production of toxins by plants as a possible important factor in succession. Numerous investigators have attributed succession of phytoplankton primarily to allelopathy as indicated in Chapter 6.

Many of my students and I became interested in various types of changes associated with plant succession in revegetating old fields in Oklahoma. These fields have been abandoned from cultivation because they are no longer profitable for farming because of low fertility. Moreover, we were interested in the causes of succession.

Booth (1941a) found that succession in abandoned fields in central Oklahoma and southeast Kansas included four stages: (1) pioneer weed, (2) annual grass, (3) perennial bunch grass, and (4) true prairie. The weed stage lasted only 2 to 3 years. The annual grass stage lasted from 9 to 13 years and was dominated by prairie threeawn, *Aristida oligantha*. The perennial bunch grass stage was dominated by little bluestem, and this stage was still present 30 years after abandonment. This was the oldest abandoned field studied by Booth so he was not able to ascertain how long a period is required for the return of the true prairie, which in

206

central Oklahoma is dominated by little bluestem, big bluestem, switchgrass, and Indiangrass.

In a study in the southern Great Plains, Savage and Runyon (1937) found that none of the abandoned fields possessed a cover comparable in composition with the climax of the region, even after 40 years. Tomanek *et al.* (1955) reported that the climax composition had not been attained in a field in central Kansas that had been abandoned for 33 years. Some fields abandoned over 30 years still have almost a pure stand of *Aristida oligantha* (Rice, 1976).

Two problems concerning old-field succession have been of major concern: (1) Why the weed stage is replaced so rapidly by a small depauperate species like *Aristida oligantha* when the weed stage is dominated by such robust plants as *Helianthus annuus, Ambrosia psilostachya, Erigeron canadensis, Chenopodium album, Sorghum halepense, Digitaria sanguinalis, Bromus japonicus*, etc. and (2) why the annual grass (*Aristida oligantha*) and perennial bunch grass stages remain so long before the true prairie returns.

A. Allelopathy and the Rapid Disappearance of the Pioneer Weed Stage

A commonly generalized explanation for the successional changes is that each stage increases the supply of minerals and organic matter and improves soil structure and water relationships thus making the listed conditions more conducive to the incoming than the outgoing species. Such a generalized story cannot explain why the first stage is replaced so rapidly by *Aristida oligantha*, because there is considerable evidence that *A. oligantha* thrives under conditions of low fertility and low water supply, which would not support most of the species in the pioneer weed stage (Rice, 1971a). Preliminary studies indicated that some pioneer weed species from stage 1 produce substances that inhibit the growth of their own seedlings and seedlings of other species in that stage, but not seedlings of *A. oligantha*. I hypothesized, therefore, that several species of stage 1 eliminate species of that stage by the production of toxins and that *A. oligantha* invades after that because its growth is not inhibited by the substances that are toxic to the pioneer species. Moreover, it grows in soil which is still so infertile it will not support species that come in still later in succession.

One of the earliest species investigated was Johnsongrass, *Sorghum halepense*. When Johnsongrass is prominent in a cultivated field before abandonment, it remains important in the early stages of succession. It often occurs in almost pure stands for protracted periods, suggesting that perhaps something more than an excellent ability to compete for light, minerals, and water might be involved. Preliminary work indicated that extracts of the rhizomes of Johnsongrass and the soil in the rhizosphere of that species inhibited growth of the primary root of rice seedlings. (Abdul-Wahab, 1964). More comprehensive experiments were performed subsequently to determine the ability of Johnsongrass to inhibit growth of several species of plants with which it is associated in abandoned fields (Abdul-Wahab and Rice, 1967).

The amount of plant material that would be added to soil by the death of Johnsongrass at the time of its peak standing crop in an old field was determined. The air-dry weight was found to be 3.65 tons of leaves and culms per acre and 2.4 tons of rhizomes per acre to a depth of 17 cm (the depth of plowing). This amounted to 1.85 g of leaves and culms per 454 g of soil and to 1.2 g of rhizomes per 454 g of soil to the depth of plowing. These values were used in determining the effects of decaying Johnsongrass residues on seed germination and seedling growth of Johnsongrass and six other weed species often associated with it.

Decaying Johnsongrass residues inhibited seed germination of most of the seven species. Both decaying rhizomes and leaves markedly inhibited seed germination of *Amaranthus retroflexus* and *Sorghum halepense*. Decaying rhizomes inhibited seed germination more than decaying leaves. Both decaying rhizomes and leaves inhibited seedling growth of all species except *Aristida oligantha*, the dominant of the second successional stage (Table 41). Even growth of this species was reduced by residue that had decayed for 6 months. Decaying leaves inhibited seedling growth of most of the test species more than decaying rhizomes did. Johnsongrass residues that had decayed for 6 months generally inhibited seedling growth more than did recently added material.

To determine the effects of root and rhizome exudates of Johnsongrass on seedling growth, roots of 14-day-old seedlings of *Amaranthus retroflexus, Setaria viridis,* and *Bromus japonicus* were placed in glass vials through which culture solution was circulated. The vials for the control plants were connected to a pot containing just quartz sand, and the test vials were connected to a pot containing quartz sand in which Johnsongrass was growing. A mineral solution was allowed to drip from a supply reservoir into the pot containing Johnsongrass or into the control pot. These solutions were allowed to pass through the vials by gravitational force and then into collecting reservoirs. The solutions were then pumped back into the supply reservoirs so the cycle could continue. The exudate reduced growth of *Amaranthus retroflexus* and *Setaria viridis,* but not the growth of *Bromus japonicus.*

Hence, Johnsongrass produces a toxin or toxins that exude from living roots and rhizomes and are released by decay of tops and rhizomes. These toxins inhibit seed germination and seedling growth of several pioneer species of weeds from revegetating old fields. However, they are generally less toxic to *Aristida oligantha,* the dominant of the second stage of succession. Chlorogenic acid, *p*-coumaric acid, and *p*-hydroxybenzaldehyde were the main plant inhibitors present in the leaf and rhizome extracts.

The common sunflower, *Helianthus annuus,* is often the most prominent plant

TABLE 41. Effect of Decaying Johnsongrass Material on Growth of Species of Plants Often Associated with Johnsongrass[a]

		Oven-dry weight (mg)		
Plant name	Experiment number	Control	Rhizome	Leaf
Amaranthus retroflexus	1	86.2	8.6[c,d]	51.7[c,d]
	2	85.0	7.1[c,d]	48.0[c,d]
	3[b]	250.5	110.0[c,d]	178.5[c,d]
Aristida oligantha	1	13.7	12.6[d]	7.4[c,d]
	2	14.0	12.0	8.5
	3[b]	50.5	23.0[c,d]	37.0[c,d]
Bromus japonicus	1	13.9	8.5[c]	7.8[c]
	2	19.3	10.3[c]	10.0[c]
	3[b]	55.5	31.0[c]	37.5[c]
Bromus tectorum	1	10.7	9.0[c,d]	7.3[c,d]
	2	22.5	11.5[c,d]	16.0[c,d]
	3[b]	72.0	43.5[c]	40.5[c]
Digitaria sanguinalis	1	58.3	24.4[c,d]	6.7[c,d]
	2	60.0	23.0[c]	6.9[c]
	3[b]	482.0	190.0[c]	277.0[c]
Setaria viridis	1	42.0	6.9[c]	6.1[c]
	2	74.7	8.5[c]	41.2[c]
	3[b]	328.0	124.0[c,d]	261.5[c,d]
Sorghum halepense	1	70.1	46.2[c,d]	16.0[c,d]
	2	195.0	70.5[c]	66.1[c]
	3[b]	272.0	138.0[c]	173.0[c]

[a] Modified from Abdul-Wahab and Rice (1967).
[b] Plant material mixed with soil and left to be decayed for 6 months.
[c] Dry weight significantly different from the control.
[d] Significant difference between rhizome and leaf treatments.

in the initial weed stage of many revegetating old fields. It strongly influences the patterning of the vegetation in such areas due to its strong allelopathic effects (see Chapter 5, Section II). It also appears to have a marked influence on succession. Soil near sunflower, decaying leaves of sunflower, leachate of leaves, and root exudate inhibited growth of seedlings of several pioneer species but did not inhibit *Aristida oligantha* in any test (see Tables 33 to 36, Chapter 5).

Digitaria sanguinalis is often a very important species in old fields during the first year after abandonment. Initial experiments using aqueous extracts of various organs of crabgrass demonstrated that this species is quite toxic to several associated species from old fields. A project was undertaken, therefore, to determine the allelopathic potential of crabgrass (Parenti and Rice, 1969).

Field measurements indicated that thick stands of crabgrass in the field averaged over 1 g of air-dry weight of tops and roots per 454 g of soil to the depth of plowing. Seeds of crabgrass and seeds of four other species usually associated with it in old fields were planted in soil containing 1 g of air-dried whole plant material per 454 g of soil or 1 g of washed air-dried peat moss per 454 g of soil for controls. Decaying crabgrass did not inhibit seed germination or seedling growth of any species.

In a test against four pioneer species and *Aristida oligantha,* root exudate of crabgrass inhibited seedling growth of all except one species (Table 42). Even *Aristida oligantha* was inhibited by the root exudate of crabgrass. Chlorogenic, isochlorogenic, and sulfosalicylic acids were identified in whole plant extracts. Sulfosalicylic acid was found only in fresh extracts, however.

Crabgrass is a very important species during the first year of the pioneer weed stage, and the pronounced inhibition of seedlings of several important species of that stage by root exudate of crabgrass probably helps eliminate some of the pioneer species. This may be true particularly in the case of *Amaranthus retroflexus.* This species is markedly inhibited by crabgrass as well as by *Helianthus annuus* and *Sorghum halepense,* and it disappears from old fields during the first year. Crabgrass is also one of the first species of the weed stage to be lost, and this is probably due to the pronounced sensitivity of its seedlings to inhibitors produced by other important pioneer species (Abdul-Wahab and Rice, 1967; Wilson and Rice, 1968).

The failure of *Aristida oligantha* to invade old fields immediately after abandonment has always been curious, because the seeds are widely dispersed (Rice *et al.,* 1960) and the plants are able to grow well in infertile soil. Possibly

TABLE 42.   Effects of Crabgrass Root Exudate on 6-Week-Old Plants[a]

Plant name	Experiment number	Mean oven-dry weight (g ± standard error)	
		Control	Test
Amaranthus retroflexus	1	9.49 ± 0.96	5.45 ± 0.56[b]
	2	9.45 ± 0.84	5.32 ± 0.51[b]
Ambrosia artemisiifolia	1	4.98 ± 0.50	4.32 ± 0.62
	2	4.83 ± 0.27	4.33 ± 0.45
Aristida oligantha	1	4.26 ± 0.30	2.18 ± 0.25[b]
	2	4.32 ± 0.30	2.35 ± 0.32[b]
Bromus japonicus	1	11.51 ± 0.79	7.80 ± 1.21[b]
	2	11.57 ± 0.75	8.00 ± 1.00[b]
Helianthus annuus	1	8.05 ± 0.75	4.05 ± 0.74[b]
	2	8.23 ± 0.61	3.95 ± 0.52[b]

[a] Modified from Parenti and Rice (1969).
[b] Dry weight significantly different from the control.

crabgrass keeps it out, because root exudate of *Digitaria* strongly inhibited seedling growth of *A. oligantha*. This was unlike the situation with other pioneer species, because they inhibited the growth of species of stage 1, and not *A. oligantha*, in most tests. The rapid disappearance of crabgrass from the weed stage would prevent it from having any effect on the invasion of *A. oligantha* 2 or 3 years after abandonment.

Western ragweed, a characteristic species found in the first stage of old-field succession, persists through the later stages with less cover. Neill and Rice (1971) studied chemical interactions between western ragweed and numerous associated species in relation to patterning of vegetation (see Chapter 5, Section II) and to succession.

Soil that was collected adjacent to western ragweed plants in the field in July either stimulated growth of associated pioneer species from stage 1 of succession or had no effect. On the other hand, soil collected similarly in January inhibited seedling growth of the same species or had no effect. Growth of *Aristida oligantha* was not affected by soil collected at either date.

Seed germination of two species and seedling growth of three species (six stage 1 species tested) were inhibited by decaying leaves of western ragweed, if the leaves had over-wintered on the parent plant. Growth of *Aristida oligantha* was inhibited also.

Leaf leachate of western ragweed inhibited the growth of some pioneer species when it was used to water seedlings growing in soil. One of six stage 1 species was stimulated by the leachate, as was *Aristida oligantha*.

Growth of five of the six stage 1 species was inhibited by root exudate of western ragweed. Growth of *Aristida oligantha* was stimulated in one experiment and reduced in another.

Western ragweed probably helps eliminate *Amaranthus retroflexus*, *Bromus japonicus*, *Erigeron canadensis*, and *Haplopappus cilliatus* from the first stage of old-field succession. On the other hand, growth of *Aristida oligantha*, the dominant of the second successional stage, is not affected by western ragweed or is stimulated.

Euphorbia supina is sometimes a common species in revegetating old fields (Drew, 1942). Often it is prominent only in limited parts of such old fields. Brown (1968) found that the root exudate of *E. supina* inhibited seedling growth of *Erigeron canadensis* but not that of *Bromus japonicus* and *Helianthus annuus*. It also slightly inhibited growth of *Aristida oligantha*.

Decaying whole plant material inhibited the seedling growth of *Bromus tectorum*, whereas it stimulated the growth of *Bromus japonicus*. Growth of *Aristida oligantha* was not affected.

It appears that *Euphorbia supina* exerts sufficient allelopathic activity against some pioneer weeds to help eliminate them from stage 1. The inhibitory effect against *Aristida oligantha* is slight enough so that it would not retard the entry of

A. oligantha into the old fields after most of the pioneer species have been eliminated.

Another species of *Euphorbia*, which is often prominent in the pioneer weed stage of some old fields, *E. corollata*, produces gallic acid and tannic acid in large amounts (Rice, 1965a), and Olmsted and Rice (1970) found that both of these compounds were inhibitory to growth of species from stage 1 but not to that of *Aristida oligantha*.

Several phenolic compounds were identified as the plant inhibitors that are produced by the allelopathic species, Johnsongrass, sunflower, crabgrass, *Euphorbia corollata*, and *Euphorbia supina* (Rice, 1965ab, 1969; Abdul-Wahab and Rice, 1967; Wilson and Rice, 1968; Parenti and Rice, 1969). Several of the pure compounds were tested against growth of species from the first stage of succession and *Aristida oligantha* (Floyd and Rice, 1967; Olmsted and Rice, 1970).

Five concentrations of six phenolic inhibitors were selected for testing against *Amaranthus retroflexus* and *Bromus japonicus* from stage 1 and against *Aristida oligantha* (Olmsted and Rice, 1970). The six inhibitors involved were chlorogenic acid, *p*-coumaric acid, gallic acid, *p*-hydroxybenzaldehyde, isochlorogenic acid, and tannic acid. At least one concentration of each phenolic inhibited growth of both species from stage 1, except for *p*-coumaric acid, which only reduced growth of *Amaranthus retroflexus*. The highest concentration of *p*-coumaric acid slightly inhibited growth of *Bromus japonicus*, but the reduction was not statistically significant. No toxin inhibited growth of *Aristida oligantha* in the concentrations tested. The only statistically significant effect on *A. oligantha* was a stimulation of growth by one concentration of chlorogenic acid (Table 43). The results with chlorogenic acid were representative of the kinds of results obtained with other inhibitors.

The evidence indicates that several important species in the first stage of old-field succession in central Oklahoma produce toxins that inhibit the growth of several other species from that stage and sometimes themselves. However, the toxins do not inhibit growth of *Aristida oligantha* in most tests. Moreover, the evidence is strongly supported by tests with pure compounds of the identified inhibitors. The evidence supports the hypothesis, therefore, that species of stage 1 are eliminated rapidly because of toxins produced by several of them that are inhibitory to species of stage 1 but not to *Aristida oligantha*, the dominant of stage 2. Moreover, *A. oligantha* grows well and reproduces in soil that is still so infertile that it will not support species that come in later in succession.

B. Allelopathy and the Slowing of Succession Starting with Stage 2

The slow invasion of abandoned fields by climax grasses, even when such species completely surround the fields, has always been puzzling. Rice *et al.*

TABLE 43. Effect of Different Concentrations of Chlorogenic Acid on Seedling Growth[a]

Species	Experiment	Control	Mean oven-dry weight (mg)					
			0.83×10^{-7} M	0.83×10^{-6} M	0.83×10^{-5} M	0.83×10^{-4} M	0.83×10^{-3} M
Amaranthus retroflexus	1	90.1	76.6[b]	78.0[b]	77.5[b]	75.7[b]	41.3[c]
	2	78.8	68.4[b]	67.3[b]	67.2[b]	64.7[b]	30.9[c]
Aristida oligantha	1	31.2	30.4	29.8	30.2	35.8[c]	31.6
	2	32.9	33.2	32.0	32.0	37.9[c]	32.1
Bromus japonicus	1	28.4	25.6	25.9	24.4	28.9	16.6[c]
	2	28.5	25.9	24.2	24.4	30.6	16.2[c]

[a] Modified from Olmsted and Rice (1970).
[b] Differs from respective control mean at 5% level or better.
[c] Differs from all other means in respective series at 5% level or better.

(1960) found that viable fruits of one of the climax species in central Oklahoma, little bluestem, are not commonly dispersed much over 6 ft from the parent plant. Even this movement, however, would accomplish a much faster invasion of old fields than occurs. Apparently ecesis fails to occur often even when the fruits of the climax species are present.

Investigations by Daniel and Langham (1936), Finell (1933), Chaffin (no date), and Harper (1932) indicated that old fields in Oklahoma were generally low in nitrogen and phosphorus. Rice et al. (1960) studied the nitrogen and phosphorus requirements of three species, which come in at different stages of succession in revegetating old fields starting with stage 2. They found the apparent order of the three species based on increasing requirements for nitrogen and phosphorus to be as follows: (1) prairie threeawn, (2) little bluestem, and (3) switchgrass. This is the order in which the three species invade abandoned fields. It appears, therefore, that the relative requirements for nitrogen and phosphorus are of considerable importance in determining the order of establishment of various species of plants in abandoned fields in central Oklahoma. If so, any factors that would regulate the rate of formation or accumulation of available nitrogen or phosphorus in such an area would affect the rate of succession.

Sources of gain of nitrogen in an abandoned field, which is not fertilized, include nitrogen fixation by lightning; by blue-green algae; by free-living bacteria, such as *Azotobacter, Clostridium,* and *Enterobacter;* and by symbiosis by *Rhizobium.* Organic nitrogenous compounds must be continuously decomposed to ammonium nitrogen because available nitrogen in the soil at any time is rarely sufficient to support the vegetation throughout a growing season. Many types of microorganisms are involved in ammonification, however, so the only significant limiting factors in this process would probably be the amount of organic nitrogenous compounds present. The rate of nitrogen fixation regulates the rate of increase of organic nitrogenous compounds in soils low in total nitrogen and thus regulates the amount of available nitrogen. Factors affecting the survival or metabolism of any of the nitrogen-fixing organisms would probably affect competition between plants with different nitrogen requirements, therefore, and thus the rate of succession in infertile old fields.

Several investigators demonstrated that certain microorganisms in the soil inhibited *Azotobacter* and *Rhizobium* (Konishi, 1931; Iuzhina, 1958; Van der Merwe et al., 1967). Others found that seeds and other plant parts inhibited *Rhizobium* (Thorne and Brown, 1937; Bowen, 1961; Fottrell et al., 1964), and Elkan (1961) reported that a nonnodulating soybean strain decreased the number of nodules produced on its normally nodulating near-isogenic sister strain when inoculated plants of the two types were grown together in nutrient solution. Beggs (1964) reported a practical problem resulting from the failure of oversown, inoculated white clover to nodulate properly and become established in large areas of *Nasella* and *Danthonia* grasslands in New Zealand. Many types of fertilization were tried, including the addition of all likely trace elements, with no improvement in nodulation and establishment. Often no nodules formed at all, and those which did appear to be ineffective as suppliers of nitrogen. Treatment of test plots with formalin or the turf killer sodium dichlorpropionate, or trichloroacetate, resulted in excellent nodulation and establishment of the white clover. Beggs concluded that the failure of clover to nodulate in untreated areas was caused by some growth-inhibitory factor or factors in the soil, probably produced by certain soil microflora. He pointed out, however, that it is very difficult to separate turf-killing effects from possible control of growth inhibiting organisms such as fungi. He emphasized that fertilizer experiments, in areas where moisture was abundant, indicated that competition between higher plants in the traditional sense was not involved. The evidence suggested that the failure of white clover to nodulate in the untreated grasslands might have been caused by inhibitors produced by the grasses.

Available evidence caused me to hypothesize that the low nitrogen-requiring early plant invaders of abandoned fields might inhibit growth of the nitrogen-fixing bacteria and blue-green algae. This would give such plants a selective advantage in competition over plants with higher nitrogen requirements and could conceivably slow down plant succession.

1. Inhibition of Nitrogen-Fixing Bacteria.   Extracts of 24 species of plants that are important in revegetating old fields were tested for growth inhibition of one strain of *Azotobacter chroococcum,* one of *A. vinelandii,* one of *Rhizobium leguminosarum,* and one of *Rhizobium* sp. (Rice 1964, 1965b,c). Extracts of at least one organ of each of 12 species inhibited growth of most of the test bacteria (Rice, 1964, 1965b,c). The inhibitors were relatively stable against autoxidation and decomposition by microorganisms (Rice, 1964). Extracts of some plants reduced nodulation of inoculated legumes in soil, and living plants of some species reduced nodulation of inoculated legumes when growing with the legumes in sand culture (Rice, 1964). *Ambrosia psilostachya, Aristida oligantha, Bromus japonicus, Digitaria sanguinalis, Euphorbia supina,* and *Helianthus annuus* were among the species that strongly inhibited the nitrogen-fixing bacteria, and these species are among the most prominent ones in the first two stages of succession in our old fields. These species were investigated further relative to their effects on nodulation and nitrogen-fixing ability of legumes (Rice, 1968, 1971a,b).

Seedlings of *Ambrosia psilostachya* and *Helianthus annuus* were transplanted from a field near Norman, Oklahoma, into prairie loam soil. Red kidney bean seeds were inoculated and planted in pots with ragweed or sunflower; other red kidney bean seeds were planted in separate pots as controls.

The nodule number was reduced greatly on bean plants growing with *H. annuus,* as compared with the number on controls. However, there was no reduction in nodule number on bean plants growing with *A. psilostachya*. Nodules on control plants were bright pink in color, whereas nodules on the test bean plants were small and grayish in color, even when growing with *A. psilostachya*. Moreover, control bean plants were dark green in color, whereas the test bean plants were chlorotic. The average fresh weight of test bean plants was less than that of the controls, and the primary leaves abscised early. It has been recognized for many years that nodules of legumes require hemoglobin to be effective in nitrogen fixation (Alexander, 1977), and test bean plants growing with both *H. annuus* and *A. psilostachya* appeared to be nitrogen deficient. The data indicated that sunflower plants inhibited nodulation of legumes in soil and suggested that both *H. annuus* and *A. psilostachya* interfered with the nitrogen-fixing ability of the nodules.

Tumors occurred at the bases of the hypocotyls of the bean plants growing with *H. annuus* similar to those produced after application of 2,4-dichlorophenoxyacetic acid or other related growth substances. The presence of tumors, along with other symptoms, indicated that a chemical was entering the bean plants as a result of association with the inhibitor plants.

Because bean plants growing with *A. psilostachya* appeared to be nitrogen-deficient, even though there was no change in nodule number, the experiment was repeated. Mulder (1954) reported that inoculated, molybdenum-deficient *Trifolium repens, T. pratense,* and *Medicago sativa* plants had more nodules than control plants, and the nodules were smaller and yellow or brown-gray instead of pink as on the controls. The experiment with *Ambrosia psilostachya* was expanded, therefore, to determine if the change in nodule size and color on bean plants growing with *A. psilostachya* was due to a deficiency of molybdenum or other trace elements resulting from the precence of the inhibitor plant in the same pot with the legume.

Very young seedlings of *A. psilostachya* were transplanted from a field as before into a similar substrate. As soon as the *A. psilostachya* plants were well-established, inoculated red kidney bean seeds were planted in all the pots as previously described. At this time, the controls were divided into three sets. Set A received no further treatment, set B had 1 ppm of molybdenum (based on weight of soil in pot) added in solution, and set C had 2 ml of Hoagland and Arnon's (1950) supplementary trace solution (B, Mn, Zn, Cu, Mo) added per pot. The test pots were also divided into three sets with one set receiving no further treatment, a second set receiving 1 ppm of Mo, and the third set receiving 2 ml per pot of the supplementary trace solution. The bean plants were watered when needed with distilled water and were grown under greenhouse conditions.

There was a highly significant reduction in nodule number in all test sets compared with all control sets (Table 44), but there were no significant differences in nodule numbers between test sets. Nodules in all test sets were minute and gray in color in contrast to the large bright pink nodules on the

TABLE 44. Effects of Living *Ambrosia psilostachya* Plants on Nodulation of Red Kidney Bean Plants Growing with Them in Soil[a]

Treatment	Bean plants (number)	Average fresh weight of bean plants (g ± standard error)	Average nodule number (± standard error)	Average μg heme per plant (± standard error)
Control A (beans only)	30	27.2 ± 0.9	303.4 ± 23.5	224.0 ± 23.9
Control B plus 1 ppm Mo	30	26.7 ± 1.1	412.9 ± 31.0[c]	
Control C plus trace solution	30	25.3 ± 0.9	364.8 ± 19.4[c]	
Ambrosia psilostachya	40	14.0 ± 0.6[b]	223.4 ± 13.6[d]	130.8 ± 14.1[c]
A. psilostachya plus 1 ppm Mo	40	13.1 ± 0.6[b]	218.6 ± 13.6[d]	
A. psilostachya plus trace solution	38	13.7 ± 0.6[b]	236.3 ± 11.6[d]	

[a] From Rice (1968).
[b] Difference from each control mean significant at better than 0.001 level.
[c] Difference from Control A significant at 0.05 level or better.
[d] Difference from each control mean significant at 0.01 level or better.

control plants in all sets. It appears, therefore, that the change in size and color of the nodules on bean plants growing with *A. psilostachya* was not due to a deficiency of molybdenum or other trace elements. According to Bould and Hewitt (1963), 1 ppm of added molybdenum should have been sufficient for adequate plant growth, even if the soil had none in the beginning. The bean plants growing with *A. psilostachya* appeared to be very nitrogen-deficient, and the heme content of the nodules on test plants was reduced (Table 44).

Stewart (1966) and Virtanen *et al.* (1947) reported that the amount of nitrogen fixation is directly proportional to the hemoglobin content of the nodules. Therefore, test plants probably fixed much less nitrogen than the control plants because of their lowered nodule numbers and reduced hemoglobin content. The average fresh weight of the test bean plants in all sets was significantly lower than the average weights of plants in all the control sets (Table 44), which might have been due, at least in part, to a reduction in the amount of nitrogen fixation.

It is possible that inhibition of nodulation of legumes by living inhibitor plants was due to competition for minerals, water, or light, when the inhibitor plants and legumes were grown in the same pots. The stairstep device was used, therefore, to eliminate competition for the above factors and to determine if exudates of species previously found to be inhibitory to *Rhizobium* would affect nodulation of inoculated legumes. The solution used was Hoagland and Arnon's No. 1 solution with one-tenth the usual amount of nitrogen. The solution was replenished daily to maintain the desired water and mineral levels.

The six inhibitor species under study were tested against three species of legumes, red kidney bean, *Lespedeza stipulacea* (Korean lespedeza), and white clover. Red kidney bean was used because it is easily grown in the laboratory, grows rapidly, and nodulates well. Korean lespedeza is the most common legume found in our abandoned fields. White clover is found fairly often in our abandoned fields also, in addition to being a very common lawn plant.

The exudates of all inhibitor species reduced the average nodule number on all three legume species (Table 45). In some instances there was almost a complete elimination of nodulation. Nodules on the test legumes generally lacked the bright pink color of those on controls. In all experiments except two, fresh weight increase of the legumes was reduced by the exudate from inhibitor species. At least part of the poor growth probably was due to the low supply of nitrogen available to those plants that did not nodulate well, because the culture solution used had only one-tenth the usual amount of nitrogen in a complete Hoagland and Arnon's solution. There may possibly have been a direct effect of the exudate on the growth of the legumes in addition to the effect on nodulation.

Based on clip quadrats in field stands of the inhibitor species, it was determined that a mature stand of *H. annuus* or *A. psilostachya* produces more than 1 g air-dry weight of leaves per pound of soil (454 g) to the depth of plowing. The other four species under study were found to produce considerably over 1 g air-dry weight of whole plant material per pound of soil to the same depth. It was decided, therefore, to determine what effect 1 g of air-dried plant material of each species per pound of soil would have on the nodulation of inoculated plants of the same three legumes. Only leaves of *H. annuus* and *A. psilostachya* were used, but entire plants of the other four species were used. One gram of milled peat was added per pound of soil in the controls to keep the organic matter content the same. The legume seeds were inoculated before planting.

Decaying material of all the forbs reduced the nodule number on all three legume species, except one, in all experiments. Moreover, the nodules were often smaller and lacked the bright pink color of the control nodules. The decaying material from the grass species did not reduce the nodule number significantly in any experiment, despite the fact that exudates of all the grass species significantly reduced the nodule numbers on most of the test legumes.

Experiments were initiated next to determine whether nodulation of inoculated legumes growing in soil is affected by leachates (artificial raindrip) from the foliage of any of the six inhibitor species involved in the previous nodulation experiments. The same three legumes were used in this investigation also (Rice, 1971b).

The nodule numbers of red kidney beans and white clover were not reduced by

TABLE 45. Effects of Root Exudates of Inhibitor Species on Nodulation of Legumes[a]

| | Average nodule number (± standard error) | | | | | |
| Inhibitor species | Red kidney bean | | Korean lespedeza | | White clover | |
	Control	Test	Control	Test	Control	Test
Forbs						
Ambrosia psilostachya	180.6 ± 10.0	62.1 ± 3.8[b]	8.1 ± 0.3	3.9 ± 0.2[b]	8.2 ± 0.6	4.3 ± 0.4[b]
Euphorbia supina	198.4 ± 11.2	181.4 ± 16.4	14.0 ± 0.5	6.5 ± 0.5[b]	5.5 ± 0.3	4.2 ± 0.4[b]
Helianthus annuus	298.6 ± 17.4	22.5 ± 2.8[b]	17.0 ± 1.5	2.3 ± 0.6[b]	6.5 ± 0.6	0.4 ± 0.1[b]
Grasses						
Aristida oligantha	214.3 ± 12.3	145.5 ± 7.7[b]	6.4 ± 0.2	5.5 ± 0.3[b]	11.1 ± 0.7	4.3 ± 0.4[b]
Bromus japonicus	129.8 ± 5.1	127.7 ± 5.5	9.9 ± 0.3	6.5 ± 0.3[b]	5.9 ± 0.8	2.3 ± 0.4[b]
Digitaria sanguinalis	174.1 ± 7.7	109.2 ± 6.3[b]	14.1 ± 0.6	8.3 ± 0.5[b]	19.5 ± 1.0	3.5 ± 0.4[b]

[a] Modified from Rice (1968).
[b] Difference from corresponding control significant at the 0.01 level or better.

the leachates of any of the inhibitor species. However, the leachate of *Euphorbia supina* leaves increased the nodule number of bean plants in one experiment. Leachates of all inhibitors species, except *Ambrosia psilostachya* and *Digitaria sanguinalis,* reduced the nodule number of Korean lespedeza in most experiments.

In the two experiments in which the hemoglobin content was quantitatively determined in test and control nodules, *Helianthus annuus* leachate reduced the mean hemoglobin content 36% per plant in Korean lespedeza; and *Euphorbia supina leachate* reduced the mean hemoglobin content 24.3% per plant in white clover. It was notable that in the latter experiment, the nodule number was not significantly affected by the leachate.

It is significant that the leachates inhibited nodulation in Korean lespedeza because it is the most important legume in our revegetating old fields. Even though the nodule numbers of the beans and clover were not reduced by the leachates, the nodules on test plants usually appeared to be smaller and gray in color in contrast to the brighter pink nodules on control plants. There is little doubt, therefore, that many of the test plants fixed less nitrogen than the control plants because of the reduced nodule number, the reduced hemoglobin content, or both. The pH of the soil is known to have a marked effect on the growth of *Rhizobium* and nodulation of legumes (Alexander, 1977). The leachates did not affect soil pH, however, so they had to exert their effects through another mechanism.

Probably the combined effects of leaf leachate, root exudate, and decaying material on symbiotic nitrogen fixation would be greater than the individual effects demonstrated.

All six inhibitor species involved in the nodulation experiments were found to be inhibitory to free-living nitrogen-fixing bacteria in addition to being inhibitory to *Rhizobium* (Rice, 1964, 1965b,c). Leuck and Rice (1976) found that, of 28 rhizosphere bacteria isolated from roots of *A. oligantha,* 5 were antagonistic toward one or more test strains of *Azotobacter* and *Rhizobium,* and 2 stimulated growth of one or more of the nitrogen-fixing test strains. Overall, 14 instances of inhibition and 3 of stimulation were recorded. This allelopathic activity of rhizosphere bacteria of *A. oligantha* against nitrogen-fixing bacteria adds an additional dimension to the direct effect of *Aristida* on biological nitrogen fixation.

All six species are very important in the pioneer stages of old-field succession in Oklahoma and adjacent states. It appears likely, therefore, that these species play a prominent role in reducing the rate of addition of nitrogen to abandoned fields and thus slow the rate of succession.

Gallic and tannic acids are produced by several species of *Euphorbia* (Rice, 1965a,b, 1969) and *Rhus copallina* (Nierenstein, 1934), and these compounds are very inhibitory to *Azotobacter* and *Rhizobium.* Experiments were performed, therefore, to determine whether these two compounds would inhibit nodulation of legumes, if they could be extracted from soil under plants that produce them, and whether resistant strains of *Rhizobium* could be selected that would cause effective nodulation in the presence of the inhibitors (Blum and Rice, 1969). Initial experiments in sand culture demonstrated that concentrations of both compounds as low as 10^{-8} M produced highly significant reductions in nodule numbers on heavily inoculated red kidney bean plants. Subsequent experiments using red kidney bean plants in soil indicated that concentrations of 33 to 400 ppm of tannic acid reduced the mean nodule number. The amount of hemoglobin was reduced in each case also. At least 400 ppm of tannic acid had to be added to the soil before any could be recovered immediately after adding it. In spite of this, the nodule number was reduced even with 33 ppm indicating that tannic acid remains biologically active even after being tightly bound in the soil.

Both gallic and tannic acids were recovered from soil under *Euphorbia supina* and *Rhus copallina* in the field. The amount of tannic acid in the top 5 cm of soil under *Rhus copallina* ranged from 600 to 800 ppm throughout the year, and this inhibitor was found to a depth of 75 cm in the soil, indicating that it is stable enough to remain for long periods of time and to leach downward in the soil. Soil taken from under *R. copallina* during all seasons of the year inhibited nodulation and hemoglobin formation in nodules of bean plants (Blum and Rice, 1969).

Tannic acid resistant strains were obtained by growing *Rhizobium phaseoli* in yeast and soil extract-mannitol broth with gradually increasing concentrations of tannic acid. Resistant and nonresistant types were incubated at 31°C for 10 days in yeast and soil extract-mannitol broth to use in inoculation of bean seeds. Red kidney bean seeds inoculated with each type were planted in control soil and similar soil to which 400 ppm of tannic acid were added. Fresh weight, nodule number, and quantity of hemoglobin were determined 4 weeks after planting. The resistant strain was not as effective in nodulation as the nonresistant strain, but it was as effective in the presence of tannic acid as without it. The resistant strain was considerably less effective also in initiating production of hemoglobin in nodules. Even though the resistant strain was just as effective in the presence of tannic acid as without it, it was no more effective in the soil with tannic acid than the nonresistant strain.

Schwinghamer (1964, 1967) found that strains of *Rhizobium* resistant to antibiotics were less effective in inducing nodulation than the original strains, and he suggested that such reduction in effectiveness was due in part to morphological changes of the bacterial walls. Therefore, even though strains of *Rhizobium* evolve that are resistant to various inhibitors, the failure to be effective in nodulation and nitrogen fixation would probably decrease the amount of nitrogen fixed in old fields, thus slowing succession.

2. Inhibition of Nitrogen-Fixing Blue-Green Algae. The ability of certain blue-green algae to fix nitrogen has been known for many years (Alexander, 1977). Shields and Durrell (1964) reviewed the literature on soil algae and indicated the importance and high frequency of blue-green algae on prairie and desert soils of Oklahoma and the southwestern United States. Booth (1941b) reported that blue-green algae, primarily *Nostoc,* which is an important nitrogen-fixing genus, may form a 32% cover between bunches of grass in the third stage of succession.

Because of the importance of nitrogen in succession in infertile old fields, experiments were performed to determine if seed plants inhibit potential nitrogen-fixing blue-green algae. Eight species of seed plants were selected for study, and these were chosen to represent the various stages of old-field succession (Parks and Rice, 1969). Samples were collected from the top one-half inch of soil from sites in which each species of seed plant occurred in almost a pure stand. The populations of potential nitrogen-fixing algae were very low in cultures made from soil samples taken near *Chenopodium album, Ambrosia psilostachya, Helianthus annuus, Sorghum halepense,* and *Rhus glabra.*

Soil samples were collected next at intervals of 1 ft for a distance of 3 ft on the north and south sides of sunflower plants growing in relatively bare areas of a field plowed a few months previously. *Anabaena, Nostoc,* and *Schizothrix* (species of which are usually nitrogen fixers) increased in prominence in the cultures with increase in distance from the sunflower plants. Using a chlorophyll extraction and quantitation procedure (Parks and Rice, 1969) to measure algal growth, it was found that there was virtually no difference in growth on the north and south sides of the sunflower plants suggesting that shading was not involved. There was an increase in total algal growth, however, with each increase in distance from the stems of sunflower. Thus, the evidence indicated that *Helianthus* inhibited algal growth, particularly of some of the possible nitrogen-fixing blue-green algae.

Plants of the eight test species were collected in June from the same fields in which soil samples were collected. These were dried and finely ground after being separated into roots, leaves, stems, etc. Ten test flasks were prepared for each plant part of each species containing 15 ml of Modified Bristol's Solution (Bold, 1949) and 0.2 g of ground plant material per flask. All flasks were steamed in an autoclave for five minutes at 100°C. This precaution against aerial algal contamination was carried out in all experiments which involved unialgal cultures. A 0.5-ml aliquot of a well-dispersed unialgal culture of *Lyngbya* sp. (Indiana Culture Coll. No. 488) was inoculated into each of the test solutions. Ten control cultures, which contained no plant material, were prepared. The cultures were incubated for 10 days, after which they were harvested and analyzed for growth by the chlorophyll extraction procedure. Blanks for the dried plant material were obtained and subtracted from the total chlorophyll *a* content of the harvested material. Some plant parts stimulated the growth of *Lyngbya,* others inhibited growth, and others had no effect. All test plants except little bluestem, a member of the climax prairie, possessed at least one plant part which inhibited growth of *Lyngbya.*

A similar experiment was conducted involving *Anabaena* sp. (Indiana Culture Coll. No. B380), a possible nitrogen-fixing species. Results were similar to those with *Lyngbya*. At least one plant part of all species, including little bluestem, inhibited the growth of *Anabaena*. The inhibition of this possible nitrogen-fixing alga by the roots of little bluestem was especially interesting, and it may help explain in part why the bunchgrass stage of old-field succession dominated by little bluestem remains for such a long time in revegetating old fields.

Leaf leachate of one species inhibited the growth of *Anabaena* and *Lyngbya*, and root exudates of several species inhibited growth of possible nitrogen-fixing blue-green algae.

Eight phenolic inhibitors produced by plants involved in this study and inhibitory to seed plants and bacteria were tested against pure cultures of *Lyngbya* and *Anabaena*. The eight phenolics were chlorogenic acid, *p*-coumaric acid, gallic acid, *p*-hydroxybenzaldehyde, isochlorogenic acid, α-naphthol, scopoletin, and tannic acid. All phenolic compounds tested inhibited the growth of *Anabaena* and *Lyngbya* in concentrations of 0.66×10^{-3} M. Tannic acid was the only compound inhibitory to *Lyngbya* in a concentration of 0.66×10^{-5} M whereas chlorogenic acid, *p*-coumaric acid, gallic acid, α-naphthol, and tannic acid were significantly inhibitory to the growth of *Anabaena* at this concentration. Tannic acid was inhibitory to *Anabaena* even at the 0.66×10^{-7} M concentration. These experiments indicated that *Anabaena* sp., a possible nitrogen-fixing species, was much more sensitive to some of the known phenolic inhibitors than *Lyngbya* sp., a non-nitrogen-fixing species.

Blum and Rice (1969) found almost 46,000 ppm of tannic acid in duff under *Rhus copallina* in fall samples and 600–800 ppm in the top 5 cm of soil under this species in the field. Moreover, they found that 30 ppm added to soil originally free of tannic acid greatly reduced the nodule number of heavily inoculated bean plants growing in the soil. A 0.66×10^{-5} M solution of tannic acid contains approximately 10 ppm of that compound, and a 0.66×10^{-7} M solution contains about 0.1 ppm of tannic acid. Both of these concentrations greatly inhibited *Anabaena* sp., and a 0.66×10^{-5} M solution markedly inhibited *Lyngbya* sp.

All phenolics tested against *Anabaena* and *Lyngbya* are produced by various test species involved in this experiment and escape from plants in various ways. It is likely, therefore, that variations in field populations of algae near certain plants were caused, at least in part, by phenolic inhibitors in the soil.

Certain weeds from stage 1 of old-field succession and *Aristida oligantha* from stage 2 were inhibitory in several ways to possible nitrogen-fixing soil algae, whereas little bluestem from stage 3 and from the climax was not inhibitory in most tests. These results complement the previous findings of inhibition of nitrogen-fixing bacteria and inhibition of effective nodulation of legumes by many of the same plant species.

Up to this point, the evidence concerning inhibition of nitrogen fixation in the early successional stages was indirect. Kapustka and Rice (1976, 1978a), measured nitrogen-fixation rates in the soil of the pioneer weed stage, the annual grass stage, and the climax prairie using the acetylene reduction technique. The nitrogen-fixation rate was about 4 times higher in the climax than in the pioneer weed stage and about 5 times higher in the climax than in the annual grass stage. Phytotoxins isolated from plants of the early successional stages inhibited growth of N-fixing bacteria isolated from soils of these stages, even in very low concentrations.

Inhibition by pioneer weeds and *Aristida oligantha* of biological nitrogen-fixation helps keep the old-field soil low in nitrogen for a prolonged period. Therefore, those plants that have higher nitrogen requirements are not able to invade and succession is slowed for a lengthy period.

Initially, it was thought that rapid nitrification was necessary for the proper growth of the vegetation and that inhibition of the process by the early invaders would help slow succession. Surprisingly, some climax species inhibited nitrification more than early invaders. Subsequently, Rice and Pancholy (1972, 1973) determined the levels of ammonium nitrogen and nitrate nitrogen and the numbers of the chief nitrifying organisms, *Nitrosomonas* and *Nitrobacter,* in two old-field successional stages and the climax in three ecosystems, the tallgrass prairie, the post oak-blackjack oak forest, and the oak-pine forest in Oklahoma. In all ecosystems, the amount of ammonium nitrogen increased from a low in the first successional stage to a high in the climax, whereas the amount of nitrate decreased from a high in the first successional stage to a low value in the climax (see Chapter 9). Moreover, concentrations of nitrifiers were high in the first successional stage and low in the climax. It was obvious from general soil data that the low rates of nitrification in the climax plots were not due to differences in pH, texture, or amounts of organic matter. It was inferred, therefore, that populations of nitrifiers were reduced by toxins in the climax plots, so that ammonium nitrogen was not oxidized to nitrate as rapidly as in the successional stages. The data therefore supported the hypothesis that many soils under climax vegetation are low in nitrate because of an inhibition of nitrification by climax plant species. A reduction of nitrification would help conserve nitrogen in the ecosystem because the ammonium ion is positively charged and is adsorbed by the negatively charged colloidal micelles in the soil. The nitrate ion is, of course, negatively charged and is readily leached below the depth of rooting or is washed away into streams. Moreover, nitrate has to be reduced back to ammonium nitrogen before it can be used by the plant and this is an energy-requiring process. Thus, inhibition of nitrification would conserve both nitrogen and energy.

Rice and Pancholy (1973) reported that condensed tannins and breakdown products of hydrolyzable tannins may be important in inhibiting nitrification. In 1974, they presented evidence that many phenolic acids, flavonoids, and coumarin derivatives may be important inhibitors of the process also.

Let us now look at the completion of old-field succession to the climax. Eventually, the nitrogen concentration increases to the point where some later species can invade. This probably causes less inhibition of nitrogen fixation and more inhibition of nitrification. The higher rate of nitrogen fixation would speed up the addition of nitrogen and the slowing of nitrification would slow the loss of nitrogen. Thus, climax species would eventually be able to invade. Once the more robust climax species are present in a relatively closed-type of vegetation, competition probably prevents pioneer weeds from reinvading. Rice and Parenti (1978) found, however, that at least one pioneer weed, *Amaranthus palmeri,* was markedly inhibited by decaying prairie litter. Thus, allelopathy may aid in keeping pioneer weeds from reinvading also.

3. Shortening the Time Period of Old-Field Succession in Oklahoma.

The data presented above suggest that one should be able to speed up old-field succession from the *Aristida oligantha* stage by fertilizing with nitrogen or nitrogen and phosphorus. To test this possibility, an area dominated by *A. oligantha* was located in McClain County, Oklahoma (Rice, 1976). The field was abandoned from cultivation for about 31 years, but was still covered primarily with *A. oligantha* (Table 46, before fertilizing). The field was heavily grazed, and this no doubt helped to keep it in the *A. oligantha* stage. It was fenced and plots were located and sampled in October. The following March, three randomly selected plots were kept as controls, three plots were fertilized with different levels of nitrogen as ammonium nitrate, and three were fertilized with different levels of nitrogen and phosphorus, using ammonium phosphate (16-20-0).

The frequency and biomass of *A. oligantha* at the end of the first growing season following fertilization were considerably reduced in the nitrogen plus phosphorus plots and remarkably reduced in the high N + P plot (Table 46). The frequency and biomass of *Bothriochloa saccharoides* were somewhat reduced in these plots also. On the other hand, the frequency and biomass of little bluestem plus *Andropogon ternarius* were considerably increased in the high N + P plot. Little bluestem is the chief dominant in the perennial bunchgrass stage of succession and in our climax prairie. *Andropogon ternarius* is relatively common in the bunchgrass stage and climax prairie in the eastern part of the tallgrass prairie where the soil is quite sandy. It was grouped with little bluestem because of its similar ecological position in succession and similar growth habit and because it was only scattered intermittently in the test plots. Two prominent forbs in the climax prairie, *Ambrosia psilostachya* and *Aster ericoides,* were much increased in biomass in the N + P plots also. High N + P speeded up succession to the point where the vegetation progressed from the *A. oligantha* stage to a late perennial bunchgrass or early prairie stage in one growing season. By the end of the second growing season after fertilization, *A. oligantha* was completely gone from the high N + P plot, and the frequency and biomass of *S. scoparium* plus *A.*

TABLE 46. Effects on Succession of Addition (in March) of N and P to *Aristida oligantha* Plots[a]

Application time and plots	Criterion	Aristida oligantha	Schizachyrium scoparium plus Andropogon ternarius	Ambrosia psilostachya	Aster ericoides
October before application of fertilizer					
All	Biomass (g/m^2)	100.0	4.7	3.9	1.2
	Frequency (%)	100.0	40.0	64.5	42.2
First October after fertilization					
Controls	Biomass	86.8	28.8	13.6	0.8
	Frequency	100.0	60.0	73.3	40.0
N + P	Biomass	16.8	80.8	81.2	8.0
(High)[b]	Frequency	40.0	80.0	100.0	40.0
Second October after fertilization					
Controls	Biomass	76.4	21.6	6.8	2.0
	Frequency	100.0	53.3	73.3	73.3
N + P	Biomass	0.0	110.8	34.8	10.0
(High)[b]	Frequency	0.0	100.0	100.0	60.0

[a] Modified from Rice (1976).
[b] Amounts of N and P added (lb/acre): high—N, 179.2; P$_2$O$_5$, 224.

ternarius were very high. Thus in two growing seasons, succession progressed from the annual grass stage to an excellent prairie sod. The results of these field studies strongly supported the hypotheses discussed previously in this chapter.

Nitrogen fertilization alone stimulated growth of little bluestem and the climax forbs and reduced growth of *A. oligantha* by the end of the second growing season. The results were not as striking, however, as they were with the combined N + P addition.

II. OLD-FIELD SUCCESSION IN AREAS OTHER THAN OKLAHOMA

Argemone mexicana grows in abundance in abandoned lands at Rajkot, India, but is gradually being replaced by *Echinops echinatus* and *Solanum surattense* (Sarma, 1974b). Sarma found that aqueous extracts of shoots of *Echinops* and *Solanum* strongly inhibited seed germination of *A. mexicana*. He suggested, therefore, that allelopathy possibly plays a role in the replacement of *Argemone*.

Gant and Clebsch (1975) pointed out that *Sassafras albidum* is a tree that invades abandoned fields in the early stages of plant succession and, unlike many members of the early seral stages, maintains itself into the mature forest stage. They suspected that allelopathy might be involved in the total interference mechanism, which makes this possible. Field studies revealed that 10 plant species consistently occurred exclusively outside of clump canopies of sassafras, and seven other species predominated beneath the canopy. Sufficient incoming and buried seeds were available to support a richer and more abundant understory flora than was found. β-Pinene, α-phellandrene, eugenol, safrole, citral, and (+)-camphor were isolated within and outside of sassafras stands from leaves, litter, soil, and roots. Germinating seeds watered with aqueous leachates of leaves, litter, and canopy washings showed varying degrees of radicle reduction during germination. Germination on soil disks from beneath sassafras stands was significantly lower in four of the test species. In seeds overwintered in sassafras litter, germination was significantly lower in all test species. They found a positive correlation also between α-phellandrene concentration and the reduction in radicle growth in *Acer negundo* and *Ulmus americana*. Gant and Clebsch pointed out that sassafras has different methods of releasing phytotoxins into the environment at various times of the year. Thus, sassafras has a continuous chemical influence on its surroundings.

Hedge bindweed (*Convolvulus sepium*) is a common component of the initial vegetation occupying abandoned cropland on the Piedmont of New Jersey (Quinn, 1974). Quinn found that the growth of *Danthonia sericea* in replicate plots in a transplant garden was inversely related to the occurrence and abundance of the hedge bindweed. Moreover, the difference in growth developed over a span of time when bindweed was dormant and was not shading *D. sericea* or competing with it for moisture or nutrients. Soil from patches of bindweed and leachates of flats containing bindweed inhibited seed germination and seedling growth of many species often associated with bindweed. Autotoxicity was statistically significant in all controlled environment studies. Quinn concluded that the allelopathic effects of bindweed on associated species could contribute to its detrimental effect as a weed in crop plants and to its dominance for several years following abandonment. Eventually, autotoxic effects override the beneficial effect of its inhibition of other species, and bindweed declines in importance as shading and other competitive stresses increase.

Ragweed (*Ambrosia artemisiifolia*) and wild radish (*Raphanus raphanistrum*) are the chief pioneer species to invade fields in the Piedmont of New Jersey abandoned after spring plowing (Jackson and Willemsen, 1976). They usually remain dominant for only 1 year after which they are replaced by *Aster pilosus* as the dominant species but with a diverse group of subordinate species. Jackson and Willemsen found that ragweed and wild radish failed to become re-established in plots cleared of second stage perennial vegetation dominated by aster, in spite of the large number of seeds of these primary invaders present in the soil. Soil studies indicated that the pattern of succession was not due to mineral or physical properties of the soil. Field soil from the second stage of succession inhibited growth and germination of ragweed and wild radish while soil from the first stage had no effect. Root exudate of ragweed and shoot extracts of ragweed and aster inhibited seed germination and seedling growth of early invaders of abandoned fields. The authors concluded that the vegetational change from the first to the second successional stage may be mediated, at least in part, by allelopathy.

Ambrosia cumanensis, a ragweed that is widely distributed in the semiarid and warm-humid zones of Mexico, is a very important pioneer species in secondary succession in the Tuxtlas region of Veracruz (Anaya and Del Amo, 1978). It exists in very dense and practically pure populations, suggesting that this species may produce substances allelopathic to potential competitors. Numerous tests were run, therefore, to determine if *A. cumanensis* does have allelopathic potential. Seven associated pioneer species and one species from an advanced stage of succession were chosen for the tests. Aqueous extracts of leaves and roots of ragweed inhibited the growth of several test species, and leachates of the leaves inhibited some species and stimulated others. Aqueous extracts of the top 10 cm of soil taken from stands of ragweed in July when the ragweeds were flowering were inhibitory to most test species. Soil extracts taken in December when the ragweeds were vegetative were less inhibitory and, in some cases, stimulatory. Decomposing leaves and roots were highly inhibitory to most test species in nonsterile soil. Residues added to sterile soil stimulated growth of several test species, but inhibited the growth of others. *Ambrosia* was autotoxic in most tests. The authors concluded that the allelopathic effects of ragweed against other species give it an advantage over associated species, and its autotoxic effects help control its own population. Del Amo and Anaya (1978) tested seven ses-quiterpene lactones isolated from A. *cumanensis* against root and shoot growth and seed germination of five test species. All of them inhibited both root and shoot growth of most test species at 250 ppm and most phytotoxins inhibited at least some test species at 100 ppm. Effects on seed germination were variable with pronounced cases of both inhibition and stimulation.

Piqueria trinervia, an herbaceous perennial, is frequently a pioneer of second-ary succession in old fields in the tropical and temperate zones of Mexico (González de la Parra *et al.*, 1981). It contains two monoterpenes, piquerols A and B, and numerous monoterpenes known to be allelochemics. Moreover, this species frequently occurs in pure stands, suggesting the possibility of allelopathic action. Bioassays were carried out against nine plant species, most of which are important in old-field succession in Mexico. Extracts of leaves and roots of *P. trinervia* and solutions of piquerol A and B inhibited seedling growth of all test species. It was demonstrated, moreover, that the results were not due to osmotic effects. Leaf extracts were more inhibitory than root extracts and piquerol A was slightly more inhibitory than piquerol B.

Lodhi (1979b) found that *Salsola kali* invades mine-spoil material in North Dakota in an early successional stage but disappears rapidly. Decomposing resi-due of *Salsola* inhibited growth of selected associated species as well as *Salsola* itself. Five phenolic acids and quercetin were identified as phytotoxins that are produced in leaves of *Salsola*. The autotoxic effects may be at least partially responsible for the rapid disappearance of *S. kali*.

Removal of the aboveground remains of the corn plants at the time of aban-donment of a corn field from cultivation increased the productivity, species richness, and numbers of nonannuals in the 5 years followed abandonment, as compared with control plots from which the corn stover was not removed (Lodhi, 1981). He found that ferulic and *p*-coumaric acids, which were prominent in the corn stover, were more highly concentrated in the soil of the noncleared plots than in the cleared plots for 4 years after abandonment. He attributed the in-creased productivity in the cleared plots, therefore, to the lower concentration of allelochemics. Removal of the corn stover also increased the soil mineralization of Ca, Mg, K, NH_4, and NO_3; and this encouraged species having higher mineral requirements to rapidly invade the cleared field.

A whole-community investigation of allelopathy in an old field in Illinois was undertaken by comparing bioassay results with association patterns in the field (Stowe, 1979). The seven most abundant species in the field were tested against each other by several bioassays, and many cases of significant inhibition were found. When the distribution patterns of species in the field were statistically compared with the results of the bioassays, no significant correlations were found. Stowe concluded ''that the types of allelopathy that were tested by these bioassays were not demonstrably effective under field conditions. . . ." This is, unfortunately, a rather sweeping conclusion based on research in one old field in an area with high precipitation. One has to agree that he showed no significant correlations in his investigation, but that is all that can be concluded.

Dominance-diversity curves are geometric during the first few years of old-field succession in forest (Bazzaz, 1975) and prairie areas (Kapustka and Moleski, 1976), and gradually change to lognormal as more species are added to the community. These authors pointed out that a geometric distribution is indicative of a niche-preemptive situation, which exists when there is a strong dominance by species with allelopathic compounds or other effective interference methods.

III. ALLELOPATHY IN URBAN PLANT SUCCESSION IN JAPAN

Urbanization around large cities in Japan has greatly changed the plant communities because of the creation of bare areas or serious disturbance of natural ecosystems (Numata *et al.*, 1973, 1974, 1975). Weed succession on urban waste land is similar to that in old fields in some parts of the United States, with *Ambrosia artemisiifolia* being the first-year dominant followed by *Solidago altissima* and *Erigeron* spp. for a few years and, next, by *Miscanthus sinensis*. Numata and his colleagues found that *S. altissima* and *Erigeron annuus* both produce polyacetylenic methyl esters (one in *Solidago* and three in *E. annuus*) that inhibit seed germination of *A. artemisiifolia*, *M. sinensis*, and a species of *Tagetes*, and the growth of rice seedlings. The phytotoxin found in *Solidago* was also extracted from soil in a stand of the species, and the concentration present was sufficient to regulate germination and growth of associated species. The polyacetylenic compounds did not inhibit several soil bacteria even in concentrations up to 1000 ppm, although they inhibited seed germination and seedling growth at 5 ppm. The inhibitory action against seedlings could not be decreased by addition of gibberellic acid or indoleacetic acid (IAA). A very dilute solution (5 ppm) of three of the phytotoxins inhibited growth of *A. artemisiifolia* in soil. Because polyacetylenic methyl esters and related compounds are widely distributed in members of the Compositae, six additional such compounds were synthesized and tested against the growth of rice seedlings. All had strong inhibitory effects in concentrations of 10 to 20 ppm.

Numata *et al.* (1975) found that a water extract of underground parts of *Artemisia princeps*, a species present in the *Miscanthus* stage of succession, strongly inhibited germination of *A. artemisiifolia* and growth of rice seedlings. Caffeic acid and methyl caffeate were identified in the aqueous root extracts and found to be extremely inhibitory to growth of rice seedlings. It was suspected, however, that the methyl ester was an artifact of extraction.

Kobayashi *et al.* (1980) found that the roots of *Solidago altissima* contain 250 to 400 ppm of the C_{10}-polyacetylene, *cis*-dehydromatricaria ester (*cis*-DME), previously identified by Numata *et al.* (1973). They found that soil under a stand of the *Solidago* contained 6 ppm of *cis*-DME plus *trans*-DME. Both compounds were found to inhibit growth of rice seedlings. The ratio of the *cis*-DME to *trans*-DME in the soil was 3:2, but no *trans*-DME was found in the roots of *Solidago*. Various environmental factors such as light and pH can isomerize the *cis*-DME to *trans*-DME, however.

Three C_{10}-polyacetylenes were identified from methanol extracts of *Erigeron annuus*, *E. canadensis*, *E. floribundus*, and *E. philadelphicus* (Kobayashi *et al.* (1980). These were the *cis*- and *trans*-matricaria ester and the *cis*-lachnophyllum ester. All were inhibitory to seed germination of *Ambrosia artemisiifolia* and seedling growth of rice at a concentration of 5 ppm or above. The *trans*-matricaria ester was less toxic than the others.

Kobayashi *et al.* concluded that the dominance of *Solidago altissima* and *Erigeron* spp. in the second stage of secondary succession is probably due to their production of acetylenic compounds, which strongly inhibit growth of many other plant species. They suggested also that the relatively short period of occupation by *S. altissima* and *Erigeron* spp. may be a consequence of the accumulation of such polyacetylenes in the soil to the point where they are toxic to these species also.

Although it does not relate directly to allelopathy, it is noteworthy that *cis*-DME is highly toxic to at least some nematodes that are parasitic on plants (Saiki and Yoneda, 1981).

8

Allelopathy and the Prevention of Seed Decay before Germination

I. DIRECT PRODUCTION OF MICROBIAL INHIBITORS BY SEED PLANTS

Probably one of the most critical points in the life cycle of many plants is seed germination. This is certainly true of many annual plants that include most cultivated plants, and it is often true in the initial establishment and spread of many perennial plants. Lieth (1960) emphasized that reproduction from seed is very common even in a perennial grassland. It seems surprising therefore that little research has been done in the past decade in this important area.

Almost every person who has planted seeds has experienced the loss of some due to decomposition by microorganisms. The incidence of decay is influenced by environmental conditions in addition to internal conditions of the seeds. Probably most seeds, which do not germinate rapidly after landing in soil, would be decomposed before germination if they did not contain or produce microbial inhibitors, namely, phytoncides. In fact, this may be the most consistent and important ecological role of allelopathy affecting annual plants and many perennial plants in natural environments. According to Horton and Kraebel (1955), seeds of many annual herbs that are important in the fire cycle in California chaparral lie dormant for as long as 40 to 50 years. Many seeds lie dormant in the soil for several years, even in humid and superhumid areas. It is unlikely that seeds could remain in soil for even 2 years in many areas without being decomposed if they did not contain microbial inhibitors. Such a mechanism may not be as critical in crop plants because they have been selected over a long period of time for rapid seed germination. Such seeds probably germinate rapidly enough generally that decomposition is not possible. Another interesting aspect of seed selection in cultivated plants is that dormancy mechanisms and germination inhibitors have generally been eliminated. There is much clear evidence that many germination inhibitors in seeds and fruits are compounds that are microbial inhibitors also. This evidence will be discussed later in this chapter.

Unfortunately, there is little direct evidence to support my suggestion concerning the prevalence of microbial inhibitors in seeds, but there is much indirect evidence, which will be discussed briefly. Even though direct evidence for the role of allelopathy in the prevention of seed decay is not great, the importance of the phenomenon warrants a separate chapter.

Ferenczy (1956) stated that Maksimov, in his plant physiology text published in 1948 in the U.S.S.R., mentioned that certain seeds in the swollen state are capable of germinating for several years in the soil because they discharge antimicrobial substances into their environment. Evenari (1949) pointed out that seeds of *Brassica oleracea* contain a microbial inhibitor belonging to the mustard oils and that the resistance of crucifers to clubroot disease, caused by *Plasmodiophora brassicae,* is attributed to their content of such oils. He stated further that mustard oils and their vapors are highly toxic to different fungi in concentrations as low as 10 ppm. He reported also that essential oils have been used for a long time as disinfectants because of their pronounced antimicrobial action. Evenari stated that Focke discovered in 1881 that plants producing essential oils are protected against attacks by parasitic fungi. Another group of compounds found in many seeds, and toxic to numerous microorganisms, is the alkaloid group (Evenari, 1949). Many unsaturated lactones are potent antimicrobial compounds, and these are common in seeds. Parasorbic acid found in seeds of *Sorbus aucuparia,* anemonin, protoanemonin, digoxigenin, gitoxigenin, and strophantidin are unsaturated lactones that are strongly antimicrobial (Evenari, 1949). Evenari listed many species and plant parts that were demonstrated to produce or contain seed germination inhibitors. He stated that these toxins were effective against many processes; therefore, it is probable that they affect one or more very basic reactions common to all living organisms.

Many types of phenolic compounds also occur in fruits and seeds, both as aglycones and as glycosides (Feenstra, 1960; Harborne, 1964; Harborne and Simmonds, 1964; Henis *et al.,* 1964; Lane, 1965; Harris and Burns, 1972). These include simple phenols such as catechol in fruits of *Psorospermum* and *Citrus* (Harborne and Simmonds, 1964); glycosides of simple phenols such as phlorin, a glycoside of phloroglucinol in *Citrus* fruits (Harborne and Simmonds, 1964); phenolic acids such as caffeic and chlorogenic acids in the seeds of sunflower (Lane, 1965); and glycosides of phenolic acids such as glucosides of *p*-coumaric and caffeic acids, which occur in seeds of *Linum* (Harborne, 1964). Other phenolics are free flavonoids such as tricin in the seeds of *Orobanche,* quercetin and myricetin in *Trifolium* seeds, casticin in the seeds of *Vitex* (Harborne, 1964), and anthocyanidins, which are widely distributed in fruits (Harborne and Simmonds, 1964); glycosides of flavonoids such as rutin in the seeds of *Brassica*, apigenin in celery seeds, two glycosides of kaempferol in *Solanum* seeds, and numerous flavonoid glycosides in fruits of many species (Harborne, 1964); and tannins, both hydrolyzable and condensed, which are widespread in plants (Bate-Smith and Metcalfe, 1957) and which have been specifically identified in fruits and seeds (Henis *et al.,* 1964; Harris and Burns, 1970, 1972). Varga and Koves (1959) identified several phenolic acids and gallotannins in dried fruits of 24 species of plants. Sycamore fruits contain the coumarin derivatives, scopolin and scopoletin (Al-Naib and Rice, 1971), and these have been shown to inhibit some microorganisms (Parks and Rice, 1969). Many similar data are available, but these examples should document the widespread occurrence of phenolics in seeds and fruits.

Much evidence indicates that phenolic compounds of several types are important in the resistance of plants to infection by fungal, bacterial, and viral diseases (Schaal and Johnson, 1955; Kuć *et al.,* 1956; Cadman, 1959; Byrde *et al.,* 1960; Hughes and Swain, 1960; Farkas and Kiraly, 1962; Gardner and Payne, 1964). The evidence does not related directly to the prevention of seed decay, but it certainly has an important indirect relation. The compounds have to be inhibitory to bacteria and fungi to prevent diseases caused by these organisms; thus, they might also inhibit the growth of organisms involved in seed decay.

Additional indirect evidence also suggests that phenolics present in seeds and fruits might prevent seed decay. Representative compounds of all the types listed above inhibit the growth of several special of *Rhizobium, Nitrobacter,* and *Nitrosomonas* and nitrification in soil (Rice, 1965a,b,c, 1969; Rice and Pancholy, 1972, 1973, 1974). Clearly, this is very indirect evidence because the bacteria involved are not decay organisms. *Rhizobium* is not appreciably affected, however, by several antibiotics commonly used for medicinal purposes (Rice, unpublished) and yet is strongly inhibited by many of the phenolic compounds. Three sensitivity disks of one of the tannins from *Euphorbia supina* virtually prevented the growth of *Rhizobium* over an entire 10-cm Petri dish (Rice, 1969). Moreover, dishes containing three disks of this tannin never had any contaminants on them even though the dishes were opened many times in the laboratory over a period of several months (Rice, unpublished). Control dishes rapidly supported a dense growth of many bacterial and fungal contaminants.

Knudson (1913) found that *Aspergillus niger* and *Penicillium* sp. use tannic acid, a hydrolyzable tannin, as a carbon source. He found only five other species of *Penicillium* and one of *Aspergillus* that grew in the presence of a 5% solution of tannic acid, even with other carbon sources present. No other fungi were found that grew in such a solution at all.

Benoit and his colleagues furnished further evidence for the effectiveness of tannins in preventing decay of organic material (Benoit and Starkey, 1968a,b; Benoit *et al.*, 1968). Benoit *et al.* (1968) found that purified wattle tannin reduced decomposition of whole plant material from rye plants harvested at young, intermediate, and mature stages of growth by approximately 50% in a given time period. Benoit and Starkey (1968b) reported that purified wattle tannin slightly inhibited decomposition of polygalacturonic acid and pectin but markedly inhibited decomposition of a plant hemicellulose preparation and cellulose. In a related project, Benoit and Starkey (1968a) found that wattle tannin reduced activity of polygalacturonase, cellulase, and urease. They concluded that inactivation by tannins of exoenzymes of microorganisms concerned with decomposition of large molecular weight compounds is an important part of the inhibitory effect of tannins on the decomposition of plant residues. This would apply also to the decay of seeds. This could obviously be a very beneficial effect in the case of tannin-containing seeds, but it could be detrimental in decomposition and mineral cycling of high tannin-containing dead organic material. The latter phenomenon is probably an important ecological role of allelopathy, but no research has been done on this effect of tannins in natural ecosystems.

The evidence presented in the past several paragraphs, although often indirect, does support the hypothesis that various types of phenolic compounds, when present in seeds, help prevent seed decay before germination. They probably play a similar role in many plants in natural areas, even when present only in the fruits.

Ferenczy (1956) performed a series of experiments directly concerned with microbial inhibition by intact seeds and fruits. Petri dishes containing a suitable solid medium were inoculated with a suspension of a test bacterium. After the surface of the medium was dry, a selected seed or fruit was half sunk into the culture medium. After 20 hours at 30°C, the dishes were examined for zones of inhibition of the test bacteria. Six different species of bacteria, which are common saprophytes in soil, but some of which are also parasitic on plants, were selected for tests: *Bacillus cereus* var. *mycoides*, *B. megaterium* (Strain 208), *B. subtilis*, *Aerobacter aerogenes*, *Erwinia carotovora*, and *Xanthomonas malvacearum*. Seeds, and sometimes fruits, of 512 species of plants belonging to 88 families were tested against all 6 species of bacteria. Seeds or fruits of 52 species belonging to 19 families were found to contain antibacterial compounds. These compounds were localized in the seed coat and in the external layer of the fruits, except in the species of *Fraxinus* where they occurred only in the embryo. Ferenczy stated that many seeds known to contain bacterial inhibitors, such as those of the Cruciferae, did not give positive results in his tests. Perhaps some chemical or mechanical activity has to occur in such cases before the toxins are released, or perhaps the toxins are produced only after microbial action on tissues of the seeds occurs (Farkas and Kiraly, 1962). Wright (1956) reported that white mustard seed (*Brassica hirta*) produced zones of inhibition on agar inoculated with *Bacillus subtilis* if the seed coats were not removed. In addition, these seeds produced a diffusible antibacterial substance that was not present in the seed coats. Ferenczy (1956) found that some of the seeds and fruits inhibited fungi also; however, no comprehensive tests were made against fungi.

Patrick and Koch (1958) reported that antifungal compounds are sometimes produced during decomposition of some plant residues in soil. If these are produced in the vicinity of seeds, they could presumably prevent seed decay.

Nickell (1960) tabulated all species of vascular plants reported to inhibit any or all of the following: gram-positive bacteria, gram-negative bacteria, fungi, mycobacteria, protozoa, phages, viruses, and yeasts. Yeasts, of course, are fungi but they are often listed separately in tests for antimicrobial substances. He found reports of active species in 157 families. Considering only those species in which the seeds were tested, 50 species in 23 families had antimicrobial activity. Often other plant parts of several species in the same genera were found to have antimicrobial activity, but the seeds were not tested. Unfortunately, this was true of most genera. Probably many species that have antimicrobial activity in parts other than seeds also have similar activity in the seeds.

Bowen (1961) reported that seed coats of sterilized seeds of the legumes *Centrosema pubescens* and *Trifolium subterraneum* inhibited *Rhizobium* and other bacteria. Fottrell *et al.* (1964) reported that myricetin, a flavonol, occurs in legume seeds and is toxic to *Rhizobium*. Work in my laboratory has demonstrated that myricetin is also inhibitory to other bacteria.

Campbell (1964) studied the viability of honey locust, *Gleditsia triacanthos*, seeds in relation to age. He planted 20 seeds each year under similar conditions and checked on germination. When the seeds were 6 years old, four seeds failed to germinate and mold grew profusely on them. Hardly any mold grew on the sixteen seeds that germinated, even though they were in close proximity to the seeds that molded. He found that very little mold grew on the seeds that germinated, for 18 years after collecting them. After that, abundant mold grew on all seeds. It appeared that an antifungal compound was present in viable seeds that slowly decreased in amount with passing time. Dr. O. J. Eigsti (personal communication) observed that viable seeds of watermelon (*Cucurbita*) did not mold, whereas the nonviable ones molded readily. It is possible that in some, or perhaps many species, living embryos produce microbial inhibitors as Ferenczy (1956) suggested for *Fraxinus* spp. In other cases, the seeds might decay before they are able to germinate if microbial inhibitors are not present in the seed coat, whereas those that have such inhibitors germinate successfully.

Henis *et al.* (1964) reported that aqueous extracts of carob (*Ceratonia siliqua*) pods inhibited *Cellvibrio fulvus, Clostridium cellulosolvens, Sporocytophaga myxococcoides,* and *Bacillus subtilis.* The carob pod was rich in condensed tannins.

Smale *et al.* (1964) tested extracts of fruits (and other parts) of 125 species in 47 families for antibacterial activity against *Agrobacterium tumefaciens, Erwinia amylovora, E. carotovora, Pseudomonas syringae, Xanthomonas phaseoli, X. pruni,* and *X. vesicatoria;* and for antifungal activity against *Aspergillus* sp., *Collectotrichum lindemuthianum, Endothia parasitica,* and *Monilia fruticola.* Many of the extracts inhibited one or more of the test microorganisms. Some inhibited only fungi, some inhibited only bacteria, and some both.

Lane (1965) investigated the dormancy mechanism of native *Helianthus annuus* seeds. He suspected initially that some water-soluble inhibitor might be responsible for the dormancy. Therefore he devised an apparatus to leach the fruits and/or seeds. After leaching the seeds, he found that they still retained their dormancy but molded and decayed rapidly when attempts were made to germinate them. On the other hand, if he exposed them to low temperatures for several weeks without any leaching, they did not mold and germinated well. Thus, these seeds contain water-soluble antimicrobial toxins. Leaching of the fruits, with or without the seeds in them, caused them to mold and decay rapidly also when exposed to the conditions used for seed germination.

Harris and Burns (1972) investigated factors affecting preharvest seed molding of grain sorghum. They observed the severity of seed molding of 49 hybrids at two locations in Georgia and rated the severity of molding on a scale of 1 to 5, with 1 showing little evidence of molding and 5 showing obvious grain deterioration. They found that consistent differences occurred between hybrids at the two locations, so correlations were run between tannin content of the seeds and the index of molding. The coefficient of correlation between tannin content and index of molding was -0.89 at one locality and -0.87 at the other. Moreover, 77.5% of the variability among hybrids for seed molding was accounted for by this relationship. Thus, it was concluded that high tannin content of the seed was the dominant inhibiting factor controlling preharvest molding.

Obviously, prevention of preharvest molding or decay of seeds is very important in agriculture and in natural ecosystems in certain climates, and this is a different aspect of prevention of seed decay than was previously emphasized. The importance of tannins in the case of grain sorghum suggests that similar microbial inhibitors are effective in prevention of both preharvest and postharvest seed decay.

Unfortunately, the only recent research concerning the production of microbial inhibitors by seeds or fruits, of which I am aware, was that of Norman Koehn (personal communication, 1980). He bioassayed numerous intact seeds and fruits of both crop and noncrop plants for growth inhibition of many common species of soil bacteria and fungi. He found that some of the seeds and fruits of crop plants and most of those of noncrop plants tested produced microbial inhibitors, at least against bacteria (Norman Koehn, personal communication, 1980). Thus, it appears that this is a fruitful phase for future research of allelopathy.

II. PRODUCTION OF MICROBIAL INHIBITORS IN SEED COATS
BY SOIL MICROORGANISMS

Wright (1956) inoculated seeds of white mustard, wheat (*Triticum*), and/or pea (*Pisum*) with either *Trichoderma viride, Penicillium frequentans, P. gladioli, Streptomyces griseus, S. venezuelae,* or *S. aureofaciens* and planted the inoculated seeds in two types of soil. *Trichoderma viride* produces the antibiotic gliotoxin in liquid culture, *Penicillium frequentans* produces frequentin, *P. gladioli* produces gladiolic acid, *Streptomyces griseus* produces streptomycin, *S. venezuelae* produces chloromycetin, and *S. aureofaciens* produces aureomycin. After a germination period of 6 or 7 days, the seed coats of mustard and pea seeds were removed and extracted, and the whole wheat seeds were extracted. The extracts were tested for antibiotic activity and the antibiotics were identified.

All three types of seeds inoculated with *Trichoderma viride* contained gliotoxin (Table 47). Pea seeds inoculated with *Penicillium frequentans* had antifungal activity and yielded 1 to 1.5 μg frequentin per seed coat. Pea seeds inoculated with *Penicillium gladioli* were found to contain gladiolic acid, but the amount was not determined. The three actinomycetes, *Streptomyces griseus, S. venezuelae,* and *S. aureofaciens,* failed to produce antibiotics when pea seeds were inoculated with them. When uninoculated pea seeds were planted in soil containing a natural strain of *Trichoderma viride,* the seed coats had both antifungal and antibacterial activity and were found to contain gliotoxin.

The production of antibiotics in uninoculated seeds in soil containing antibiotic-producing microorganisms adds still another dimension to the phenomenon of

TABLE 47.   Production of Gliotoxin[a] in Wheat, Mustard, and Pea
Seeds Inoculated with *Trichoderma viride*[b]

Seed	R_f value of active extract	Weight of gliotoxin/seed coat (μg)
Wheat	0.84	0.02–0.03
Mustard	0.85	0.25
Pea	0.82	4.0

[a] R_f value of gliotoxin 0.85.
[b] From Wright (1956). Reproduced by permission of Cambridge University Press.

prevention of seed decay before germination. As previously indicated, Wright (1956) found that mustard seed coats contain natural microbial inhibitors in addition to those produced by associated microorganisms.

III. CONCLUSIONS

The foregoing discussion indicates the important survival value of microbial inhibitors in seeds, and strong evidence also could be presented for the presence and importance of such inhibitors in the outer parts of certain vegetative reproductive structures such as the scales of bulbs and the ''peelings'' of tubers. Undoubtedly, strong evolutionary pressures operate in the selection of seeds and other reproductive structures that contain effective antimicrobial toxins.

9

Allelopathy and the Nitrogen Cycle

I. THE NITROGEN CYCLE AND PHASES KNOWN TO BE AFFECTED BY ALLELOPATHY

The nitrogen cycle will be discussed in a generalized fashion in order to point out which phases have been shown to be affected by allelopathy. The cycle involves the sources of gain of nitrogen, mineralization of organic nitrogen, and losses of nitrogen.

In an unfertilized area, the sources of gain of N in the soil are biological N_2 fixation, nitrogen fixation by lightning, and addition of organic matter. The major steps involved in mineralization of organic nitrogen are aminization, ammonification, and nitrification. The chief losses are plant and animal removal, leaching, volatilization, and erosion.

It is obvious from Molisch's definition that allelopathy could affect only those phases of the nitrogen cycle that involve plants or microorganisms. Thus, in relation to sources of gain of nitrogen, biological N_2 fixation, and, to some extent, the addition of organic matter could be affected. There is considerable evidence in the literature concerning allelopathic effects on biological N_2 fixation (Chapters 2, 3, 7 and Section II of this chapter), but there is virtually no literature directly related to the amounts of organic matter added to soil.

It is doubtful that allelopathy has any appreciable effect on aminization and ammonification because these processes are carried out by a very heterogenous group of soil organisms (Harmsen and Kolenbrander, 1965; Brady, 1974). This is no doubt a very important evolutionary development because the heterogeneity of the organisms involved makes it virtually impossible for most environmental factors (including allelopathy) to completely stop these processes, and they are essential in the cycling of N in ecosystems.

On the other hand, nitrification is apparently carried out chiefly by two genera of bacteria, *Nitrosomonas,* which oxidizes ammonium to nitrite, and *Nitrobacter,* which oxidizes nitrite to nitrate (Brady, 1974; Alexander, 1977). It should not be surprising, therefore, that much evidence exists indicating that allelopathy strongly inhibits nitrification in many ecosystems (Chapter 7 and Section III of this chapter).

In considering the losses of N from soil, it appears likely that allelopathy could affect losses due to plant and animal removal, leaching, and volatilization. Allelopathy has definitely been shown to affect mineral uptake of plants (Chap. 13, V) and productivity (Chapters 2, 3, 5, 6, 7), and thus the loss of minerals due to plant removal by cropping or animal removal would be altered. Unfortunately, no specific data are available concerning this possible effect. There is much evidence that concentrations of nitrate are lower in climax ecosystems than in successional stages or crop lands and that losses of N due to leaching and runoff are therefore lower in climax areas (Section III of this chapter). Moreover, there is much evidence that the low nitrate concentrations in the climax areas are due to inhibition of nitrification by climax plants (Section III).

There is considerable loss of N from some soils due to denitrification resulting in volatilization of the nitrogen (Bracken and Greaves, 1941; Allison, 1955). Denitrifying bacteria are responsible for the reduction of nitrite or nitrate to N_2, NO, or N_2O, and the active species are largely limited to the genera *Pseudomonas, Bacillus,* and *Paracoccus* (Alexander, 1977). It appears likely that allelopathy could affect denitrification directly through an effect on denitrifying bacteria and indirectly through its effect on nitrification, which would determine the amount of substrate available for the denitrifying organisms. Unfortunately, there is no information available on the effects of allelopathy on the denitrifying bacteria.

II. ALLELOPATHIC EFFECTS ON NITROGEN FIXERS AND NITROGEN FIXATION

Nineteen genera of bacteria and 18 of blue-green algae have been verified to fix nitrogen asymbiotically (Alexander, 1977). The bacterial genus *Rhizobium* is responsible for symbiotic N_2 fixation in the nodules of legumes, and the actinomycete *Frankia* is responsible for symbiotic N_2 fixation in the nodules of nonlegumes (Becking, 1970; Alexander, 1977; Callaham *et al.,* 1978). Callaham *et al.* (1978) identified the organism from the nodules of the nonlegume *Comptonia peregrina* (Myricaceae) as an actinomycete. They were able to grow it in pure culture, reinoculate other plants, and verify N_2 fixation.

Considerable research has been done on the allelopathic effects of several types of plants and microorganisms on *Azotobacter,* and a small amount on *Clostridium* and *Enterobacter.* I have found no publications concerning allelopathic effects against any of the remaining sixteen verified bacterial genera of asymbiotic N_2 fixers. Much research has been reported on allelopathic effects against *Rhizobium*, some on effects against the N_2-fixing blue-green algae, and a small amount on effects against nodulation and symbiotic N_2 fixation by nonlegumes.

A. Inhibition of Asymbiotic Nitrogen Fixers

Asymbiotic N_2 fixation varies greatly in amount with estimates in various ecosystems ranging from less than 1 kg/ha/yr to 100 kg/ha/yr (Steyn and Delwiche, 1970; Paul *et al.*, 1971; Rychert and Skujins, 1974; Kapustka and Rice, 1976). This input can obviously be very important, therefore, in at least some ecosystems. The highest estimates have been based on acetylene reduction measurements in desert rain crusts made up of blue-green algae and lichens (Rychert and Skujins, 1974).

Iuzhina (1958) reported that many bacteria, fungi, and actinomycetes isolated from soils of the Kola Peninsula in the U.S.S.R. inhibited the growth of *Azotobacter.*

Rice (1964, 1965b,c) tested 24 species of plants from revegetating old fields against the growth of *Azotobacter chroococcum* and *A. vinelandii.* Water extracts of at least one organ of each of 12 species inhibited growth of one or both of these N_2 fixers.

Chan *et al.* (1970) reported that *Azotobacter chroococcum* (strain 536), a common free-living nitrogen fixer in the soil, was inhibited by *Pseudomonas* sp., a rhizosphere isolate, and several other species of *Pseudomonas. Pseudomonas* is a very common bacterium in soil and in the rhizosphere.

In the eastern part of Oklahoma and throughout the eastern United States, broomsedge, *Andropogon virginicus,* invades old fields 3 to 5 years after abandonment and remains for many years, sometimes in almost pure stands (Rice, 1972). These fields are generally very low in nitrogen (Chapter 7), so I hypothesized that broomsedge might inhibit nitrogen-fixing bacteria. Sterile aqueous extracts of roots and shoots inhibited two test species of *Azotobacter.*

Aristida adscensionis is a common grass species in abandoned lands at Rajkot, India (Murthy and Ravindra, 1974). Because of previous work indicating that *A. oligantha* is inhibitory to nodulation of legumes and hemoglobin formation in the nodules, Murthy and his colleagues decided to determine if *A. adscensionis* inhibits N_2-fixing bacteria (Murthy and Ravindra, 1974, 1975; Murthy and Nagodra, 1977). They found that the shoot and root extracts and leachates of *Aristida* residues strongly inhibited *Azotobacter.*

Twenty-eight distinct bacterial isolates from the rhizosphere of *Aristida oligantha,* the dominant species in the second stage of old-field succession in Oklahoma, were tested for activity against three strains of *Azotobacter* (Leuck and Rice, 1976). Cell-free media from five of the isolated bacteria inhibited the growth of at least one strain of *Azotobacter*. This might increase the depression in the rate of N_2 fixation in the *A. oligantha* stage of succession and thus help to prolong that stage (Chapter 7).

Kapustka and Rice (1976) found that seven phenolic acids, identified as allelopathic agents, inhibited the growth of three free-living N_2 fixers, *Azotobacter chroococcum*, *Enterobacter aerogenes*, and *Clostridium* sp. Aqueous soil extracts from the pioneer weed stage, the annual grass stage, and the climax prairie significantly inhibited growth of *E. aerogenes* but stimulated growth of *A. chroococcum*.

In spite of the apparent importance of blue-green algae in N_2 fixation in rice cultivation and some natural ecosystems, very little research has been done on allelopathic effects against these organisms in terrestrial habitats. There has been a moderate amount of research done on allelopathic effects of other algae against some of the blue-greens, but not in relation to their N_2-fixing potential (Chapter 6).

Certain weeds from stage 1 of old-field succession in Oklahoma and *Aristida oligantha* from stage 2 inhibited nitrogen-fixing blue-green algae, whereas little bluestem from stage 3 and the climax did not inhibit them in most tests (Parks and Rice, 1969; Chapter, 7, Section I). These results complemented other evidence, which indicated inhibition of N_2-fixing bacteria by many of the same species. The combined effects may result in a slowing of the rate of addition of N to infertile old fields and thus to the slowing of succession. This could explain why the intermediate stages remain so long.

Nitrogen fixation by desert algal crusts probably constitutes a major inputs of nitrogen into desert ecosystems (Rychert and Skujins, 1974). These investigators determined that blue-green algae–lichen crusts from *Atriplex*, *Eurotia*, and *Artemisia* dominated sites in the Great Basin Desert had a laboratory potential of fixing nitrogen at rates up to 84 g of N ha^{-1} hour^{-1}. They found that the rates were reduced under canopies of the desert shrubs and demonstrated that water extracts of leaves of *Atriplex*, *Eurotia*, and *Artemisia* markedly inhibited N_2 fixation. Thus it appears that allelopathy plays an important role in the nitrogen economy of desert ecosystems.

Rice *et al.* (1980) found that four of five phenolic compounds present in decomposing rice straw markedly inhibited the growth of *Anabaena cylindrica*, a N_2-fixing blue-green alga (Chapter 2, Section II). All except two of them inhibited N_2 fixation (acetylene reduction) by this alga also. A combination of all five compounds was particularly effective in inhibiting growth and nitrogen fixation. This suggests a synergistic action and seems particularly significant because the five compounds always occur together in decomposing rice residues.

B. Inhibition of Symbiotic Nitrogen Fixers

1. Legumes.   The amounts of N_2 fixed symbiotically by *Rhizobium* can be high in agroecosystems, especially with alfalfa (Brady, 1974). The reported amounts fixed by *Rhizobium* in alfalfa range from 125–335 kg/ha/year (Alexander, 1977). The amounts are much lower, however, in natural ecosystems with estimated values ranging from about 0.2 to 1.4 kg/ha/year in ecosystems with nodulated legumes (Vlassak *et al.*, 1973; Kapustka and Rice, 1978b). The recognized importance of symbiotic nitrogen fixation by *Rhizobium* has caused many persons to research possible allelopathic effects against this organism.

Konishi (1931) reported that several bacteria isolated from soil were inhibitory to various species of *Rhizobium,* and generally these were aerobic, rod-shaped, gram positive, and non-spore forming. He also experimented with several species of known bacteria and found that two common soil bacteria, *Bacillus subtilis,* and *Bacillus megaterium,* inhibited *Rhizobium* from both alfalfa (*Medicago*) and pea (*Pisum*) nodules.

When Konishi (1931) used 4-week-old liquid cultures of the alfalfa *Rhizobium* and *Bacterium coli* (apparently *Escherichia coli*), which were grown together, to inoculate tubes containing alfalfa plants in agar, *Bacterium coli* inhibited or completely prevented nodule formation. *Bacillus fluorenscens* and *Bacterium aerogenes* slightly inhibited nodulation in similar experiments. The design of the experiments was such that it was not possible to tell whether the inhibition of nodulation was due to the inhibition of *Rhizobium,* to some more direct effect on the nodulation process, or to both.

Thorne and Brown (1937) found that most legume-nodule bacteria were able to grow in freshly expressed juices of their host plants, but such juices were bactericidal to other species of root-nodule bacteria. Bowen (1961) found that seeds of *Centrosema pubescens* and *Trifolium subterraneum* were inhibitory to *Rhizobium* when the seed coats were sterilized and the seeds were placed on Petri plates inoculated with *Rhizobium*. Fottrell *et al.* (1964) reported that the flavonoid myricetin occurs in some legume seeds and is toxic to *Rhizobium.*

Elkan (1961) found that a nonnodulating, near-isogenic soybean strain significantly decreased the number of nodules produced on its normally nodulating sister strain when inoculated plants of the two types were grown together in nutrient solution. Nodulation in ladino clover was inhibited also by the mutant soybean. Both results suggested the exudation by the roots of the nonnodulating strain of substances inhibitory to *Rhizobium,* to the nodulating process, or to both. Visona and Pesce (1963) and Visona and Tardieux (1964) performed a series of experiments that demonstrated that there are microorganisms in the rhizosphere of red clover and alfalfa, which are antibiotic to *Rhizobium.*

Beggs (1964) reported that oversown, inoculated white clover failed to nodulate properly and become established in large areas of *Nasella* and *Danthonia* grasslands in New Zealand (Chapter 7, Section I). Subsequent research caused him to conclude that the failure to nodulate was due to some toxin in the soil.

Twelve of 24 species of plants important in old-field succession in Oklahoma inhibited the growth of *Rhizobium* and nodulation and hemoglobin formation in legumes (Rice, 1964, 1965b,c, 1968, 1969, 1971b; Chapter 7, Section I). Blum and Rice (1969) reported that strains of *R. phaseoli* could be selected that were resistant to tannic acid, an inhibitor produced by several of the old-field species. The resistant strains had a much lower N_2-fixing ability, however.

Van der Merwe *et al.* (1967) and Hattingh and Louw (1969a) isolated 1091 bacteria from the rhizoplane of inoculated clovers, *Trifolium repens*, *T. pratense*, and *T. subterraneum*. Eighty-three of these isolates inhibited the growth of two *Rhizobium trifolii* strains. The inhibitory bacteria belonged to the genera *Pseudomonas*, *Xanthomonas*, *Flavobacterium*, *Achromobacter*, *Alcaligenes*, *Erwinia*, *Aerobacter*, *Bacillus*, *Streptomyces*, *Nocardia*, *Corynebacterium*, *Arthrobacter*, and *Brevibacterium*. The strongest antagonists and the greatest number belonged to *Pseudomonas*. Van der Merwe *et al.* (1967) identified the inhibitor produced by the strongest antagonist, *Pseudomonas* sp. (Strain W78), as 2,4-diacetylphloroglucinol.

Hussain and Mallik (1972) tested 28 species of fungi and 18 species of bacteria isolated from the rhizosphere of *Trifolium alexandrinum* against *Rhizobium trifolii*, the root nodule bacterium of *T. alexandrinum*. Two species of fungi and two of bacteria were antagonistic and two species of bacteria were stimulatory to the growth of *R. trifolii*. Mallik and Hussain (1972) tested 41 species of fungi, 66 species of bacteria, and 24 species of actinomycetes from the rhizosphere of lucerne (*Melilotus alba*) against the growth of two isolates of *Rhizobium meliloti* from lucerne nodules. Five percent of the fungi, 14% of the bacteria, and 21% of the actinomycetes exerted antagonistic effects, whereas 7% of the fungi and no bacteria or actinomycetes were stimulatory.

Sterile aqueous extracts of roots and shoots of broomsedge inhibited the growth of two species of *Rhizobium* (Rice, 1972). Small amounts of decaying shoots of broomsedge (lg/454g of soil) inhibited the growth and nodulation of the two most important species of legumes in old-field succession in eastern Oklahoma, *Lespedeza stipulacea* and *Trifolium repens*. Broomsedge competes vigorously and grows well on soils of low fertility; thus, inhibition of nodulation of legumes could keep the nitrogen supply low and give broomedge a selective advantage in competition over species that have higher nitrogen requirements. Interference of broomsedge against other species could help explain why it invades old fields in 3 to 5 years after the land is abandoned from cultivation, and it remains in almost pure stands for many years.

Plants of the Faboideae and Mimosoidae generally nodulate under proper conditions, and the majority of the legumes that do not nodulate are restricted to the Caesalpinoideae (Rao *et al.*, 1973). These investigators decided to test extracts of nonnodulating *Cassia fistula* and *C. occidentalis* of the Caesalpinoideae and nonnodulating *Leucaena leucocephala* of the Mimosoideae against *Rhizobium* from six genera and species of legumes. All strains of *Rhizobium* were inhibited by extracts of all the nonnodulating species. Pieces of roots of the nonnodulating species also inhibited *Rhizobium*.

Shoot and root extracts of *Aristida adscensionis* and leachates of its residues

(Section II,A of this chapter) inhibited *Rhizobium* isolated from nodules of *Indigofera cordifolia*, a common associate of *A. adscensionis* in India (Murthy and Ravindra, 1974, 1975; Murthy and Nagodra, 1977). Numbers of nodules on *Indigofera* in association with *Aristida* in the field were significantly lower also than on *I. cordifolia* growing where *Aristida* was absent. They suggested that inhibition of nitrogen fixation might prevent species with higher nitrogen requirements from invading these fields and thus allow *Aristida* to remain dominant for a prolonged period of time.

Schenck and Stotsky (1975) reported that volatile compounds released by germinating seeds of five crop plants and slash pine (*Pinus elliottii*) stimulated the growth of several species of fungi and of many gram-positive and gram-negative bacteria. *Rhizobium japonicum* was among the bacteria stimulated. Effects on the growth of the various organisms were apparent only during the first 3 or 4 days after planting the seeds, and killed or dried seeds had no effect.

Leuck and Rice (1976) found that three of 28 bacterial isolates from the rhizosphere of *Aristida oligantha*, the dominant in the second stage of old-field succession in Oklahoma, either inhibited or stimulated growth of one or more of the three test species of *Rhizobium*. They suggested that this might add an additional dimension to the direct effect of *A. oligantha* on N_2 fixation and help prolong the annual grass stage (Chapter 7, Section I).

Five phenolic compounds that were present in decomposing rice straw and sterile extracts of decomposing rice straw in soil were very inhibitory to the growth of three strains of *Rhizobium* (Rice *et al.*, 1981). The effects were additive and in several instances synergistic. The phenolic compounds also reduced nodule numbers and hemoglobin content of the nodules in two bean varieties. Extracts of decomposing rice straw in soil significantly reduced N_2 fixation in Bush Black Seeded Beans. This may help explain the great reduction in soybean yields in Taiwan following rice crops when the rice stubble is left in the field (Chapter 2, Section II).

2. Nonlegumes.   Some 17 genera of nodulated nonlegumes have been confirmed as symbiotic N_2 fixers (Farnsworth and Clawson, 1972; Torrey, 1978; Heisey *et al.*, 1980). Relatively large amounts of nitrogen have been calculated to be fixed by several of these actimomycete-nodulated plants: *Alnus*, 40 to 362 kg/ha/year, *Casuarina*, 58 to 200, and *Hippohaë*, 15 to 179 (Torrey, 1978). Thus, several of these species are of potential use in forestry, both as forest products and as fixers of N_2 in the ecosystems (Tarrant and Trappe, 1971; DeBell and Radwan, 1979; Gordon and Dawson, 1979). Some of the symbiotic N_2-fixing nonlegumes are important during early succession on recently deglaciated areas in Alaska, which are low in N (Crocker and Major, 1955). *Dryas drummondii* is a prominent early colonizer in such areas; *Shepherdia canadensis* occurs also in the early stage but is less prominent. *Alnus* usually follows and soon becomes dominant. It appears, therefore, that allelopathic effects against some of the symbiotic nonlegumes could be very important in some ecosystems.

Jobidon and Thibault (1981, 1982) noted the allelopathic effects of balsam poplar against green alder (see Chapter 3, Section I,A). They found that aqueous extracts of leaf litter and buds and fresh leaf leachates of balsam poplar inhibited seed germination and radicle and hypocotyl growth of green alder. Root hair development was severely inhibited and necrosis of the radicle meristems occurred.

Jobidon and Thibault (1981) suggested that the inhibition of root hair development might be important in the infection of the alder by the actimomycete *Frankia*, because it has been clearly demonstrated that *Frankia* penetrates the root hair to initiate nodule formation (Lalonde and Quispel, 1977). Subsequent investigation demonstrated that the average number of nodules per alder plant after treatment with balsam poplar extracts was only 51% that of control plants (Jobidon and Thibault, 1982). Moreover, there was a 62% decrease in nitrogen fixation (acetylene reduction) by alder seedlings that were treated with the most concentrated bud and leaf litter extracts of balsam poplar. Foliar N content of both nodulated and unnodulated alder seedlings was significantly lower in extract-treated plants than in controls.

The striking results of Jobidon and Thibault suggest a need for more research on allelopathic effects of several plant species on symbiotic N_2 fixation by nonlegumes. They also call into question the proposals for mixed plantations of alders and poplars.

III. INHIBITION OF NITRIFICATION

A. Theoretical Basis for Selection in Natural Ecosystems of Species Inhibitory to Nitrification

My early training in plant physiology caused me to begin research on the role of allelopathy in old-field succession with the concept that nitrification is required to make nitrogen available for the plant growth (Chapter 7, Section I,B). If so, inhibition of nitrification would give pioneer plants with low nitrogen requirements a selective advantage in competition with the later invaders, which have higher nitrogen requirements and, thus, slow succession. Pioneer species did, indeed, inhibit nitrifying bacteria in addition to inhibiting nitrogen-fixing bacteria. However, climax species inhibited nitrification more strongly than did pioneer species, but hardly inhibited nitrogen-fixing bacteria at all (Rice, 1964). These results caused me to re-evaluate the necessity of nitrification for the growth of plants.

Ammonium ions are positively charged and are therefore adsorbed on the negatively charged colloidal micelles, thus preventing leaching below the depth of rooting due to percolating water. On the other hand, nitrate ions are negatively charged and are repelled by the colloidal micelles in the soil. Thus, they readily leach below the depth of rooting or are easily carried away in surface drainage. It would appear from these facts that inhibition of nitrification would help to conserve nitrogen.

If plants take up nitrate ions, they have to reduce these ions to nitrite and then to ammonium ions before the nitrogen can become involved in reactions leading to the formation of amino acids and subsequently to other nitrogenous organic compounds. The reduction of nitrate ions to ammonium ions requires energy; thus, inhibition of nitrification would conserve energy.

The conservation of energy and of nitrogen resulting from the inhibition of nitrification would appear to be strong forces during succession in the selection of plant species inhibitory to nitrification. If nitrification is inhibited, this would mean that ammonium nitrogen would be the chief form of available nitrogen in later successional stages and in climax ecosystems. The next item of importance, therefore, concerns the ability of plants to use ammonium nitrogen, especially noncrop plants.

There is growing evidence that many plant species, probably most, can use ammonium nitrogen as effectively or more so than nitrate nitrogen in ecologically meaningful concentrations (Allison, 1931; Addoms, 1937; Tam and Clark, 1943; Cramer and Myers, 1948; Cain, 1952; Swan, 1960, Oertli, 1963; Pharis *et al.*, 1964; Nielsen and Cunningham, 1964; McFee and Stone, 1968; Ferguson and Bollard, 1969; Shen, 1969; Gamborg and Shyluk, 1970; Moore and Keraitis, 1971; Christersson, 1972; Gigon and Rorison, 1972; Weissman, 1972; Smith and Rice, 1983). Thus, inhibition of nitrification makes good biological sense.

The evolution of only two genera of bacteria primarily responsible for nitrification suggests also that this process is not a necessary link in the N cycle in many ecosystems. There is growing evidence to support this possibility.

It is noteworthy that many ecologists, physiologists, and soil scientists still assume that nitrification is essential in most ecosystems. Virtually all authors of texts on soils still state that nitrification is the weak link in the N cycle. This probably stems from the fact that crop plants were selected for many years to grow and produce well on nitrate as the chief N fertilizer. Moreover, much of our knowledge of plant physiology is based on research on crop plants, which has no doubt skewed our information in this important area.

B. Evidence for Inhibition of Nitrification

Leather (1911) reported low nitrification under perennial grass in India, indicating an inhibition of the nitrifying bacteria by certain perennial grass species. Russell (1914) reported that cropped soil had a much lower total nitrate content than uncropped similar soil, even when the amount taken up by plants was included, and suggested that the lower amount was due to a diminished production in the presence of plants. Lyon *et al.* (1923) found that maize, wheat, and oats markedly depressed nitrate production. They suggested that the reason for the recovery of smaller amounts of nitrates from the soil and plants was that plants liberate carbonaceous matter into the soil, which favors the development of nitrate-consuming organisms. They further suggested that the nitrates are converted by these microorganisms into other compounds and that differences in the composition of the exudate from the roots of plants might account for the differences they cause in the disappearance of nitrates. They stated that plants high in nitrogen might liberate substances less rich in carbohydrates, which might be less encouraging to the nitrate-consuming bacteria. They found that addition of ground organic matter containing different percentages of nitrogen to the soil resulted in differences in amounts of nitrates, which could subsequently be leached from the soil. No data were given, however, on amounts of ammonium nitrogen in the various soils to determine if nitrification may have been reduced, and no tests were made to determine if the various types of organic matter may have contained different amounts of substances inhibitory to nitrifying bacteria.

Richardson (1935, 1938) studied the nitrogen cycle in grassland soils at Rothamsted Experimental Station in England. He found that the concentration of ammonium nitrogen was several times greater than that of nitrate nitrogen in the soil, and the ratio of ammonium to nitrate nitrogen increased with the age of the sward. He reported also that the grasses and other plants absorbed the ammonium nitrogen as readily as nitrate and that much of the nitrogen was taken up as ammonium.

Theron (1951) investigated problems that arose in South Africa when grasses were used as a means of rehabilitating wornout soils. Normally, a luxuriant growth of grass took place during the first season after the sward was established, but growth soon deteriorated. By the third or fourth season, the sward was so poor that it hardly afforded any grazing and appeared to be valueless as a rebuilder of soil. The growth of the grass could be increased again by plowing the soil, or by the application of nitrogenous fertilizers. In any event, the sequence was soon repeated. Theron found ample quantities of nitrates in the soil solution during the period of luxuriant growth, but during the intervening periods of poor growth, little or no nitrate was found. Exchangeable ammonium was still present in small quantities.

Theron ran numerous lysimeter experiments in which some lysimeters contained crop plants, some contained a local perennial grass species, and some were left fallow. An example of his usual results were those obtained using annual millet as the crop plant and *Hyparrhenia* sp. as the perennial grass. Large quantities of nitrogen were mineralized during the first year in all three types of lysimeters. Under the millet, virtually a steady state was reached with respect to both yield and the nitrogen lost to the crop and to the percolate that year. The nitrate in the percolate of the cropped soil had cyclic changes associated with the growth and maturation of the millet. By the time the second crop had matured, the nitrate content had fallen to a very low concentration. However, it again increased to a high value during the ensuing winter. Virtually no nitrates were found, however, in the percolate of the perennial grass plots at any time after the first year of growth. In the fallow soil, nitrogen was freely mineralized consistently. These results illustrate the very important point that nitrification took place actively from the second season on throughout the entire winter in the cropped and fallow soil but not in the soil under perennial grass, even though the grass was dormant from May to September. According to Theron, the soil remained equally moist in all plots and other external conditions were similar.

Theron argued emphatically that the continued low concentration of nitrate in the percolate under perennial grass was not due to the consumption of nitrate by microorganisms because of their stimulation by carbonaceous matter from the roots of the grass, as suggested by Lyon *et al.* (1923). He felt this explanation was untenable for the following reasons: (1) The amount of carbonaceous material required to bring about the results was too great to be excreted by the roots of the grass; (2) nitrate reappeared in the cropped soil immediately after the maturity and death of the crop and continued high until the growth of the subsequent crop, even though the total supply of carbonaceous material from the dead roots was higher than at any other time; (3) nitrate did not appear in the soil under grass even when the roots were dormant and could not have excreted sufficient carbonaceous material for nitrates to be reassimilated by microorganisms; (4) when such carbonaceous material was added to soil, both nitrate and ammonium were assimilated by microorganisms. Although little or no nitrate occurred under the grass, ammonium nitrogen was generally present in larger amounts than were usually present in cultivated soil. Theron grew sunflowers in small quantities of soil and found that ammonium accumulated in the soil when the plants were maturing even though it was entirely absent during the early stages of growth.

Theron concluded, therefore, that perennial grasses and other actively growing plants interfere only with nitrification and not with ammonification. He concluded further that the inhibition of nitrification is probably due to bacteriostatic excretions by living roots; he suggested that only very minute quantities of the inhibitors would probably be necessary to inhibit the rather sensitive nitrifying bacteria.

Eden (1951) found that grassland (patana) soils in Ceylon are extremely low in nitrate and the low nitrification rate lasts for several years after breaking the land for tea cultivation. This is the time when one would normally expect an increase in nutrients from decomposing vegetation.

In a study of the rate of nitrification in soils taken from plant communities representing various stages of secondary grassland succession on the Transvaal Highveld in South Africa, Stiven (1952) found that soils taken from the *Trachypogon plumosus* grassland climax community consistently lagged in the production of nitrate. When *T. plumosus* plants with their surrounding soil cores were placed in pots and the aerial parts of the plants were removed, little regeneration of the top occurred. Little or no nitrate was found in water which percolated through the soil, and no weeds seemed to grow in the pots. Stiven hypothesized that the roots were secreting toxic materials that inhibited the activity of the nitrifying bacteria. He found that distilled water extracts of the roots inhibited *Escherichia coli, Bacillus subtilis, Staphylococcus aureus,* and *Streptococcus haemolyticus,* but he did not test the extract against the nitrifiers, probably because of the difficulty of culturing them.

Mills (1953) reported that the fallow soil in Uganda, Africa, may have nitrate concentrations as high as 200 ppm (even 300 ppm in the top 2.5 cm), 10 ppm on shaded and/or mulched soil, and something in between in cropped soil. After resting 3 years under elephant grass (*Pennisetum purpureum*), *Chloris,* or *Paspalum,* the nitrate content, which was tested to a depth of 183 cm, was found to be virtually zero. After the area was plowed, the rate at which nitrate accumulation occurred in the surface horizons varied with the species of grass previously present. It accumulated very slowly in *Paspalum* plots, slightly faster in *Chloris* plots, and still faster in elephant grass plots. The rate was still relatively slow, however, in the elephant grass plots.

Greenland (1958) found almost no nitrate throughout the year in permanent grassland plots at Ejura in Ghana, Africa. On the other hand, he found that the amounts of nitrate under temporary or successional grass plots were greater than under the climax grasses, but lower than in cropland. He pointed out that the cause of low nitrate concentrations was not likely due to nitrate absorption, because he found that little microbial nitrate absorption took place under crops. In addition, soil samples taken from the permanent grassland plots and incubated in the laboratory showed a high rate of nitrification in spite of a very high carbon/nitrogen ratio of over 20. Greenland concluded, therefore, that the low level of nitrate under grassland was due to suppression of mineralization due to an excretion of the plant roots that was toxic to the nitrification process.

Nye and Greenland (1960) reviewed research in many African areas and vegetation types and reported that irrespective of the C/N ratio of the soil, its pH, or moisture regime, very little nitrogen is found in the soil while the dominant vegetative cover is a grass.

Meiklejohn (1962) found that grassland soils in Ghana contained few ammonium oxidizers and very few or no nitrite oxidizers; she stated that the lack of available nitrogen in these soils seemed to be due mainly to the absence of bacteria that were able to oxidize nitrite to nitrate. In a later project near Salisbury, Rhodesia, she reported that soils under native grass contained few nitrifiers but that the same soils, when cleared and planted with crops, contained many more nitrifiers (Meiklejohn, 1968). Soils under improved grass pastures and under two legumes contained about 100 times as many nitrifiers as soils under native grass.

Robinson (1963) reported that New Zealand tussock-grassland soil is typically low in nitrifiers, and no nitrification took place in 50 days in the laboratory when the soil was percolated with ammonium sulfate solution. However, when the tussock grassland soil was inoculated with a small amount of garden soil and percolated with ammonium sulfate solution, nitrification began immediately without a lag phase. Robinson concluded, therefore, that the lack of nitrification in the uninoculated soil was due to a deficiency of nitrifiers and not due to toxins produced by the grasses. He argued that the low populations of nitrifiers in the tussock-grassland soils occurred because of the very low amounts of ammonium available in the infertile soils. He argued also that the few nitrifiers that persisted were strains that had lost the ability to carry on nitrification even when ammonium became available. His argument is weak, however, because the same unresponsive small populations of nitrifiers could have resulted from toxin production by the grasses (see Chapter 7, Section I,B).

Boughey et al. (1964) reported that two species of *Hyparrhenia,* grasses abundant in the Rhodesian high-veld savanna, secrete a toxin that suppresses the growth of nitrifying bacteria. Warren (1965) found that the populations of nitrifiers in the climax purple-veld of South Africa were much lower than in successional stages. During the greater part of the year, the nitrite oxidizers were almost completely absent in the purple-veld soil and were only slightly higher in *Hyparrhenia* soils. Moreover, there was a gradual decrease in nitrite and nitrate with succession. Munro (1966a,b) found that root extracts of several climax species from the Rhodesian high-veld were more inhibitory to nitrification than several seral species investigated. The evidence is strong, therefore that inhibition of nitrification increases with succession in grassland areas in Africa and that inhibition is very pronounced in the climax grasslands.

Neal (1969) investigated the effects on nitrification *in vitro* on aqueous root extracts of six climax grass species and eight species of grasses or forbs that increase in importance or invade overgrazed grasslands, in Alberta, Canada. Inhibitors of both *Nitrosomonas* and *Nitrobacter* were found in root extracts of grasses and forbs, which commonly increase on, or invade overgrazed grasslands. *Stipa comata* was the only climax grass that inhibited nitrification appreciably. In general, *Nitrobacter* was inhibited more than *Nitrosomonas*. These results are obviously contradictory to those obtained in African grasslands and to most results obtained in North American grasslands.

Dommergues (1954) did a microbiological study of five forest soil types from central and eastern Madagascar. He found that ammonification was higher and nitrification was lower in the forest soils than in cultivated soils. In a subsequent study of dry tropical forest soils in Senegal, Dommergues (1956) stated that nitrification in dry tropical forest soils is more active than in dense, humid forest soils. Nevertheless, nitrification increases greatly in the dry forest soils on clearing and cultivating.

Jacquemin and Berlier (1956) reported that the nitrifying power in forest-covered soils of the lower Ivory Coast of Africa was low and that it increased on clearing. Berlier et al. (1956) compared the biological activity of forest and savanna soils on the Ivory Coast and found that the nitrifying activity in savanna soils was practically zero, or in other words, still lower than in the forest soils. Ammonification was higher also in the forest soils than the savanna soils. Nye and Greenland (1960) reported that the numbers of nitrifiers are exceptionally low in the Moist Evergreen Forest covered soils of Africa. In contrast to the low rate of nitrification in such soils, ammonification proceeds rapidly.

Data cited by Russell and Russell (1961) and Weetman (1961) indicated that the rate of nitrification is very low generally under spruce and several other conifers and also in forests in general with a mor type of mulch. In Finland, Viro (1963) reported that most of the available nitrogen in the humus layer from stands in which spruce was dominant, in pure pine stands, and in pure spruce stands was ammonium nitrogen with the amount of nitrate nitrogen being very low.

Smith et al. (1968) found an 18-fold increase in *Nitrosomonas* and a 34-fold increase in *Nitrobacter* after clear-cutting a forest ecosystem in Connecticut. They stated that the evidence indicated that the pronounced increase in numbers of nitrifiers was due, not to changes in physical conditions, but to elimination of uptake of nitrate by the vegetation or to a reduced production of substances inhibitory to the autotrophic nitrifying population. Likens et al. (1969) reported a 100-fold increase in nitrate loss in the same ecosystem after cutting.

Thus, the limited evidence available at this point in time indicated that numbers of nitrifiers and rates of nitrification are low in at least some climax forest ecosystems. Moreover, the number of nitrifiers and rates of nitrification increase with disturbances in the ecosystems.

On the basis of this evidence, it was hypothesized that inhibition of nitrification increases during succession and is high in climax ecosystems. Three stands representing two stages of old-field succession and the climax were selected in each of the following vegetation types in Oklahoma: oak–pine forest, post oak–blackjack oak forest, and tall grass prairie (Rice and Pancholy, 1972). Stands in the tall grass prairie area were located near Norman in Cleveland and McClain Counties, the post oak–blackjack oak area stands were in Hughes County southeast of Wetumka, and the oak–pine stands were in Latimer County north of Wilburton. Average annual precipitation in the study areas varied from 44 inches in the oak–pine area to 38 inches in the post oak–blackjack oak area to 33 inches in the tall grass prairie area (Gray and Galloway, 1959). The oak–pine plots were located in the Enders-Conway-Hector soil association (Gray and Galloway, 1959) in the Ultisols order (Gray and Stahnke, 1970); the post oak–blackjack oak plots were in the Darnell-Stephenville association in the Alfisols order, and the tall grass prairie plots were in the Renfrow-Zaneis-Vernon association in the Mollisols order.

All plots had a sandy loam soil except the first successional stage in the tall grass prairie area which had a sandy clay loam soil. Moreover, the pH was virtually the same in all three plots of each vegetational type. The amount of organic carbon often varied considerably in the different plots with the only consistent trend in relation to succession occurring in the tall grass prairie plots.

The first successional stage investigated (P_1) was in the first year after abandonment from cultivation in the oak–pine and tall grass prairie areas and in the second year in the post oak–blackjack oak area. All P_1 plots were in the pioneer weed stage of succession, stage 1 of Booth (1941a) (Chapter 7, Section I). The second successional stage (P_2) was in the sixth year after abandonment in the tall grass prairie area and was dominated by *Aristida oligantha,* which represented stage 2, the annual grass stage, of Booth. The P_2 plot in the post oak-blackjack oak area was in the eighth year after abandonment and was dominated by *Ambrosia psilostachya, Andropogon virginicus, Aristida oligantha,* and *Lespedeza stipulacea.* The P_2 plot in the oak–pine area was abandoned from cultivation for 25 years and was dominated by *Andropogon virginicus;* a few pine and oak seedlings were present. The climax prairie (P_3) was dominated by *Andropogon gerardii* and *Schizachyrium scoparium* with *Panicum virgatum* and *Sorghastrum nutans* as important secondary sepcies. The climax post oak–blackjack oak stand was dominated by *Quercus marilandica* and *Q. stellata* with ground cover primarily of *S. scoparium.* The climax oak–pine stand was dominated by *Pinus echinata* and *Quercus stellata.* This stand had virtually no herbaceous ground cover. There were a few very small and sparse patches consisting of *A. gerardii, S. scoparium, Desmodium laevigatum,* and *Tephrosia virginiana.*

Ten soil samples were taken from the 0 to 15 cm and 10 from the 45 to 60 cm depths in each plot every other month for a full year. These were analyzed for ammonium nitrogen and nitrate nitrogen. Eight evenly distributed soil samples were taken from the 0 to 15 cm depth in each plot every other month for a year. These were analyzed for numbers of *Nitrosomonas* and *Nitrobacter.*

The concentration of ammonium nitrogen was lowest in the first successional stage, intermediate in the second successional stage, and highest in the climax stand (Tables 48 and 49). Moreover, the differences in concentration between P_1 and P_2, P_2 and P_3, and P_1 and P_3 were generally statistically significant. This trend was remarkably consistent throughout all sampling periods, all vegetation types, and both sampling depths in the soil. The concentration of nitrate was highest in the first successional stage, intermediate in the second successional stage, and lowest in the climax stand at both sampling depths in all vegetation

TABLE 48.   Amounts of Ammonium and Nitrate Nitrogen in 0 to 15 cm Level of Research Plots[a,b]

Source and date	NH_4^+ (ppm)			NO_3^- (ppm)		
	P_1	P_2	P_3	P_1	P_2	P_3
Tall grass prairie						
March, 1971	3.28	3.98	4.60[d,e]	4.42	1.42[c]	1.50[e]
May, 1971	2.87	3.71[c]	4.86[d,e]	3.11	1.98[c]	1.39[d,e]
July, 1971	0.94	1.45[c]	2.21[d,e]	4.12	2.70[c]	1.77[d,e]
September, 1971	1.68	5.63[c]	6.68[e]	4.27	2.78[c]	1.78[d,e]
November, 1971	2.82	6.69[c]	7.83[d,e]	2.42	0.91[c]	0.20[d,e]
January, 1972	2.08	3.10	4.35[e]	2.49	1.31[c]	0.57[d,e]
Post Oak–Blackjack Oak						
March, 1971	2.17	3.09[c]	4.49[d,e]	2.13	1.62	1.14
May, 1971	2.24	3.97[c]	5.45[d,e]	2.78	1.53[c]	0.86[d,e]
July, 1971	1.99	2.90[c]	4.36[d,e]	3.43	1.83[c]	1.06[d,e]
September, 1971	4.93	4.27	2.80[e]	3.21	1.65[c]	1.05[d,e]
November, 1971	2.87	4.41[c]	4.90[e]	1.67	0.88[c]	0.42[e]
January, 1972	1.36	3.04[c]	3.02[e]	1.89	1.04[c]	0.73[d,e]
Oak–Pine						
March, 1971	2.40	6.43[c]	7.39[e]	3.92	1.25[c]	1.82[d,e]
May, 1971	3.36	6.24[c]	7.13[e]	3.09	2.03[c]	1.52[d,e]
July, 1971	2.62	3.89[c]	4.72[d,e]	3.72	2.40[c]	1.62[d,e]
September, 1971	2.97	3.15	3.01	2.18	1.44	0.07[d,e]
November, 1971	1.63	3.91[c]	4.37[e]	2.62	2.01[c]	0.79[d,e]
January, 1972	1.00	3.85[c]	5.43[e]	2.59	0.92[c]	0.35[d,e]

[a] From Rice and Pancholy (1972).
[b] Each number is average of ten analyses.
[c] Difference between P_1 and P_2 significant at 0.05 level or better.
[d] Difference between P_2 and P_3 significant at 0.05 level or better.
[e] Difference between P_1 and P_3 significant at 0.05 level or better.

types and in virtually all sampling periods (Tables 48, 49). Again, the differences between P_1 and P_2, P_2 and P_3, and P_1 and P_3 were usually statistically significant. The consistently high concentrations of ammonium nitrogen and very low concentrations of nitrate in the climax stands throughout all vegetation types, all sampling periods, and both sampling levels were particularly striking.

The number of *Nitrosomonas* per gram of soil was highest in the first successional stage, intermediate in the second successional stage, and lowest in the climax throughout all sampling periods and all vegetation types (Table 50). The number of *Nitrobacter* per gram of soil was generally considerably higher in the first and second successional stages than in the climax (Table 50). In the post oak–blackjack oak and oak–pine areas, the number in the first successional stage was highest with the second successional stage generally having an intermediate

TABLE 49. Amounts of Ammonium and Nitrate Nitrogen in 45 to 60 cm Level of Research Plots[a,b]

Source and date	NH_4^+ (ppm) at 45–60 cm			NO_3^- (ppm) at 45–60 cm		
	P_1	P_2	P_3	P_1	P_2	P_3
Tall grass prairie						
March, 1971	1.37	3.40[c]	4.12[e]	4.63	3.68[c]	1.16[d,e]
May, 1971	1.47	1.89[c]	2.48[d,e]	2.49	2.08[c]	1.36[d,e]
July, 1971	0.76	0.97	1.27	2.67	2.12[c]	1.44[d,e]
September, 1971	3.32	5.51	5.67[e]	2.66	2.25[c]	1.49[d,e]
November, 1971	3.18	5.39[c]	5.60[e]	0.85	0.31[c]	0.13[e]
January, 1972	1.88	2.95	2.75[e]	1.06	0.92	0.18[d,e]
Post Oak–Blackjack Oak						
March, 1971	1.65	4.54[c]	5.19[d,e]	2.08	1.63[c]	0.94[e]
May, 1971	0.80	1.12	2.65[d,e]	1.98	1.40	1.18[e]
July, 1971	0.87	1.26	2.14[d,e]	2.18	1.66[c]	0.96[d,e]
September, 1971	3.64	3.01	2.79	2.08	1.63[c]	0.80[d,e]
November, 1971	1.78	2.95[c]	3.22[e]	0.54	0.45	0.22[e]
January, 1972	0.93	1.20	1.55[e]	0.88	0.64[c]	0.60[e]
Oak–Pine						
March, 1971	2.79	5.23[c]	7.06[d,e]	2.23	1.00[c]	2.28[d]
May, 1971	1.99	5.37[c]	6.05[e]	2.08	1.82	1.58[e]
July, 1971	1.89	3.46[c]	4.20[d,e]	2.19	2.12[c]	1.06[d,e]
September, 1971	1.05	1.57	2.10[e]	0.00	0.37	0.00
November, 1971	1.33	1.61	1.82[e]	1.39	1.28	0.39[d,e]
January, 1972	0.98	1.31	2.01[d,e]	0.14	0.06	0.04

[a] From Rice and Pancholy (1972).
[b] Each number is average of ten analyses.
[c] Difference between P_1 and P_2 significant at 0.05 or better.
[d] Difference between P_2 and P_3 significant at 0.05 or better.
[e] Difference between P_1 and P_3 significant at 0.05 or better.

number. In the prairie area, the first and second successional stages generally had similar numbers of *Nitrobacter*. The particularly striking feature was the very low number of *Nitrobacter,* which occurred in the climax stands in all vegetation types and all sampling periods, with the exception of June, in the tall grass prairie. The number was often zero.

The inverse correlation between the concentration of nitrate and the concentration of ammonium in all plots was striking. The concentration of ammonium increased from a low in the first successional stage to a high in the climax, whereas the concentration of nitrate decreased from a high in the first successional stage to a low value in the climax. Moreover, the numbers of nitrifiers were high in the first successional stage and low in the climax. Thus, some factor or factors reduced the populations of nitrifiers during succession resulting in an apparent reduction in the rate of oxidation of ammonium to nitrate. It was obvious from the general soil data (Rice and Pancholy, 1972) that the low rates of nitrification in the climax plots were not due to pH or textural differences. Moreover, the lack of definite trends in amounts of organic carbon in relation to succession in the oak–pine and post oak–blackjack oak areas indicated that the quantity of organic carbon was not responsible for the low rate of nitrification in the climax plots. These facts, along with the previous discovery (Rice, 1964) that the climax species investigated were very inhibitory to nitrification, caused us to infer that the climax plants reduced the rates of nitrification in the three eco-systems involved. It appears that inhibition of nitrification started during succession and increased in intensity as succession proceeded toward the climax.

Data obtained during a second growing season were similar to those reported here, and thus supported the same conclusions (Rice and Pancholy, 1973).

TABLE 50. Numbers (MPN) of Nitrifiers in 0 to 15 cm Level of Research Plots[a,b]

Source and date	Nitrosomonas/g soil			Nitrobacter/g soil		
	P_1	P_2	P_3	P_1	P_2	P_3
Tall grass prairie						
April, 1971	111	42[c]	140	25	25	25
June, 1971	3012	525[c]	147[d,e]	347	417	280
August, 1971	334	158	50[e]	62	72	3[d,e]
October, 1971	817	51	37	24	24	24
December, 1971	177	26[c]	32[e]	23	23	18
February, 1972	127	51	51	110	270[c]	43[e]
Post Oak–Blackjack Oak						
April, 1971	186	47	22	116	19[c]	5[d]
June, 1971	4470	710[c]	32[e]	22	8	0[e]
August, 1971	303	207	64[e]	51	1[c]	0[e]
October, 1971	216	292	15	50	321	15
December, 1971	126	78	15[e]	19	12	9
February, 1972	50	15	14	132	41	6[d,e]
Oak–Pine						
April, 1971	2770	65[c]	7[d,e]	248	9[c]	5[e]
June, 1971	3365	188[c]	8[d,e]	951	29[c]	0[d,e]
August, 1971	36,286	964[c]	24[d,e]	384	45[c]	0[d,e]
October, 1971	1588	315[c]	5[d,e]	661	40[c]	30[e]
December, 1971	783	451[c]	30[d,e]	192	50[c]	8[d,e]
February, 1972	3950	57[c]	0[d,e]	205	37[c]	0[d,e]

[a] From Rice and Pancholy (1972).
[b] Each number is average of four determinations at different locations.
[c] Difference between P_1 and P_2 significant at 0.05 level or better.
[d] Difference between P_2 and P_3 significant at 0.05 level or better.
[e] Difference between P_1 and P_3 significant at 0.05 level of better.

All the dominant herbaceous and tree species from the intermediate and climax areas produced considerable amounts of condensed tannins, and the concentration of these tannins in the top 15 cm of soil was always higher in the climax stand than in the intermediate successional stage in each vegetation type (Rice and Pancholy, 1973). Gallic and ellagic acids, which result from the digestion of hydrolyzable tannins in oak species, were present in the climax oak–pine forest soil also. Condensed tannins, hydrolyzable tannins, ellagic acid, gallic acid, digallic acid, and commercial tannic acid (a hydrolyzable tannin), in very small concentrations, completely inhibited nitrification by *Nitrosomonas* in soil suspensions for 3 weeks, the duration of the tests. Moreover, the concentrations of tannins, gallic acid, and ellagic acid found in the soil of the climax research plots were several times higher than the minimum concentrations necessary to completely inhibit nitrification.

Seventeen additional potential inhibitors of nitrification were identified from eleven dominant climax species in the three ecosystems (Rice and Pancholy, 1974). These were mostly phenolic acids and flavonoids, but one coumarin compound, scopolin, was found in high concentrations in several species. The aglycones of most of the compounds completely inhibited nitrification by *Nitrosomonas* at concentration as low as 10^{-6} to 10^{-8} *M*. We found a very inhibitory compound in large quantities in the oak–pine climax soil, which appeared in all tests to be a flavonoid aglycone, but we were never able to identify it. It is likely that some, if not all, of the nitrification inhibitors identified may be important in the inhibition of nitrification in the later stages of succession and in the climax, along with the tannins.

The data strongly supported our original hypothesis that inhibition of nitrification increases during succession and is high in climax ecosystems. There have been numerous papers published on this subject since our hypothesis was made. Many have supported the hypothesis, but a few have not.

As noted previously, several investigators reported that nitrification is inhibited under several species of grasses in Africa. One of the genera involved was *Hyparrhenia,* and Purchase (1974) examined the possible inhibition of nitrification by *H. filipendula.* He reported that washings from live roots slightly prolonged the lag phase of the nitrite oxidizers in liquid cultures, and washings from decaying roots were inhibitory only if collected during the initial stages of decay. When roots were incubated in soil with added $(NH_4)_2SO_4$, they caused considerable immobilization of mineral nitrogen with a resultant depression of nitrate accumulation but did not prevent nitrification of surplus ammonia. *Hyparrhenia* grasslands were found to have very low numbers of nitrifying bacteria, but the numbers were low away from the root zone of the grass as well as in this zone. The numbers of ammonia oxidizers increased regularly with addition of ammonia, but the numbers of nitrite oxidizers increased slightly in some experiments but not in others. Purchase suggested that the lack of response of the nitrite oxidizers to added ammonia was caused by a phosphate deficiency, but no evidence was given to support this. Despite this discrepancy, he concluded "that nitrification in *Hyparrhenia* grasslands is restricted by limited availability of ammonia and that there is no convincing evidence for toxic inhibition by root secretions."

Woodwell (1974) reported a relatively high concentration of nitrate in a shallow water table under a cultivated field (4.48 mg/ml) on Long Island, New York, whereas the concentration was only 0.93 mg/liter in the water table under an abandoned field, 0.42 mg/liter under a pine forest, and 0.02 mg/liter under an oak–pine forest. He stated that "Mature systems are tight; disturbance causes leakage of nutrients."

Todd *et al.* (1975) measured the nitrate content of stream water and the nitrifying bacterial population of the upper 40 cm of soil of three Appalachian watersheds over a 22-month period. The watersheds were a fescue grass catchment, a 15-year-old white pine plantation, and a mature undisturbed hardwood forest. Monthly averages of nitrate nitrogen in stream water were 730, 190 and 3 ppb, respectively, for the three ecosystems. The respective nitrifying populations averaged 16,000, 175, and 22 per gram dry weight of soil, respectively. There was an obvious correlation between numbers of nitrifiers and nitrate contents of the streams. Thus, the authors concluded that "Nitrifying activity appears to be dependent on vegetation types and successional stage." It appears that inhibition of nitrification increased as succession progressed toward the climax hardwood forest of the region.

Vitousek and Reiners (1975) hypothesized that increases in biomass cause the decreasing losses of nutrients from ecosystems during succession because the added biomass ties up more of the nutrients. They suggested that this probably accounts for a lowering in concentration of nitrate in soil during succession instead of an inhibition of nitrification. Vitousek (1977) cited data from trenching experiments to support that hypothesis. He pointed out, however, that trenching eliminated the possibility of inhibition of nitrification by root exudates. This is a very important possibility in light of Moleski's (1976) results (see the next paragraph). Interestingly, trenching resulted in no significant changes in concentrations of elements other than nitrogen. This certainly calls into question the hypothesis that differences in ammonium nitrogen and nitrate nitrogen in trenched and untrenched plots were due to differences in uptake. Moreover, this hypothesis does not explain the increasing ratios of NH_4 to NO_3 during succession as reported by Rice and Pancholy (1972, 1973), nor the decreasing numbers of nitrifiers. Another point of significance is that the total biomass of the pioneer weed stage in Oklahoma old-field succession is always much greater than that of the succeeding annual grass stage. In spite of this fact, the concentrations of nitrate in the soil throughout the year are significantly greater in the pioneer stage than the annual grass stage.

Moleski (1976) investigated the inputs of condensed tannins into the soil in a *Quercus stellata–Q. marilandica* forest and effects of tannins on microbial populations, including the nitrifiers. As a basis for comparison, he clear-cut a part of the forest, competely trenched around the cleared area, and inserted a heavy plastic sheet to prevent reinvasion of roots from outside the plot. He measured many microclimatic and soil factors in addition to the changes in soil tannins to obtain a broad picture of factors affecting the microbial populations. In contrast to previous suggestions that condensed tannins in soil come chiefly from the throughfall, stem flow, and leaching or decay of fallen leaves (Rice and Pancholy, 1973), Moleski's evidence indicated that most come from root exudates. This result certainly makes the practice questionable of trenching to demonstrate the amount of root uptake of nutrients. Nitrate nitrogen accumulated in the clear-cut area but was not detected at all in the forest, and numbers of *Nitrosomonas* and *Nitrobacter* were higher in the cleared area. Correlation coefficients between abiotic factors and the numbers of nitrifiers indicated that these factors were more conducive to nitrification in the forest than in the clear-cut area. Moleski concluded that ''the evidence supports the hypothesis that the lower amount of nitrate in the forest was due to an inhibition of nitrification (probably by condensed tannins), the uptake of nitrate by the vegetation, or a combination, and not due to a change in microclimate.'' The overall evidence suggests that the decreasing concentrations of soil nitrate during succession probably result from a combination of increased uptake and decreased nitrification, the relative importance of each depending on the types of ecosystems involved, and the stage of succession.

Jones and Richard (1977) studied the effect of reforestation on turnover of nitrate and ammonium in Queensland, Australia, and reported that addition of both lime and pine needles to the soil suppressed the nitrifiers; immobilization of NH_4 by heterotrophic bacteria dominated. Pine needles alone stimulated fungi to immobilize ammonium nitrogen.

Lodhi (1977, 1978b) investigated soil properties in relation to the distance from the tree trunks of several forest species. He found that nitrate nitrogen had the greatest overall variability at different distances from the trunks. Ammonium nitrogen was always considerably higher than nitrate nitrogen when compared at each distance from the trunk of each species, and he stated that the low nitrate could not have been due to its uptake by intact vegetation because the soil samples were taken before active growth. Additionally, low numbers of *Nitrosomonas* and *Nitrobacter* were associated with large amounts of ammonium nitrogen in most samples. Lodhi concluded that the high ratio of NH_4 to NO_3 was due to an inhibition of nitrification and that the differences under different species were caused by variations in the type of litter.

Lodhi (1979c) found that the ratio of ammonium nitrogen to nitrate nitrogen increased with succession on mine-spoils in North Dakota. Moreover, the numbers of *Nitrosomonas* and *Nitrobacter* decreased with succession. He concluded that the vegetation inhibits nitrification and the inhibition increases with succession, thus resulting in less loss of nitrogen.

The dynamics of climax ponderosa pine stands in western North Dakota were studied to determine the influence of plant-produced chemicals on nitrification (Lodhi and Killingbeck, 1980). Low levels of nitrate nitrogen relative to ammonium nitrogen and low numbers of *Nitrosomonas* and *Nitrobacter* in the soils suggested that nitrification rates were low. The low nitrification rate could not have been due to low pH because the soils were alkaline (pH 7.25 to 7.75). Evidence suggested that the reduction in nitrate synthesis was due to the production and subsequent transfer to the soil of allelochemics toxic to *Nitrosomonas*. Several compounds inhibitory to nitrification were found in extracts from ponderosa pine needles, bark, and A-horizon soils.

Foliar leachates, leaf extracts, and bud extracts of balsam fir (*Abies balsamea*) and balsam poplar were added to ammonium percolation solution to determine their effects on nitrification in soils (Thibault *et al.,* 1982). Foliar leachates of balsam fir, a climax species, strongly inhibited nitrification, whereas leachates of balsam poplar, a transitional species, were somewhat less inhibitory. Five percent aqueous extracts of balsam fir needles and dormant buds of balsam poplar completely prevented oxidation of ammonium. Even 2% extracts virtually prevented oxidation of ammonium to nitrate, with balsam fir again being more inhibitory than balsam poplar.

Vitousek and his colleagues questioned the hypothesis that inhibition of nitrification increases during succession (Vitousek and Reiners, 1975; Vitousek, 1977; Robertson and Vitousek, 1981). Robertson and Vitousek (1981) questioned the use of pool sizes of ammonium and nitrate and numbers of nitrifiers to indicate relative rates of nitrification. This is a valid criticism because it would obviously be much more closely related to the real world if rates of nitrification could be measured in the field. Unfortunately, no such technique exists at present. They chose to measure nitrification by taking soil samples into the laboratory under ideal and completely artificial conditions and to use these results to indicate rates of nitrification in the real world. Their method completely prevents the continued addition of allelochemics, which could occur in the field and would speed the dilution and breakdown of any allelochemics present in the soil at the time of collection. This could cause the population of nitrifiers to build rapidly and cause nitrification in all samples.

I feel strongly that, if tests of pool sizes and numbers of nitrifiers are consistent throughout the year in indicating low numbers of nitrifiers and high NH_4 to NO_3 ratios, this is a more realistic indication of a low rate of nitrification in the field than laboratory incubation studies. I too have found that soil samples from plots with the consistency described above, can be incubated under ideal conditions in the laboratory with relatively good rates of nitrification occurring after a lag period of a few weeks. This result actually supports the inference that some factor or factors in the field are inhibiting nitrification, instead of disproving it.

Robertson and Vitousek (1981) investigated "potential" nitrogen mineralization and nitrification in soils from a primary sere in the Indiana Dunes and a secondary sere in the William L. Hutcheson Memorial Forest in New Jersey, using the laboratory incubation technique. They reported that "potential" nitrogen mineralization in soils from the primary sere increased through the first five stages and then leveled off. Nitrogen mineralization was relatively constant in soils from the secondary sere except that the highest rates were observed in the oldest site. "Potential" nitrification was very closely correlated with nitrogen mineralization in soils from both seres, except that low nitrification occurred in one site with substantial mineralization. They concluded that their results did not support the hypothesis that nitrification is progressively inhibited during the course of ecological succession.

As a matter of fact, one look at the amounts of organic carbon in the successional stages of each sere would have caused me to postulate their results concerning "potential" rates of nitrification under ideal conditions free of any regular additions of allelochemics. Unfortunately, Robertson and Vitousek reported numbers of nitrifiers and concentrations of nitrate and ammonium in soils of the various successional stages for only one sample time (August). The variability was so great that no consistent trend could be detected. The concentration of ammonium was significantly greater in the two oldest stages in the primary sere than in the four earlier stages and was also significantly greater in the climax forest in the secondary sere than in the three earlier stages of that sere. The oldest stage in each sere also had the lowest mean number of ammonium-oxidizing bacteria, but the difference from some of the other counts was not statistically significant in either case because of the large variability. It might be very enlightening to make careful measurements in these seres of ammonium and nitrate and numbers of nitrifiers throughout the year, with sufficient numbers of individual soil samples analyzed from each successional stage to decrease standard errors.

Considerable additional evidence from other types of investigations supports the hypothesis that nitrification is increasingly inhibited during succession. If the ratio of ammonium to nitrate does increase with succession, there should be a selection of species along the sere in favor of those which selectively absorb and grow better on ammonium. This has been reported by some investigators (Wiltshire, 1973; Haines, 1977). Whilshire (1973) found that climax grasses in Rhodesia showed a greater preference for ammonium than did earlier seral species or crop cultivars. Haines (1977) carried out innovative field investigations on the relative uptake of nitrate and ammonium by plants from different stages in a South Carolina old-field to forest sere. He found that nitrate uptake decreased, whereas ammonium uptake increased with succession.

If the hypothesis is correct, species that are adapted to a particular successional stage must be adapted to the form of nitrogen available to them. Thus, the enzymatic systems for absorbing and incorporating nitrogen should reflect the form of nitrogen used. Nitrate reductase is the enzyme involved in the first step of the reduction of nitrate into a form the plant can use. This enzyme thus controls the ability of a plant to use nitrate; therefore, species that use mostly nitrate as their nitrogen source should have relatively high levels of nitrate reductase activity (NRA). Conversely, species that use mostly ammonium as their nitrogen source should have relatively low levels of NRA.

In a study of nitrate utilization by species from acidic and calcareous soils, Havill *et al.* (1974) found that ruderal plants had much higher nitrate reductase activities than climax plants. Bate and Heelas (1975) reported that *Sporobolus pyramidalis,* a pioneer species in the Rhodesian grasslands, had a NRA 4 times that of the climax species, *Hyparrhenia filipendula.* Franz and Haines (1977) found that the NRA of vascular plants decreased during an old-field to forest succession in South Carolina. Smith and Rice (1983) measured the NRA of leaves and roots of characteristic species from different stages of an old-field to prairie sere in Oklahoma and found that the pioneer species had high NRA levels, whereas the four climax species had low NRA levels.

Overall, the weight of evidence at this time supports the hypothesis that nitrification is increasingly inhibited during succession in many seres. Nevertheless, there are some ecologists who strongly disagree with this hypothesis as I pointed out earlier. There is a definite need for the development of a technique that can accurately measure the rate of nitrification under field conditions. It appears to me that this will be necessary before the matter can be resolved satisfactorily.

Inhibition of Nitrification in Agriculture.   Russell's (1914) report on the inhibition of nitrification by crop plants was mentioned previously. There was little work done on this subject again until the 1970s, probably because of the suggested explanation of Russell's results by Lyon *et al.* (1923) involving a different mechanism, which was already an accepted one at that time.

Moore and Waid (1971) did an elegant series of experiments on the effects of exudates of ryegrass (*Lolium perenne*), wheat, Cos lettuce, salad rape (*Brassica napus* var. *arvenis*), and onion on nitrification in a clay loam soil from a cultivated site at the University Farm at Reading, England. The procedure in each experiment was as follows: (1) The soil was percolated intermittently for 28 to 30 days with ammonium and nutrient solutions until nitrification proceeded at a steady rate; (2) the soil was percolated intermittently for a desired experimental period with ammonium solution and solution which had percolated through quartz chips containing the living roots of the test species; (3) control soil was percolated intermittently during the experimental period with ammonium solution only; (4) the leachates of the test and control soils were collected at regular intervals and analyzed immediately for ammonium, nitrate, and nitrite; and (5) the amounts of ammonium disappearing from, and nitrate appearing in, the solutions that had leached through the soil columns were calculated. Corrections were made for retention in the plant containers and additions of nitrate from the plant containers and the nutrient solution. Results were expressed on a rate per day basis. Modifications of this procedure were used also in which various stages were repeated on a cyclic basis.

All test species reduced the rate of nitrification, but the effects of rape and lettuce were only temporary. Ryegrass root exudates had the most pronounced and persistent effects and reduced the rate of nitrification up to 84%. The control rate remained virtually constant over an 80-day experimental period after equilibrium was reached. Wheat root exudate also caused a pronounced and persistent decrease in rate over the 80-day experimental period, as did onion root exudate. As neither microbial immobilization of inorganic nitrogen nor denitrification appeared to be taking place in the presence of the root exudates, they concluded that the exudates contained inhibitors that retarded nitrification.

Lodhi (1981) divided an abandoned corn field in Saint Louis County, Missouri, into two plots. All the aboveground corn residues were removed from one plot, whereas the residues were left in the other plot. Removal of the corn residues caused a significant increase in biomass of the pioneer species in 4 of the 5 years following abandonment. In spite of the increased biomass in the clear-cut area, the concentrations of nitrate and ammonium in the soil of this area were significantly higher than in the area with the corn residues (Fig. 23). Moreover, the concentration of nitrate was mich higher than that of ammonium in the clear-cut area, whereas the reverse was true in the uncleared area. The numbers of *Nitrosomonas* were also always significantly higher in soil from the clear-cut plot. Five phenolic phytotoxins were identified in corn residue and in soils of both the clear-cut and the uncleared plots. Two of the inhibitors, ferulic and *p*-coumaric acids, were much higher in concentration in soil of the uncleared plot than in the clear-cut plot for 4 years after the start of the experiment. Lodhi concluded, therefore, that clear-cutting increased the rate of nitrification by removing the inhibitors of that process.

There has been increasing concern about the loss of nitrate from arable lands, both from the viewpoint of conservation of nitrogen and the prevention of contamination of ground and surface waters. Additionally, surplus nitrate is subject to denitrification and loss by volatilization, and considerable amounts of N_2O, a serious air pollutant, can be produced under some denitrifying conditions. Therefore, many compounds have been developed and tested in agriculture for the inhibition of nitrification (Boswell and Anderson, 1974; Bundy and Bremner, 1974). Huber *et al.* (1977) reviewed the literature on nitrification inhibitors as "new tools for food production." They cited many examples of large increases

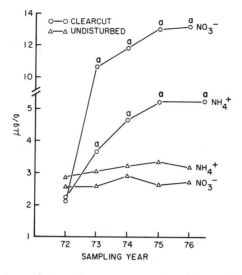

Fig. 23.   Ammonium and nitrate nitrogen concentrations during study years. (a) Differences in concentrations between clear-cut plots and undisturbed plots are significantly different at the 0.05 level in their respective sampling years after abandonment. (After Lodhi, 1981.)

in crop yields in various parts of the United States because of addition of nitrification inhibitors. They also cited many cases of increased protein and/or total nitrogen and decreased nitrate in numerous crop plants due to addition of nitrification inhibitors. They concluded that the use of nitrification inhibitors may markedly increase the efficiency of food production, reduce energy requirements for growing crops, decrease the incidence of plant disease, and reduce the pollution potential of N fertilizers.

10

Chemical Nature of Allelopathic Agents

My goal in this chapter is to discuss the different types of organic compounds that have been identified as allelochemics produced by microorganisms or higher plants. I have no intention of naming every individual compound which has been implicated in an allelopathic role. It would be a tremendous task to discuss all antibiotics and marasmins (see Chapter 1, Section II) that have been identified; thus, only certain representative ones will be discussed. The detailed biosynthesis of the various types of inhibitors will not be discussed, although this is a fascinating area. However, phytoncides and kolines from higher plants will be reviewed more thoroughly.

Most chemical inhibitors are compounds that have been termed secondary substances by Fraenkel (1959) and Whittaker and Feeny (1971), because they are of sporadic occurrence and thus do not appear to play a role in the basic metabolism of organisms. There are many thousands of such compounds, but only a limited number of them have been identified as toxins involved in allelopathy. Whittaker and Feeny (1971) stated that, with few exceptions, the secondary compounds could be classified into five major categories: phenylpropanes, acetogenins, terpenoids, steroids, and alkaloids. They pointed out further that the phenylpropanes and alkaloids originate from a small number of amino acids and the rest originate generally from acetate. As the authors pointed out, the flavonoids, of course, are hybrids in this scheme, because one ring arises from phenylalanine and the other has an acetate origin (see Section I).

The term *acetogenins* is attractive in that it conveniently includes all the diverse secondary compounds that arise from acetate. Unfortunately, it does not help much in indicating chemical similarities. I have chosen therefore to devise a system that has fourteen categories, plus a catch-all (miscellaneous). Most antibiotics, marasmins, phytoncides, and kolines, which have been identified, fit in one of the fourteen chemical categories, but a few do not. The categories are given in capital letters in Fig. 24, and the known or suspected metabolic pathways are indicated [based on Neish (1964), Brown (1964), and Robinson

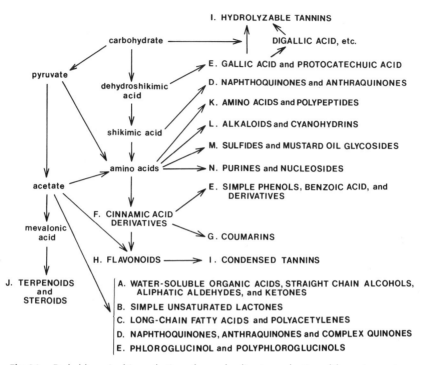

Fig. 24. Probable major biosynthetic pathways leading to production of the various categories of allelopathic agents. Letters refer to subheadings under which categories are discussed in next section of this chapter.

(1983)]. It is evident from the diagram that the inhibitors arise either through the acetate or through the shikimic acid pathway. Several types of inhibitors, which originate from amino acids, actually come through the acetate pathway. These include some of the amino acid and polypeptide inhibitors, some of the alkaloids, probably some of the sulfides, and the purines and nucleosides (Neish, 1964; Whittaker and Feeny, 1971; Robinson, 1983). The other types of inhibitors that originate from amino acids apparently arise from phenylalanine or tyrosine and these compounds are formed from shikimic acid.

I. TYPES OF CHEMICAL COMPOUNDS IDENTIFIED AS ALLELOPATHIC AGENTS

A. Simple Water-Soluble Organic Acids, Straight-Chain Alcohols, Aliphatic Aldehydes, and Ketones

Evenari (1949) pointed out that the concentrations of several organic acids such as malic, citric, acetic, and tartaric acids in fruits are often high enough to inhibit seed germination. He also stated that unripe grains of corn and unripe seeds of peas will not germinate because of the presence of acetaldehyde.

Three organic acids (malonic, citric, and fumaric) were exuded from seeds of *Pinus resinosa* and inhibit germination of the zoospores and growth of *Pythium afertile* (Agnihotri and Vaartaja, 1968). Acetaldehyde, propionic aldehyde, acetone, methanol, and ethanol were emitted as volatile growth inhibitors by beet, tomato, sweet potato, radish leaves, and carrot roots in closed systems (Dadykin *et al.*, 1970). Prutenskaya *et al.* (1970) determined that several organic acids were among the toxins produced from decomposing soybean residues. According to Gaidamak (1971), several organic acids of an aliphatic series were exuded as toxins by roots of cucumber and tomato plants. Patrick (1971) reported that acetic and butyric acids were among the toxins produced during decomposition of rye residues, and Chou and Patrick (1976) found that butyric acid was produced in decomposing corn residues. Tang and Waiss (1978) reported that salts of acetic, propionic, and butyric acids were the chief phytotoxins produced in decomposing wheat straw. Traces of isobutyric, pentanoic, and isopentanoic acids also occurred.

Many of the green, brown, and red algae also produce organic acids, aldehydes, and alcohols (Katayama, 1962; Maksimova and Pimenova, 1969). Water-soluble organic acids make up as much as 25% of the extracellular compounds produced by *Chlorella vulgaris* (Maksimova and Pimenova, 1969).

B. Simple Unsaturated Lactones

There are of course many types of lactones of varying complexity, but only the simple ones that arise from acetate are included in this category. The more complex ones, such as the coumarins and cardiac glycosides (steroids), will be discussed in later categories.

Parasorbic acid (Fig. 25) was identified from the fruits of *Sorbus aucuparia* and is very inhibitory to seed germination and seedling growth and also is antibacterial (Evenari, 1949). The aglycone of ranunculin (Fig. 25), protoanemonin, is also inhibitory to seed germination and to many bacteria. This compound is produced by several species of the Ranunculaceae (Evenari, 1949).

Several well-known antibiotics, such as patulin and penicillic acid (Fig. 25), are simple lactones (Evenari, 1949; Neish, 1964). All are of course inhibitory to at least some microorganisms, and patulin is very inhibitory to higher plants (Norstadt and McCalla, 1963). Patulin is produced by a number of fungi according to Norstadt and McCalla, and they found that *Penicillium urticae* produces large amounts when growing on wheat straw residue. Börner (1963a,b) reported that *Penicillium expansum* produces significant amounts of patulin during de-

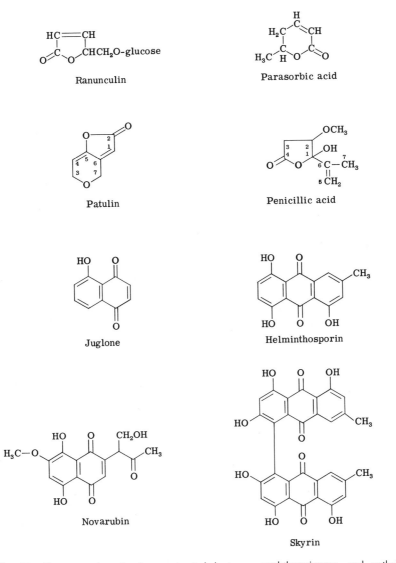

Fig. 25. Representative simple unsaturated lactones, naphthoquinones, and anthraquinones which have been implicated in allelopathy.

composition of apple root and leaf residues, and he feels this may be important in the apple replant problem as previously indicated (see Chapter 3, Section II,C). Penicillic acid is produced by *Penicillium cyclopium* (Neish, 1964) and probably by other species.

C. Long-Chain Fatty Acids and Polyacetylenes

The polyacetylenes are included here because they are apparently derived from long-chain fatty acids (Fenical, 1975; Robinson, 1983). There is rapidly expanding evidence that these two groups of compounds are very important in allelopathy.

Spoehr *et al.* (1949) found that the inhibitors from *Chlorella* (see Chapter 6) are apparently highly unsaturated fatty acids of C_{16} and C_{18} series, and they are active only after photooxidation resulting in fatty acids with 12 or fewer carbons.

Proctor (1957b) concluded that the inhibitor produced by *Chlamydomonas reinhardii*, which is very toxic to another alga, *Haematococcus pluvialis* (see Chapter 6), was probably a long-chain fatty acid or a mixture of such acids. He tested eight such acids, nonanoic, decanoic, lauric, myristic, palmitic, stearic, oleic, and linoleic, and all inhibited *Haematococcus* and other algae. McCracken *et al.* (1980) demonstrated conclusively that the toxicity of *C. reinhardii* is due to the fatty acids produced. They identified 15 fatty acids ranging in length from 14 to 20 carbons. Seven of the identified compounds were tested against four algal genera, and it was concluded that unsaturated compounds were most inhibitory and that, in general, toxicity increased with an increase in double bonds. Katayama (1962) reported that many green, brown, and red algae produce many fatty acids and that both the lower and higher fatty acids are antibacterial.

Long-chain fatty acids may also be important in allelopathic interactions in higher plants. Alsaadawi *et al.* (1983) found nine fatty acids in decomposing residue of *Polygonum aviculare* (myristic, palmitic, linolelaidic, oleic, stearic, arachidic, 11,14-eicosadienoic, heneicosanic, and behenic acids). All except myristic and heneicosanic acids also occurred in soil under decomposing *Polygonum* residue. Sodium salts of all these fatty acids inhibited seedling growth of bermudagrass in a concentration of only 5 ppm. All of them also inhibited some test strains of *Azotobacter* and *Rhizobium*.

Grümmer (1961) reported that the polyacetylene, agropyrene, is produced by *Agropyron repens* and that it is antimicrobial. Subsequently, many polyacetylenes were indentified, and most are distributed in about 12 families of dicotyledons(Krupa, 1974). The family Compositae has by far the largest number. The allelochemics produced by two strongly allelopathic pioneer plant species, *Solidago altissima* and *Erigeron annuus*, were polyacetylenes (Numata *et al.*, 1973). *S. altissima* produces *cis*-dehydromatricaria ester

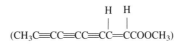

whereas *E. annuus* produces *cis*- and *trans*-matricaria ester and *cis*-lachnophyllum ester. Kobayashi *et al.* (1980) reported that three additional species of *Erigeron* produced the same polyacetylene inhibitors. They also reported that the *cis-* and *trans-*dehydromatricaria esters were found in soil of *S. altissima* communities in concentrations inhibitory to test plants.

Campbell *et al.* (1982) reported that α-terthienyl, produced by roots of *Tagetes erecta,* caused 50% mortality in seedlings of four test species in concentrations from 0.15 to 1.93 ppm. Phenylheptatriyne, produced by leaves of *Bidens pilosa,* caused 50% mortality in the same test species in concentrations ranging from 0.66 to 1.82 ppm. Obviously these compounds are very toxic to plants. Safflower (*Carthamus tinctorius*) produces two antifungal polyacetylenes, safynol and dehydrosafynol, which increase rapidly in concentration when safflower is infected with *Phytophthora drechsleri,* a root and stem rot organism (Krupa, 1974).

Marx (1969b) reported that the ectotrophic mycorrhizal fungus *Leucopaxillus cerealis* var. *piceina* produces three polyacetylenes, which are antimicrobial. Diatretyne nitrile is antifungal, whereas diatretyne amide and diatretyne 3 are antibacterial but not antifungal (see Chapter 4). Diatretyne nitrile completely inhibited germination of zoospores of *Phytophthora cinnamomi* in a concentration of 2 ppm. Several other fungi are known to produce polyacetylenes and the biological roles should be investigated (Krupa, 1974). Some of the red algae produce halogenated acetylenes, and some of these substances have been shown to cause cytotoxicity (Fenical, 1975).

D. Naphthoquinones, Anthraquinones, and Complex Quinones

Some anthraquinones and naphthoquinones are thought to be formed by head-to-tail condensation of acetate units with decarboxylation, reduction, and oxidation reactions occurring in addition to cyclization (Neish, 1964; Robinson, 1983). Other naphthoquinones are known to be produced by the shikimate pathway (Thomas and Threlfall, 1974), and some anthraquinones are known to be produced by the addition of an isoprene unit to a naphthoquinone following synthesis of the naphthoquinone by the shikimate pathway (Robinson, 1983).

Juglone, a toxin from walnut trees, was identified by Davis (1928) as 5-hydroxynaphthoquinone (Fig. 25). This is the only inhibitor produced by higher plants that is definitely known to be a naphthoquinone. Wilson and Rice (1968) found a naphthalene derivative in the leachate of leaves of *Helianthus annuus.* They suggested that it might be a derivative of α-naphthol. Novarubin (Fig. 25) is a marasmin produced by the pathogenic fungus *Fusarium solani,* and it has been isolated from various cultures of the fungus and from diseased pea plants (Owens, 1969). It causes diseased plants to wilt.

Numerous anthraquinones are produced by higher plant and by fungi (Harborne and Simmonds, 1964). I know of none, however, that has been identified as an allelopathic compound in higher plants. Several antibiotics and marasmins have been identified as anthraquinones, however (Neish, 1964; Owens, 1969). The antibiotic helminthosporin (Fig. 25) produced by the fungus *Helmintho-sporium graminium* is one example (Neish, 1964). Skyrin (Fig. 25) is a dianthra-quinone marasmin produced by the chestnut blight fungus, *Endothia parasitica;* it changes water permeability of cells at the low concentration of 10^{-6} M (Owens, 1969).

The tetracycline antibiotics, such as aureomycin, are dimeric quinones and are even more complex than skyrin (Whittaker and Feeny, 1971). Aureomycin is produced by *Streptomyces aureofaciens* (Wright, 1956).

E. Simple Phenols, Benzoic Acid, and Derivatives

The compounds listed in this category have a somewhat mixed origin. Some compounds, such as gallic (Fig. 29) and protocatechuic acids, originate directly from dehydroshikimic acid (Neish, 1964), some, such as phloroglucinol, are formed directly from acetate (Robinson, 1983), but apparently most are derived from cinnamic acid (Neish, 1964; Robinson, 1983).

This category and the cinnamic acid derivatives have been the most commonly identified allelopathic compounds produced by higher plants. As long ago as 1908, Schreiner and Reed reported that vanillin, vanillic acid, and hydroquinone (Fig. 26) are among phenolic compounds commonly produced by plants that are inhibitory to seedling growth. Hydroquinone is the aglycone of arbutin. Gray and Bonner (1948b) identified 3-acetyl-6-methoxybenzaldehyde as the toxin pro-duced in leaves of *Encelia farinosa.* Börner (1959) found that phloroglucinol and *p*-hydroxybenzoic acid are two inhibitors produced during the breakdown of phlorizin from apple root residues.

p-Hydroxybenzoic acid and vanillic acid are the most commonly identified benzoic acid derivatives involved in allelopathy. These were identified as impor-tant inhibitors in the following cases: in *Camelina alyssum* (Grümmer and Beyer, 1960); in soil (Whitehead, 1964); in residues of corn, wheat, sorghum, and oats (Guenzi and McCalla, 1966a); in cropped soil (Guenzi and McCalla, 1966b); in many soils in France (Hennequin and Juste, 1967); in soils in Taiwan (Wang *et al.*, 1967a,b); and in sugar beet flower clusters (Battle and Whittington, 1969). Guenzi and McCalla (1966a,b) found syringic acid (Fig. 26) in the same residues and soils also. Lodhi (1976, 1981) reported *p*-hydroxybenzoic acid in bottom-land forest soil and in soil with corn residues in Missouri, and he also found syringic acid in the soil of the abandoned cornfield.

Arbutin was found to be the inhibitor in *Arctostaphylos uva-ursi* (Winter, 1961) and one inhibitor in the leachate of foliaceous branches of *Arctostaphylos glandulosa,* along with hydroquinone, gallic, protocatechuic, vanillic, and *p*-hy-droxybenzoic acids (Chou and Muller, 1972). Chou and Muller also found *p*-hydroxybenzoic and syringic acids in soil under the shrub. Gallic acid is one

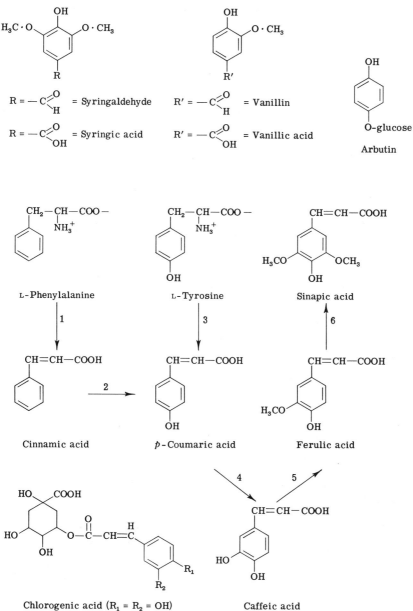

Fig. 26. Some benzoic and cinnamic acid derivatives and related compounds which have been identified as allelopathic agents. The proposed pathway of biosynthesis of cinnamic acid and its derivatives is shown. (After Neish, 1964.)

inhibitor produced by *Euphorbia corollata* (Rice, 1965a), *Eurphorbia supina* (Rice, 1969), and *Eucalyptus camaldulensis* (del Moral and Muller, 1970). Gentistic acid (2,5-dihydroxybenzoic acid) is one of the toxins produced by *Celtis laevigata* (Lodhi and Rice, 1971) and *Eucalyptus globulus* (del Moral and Muller, 1969).

Hattingh and Louw (1969b) isolated a strain of *Pseudomonas* from the rhizoplane of *Trifolium repens,* which inhibited development of clover seedlings. They suspected that the inhibitor was 2,4-diacetylphloroglucinol, but did not confirm it. Sulfosalicylic acid (2-hydroxy-5-sulfobenzoic acid) is one of the inhibitors produced by crabgrass (Parenti and Rice, 1969), and phenolcarbonic acid is a potent inhibitor exuded from the roots of tomatoes (Gaidamek, 1971).

At least one blue-green alga, *Calothrix brevissima,* and numerous genera and species of red algae produce brominated phenols (Fenical, 1975). Many of these compounds are antibacterial and some inhibit the growth of algae.

Most brown algae produce polyphloroglucinols; these are very inhibitory to many species of algae (Craigie and McLachlan, 1964; Ragan and Craigie, 1978).

F. Cinnamic Acid and Derivatives

These compounds are clearly derived from phenylalanine or tyrosine (Fig. 26) through the shikimic acid pathway (Neish, 1964), and they are widespread in higher plants. They have been implicated in many cases of allelopathy since Schreiner and Reed (1908) demonstrated that cinnamic acid (Fig. 26), *o*-coumaric acid, and *o*-hydrocoumaric acid were among several organic compounds produced by plants, which are inhibitory to seedling growth. Bonner and Galston (1944) identified *trans*-cinnamic acid as the toxin produced by roots of guayule. There are many examples in which cinnamic acid derivatives were implicated as allelopathic agents, and a few of these are given below. Chlorogenic and caffeic acids (Fig. 26) are fungistatic agents produced by potatoes in response to inoculation with *Helminthosporium carbonum* (Kuć *et al., 1956*), and *p*-hydroxyhydrocinnamic acid is an inhibitor produced in the breakdown of phlorizin from apple residues (Börner, 1959). Several cinnamic acid derivatives were identified as germination inhibitors in many species and families (van Sumere and Massart, 1959); several cinnamic acid derivatives including chlorogenic acid were identified as growth and germination inhibitors in many dry fruits (Varga and Köves, 1959); ferulic acid is one inhibitor produced by *Camelina alyssum* (Grümmer and Beyer, 1960); *p*-coumarylquinic (Fig. 26) and chlorogenic acids were identified in apple leaves (Williams, 1960) and may be involved in the apple replant problem. Whitehead (1964) identified *p*-coumaric and ferulic acids in several soils; chlorogenic acid is one inhibitor in *Galium mollugo* (Kohlmuenzer, 1965a); chlorogenic and isochlorogenic acids are the chief inhibitors in *Helianthus annuus* (Rice, 1965a); and chlorogenic acid, isochlorogenic acid, and a glucose ester of caffeic acid are toxins produced by *Ambrosia psilostachya* and *A. artemisiifolia* (Rice, 1965c).

Ferulic and *p*-coumaric acids are among the inhibitors present in residues of corn, wheat, sorghum, and oats and in soil under these crops (Guenzi and McCalla, 1966a,b); chlorogenic and *p*-coumaric acids are inhibitors in *Sorghum halepense* (Abdul-Wahab and Rice, 1967); *p*-coumaric and ferulic acids were found in many cultivated soils (Hennequin and Juste, 1967); a glucose ester of ferulic acid is one of six phenolic inhibitors produced by *Bromus japonicus* (Rice and Parenti, 1967); and *p*-coumaric and ferulic acid were found in soils in Taiwan (Wang *et al.*, 1967a,b). *p*-Coumarylquinic and chlorogenic acids are inhibitors in fog drip from *Eucalyptus globulus* (del Moral and Muller, 1969); ferulic and *p*-coumaric acids are two of the inhibitors present in flower clusters of sugar beet (Battle and Whittington, 1969); chlorogenic and isochlorogenic acids are inhibitors in *Digitaria sanguinalis* (Parenti and Rice, 1969); caffeic, chlorogenic, *p*-coumaric, and ferulic acids are among the toxins in the leaf leachate of *Eucalyptus camaldulensis* (del Moral and Muller, 1970); chlorogenic acid, isochlorogenic acid, neochlorogenic acid, band-510, and *o*-coumaric acid are toxins produced by *Platanus occidentalis* (Al-Naib and Rice, 1971); and ferulic, caffeic, and *p*-coumaric acids are toxins in *Celtis laevigata* leaves (Lodhi and Rice, 1971). Several cinnamic acid derivatives are produced during decomposition of corn and rye residues (Chou and Patrick, 1976). *p*-Coumaric and ferulic acids are inhibitors in *Sporobolus pyramidatus* (Rasmussen and Rice, 1971), chlorogenic acid was found in the leachate of leaves of *Arctostaphylos glandulosa,* and *p*-coumaric, ferulic, and *o*-coumaric acids were found in soil under the plant (Chou and Muller, 1972). Isochlorogenic acid is a mixture of 4,5-, 3,4-, and 3,5-dicaffeoylquinic acids (Corse *et al.,* 1965). Apparently, it sometimes contains a small amount of monoferuloyl-mono-caffeoylquinic acid also.

Many other more recent examples could be cited in which cinnamic acid derivatives have been identified as allelopathic agents, but the examples given should suffice. It is noteworthy that only *p*-coumaric and ferulic acids have generally been found in appreciable concentrations in soil under plants that have been demonstrated to be allelopathic due presumably to the production of cinnamic acid derivatives. Lodhi (1976) also found caffeic acid in appreciable concentrations in soils in bottomland forests in Missouri. He stated that caffeic, ferulic, *p*-coumaric, and *p*-hydroxybenzoic acids were the most persistent allelochemics in those soils. It appears, therefore, that most of the cinnamic acid derivatives other than caffeic, ferulic, and *p*-coumaric acids are only temporarily effective as allelopathic agents.

G. Coumarins

The coumarins are lactones of *o*-hydroxycinnamic acid (Robinson, 1983). Various kinds of side chains may be present, with isoprenoid ones being especially common. They occur in all parts of plants and are widely distributed in the plant kingdom.

Schreiner and Reed (1908) reported that coumarin (Fig. 27) and esculin (6-

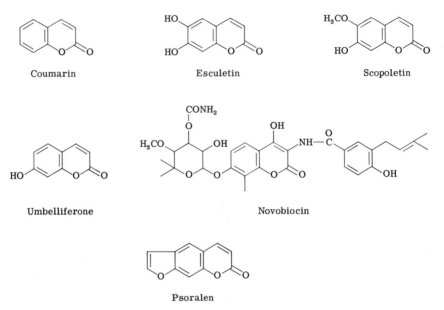

Fig. 27.   Some coumarins that have been implicated in allelopathy.

glucoside of esculetin, Fig. 27) are very inhibitory to the growth of wheat seedlings. Evenari (1949) also listed coumarin as a very potent inhibitor of seed germination. Esculin is a toxin produced by *Aesculus hippocastanum*, and coumarin is one produced by *Melilotus alba* (Winter, 1961). A glycoside of esculetin, probably esculin, is produced prominently in *Phleum pratense* roots (Avers and Goodwin, 1956). Scopoletin (Fig. 27) and/or its 7-glucoside, scopolin, have been reported as inhibitors in oat roots (Goodwin and Kavanagh, 1949; Martin, 1957), in *Galium mollugo* (Kohlmuenzer, 1965a), in *Platanus occidentalis* (Al-Naib and Rice, 1971), and in *Celtis laevigata* (Lodhi and Rice, 1971). Van Sumere and Massart (1959) listed several coumarins as inhibitors of seed germination in many species and families of plants.

Rice and Pancholy (1974) reported that scopolin is a strong inhibitor of nitrifying bacteria. It is produced by three oak species and two herbaceous species important in old-field succession in Oklahoma. Scopolin was found in soils of bottomland forests in Missouri in January and April (Lodhi, 1976).

The antibiotic novobiocin (Fig. 27) is a coumarin with some rather complex side chains (Neish, 1964). It is produced by *Streptomyces niveus*.

There are a few examples of furanocoumarins that have been identified as plant inhibitors. These are compounds with a furan ring fused with the benzene ring of coumarin (Robinson, 1983). Psoralen (Fig. 27) is produced by some species of *Psoralea* (Robinson, 1983; Baskin *et al.*, 1967) and by *Aegle marmelos* (Sinha-Roy and Chakraborty, 1976). Byakangelicin, isopimpinellin, and another unidentified furanocoumarin are produced by *Thamnosma montana* (Bennett and Bonner, 1953), and pimpinellin, bergapten, isobergapten, and angelicin are furanocoumarin allelochemics produced by seeds of *Heracleum laciniatum* (Junttila, 1976). Camm *et al.* (1976) found that the phytotoxin produced by *Heracleum lanatum* is the furanocoumarin xanthotoxin. Bishop's weed (*Ammi majus*) produces twelve furanocoumarins in the fruits, of which xanthotoxin is most concentrated (Friedman *et al.*, 1982). This compound was very inhibitory to germination of seeds of *Anastatics hierochuntica* in low concentrations.

H. Flavonoids

The flavonoids have a basic C_6—C_3—C_6 skeleton (Fig. 28) in which the A ring is of acetate origin and the B ring of shikimic acid origin (Neish, 1964). The

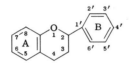

Common flavonoid skeleton

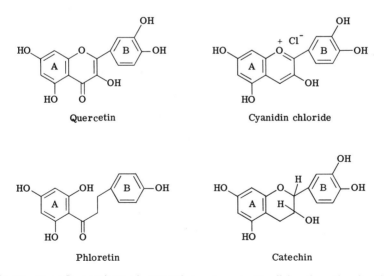

Quercetin

Cyanidin chloride

Phloretin

Catechin

Fig. 28.   Some flavonoids (Neish, 1964) that are important in allelopathy, either directly or indirectly. Catechin is thought to be involved along with flavan-3,4-diols in the synthesis of condensed tannins, and cyanidin chloride is formed in the hydrolysis of condensed tannins by concentrated HCl (Brown, 1964.)

largest group of flavonoids is characterized by having a pyran ring linking the three-carbon chain with one of the benzene rings (Robinson, 1983). There is a huge variety of flavonoids, and they are very widespread in seed plants (Harborne and Simmonds, 1964). In spite of the large number and wide distribution, only a few have been implicated in allelopathy. This may be due in part to the difficulties involved in identifying many of the flavonoids and their numerous glycosides.

Börner (1959) found that phlorizin in apple root residues inhibited the growth of apple seedlings. This compound is the 6-glucoside of phloretin (Fig. 28), and Börner found that phloretin, phloroglucinol, p-hydroxyhydrocinnamic acid, and p-hydroxybenzoic acid are produced from phlorizin during decomposition of apple root residues. All of them inhibit the growth of apple seedlings. Börner (1959) also found quercitrin, a glycoside of quercetin (Fig. 28), in apple bark but he did not test it for inhibitory activity. Williams (1960) reported that numerous glycosides of quercetin and kaempferol, and epicatechin and catechin are all present in apple residues. Possibly these compounds play roles in the apple replant problem in addition to those played by phlorizin and its breakdown products.

Kohlmuenzer (1965a) identified the flavonoid diosmetin trioside as one of the toxins produced by *Galium mollugo;* and Grümmer (1961) listed quercitrin as an inhibitor in leaves of *Artemisia absinthium.* Fottrell *et al.* (1964) found that the flavonol myricetin, which is present in some legume seeds, is inhibitory to *Rhizobium.* Rice and Pancholy (1974) found numerous flavonoids and their glycosides in several herbaceous species from the tall grass prairie and the post oak–blackjack oak forest that were inhibitory to nitrifying bacteria and to seed germination. Lodhi and Killingbeck (1980) found quercitin in ponderosa pine needles, bark, and soil under the pine and found it to be toxic to soil suspensions of *Nitrosomonas.*

Chang *et al.* (1969) reported that red clover soil sickness is caused by five isoflavonoids, which are degraded to phenolic acids after being exuded from the roots.

I. Tannins

1. Hydrolyzable Tannins.   The only logical reason for including the hydrolyzable and condensed tannins in one category is that both types possess an astringent taste and tan leather. The former contains ester linkages, which can be hydrolyzed by boiling with dilute mineral acid; this is not true of the condensed tannins. The most common hydrolyzable tannins are sugar esters of gallic acid or of gallic and hexaoxydiphenic acid (Robinson, 1983) (Fig. 29). Ellagic acid (Fig. 29) is produced from hexaoxydiphenic acid on hydrolysis (Robinson, 1983). Chebulic acid (Fig. 29) is another secondary product formed from the

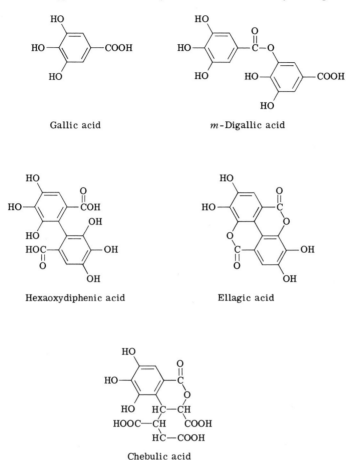

Fig. 29.   Some acid components of hydrolyzable tannins. (After Robinson, 1983.)

hydrolysis of some tannins. Some hydrolyzable tannins are complex mixtures of several phenolic acids (Robinson, 1983). Digallic (Fig. 29) and trigallic acids sometimes result from the mild hydrolysis of hydrolyzable tannins in addition to gallic acid. There are obviously many kinds of hydrolyzable tannin molecules possible, and they are very difficult to identify specifically. They are widespread in dicotyledonous plants (Bate-Smith and Metcalfe, 1957; Swain, 1965), but only a few researchers have implicated them in allelopathy.

They have been identified in the following cases: as growth and germination inhibitors in several dry fruits (Varga and Koves, 1959); as growth retarders of nitrogen-fixing and nitrifying bacteria in *Euphorbia corollata, E. supina,* and *E. marginata* (Rice, 1965a,b, 1969); as growth inhibitors of *Rhizobium* from *Rhus copallina* (Blum and Rice, 1969); as reducers of seedling growth in *Carpinus betulus* (Mitin, 1970); as inhibitors of seed germination and seedling growth in *Arctostaphylos glandulosa* (Chou and Muller, 1972); and as reducers of nitrification produced by *Quercus marilandica, Q. stellata,* and *Q. velutina* (Rice and Pancholy, 1973). Plant residues that have hydrolyzable tannins often contain either gallic or ellagic acid, or both, and sometimes digallic acid. All of these are inhibitors of nitrification in soil and occur in some forest soils in concentrations above that required to inhibit nitrification (Rice and Pancholy, 1973).

Lodhi (1976) reported that digallic, ellagic, and gallic acids were important phytotoxins in bottomland forest soils in Missouri. They were the most persistent allelochemics in the soil after caffeic, ferulic, *p*-coumaric, and *p*-hydroxybenzoic acids.

2. Condensed Tannins. Condensed tannins apparently arise by the oxidative polymerization of catechins (Fig. 28) and flavan-3,4-diols (Brown, 1964). The latter have OH-groups at the three and four positions in the pyran ring. Condensed tannins are only partially broken down by rather drastic heating with concentrated acid to release cyanidin chloride (Fig. 28), which has a bright red color, and some red-brown polymers often termed *phlobaphenes* (Robinson, 1983). Obviously, there are huge numbers of condensed tannins, just on the basis of the length of the polymer chain alone.

There are even fewer reports of the involvement of condensed tannins in allelopathy than of hydrolyzable tannins. Harris and Burns (1970, 1972) reported that tannins in the grains of certain sorghum hybrids inhibit preharvest seed germination and preharvest seed molding. They did not specify the type of tannins but work in my laboratory has shown that the grasses analyzed for tannins contain only condensed tannins (Rice and Pancholy, 1973). I have found nothing else in the literature on the chemistry of grass tannins. Mitin (1970) identified the seedling growth inhibitors in dead leaves of beech, *Fagus silvatica,* as condensed tannins.

Basaraba (1964) reported that both hydrolyzable and condensed tannins inhibited nitrification when added to soil. Rice and Pancholy (1973) found that condensed tannins are possibly important inhibitors of nitrifying bacteria in the following species of the tall grass prairie, the post oak–blackjack oak forest, and the oak–pine forest: *Andropogon gerardii, A. virginicus, Schizachyrium scoparium, Aristida oligantha, Panicum virgatum, Sorghastrum nutans, Pinus echinata, Quercus marilandica, Q. stellata,* and *Q. velutina.* Lodhi and Killingbeck (1980) identified condensed tannins in ponderosa pine needles, bark, and soil under the pine, and these compounds strongly inhibited soil suspensions of *Nitrosomonas.*

Starkey and his colleagues demonstrated that condensed tannins markedly inhibited the rate of decomposition of organic matter in soil (Benoit and Starkey,

1968a,b; Benoit *et al.*, 1968). This could be very important in mineral cycling in some ecosystems.

Somers and Harrison (1967) isolated condensed and hydrolyzable tannins from wood shavings and found that all tannin fractions inhibited germination and hyphal growth of spores of *Verticillium alboatrum,* which causes the *Verticillium* wilt disease. The main condensed tannin of highest molecular weight was most inhibitory.

J. Terpenoids and Steroids

The terpenoids and steroids have basic skeletons derived from mevalonic acid or a closely related precursor (Robinson, 1983). They consist of five-carbon isoprene or isopentane units linked together in various ways and with different types of ring closures, functional groups, and degrees of saturation. According to Robinson, the term *terpenoids* is preferred over *terpenes* because the former includes all compounds built of isoprene units, regardless of the functional groups, whereas the latter refers only to hydrocarbons. The basic types of terpenoids are the monoterpenoids (C_{10}), sesquiterpenoids (C_{15}), diterpenoids (C_{20}), triterpenoids (C_{30}), and tetraterpenoids (C_{40}).

Higher plants produce a great variety of terpenoids (Robinson, 1983), but only a few of them have been implicated in allelopathy. Microorganisms generally do not appear to produce large numbers and varieties of terpenoids, but certain fungi produce terpenoids and mixed terpenoids that are important as marasmins (Owens, 1969). Several red algae produce halogenated monoterpenes, sesquiterpenes, and diterpenes, which are antibacterial and often antialgal (Fenical, 1975).

The monoterpenoids are the major components of the essential oils of plants (Robinson, 1983), and they are also the predominant terpenoid inhibitors that have been identified from higher plants. Sigmund (1924) reported that several monoterpenoids from essential oils are toxic to seed germination and to the growh of certain bacteria. Evenari (1949) suggested that the monoterpenoids and aromatic aldehydes may be chiefly responsible for the inhibitory activity of essential oils.

Camphene, camphor, cineole, dipentene, α-pinene, and β-pinene were identified by Muller and Muller (1964) as the volatile inhibitors produced by *Salvia leucophylla, S. apiana,* and *S. mellifera* (Fig. 30). Camphor and cineole were found to be most toxic to root growth of test seedlings, and these two terpenes were later identified in the air around *Salvia* plants in the field (C. H. Muller, 1965). Asplund (1968) investigated the relationship between the structure of ten monoterpenes and the inhibition of germination of radish seeds. The compounds tested were (+)camphor, (−)camphor, (+)pulegone (Fig. 30), (−)borneol, 1,8-cineole, limonene (Fig. 30), α-phellandrene (Fig. 30), *p*-cymene, α-pinene, and

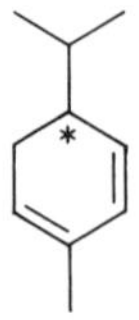

α-Phellandrene

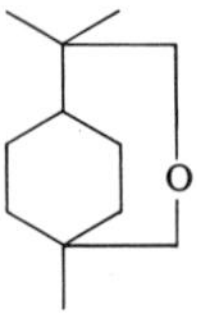

1:8 Cineole

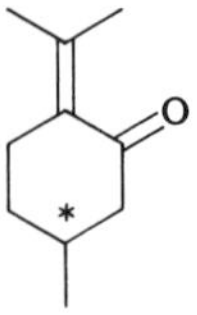

Pulegone

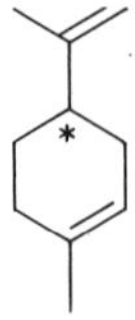

Limonene

α-Pinene

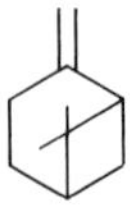

β-Pinene

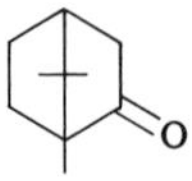

Camphor

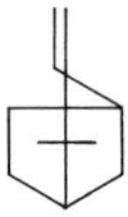

Camphene

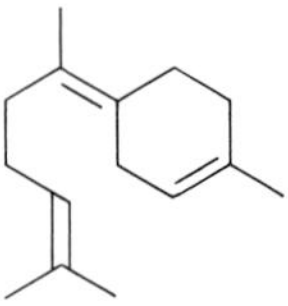

γ-Bisabolene

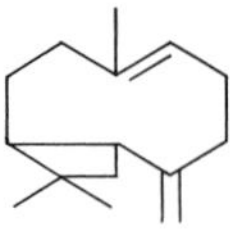

Caryophyllene

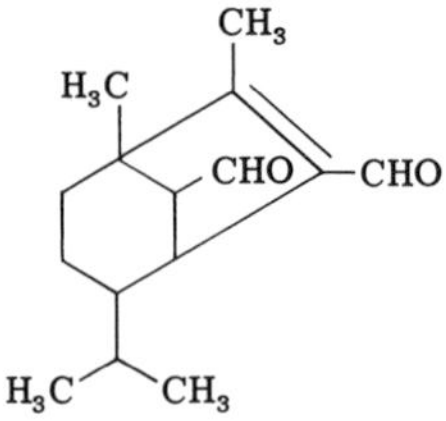

Helminthosporal

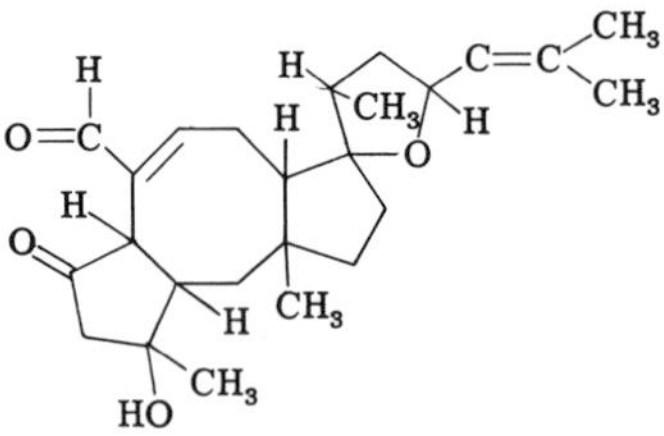

Ophiobolin

Fig. 30. Some terpenoids, steroids, and compounds with isoprenoid side chains that have been implicated in allelopathy. (Helminthosporal, ophiobolin, alternaric acid, zinniol, and ascochitine after Owens, 1969).

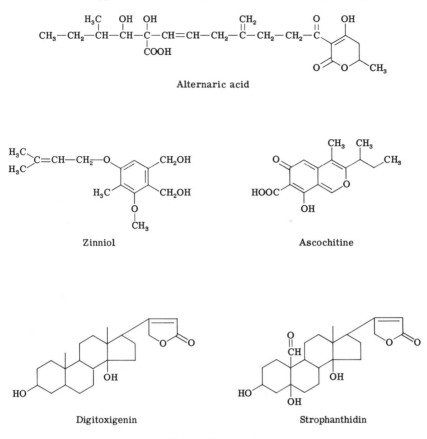

Fig. 30. (Continued)

β-pinene. Those compounds with a functional ketone group, the two camphors and pulegone, were much more inhibitory than all the others. β-Pinene was least toxic, followed by 1,8-cineole, and the rest were similar in activity.

Cineole, α-phellandrene, α-pinene, and β-pinene are volatile inhibitors produced by *Eucalyptus camaldulensis;* cineole and α-pinene are apparently most important in the allelopathic activity of this species because these were adsorbed on the soil in significant amounts (del Moral and Muller, 1970).

Artemisia absinthium produces three sesquiterpene inhibitors, β-carophyllene (Fig. 30), bisabolene (Fig. 30), and chamazulene (Grümmer, 1961). *Ambrosia psilostachya* and *Ambrosia acanthicarpa* produce several sesquiterpenes, some of which may be inhibitory (Miller *et al.*, 1968; Geissman *et al.*, 1969). This has not been confirmed. However, Neill and Rice (1971) showed that *Ambrosia psilostachya* produces volatile plant growth retarders and these might be sesquiterpenes. Abscisic acid is a sesquiterpene and a very important hormone in plants. It is also involved in allelopathy as one of the inhibitors of seed germination that is present in leaves of *Fagus silvatica* (Mitin, 1971) and in flower clusters of sugar beet (Battle and Whittington, 1969).

White and Starratt (1967) isolated a phytotoxic compound, which they named zinniol (Fig. 30), from cultures of *Alternaria zinniae*. It is not strictly a terpenoid but does have an isoprene side chain. This fungus causes leaf and stem blight of zinnia, sunflower, and marigold. Owens (1969) suggested that zinniol is possibly the marasmin responsible for the stem withering, chlorosis, and leaf-tip curling characteristic of the disease. White and Starratt found that the compound also inhibits seed germination and has weak activity against fungi and bacteria.

The sesquiterpenoid helminthosporal (Fig. 30) is a marasmin produced by the fungus *Cochliobolus sativus,* which causes common root rot of cereals (Owens, 1969). Ophiobolin (Fig. 30) is a mixed terpenoid-like compound that is a marasmin produced by *Cochliobolus miyabeanus,* which causes a leaf spot of rice (Owens, 1969). Alternaric acid (Fig. 30) is a lactone with a long terpenoid side chain, which is thought to be a secondary determinant of the early blight disease of various species of the Solanaceae caused by *Alternaria solani* (Owens, 1969). Ascochitine (Fig. 30) is a sesquiterpenoid-like compound, which was isolated from culture filtrates of *Ascochyta fabae,* the causal fungus of brown spot disease of broad bean (Owens, 1969). Application of small amounts of the toxin to coleoptiles of broad bean causes the brown necrotic spots characteristic of the disease.

Many halogenated monoterpenes, sesquiterpenes, and diterpenes are produced by numerous species of red algae, and these allelochemics are generally antibacterial and often antialgal (Fenical, 1975).

The basic steroid nucleus is the same as that of tetracyclic triterpenoids, but only two methyl groups are attached to the ring system (Robinson, 1983). Few steroids have been implicated in allelopathy. Digitoxigenin and strophanthidin (Fig. 30) are two well-known examples which have strong antimicrobial activity (Evenari, 1949). Both are aglycones of cardiac glycosides. Digitoxigenin is the aglycone of the digilanides A, B, and C produced by foxglove, *Digitalis pupurea,* and strophanthidin is the aglycone of the convallatoxin produced by *Convallaria majalis* (Robinson, 1983).

K. Amino Acids and Polypeptides

Amino acids and peptides are among the best-known constituents of living matter. Amino acids are caboxylic acids having at least one amino group, and peptides are polymers of two or more amino acid molecules connected by peptide (C–N) linkages (Robinson, 1983). Amino acids can be divided arbitrarily into two major groups: One group is found in all living systems in either the free state or condensed as peptides, whereas representatives of the second group occur in a limited number of organisms and do not occur in proteins.

There are only a few instances in which amino acids have been implicated in allelopathy, and in most instances the specific amino acids have not been identified. Thus, it is impossible to say at this time whether both groups of amino acids described above may be involved as inhibitors. In the case of rhizobitoxine, which is produced by certain strains of *Rhizobium japonicum,* it is an amino acid of the noncommon category (Owens, 1969; Owens *et al.,* 1972). Owens *et al.* (1972) identified it as 2-amino-4-(2-amino-3-hydroxypropoxy)-*trans*-but-3-enoic acid. This compound irreversibly inactivates β-cystathionase; thus, it inhibits the conversion of methionine into ethylene.

Gressel and Holm (1964) found that several free amino acids were the compounds present in seeds of *Abutilon theophrasti,* which inhibited seed germination of several crop plants. They did not identify the amino acids, however. Prutenskaya *et al.* (1970) reported that amino acids were among the phytotoxins produced during the decomposition of soybean plant residue, but they did not identify the specific amino acids. Gaidamak (1971) found that unspecified amino acids were among the phytotoxins exuded by roots of cucumbers and tomatoes. It would be interesting to know if the inhibitory amino acids involved in these last three cases were unusual amino acids that acted as antimetabolites in amino acid or protein synthesis.

Several of the known marasmins produced by pathogenic microorganisms are polypeptides and related glycopeptides (Owens, 1969). One of these is lycomarasmin (Fig. 34) (Owens, 1969). This toxin is produced by *Fusarium oxysporum* f. *lycopersicum* and causes wilting of tomato cuttings.

Other marasmins, which have been identified as polypeptides according to Owens (1969), are as follows: victorin produced by *Helminthosporium victoriae,* which causes blight in Victoria oats and its derivative cultivars *Helminthosporium carbonum* toxin; carbtoxinine, which is also produced by *H. carbonum;* and toxins A and B produced by *Periconia circinata,* which cause milo disease in certain grain sorghums. *H. carbonum* is pathogenic to certain corn hybrids.

Some marasmins, which have been identified as glycopeptides according to Owens (1969), are *Corynebacterium sepidonicum* toxin; *C. michiganense* toxins, I, II, III; and colletotin produced by *Colletotrichum fuscum,* which infects *Digitalis. Corynebacterium sepidonicum* infects potato plants and *C. michiganense* infects tomato plants.

L. Alkaloids and Cyanohydrins

The logic for including these types of compounds together is that they are derived from amino acids and contain nitrogen (Neish, 1964).

1. Alkaloids. The nitrogen of the alkaloids may be in heterocyclic rings or in side chains but, in any event, the simple aliphatic amines are not included in this category. The alkaloids as a group are distinguished generally from most

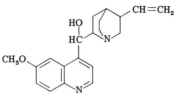

Cocaine   Physostigmine   Quinine

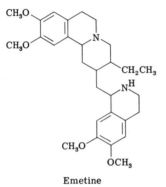

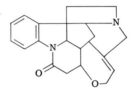

Strychnine

Emetine

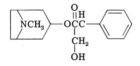

Atropine   Ephedrine

Papaverine   Codeine

Fig. 31.   Alkaloids and related compounds that have been identified as allelopathic agents. Salts and esters of fusaric and picolinic acids are alkaloids.

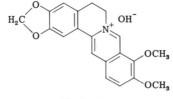

Berberine

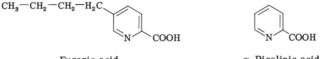

Fusaric acid   α-Picolinic acid

Fig. 31. (*Continued*)

other plant components by being basic, and they usually occur in plants as the salts of various organic acids (Robinson, 1983). Caffeine (Fig. 34) and several related compounds are often included as alkaloids but are included here with the purines because of their purine ring structures (Robinson, 1983). The alkaloids are best known for their physiological effects on man and their use in pharmacy.

Recently, there has been little work on the role of alkaloids in allelopathy. Evenari (1949) emphasized strongly, however, the importance of these compounds as seed germination inhibitors. He stated in fact that all seeds and fruits known for their high alkaloid content are strong inhibitors of seed germination and that the alkaloids are the main, if not the only, cause of inhibition. He listed the following as strong inhibitors of seed germination: cocaine, physostigmine, caffeine, quinine, strychnine, berberine, codeine, (Fig. 31), cinchonin, cinchonidin, and tropa acid. He listed narkotine, scopolamine, emetine (Fig. 31), papaverine (Fig. 31), ephedrine (Fig. 31), piperine, and atropine (Fig. 31) as weak inhibitors of seed germination.

Overland (1966) found that barley roots exude alkaloids and that gramine, which is known to be produced by barley, was inhibitory to the growth of *Stellaria media*.

Fusaric acid (Fig. 31) is a marasmin produced by many species of *Fusarium* and has been detected in infected tomato plants and wilted cotton (Owens, 1969). α-Picolinic acid (Fig. 31) also acts as a marasmin in some cases (Owens, 1969), and salts or esters of fusaric acid and α-picolinic acid are alkaloids.

2. Cyanohydrins. Tyrosine appears to be the precursor of the aglycone of dhurrin (Fig. 32), and phenylalanine appears to be the precursor of the aglycone of amygdalin and prunasin (Fig. 32) (Neish, 1964). Dhurrin occurs in sorghum (*Sorghum bicolor*) seedlings, and Conn and Akazawa (1958) found that the

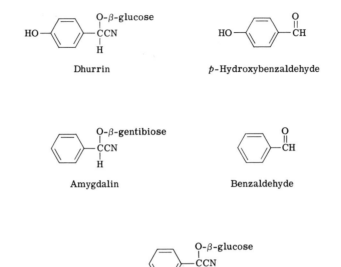

Fig. 32.   Representative cyanogenic glycosides and their hydrolysis products. (After Robinson, 1983.)

seedlings also contain enzymes that hydrolyze dhurrin to an equimolar mixture of glucose, HCN, and p-hydroxybenzaldehyde (Fig. 32). Later, Abdul-Wahab and Rice (1967) found HCN and p-hydroxybenzaldehyde among the phytotoxins produced by Johnsongrass (*Sorghum halepense*). They demonstrated that dhurrin is the source of these inhibitors in Johnsongrass.

Proebsting and Gilmore (1941) and Patrick (1955) pointed out that HCN and benzaldehyde (Fig. 32) are produced by the breakdown of amygdalin, which is present in peach root residues. They found that HCN and benzaldehyde are inhibitory to the growth of peach seedlings, but amygdalin is not.

Evenari (1949) reported that seeds of many species of the Prunaceae and Pomaceae contain large amounts of cyanogenic glucosides and that the HCN which is released slowly from these glucosides inhibits germination. He pointed out further that large quantities of HCN are released from *Crataegus* seeds just before germination and suggested that this may be the culmination of the period of after-ripening. The HCN could thus be liberated from the tissues and germination processes could proceed.

M. Sulfides and Mustard Oil Glycosides

The sulfur- or sulfur and nitrogen-containing compounds included in this category have considerable diversity but are thought to be derived from amino acids

(Robinson, 1983). The sulfides are volatile and have an offensive odor. Robinson (1983) stated that there is no conclusive evidence for the occurrence of di- and polysulfides in plants and that they probably arise through secondary transformations brought about by plant enzymes. Allicin (Fig. 33) is an example of a disulfide known to be produced enzymatically from alliin (Fig. 33) when garlic, *Allium sativum,* is crushed (Robinson, 1983). As early as 1936, McKnight and Lindegren reported that vapors from crushed garlic are bacteriocidal to *Mycobacterium cepae.* Cavallito *et al.* (1944) identified allicin as the antibacterial substance.

Mustard oils, such as allyl isothiocyanate and allyl thiocyanate (Fig. 33), are products of the hydrolysis of mustard oil glycosides such as sinigrin (Fig. 33) (Robinson, 1983). The aglycones undergo rearrangement on hydrolysis so that they often bear little resemblance to the aglycones which exist in the glycosides. According to Evenari (1949) mustard oils are produced by all organs of plants belonging to the Cruciferae and are produced in especially large amounts in the genera *Brassica* and *Sinapis.* He pointed out that mustard oils are potent inhibitors of seed germination and of microorganisms.

Bell and Muller (1973) found that large quantities of allyl isothiocyanate were liberated when leaves of *Brassica nigra* were macerated (see Chapter 5). They also found that this mustard oil was very inhibitory to seed germination in laboratory tests, but they could not demonstrate any appreciable activity under field conditions.

N. Purines and Nucleosides

The purines are best known of course as constituents of nucleic acids, and the specific ones involved in both ribonucleic acid (RNA) and deoxyribonucleic acid

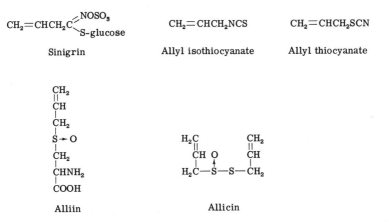

Fig. 33. Sulfides, a mustard oil glycoside and mustard oil implicated in allelopathy. (After Robinson, 1983.)

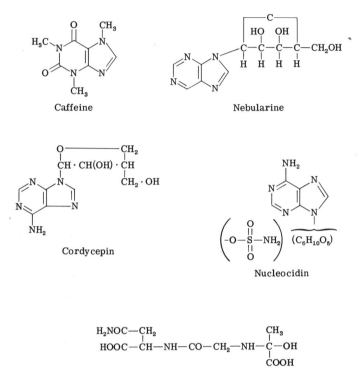

Fig. 34. A representative polypeptide (lycomarasmin-Owens, 1969), purine (caffeine), and some nucleosides (cordycepin-Bentley *et al.*, 1951; nucleocidin-Waller *et al.*, 1957; nebularine-Löfgren *et al.*, 1954) known to be involved in allelopathy.

(DNA) are adenine and guanine. When a sugar molecule is attached by a β-glycosidic bond to nitrogen in position 9 of a purine, the resulting compound is called a nucleoside (Robinson, 1983). The sugar involved in DNA is deoxyribose, and the one involved in RNA is ribose.

There are several known naturally occurring purines and nucleosides in plants in addition to those that are involved in nucleic acids (Robinson, 1983). The only ones in higher plants which have been shown to be involved in allelopathy are caffeine (Fig. 34), theophylline, paraxanthine, and theobromine from the coffee tree (Chou and Waller, 1980a,b; Rizvi *et al.*, 1981). These compounds are all purines. Evenari (1949) included caffeine as an alkaloid and pointed out that it is one of the most potent alkaloids in the inhibition of seed germination.

Several antibiotics produced by various microorganisms have been shown to be nucleosides. Among these are nebularine, cordycepin, and nucleocidin (Fig. 34). Cordycepin is produced by *Cordyceps militaris*, and the sugar involved is cordycepose, which is attached to an adenine base (Bentley *et al.*, 1951). Nucleocidin is a glycoside of adenine in which sulfamic acid is bound to the sugar moiety by an ester linkage (Waller *et al.*, 1957), and nebularine has a rather unusual 6-carbon sugar attached to the 9 position of purine (Löfgren *et al.*, 1954).

O. Miscellaneous

There are several toxins that have been implicated in allelopathy, which do not fit clearly into any of the specific categories discussed, although they usually have relationships to one or more categories. Phenylacetic and 4-phenylbutyric acids are among the phytotoxins produced during the decomposition of corn and rye residues (Chou and Patrick, 1976), and these compounds are apparently produced either directly from shikimic acid or from cinnamic acid (Robinson, 1983). Phenethyl alcohol is one of the autoantibiotics produced by the fungus *Candida albicans,* and it probably fits into the same category as the compounds mentioned in the previous sentence (Lingappa *et al.*, 1969). Tryptophol is another autoantibiotic produced by *C. albicans* (Lingappa *et al.*, 1969), and this compound possibly is produced from the amino acid tryptophan, as has been suggested in higher plants (Leopold, 1955).

Ethylene ($CH_2{=}CH_2$) is the volatile allelochemic produced by various fruits, such as apples and pears (Molisch, 1937), and by decomposing litter of *Pinus radiata* (Lill and McWha, 1976). It is derived from the amino acid methionine (Owens *et al.*, 1971).

II. UNIDENTIFIED INHIBITORS

There are many cases in which significant allelopathic mechanisms are known to be operative and in which nothing is known of the toxins involved. There are no doubt important unidentified inhibitors also in numerous instances where some inhibitors have been identified. Often, the allelochemics that have been identified are those, such as phenolic compounds, that are easily detected on chromatograms.

The increasing use of gas chromatography in conjunction with mass spectrometry is making possible the identification of many types of allelochemics, which previously remained unknown. I predict there will be a rapid increase in identification of significant allelopathic compounds. A knowledge of these compounds is very important in determining methods of escape into the environment, amounts present in the environment, amounts absorbed by affected organisms, and methods and rates of decomposition.

11

Factors Affecting Amounts of Allelopathic Compounds Produced by Plants

I. INTRODUCTION

Factors affecting the amounts of allelochemics produced by plants are extremely important in allelopathy, but research on this subject was active only during a 15-year period starting in 1957. Unfortunately, very little research has been done since that time.

Early in my own research in allelopathy, I found that inhibitor plants growing in glass houses do not produce as large quantities of inhibitors as the same kinds of plants growing out-of-doors. This suggested an important effect of light quality on the production of inhibitors. My students, colleagues, and I also became interested in the possible roles of mineral deficiency in the production of inhibitors, because we were interested in old-field succession; the old fields with which we were working had been abandoned from cultivation because of low fertility (see Chapter 7). Our early work led to the study of the effects of other stress factors on the content of various phenolic inhibitors in plants. Many other persons have investigated the effects of various factors on the phenolic content of plants because of an interest in the possible physiological roles of such compounds.

There has been a great deal of research also done on the effects of various factors on the production of commercially important antibiotics by microorganisms, but that work will not be reviewed here.

II. EFFECTS OF RADIATION

A. Light Quality

1. Ionizing Radiation. Ionizing radiation markedly increases the amount of various phenolic inhibitors in tobacco and sunflower plants (Fomenko, 1968; Koeppe *et al.*, 1970a). Fomenko (1968) found that exposure of sunflower plants to 20,000 R of ionizing radiation greatly increases the concentrations of caffeic acid and quercetin. Koeppe *et al.* (1970a) investigated the effects of 1000 R, 2500 R, and 4400 R of X-irradiation on the concentration of chlorogenic acids and scopolin in tobacco plants (*Nicotiana tabacum* var. One Sucker). Amounts of chlorogenic acid (3-*O*-caffeoylquinic acid), neochlorogenic acid (5-*O*-caffeoylquinic acid), and band 510 (4-*O*-caffeoylquinic acid) were determined separately in their project and in most of the other projects discussed. However, these values will be combined and reported as total chlorogenic acids. Plants were harvested and analyzed at 12, 21, and 29 days after irradiation.

Substantial increases in scopolin due to irradiation were found at first harvest in roots, stems, and leaves, and the amounts were dose-dependent. The concentrations remained higher in leaves and stems of plants exposed to the two highest doses throughout the test period. Total chlorogenic acids were decreased by all doses except for a temporary increase in leaves at first harvest, which was caused by the highest dose.

2. Ultraviolet Radiation. Many workers have investigated the effects of UV treatment on the phenolic content of plants since Frey-Wyssling and Babler (1957) reported that greenhouse tobacco (*N. tabacum* var. Mont Calme brun) does not produce rutin and only about one-seventh of the normal amount of chlorogenic acid (Lott, 1960; Koeppe *et al.*, 1969; del Moral, 1972; Hadwiger, 1972). Frey-Wyssling and Babler (1957) found that supplementation of greenhouse light with UV light improved the growth of greenhouse tobacco and increased the chlorogenic acid content from 0.41 to 2.52%. This approached the concentration present in control plants grown out-of-doors, i.e., 2.72%.

Lott (1960), working with the same variety of tobacco as Frey-Wyssling and Babler (1957), found that the maximum increase in concentration of chlorogenic acid achieved in open air by supplementing natural radiation with UV light was 79%. On the other hand, he achieved a maximum increase of 550% in the greenhouse by supplementing the radiation with UV light. When he removed the short UV rays (below 350 mμ), from the supplemental irradiation, the maximum increase attained in the greenhouse was 287%. He found also that supplementation of normal radiation with short-wavelength UV light gave a maximum increase in concentration of rutin of 27% in open air plants and 28.5% in greenhouse plants.

Koeppe *et al.* (1969) investigated the effects of different levels of supplemental UV irradiation on concentrations of chlorogenic acids and scopolin in tobacco (*N. tabacum* var. One Sucker) and Russian Mammoth sunflower plants. All levels of UV increased the concentrations of scopolin in old leaves, young leaves, and stems of tobacco (Table 51), but only the low UV dose increased the concentration of this compound in roots. All doses increased the concentrations of total chlorogenic acids in young leaves and stems of tobacco (Table 51), and the low dose increased the concentrations in old leaves and roots. The medium dose also increased the concentration of chlorogenic acids in old leaves. All doses markedly increased the scopolin concentration in sunflower leaves, and the two highest doses increased the concentration of chlorogenic acids in sunflower leaves.

Del Moral (1972) found that supplemental UV light markedly increased con-

TABLE 51.   Effects of Varying Supplemental UV Intensities on Concentrations of Chlorogenic Acids and Scopolin in Tobacco Plants[a]

| | Phenol concn. (μg/g fresh wt of plant part) | |
Treatment[b]	Total chlorogenic acids	Scopolin
Older leaves		
Control	657	2.1
Low UV	910	3.1
Medium UV	675	11.3
High UV	464	59.3
Younger leaves		
Control	1290	4.5
Low UV	2076	7.6
Medium UV	1615	28.1
High UV	1797	38.6
Stems		
Control	143	11.0
Low UV	290	15.9
Medium UV	229	38.7
High UV	323	34.1
Roots		
Control	222	39.4
Low UV	234	46.2
Medium UV	187	37.4
High UV	89	19.5

[a] Data from Koeppe *et al.* (1969). Reproduced by permission of Microforms International Marketing Corporation.
[b] UV in mW/ft^2: low = 1–1.5; medium = 4–5; high = 5–8.

centrations of total chlorogenic acids and total isochlorogenic acids in sunflower (*Helianthus annuus*) leaves, stems, and roots.

Hadwiger (1972) reported that psoralen, plus 4 min of 366 nm UV light, caused 2 times as much phenylalanine ammonia-lyase (PAL) activity in pea plants 3 hr after irradiation as in controls and 12 times as much PAL activity 20 hr after irradiation. Psoralen alone did not have this effect. PAL is the enzyme responsible for the formation of cinnamic acid from phenylalanine; cinnamic acid derivatives and coumarins are produced from cinnamic acid (see Chapter 10).

3. Red and Far-Red Light.   Jaffe and Isenberg (1969) demonstrated that concentrations of several phenolic compounds increased at a faster rate in potato tuber disks irradiated with red light than in disks irradiated with an equivalent dose of far-red light. The only phenolics identified were ferulic and *p*-coumaric acids.

In a study on the effect of photoperiod on concentrations of alkaloids and phenolic compounds in tobacco plants, Tso *et al.* (1970) exposed some plants on each photoperiod to 5 min of red light at the end of each day (light period) and others to 5 min of far-red light. Within each photoperiod, the plants that received the red light each day had significantly higher concentrations of total alkaloids than those that received far-red. On the other hand, plants that received far-red last each day had higher concentrations of soluble phenols, particularly of chlorogenic acid. The results concerning phenolics appear to be just the opposite from those of Jaffe and Isenberg (1969). This probably is not true, however, because the experiments were different in so many ways: (1) Jaffe and Isenberg worked with potato tuber disks; (2) the disks were irradiated with the given light quality for 24 hr each day; (3) Tso *et al.* used whole tobacco plants; and (4) the plants received only 5 min of red or far-red light at the end of each light cycle. Thus, the question concerning the relative effects of these two light qualities on the phenolic content of intact plants is still unanswered.

B. Intensity of Visible Light

Zucker (1963) found that visible light stimulates the synthesis of chlorogenic acid in potato tuber disks in water, and it stimulates synthesis of *p*-coumaryl esters in similar disks in the phenylalanine culture. A brief exposure to light of low intensity doubles the rate of synthesis of chlorogenic acid over that in darkness. Later, Zucker (1969) found that some photosynthetic product is necessary in apparently very small amounts for the synthesis of phenylalanine ammonia lyase in *Xanthium* leaf disks. A weak light for a short period suffices to produce the required material, however.

Jaffe and Isenberg (1969) found that white light at an intensity of 244

$\mu W/cm^{-2}/sec^{-1}$ was not quite as effective in stimulating the formation of lignin in peeled potato tubers as red light does at an intensity of 73 $\mu W/cm^{-2}/sec^{-1}$. They did not test other intensities of white light, so it is possible that a lower intensity white light would be more effective than the intensity used. The relationship to the question being discussed is that cinnamic acid derivatives are precursors of lignin.

C. Daylength

It appears that long days increase the concentrations of phenolic acids and terpenes in plants, regardless of the daylengths required for flowering (Taylor, 1965; Burbott and Loomis, 1967; Zucker, 1969). However, Zucker *et al.* (1965) discovered that concentrations of the chlorogenic acids increase markedly in the leaves of Maryland Mammoth tobacco (short-day plant) and *Nicotiana sylvestris* (long-day plant) just prior to the change of the meristem from the vegetative to the flowering shape. This increase occurs under short days in Maryland Mammoth tobacco and under long days in *N. sylvestris*. Taylor (1965) stated in his review that biosynthesis of anthocyanins in *Kalanchoe blossfeldiana* is also regulated by the same photoperiodic conditions that regulate flowering.

Xanthium pennsylvanicum is a striking short-day plant because it will flower if given a single long, dark period. Nevertheless, much higher concentrations of chlorogenic acid, isochlorogenic acids, flavonoid aglycones, and quercetin glycosides are produced in the leaves on very long days (Taylor, 1965).

Burbott and Loomis (1967) found that *Mentha piperita* grows better on long days and produces considerably greater concentrations of monoterpenes on long days. Under 8-hour days, temperature affects the composition of the terpenes produced; warm nights produced oxidized terpenes, such as pulegone and menthofuran, whereas cold nights favored production of the more reduced compound menthone. In long days, temperature does not affect the composition, with menthone predominating whatever the temperature.

Zucker (1969) reported that the induction of phenylalanine ammonia lyase increases in leaf disks of *Xanthium pennsylvanicum* with increases in daylength. This correlates well, of course, with the report of Taylor (1965) that many phenolics increase in leaves of *Xanthium pennsylvanicum* with increases in daylength, including several cinnamic acid derivatives.

III. MINERAL DEFICIENCIES

A. Boron

Watanabe *et al.* (1961) discovered a 20-fold increase in scopolin in leaves of tobacco plants that grew in a minus boron solution for 38 days. A few years later, Dear and Aronoff (1965) found a pronounced increase in caffeic and chlorogenic acids in leaves and growing points of boron-deficient sunflower plants. They also found that the ratio of caffeic acid to chlorogenic acid increased tenfold in the leaves and fourfold in the growing points of minus-boron plants.

B. Calcium

Loche and Chouteau (1963) reported that concentrations of scopolin increase and of chlorogenic acid decrease in leaves of tobacco plants deficient in calcium. This is the only report I have seen on this subject, and much more research is needed because calcium deficiency is very common in areas with high precipitation.

C. Magnesium

Loche and Chouteau (1963) found increases in concentrations of scopolin and decreases in chlorogenic acid in magnesium-deficient tobacco leaves just as they did in the case of calcium. Their results were supported by Armstrong et al. (1971) who found exactly the same effects in magnesium-deficient tobacco leaves. The scopolin concentration did not change, however, in Mg-deficient stems but decreased in the deficient roots. The total chlorogenic acids decreased in concentration in Mg-deficient stems and roots just as in the leaves.

D. Nitrogen

Chouteau and Loche (1965) also did some of the early research on the effects of nitrogen-deficiency on concentrations of phenolics in plants. They reported an increase in concentration of chlorogenic acid in nitrogen-deficient tobacco leaves, but did not analyze roots and stems. Shortly thereafter, Tso et al. (1967) reported a direct relationship between amounts of applied nitrogen and concentrations of chlorogenic acid and scopolin in three varieties of tobacco with an inverse relationship in the fourth. For some strange reason, only the results with the fourth variety agree with the results of Chouteau and Loche (1965) and other workers (Armstrong et al., 1970; del Moral, 1972; Lehman and Rice, 1972).

Armstrong et al. (1970) found large increases in concentrations of total chlorogenic acids and scopolin in roots, stems, and leaves in nitrogen-deficient tobacco plants (N. tabacum var. One Sucker). There was almost a fivefold increase in concentration of total chlorogenic acids in leaves and stems and of scopolin in stems. Lehman and Rice (1972) found similar large increases in concentration of total chlorogenic acids in old leaves, stems and roots of nitrogen-deficient Russian Mammoth sunflower plants (Table 52). There was a very slight decrease, however, in concentration of scopolin in old leaves and stems of nitrogen-deficient sunflower plants. This last point is not of much significance, however, because the concentrations of scopolin are low in various parts of

TABLE 52.   Concentrations of Chlorogenic Acids and Scopolin in
Nitrogen-Deficient and Control Sunflower Plants 5 Weeks from
Start of Treatment[a]

Plant organ and treatment	Phenol concn. (μg/g fresh wt of plant part)	
	Total chlorogenic acids	Scopolin
Older leaves		
Control	1139	7.2
Deficient	8884	6.4
Younger leaves		
Control	1737	—[b]
Deficient	873	—[b]
Stems		
Control	383	1.8
Deficient	3275	—[b]
Roots		
Control	303	—[b]
Deficient	490	—[b]

[a] Data from Lehman and Rice (1972).
[b] Below amounts determinable by procedure used.

sunflower plants. The increase in concentration of total chlorogenic acids in nitrogen-deficient old leaves was about eightfold, and it was about eight and a half-fold in N-deficient stems.

Del Moral (1972) reported very large increases in the concentrations of total chlorogenic and isochlorogenic acids in roots, stems, and leaves of nitrogen-deficient sunflower plants. There was about a ten and a half-fold increase in concentration of total chlorogenic acids in N-deficient plants, based on his weighted mean data, and approximately an eightfold increase in concentration of isochlorogenic acids.

The great increases in concentrations of inhibitors in plants that result from nitrogen deficiency are probably of great significance in allelopathy, because there are large areas of land deficient in nitrogen. This is extremely important in connection with allelopathic mechanisms operating in revegetation of infertile old fields (see Chapter 7).

E. Phosphorus

Loche and Chouteau (1963) reported increases in scopolin concentration and decreases in chlorogenic acid in tobacco leaves deficient in phosphorus. A large increase in the concentration of the isomers of chlorogenic acid was observed in extracts of sunflowers grown under phosphate-deficient conditions (Koeppe *et al.,* 1976). Moreover, larger amounts of phenolic compounds were leached from the living intact roots, dried roots, and tops of phosphate deficient plants than from phosphate-sufficient ones (Table 53). These leachates contained scopolin, but none of the isomers of chlorogenic acid or caffeic acid.

There is a widespread deficiency of easily soluble phosphorus in many soils (Rice *et al.,* 1960). Over one-half the soils tested in eastern Oklahoma have been found to be deficient in phosphorus, with the subsurface soils having even lower amounts than the surface soils. Available phosphorus is particularly low in old fields, which have been abandoned because of low fertility (Rice *et al.,* 1960). Thus, the pronounced increases in concentrations of inhibitors in plants resulting from phosphorus deficiency are probably very important in the allelopathic mechanisms operating during old-field succession (see Chapter 7).

F. Potassium

Chouteau and Loche (1965) reported decreased concentrations of chlorogenic acid in leaves of potassium-deficient tobacco plants. The results of Armstrong *et al.* (1971) supported this report; they found decreases in concentrations of total chlorogenic acids in roots, stems, and leaves of tobacco plants (One Sucker) maintained on a minus-potassium solution for 3 to 5 weeks. Concentrations of scopolin, however, almost doubled in leaves of potassium-deficient tobacco plants and were slightly higher in roots and stems.

TABLE 53. Quantities of Phenolic Compounds Found in Leachates of Sunflowers Grown under (+) or (−) Phosphate Conditions[a]

Leachate type	Plant age (days)	±P	Phenol concn. (mg/g dry wt of root or top)
Intact roots	34	+P	0.069
		−P	0.14
Intact roots	31	+P	0.048
		−P	0.22
Intact roots	29	+P	0.14
		−P	0.42
Dried tops	34	+P	0.14
		−P	0.22
Dried tops	30	+P	0.16
		−P	0.40
Dried roots	34	+P	0.23
		−P	0.46
Dried roots	30	+P	0.22
		−P	0.44

[a] From Koeppe *et al.* (1976). Reproduced by permission of the National Research Council of Canada from the Can. J. Bot. **54.**

Lehman and Rice (1972) found that maintenance of Russian Mammoth sunflower plants on a minus-potassium solution for 5 weeks caused marked increases in concentrations of total chlorogenic acids in young leaves and stems (Table 54). Concentrations were also increased in old leaves during the period from 1 to 4 weeks after start of treatment. Concentrations of scopolin were increased in old leaves, young leaves, and stems of potassium-deficient sunflower plants (Table 54). In fact, the concentration in potassium-deficient old leaves was more than doubled, 5 weeks after the start of treatment.

G. Sulfur

Lehman and Rice (1972) reported that the concentrations of total chlorogenic acids were substantially increased in old leaves, young leaves, stems, and roots of Russian Mammoth sunflower plants grown in a minus-sulfur solution for 5 weeks (Table 55). The concentrations of scopolin were slightly increased also in old leaves and roots of the sulfur-deficient plants, but slightly decreased in stems.

The increases in concentrations of total chlorogenic acids resulting from sulfur deficiency were surprisingly great, ranking second in amount only to the extremely large increases resulting from nitrogen deficiency.

TABLE 54. Concentrations of Chlorogenic Acids and Scopolin in Potassium-Deficient and Control Sunflower Plants 5 Weeks after Start of Treatment[a]

Plant organ and treatment	Phenol concn. (μg/g fresh wt of plant part)	
	Total chlorogenic acids	Scopolin
Older leaves		
Control	1139	7.2
Deficient	832	16.5
Younger leaves		
Control	1737	—[b]
Deficient	2001	1.1
Stems		
Control	383	1.8
Deficient	1458	2.1
Roots		
Control	303	—[b]
Deficient	178	—[b]

[a] Data from Lehman and Rice (1972).
[b] Below amounts determinable by procedure used.

TABLE 55. Concentrations of Chlorogenic Acids and Scopolin in
Sulfur-Deficient and Control Sunflower Plants 5 Weeks after the
Start of Treatment[a]

	Phenol concn. (μg/g fresh wt of plant part)	
Plant organ and treatment	Total chlorogenic acids	Scopolin
Older leaves		
Control	1139	7.2
Deficient	4399	7.7
Younger leaves		
Control	1737	—[b]
Deficient	4272	—[b]
Stems		
Control	383	1.8
Deficient	1192	0.6
Roots		
Control	303	—[b]
Deficient	464	1.8

[a] Data from Lehman and Rice (1972).
[b] Below amounts determinable by procedure used.

H. General Mineral Deficiency in an Ecosystem

Mature leaves of abundant trees in rain forest vegetation on infertile acid white-sand soils of the Douala-Edea Reserve, Cameroon, contained about 2 times the concentration of phenolic compounds found in similar rain-forest vegetation on fertile lateritic soils of the Kibale Forest, Uganda (McKey *et al.*, 1978). There was substantial evidence that the high phenolic content of the leaves in the Douala-Edea forest serves as a feeding deterrent to herbivores, and it probably increases the allelopathic potential of at least some species. There is little doubt that the effects of soil fertility on concentrations of secondary compounds in plants play many important roles in ecosystems.

IV. WATER STRESS

All the factors discussed so far that result in increased concentrations of inhibitors represent stress conditions to the plants. Water stress is certainly a very obvious stress condition, but very little has been done to determine its effect on the inhibitor content of plants.

Del Moral (1972) used NaCl in the culture solution to cause water stress of

TABLE 56. Effects of Stress Factors on Concentrations of Total Chlorogenic Acids and Total Isochlorogenic Acids in Sunflower Plants[a]

	Phenol concn. (μg/g dry wt of plant)[b]	
Stress applied	Total chlorogenic acids	Total isochlorogenic acids
None, control	43	135
UV Light	113	203
$- H_2O$	258	320
UV; $- H_2O$	455	512
$-$ Nitrogen	458	1065
$-$ Nitrogen; UV	310	375
$-$ Nitrogen; $- H_2O$	645	2185
$-$ Nitrogen; $- H_2O$; UV	546	979

[a] Data from del Moral (1972).
[b] Weighted mean of leaf, stem, and root tissues.

sunflower (*H. annuus*) plants. The osmotic potential of the solution in the drought stress vessels ranged from -4.0 to -4.3 atmospheres during a 24-hr test period. After 31 days of treatment, drought stress resulted in substantial increases in concentrations of total chlorogenic and isochlorogenic acids in roots, stems, and leaves over amounts in control plants.

Del Moral also tested the effects of combinations of stress factors and found

TABLE 57. Effect of Chilling Temperatures on Concentrations of Chlorogenic Acids and Scopolin in Tobacco Plants 24 Days after Start of Treatment[a]

	Phenol concn. (μg/g fresh wt of plant part)	
Organ and treatment	Total chlorogenic acids	Scopolin
Older leaves		
Control	1204	9.4
Chilled	3812	12.2
Younger leaves		
Control	1714	4.0
Chilled	4387	1.1
Stems		
Control	205	109.0
Chilled	965	96.5
Roots		
Control	2460	555.0
Chilled	860	178.0

[a] Data from Koeppe et al. (1970b).

that a combination of water stress and exposure to supplemental UV light increased the concentrations of total chlorogenic and isochlorogenic acids, more than either factor alone, with normal nitrogen (Table 56). With nitrogen deficiency however, the stimulatory effects of drought plus UV light were less than with drought alone. The greatest increases in concentrations of total chlorogenic and isochlorogenic acids, on a whole-plant basis, resulted from a combination of drought stress and nitrogen deficiency. This combination resulted in a 15-fold increase in concentration of total chlorogenic acids and a 16-fold increase in concentration of total isochlorogenic acids.

The synergistic effects of stress factors are particularly important because they generally occur in combinations under field conditions. Conditions that cause low fertility in soils, such as excessive erosion, often result in soils with lower infiltration rates and thus in soils that are often deficient in available water. These combinations would increase, therefore, the allelopathic potentials of inhibitory species.

α-Pinene concentration increased in water-stressed loblolly pine (*Pinus taeda*), whereas β-pinene, myrcene, and limonene decreased (Gilmore, 1977). Camphene concentration increased with soil-moisture stress in one test year and decreased during the second test year.

V. TEMPERATURE

Martin (1957) found that about seven and one-half times as much scopoletin exuded from roots of oak plants in 72 hours at 30°C as in 135 hours at 19°C. Obviously, this does not necessarily relate directly to the amount of scopoletin produced, but it does relate to the intensity of any allelopathic effect resulting from the scopoletin.

Koeppe *et al.* (1970b) maintained tobacco plants (One Sucker) on temperatures of either 32°C (control) or 8° to 9°C (chilled) during a 16-hr light period each day, and all plants were subjected to a dark period temperature of 15 to 16°C each day. Chilling increased the concentrations of total chlorogenic acids markedly in old leaves, young leaves, and stems, but it decreased the concentration in the roots (Table 57). Chilling also increased the concentration of scopolin slightly in old leaves, but it decreased the concentration substantially in young leaves and roots. The net effect was an increase in inhibitory material in the form of chlorogenic acids.

VI. ALLELOPATHIC AGENTS

Commercial phytocides and allelopathic agents represent stress factors for affected plants, and, thus, it is of value to determine what effects these factors have on the inhibitor content of affected plants.

Dieterman *et al.* (1964a) discovered that tobacco plants sprayed with a 1000 ppm concentration of 2,4-dichlorophenoxyacetic acid (2,4-D) had a 31-fold increase in concentration of scopolin in the leaves, a 28-fold increase in the stems, and over a 4-fold increase in the roots 30 days after spraying. Dieterman *et al.* (1964b) found similar but smaller increases in Russian Mammoth sunflower plants 120 hours after spraying with a 1000 ppm concentration of 2,4-D. Winkler (1967) found that spraying tobacco plants with maleic hydrazide also substantially increased the concentrations of scopolin in the various plant parts.

Einhellig *et al.* (1970) treated tobacco (One Sucker) and Russian Mammoth sunflower seedlings with 1×10^{-4} and 5×10^{-4} M solutions of scopoletin by immersing the roots in the solutions. The plants were harvested 11 days after the start of the treatment, and the tops and roots were analyzed separately for concentrations of scopolin and chlorogenic acids. The 10^{-4} M treatment caused significant increases in concentrations of scopoletin and scopolin in both roots and shoots of tobacco (Fig. 35, 36), and the 5×10^{-4} M treatment caused dramatic increases in the same inhibitors. No significant changes in concentrations of total chlorogenic acids occurred, however. The results were similar in sunflower for all three inhibitors. Scopolin and scopoletin concentrations are usually very low in sunflower plants even after exposure to stress conditions. In the present experiments, however, the concentration of scopolin reached almost 150 μg/g fresh weight in the shoots of the 5×10^{-4} M treated plants and almost 100 μg/g in the roots.

As was previously pointed out, Hadwiger (1972) found 12 times as much PAL activity in pea plants 20 hours after treating with psoralen and irradiating with UV light as in control plants. UV light alone did not have any effect at the wavelength (366 nm) and intensity used.

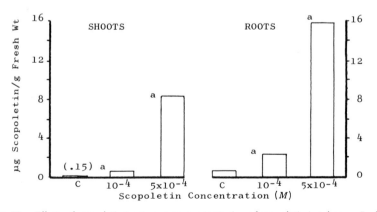

Fig. 35.   Effects of scopoletin treatment on concentration of scopoletin in tobacco. Each bar is mean of 10 samples. (C) Control; (a), significantly different from all others at 1% level. (From Einhellig *et al.*, 1970.)

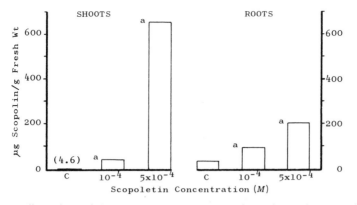

Fig. 36. Effects of scopoletin treatment on concentration of scopolin in tobacco. Each bar is mean of ten samples. (C) Control; (a), significantly different from all others at 1% level. (From Einhellig *et al.*, 1970.)

Sarkar and Phan (1974) reported that the amount of total phenols in carrot roots increased fivefold after a 3-day exposure to ethylene, and sevenfold after a 7-day exposure. Moreover, at least four new phenols were produced, which do not occur normally in carrot tissues. This mechanism could certainly enhance the effect of ethylene as an allelopathic agent.

The pronounced increases in the inhibitor content of plants exposed to a great variety of stress conditions make it appear likely that most allelopathic agents would stimulate the production of toxins by plants affected by the agents. Certainly, more research is needed on the effects of allelochemics on the production of phytotoxins by plants.

VII. AGE OF PLANT ORGANS

Koeppe *et al.* (1969) found that concentrations of scopolin and chlorogenic acids in leaves of tobacco plants varied with the ages of the leaves, even in control plants. Because of these results, Koeppe *et al.* (1970c) decided to determine the effect of tissue age on the concentrations of chlorogenic and isochlorogenic acids in native sunflower plants (*H. annuus*) that were collected in old fields near Norman, Oklahoma. Concentrations of total chlorogenic acids increased with an increase in age of leaves to node 6 (Table 58) after which they declined. On the other hand, concentrations of total isochlorogenic acids decreased with increasing age of leaves from the apex. Concentrations of both total chlorogenic acids and total isochlorogenic acids in stems decreased with increasing age from the apex to node 5 or node 6 after which the concentrations began to increase again very slowly with age.

TABLE 58.   Effect of Age on Concentrations of Chlorogenic and Isochlorogenic Acids in Leaves of the Native Sunflower, *Helianthus annuus*, in a Field[a]

| Harvest time and leaf position | Phenol concn. (μg/g fresh wt of plant part) | |
	Total chlorogenic acids	Relative isochlorogenic acids
May 2[b]		
Apex	2465	157.3
Nodes 3,4	3094	109.2
Nodes 5,6	3894	66.4
May 24[b]		
Apex	1892	182.5
Nodes 3,4	1960	146.0
Node 6	2603	159.2
Node 8	2445	100.8
Node 10	2082	88.2
Node 12	925	30.6
Node 14	863	28.4

[a] Data from Koeppe *et al.* (1970c). Reproduced by permission of Microforms International Marketing Corporation.
[b] May 2, plants 16–24-cm tall; May 24, plants 48–64-cm tall.

Koeppe *et al.* (1970b) discovered that the concentrations of scopolin and total chlorogenic acids decreased with age of leaves in tobacco but, due to the increase in size with age, the total amounts of these compounds increased with age in the leaves.

Woodhead (1981) measured the phenolic acid concentration in leaves from the same nodes in 10 cultivars of sorghum (*S. bicolor*) from the early seedling stage to heading, under field conditions. She found that the concentration in all cultivars decreased with age in healthy plants, with the decrease being particularly great after 28 days of age. The phenolic concentration increased again, however, at heading time to about the same level found in young plants.

VIII. GENETICS

All persons who work on allelopathy become aware very quickly that plants of the same species growing close together vary greatly in their allelopathic effects. Obviously, this could be due in part to differences in the microhabitats and thus to differences in stress conditions. It is logical to assume that genetics must play an important role also in determining amounts of inhibitors produced by a given plant, and the sensitivity of the plant to the stress factors discussed above.

Putnam and Duke (1974) screened 526 accessions of cucumber and found several that were strongly allelopathic to two indicator weeds (see Chapter 2, Section III). Moreover, Fay and Duke (1977) found several accessions of oats, which were allelopathic to an indicator weed. James Spruell (personal communication) has screened approximately 500 accessions of wheat for allelopathic potential and has found several accessions with considerable promise for breeding programs.

The allelopathic effects of sorghum are caused primarily by certain phenolic compounds (Burgos-Leon *et al.*, 1980), and Woodhead (1981) reported that the phenolic acid concentrations of leaves from the same nodes differed considerably among 10 cultivars of *S. bicolor* grown in similar conditions. Unfortunately, no tests were made of the allelopathic potentials of the 10 cultivars.

There are large differences in concentrations of many terpenoids in *Juniperus scopulorum* due to genotypic variation (Adams and Hagerman, 1977; Adams, 1979). Such differences could cause pronounced changes in the allelopathic potential of *J. scopulorum,* but no evidence is available on this subject.

A species such as *Ambrosia psilostachya,* which invades old fields in Oklahoma in the pioneer weed stage, almost disappears before the *Aristida oligantha* stage (second stage) becomes predominant, reappears in the perennial bunchgrass stage, and remains in the climax prairie, may vary greatly genetically in the different stages. The plants of this species, which appear in the pioneer weed stage, are very inhibitory to the growth of several species of that stage and probably help eliminate the pioneer weed stage rapidly (Neill and Rice, 1971). It is doubtful that plants of this species, which persist into the climax, exert the same allelopathic impact, but only additional research can determine whether this suggestion is correct.

There is an urgent need for research on the genetics of allelopathy. Individuals interested in both genetics and allelopathy could make important contributions to both areas by combining them.

IX. PATHOGENS AND PREDATORS

Considerable evidence was presented in Chapter 4, Section V, that infection by many pathogens causes marked increases in concentrations of phenolics and other types of chemical compounds in plants. Woodhead (1981) found that various sorghum cultivars infected with sorghum downy mildew (*Sclerospora sorghi*) or rust (probably *Puccinia purpurea*) had increased the concentrations of phenolics under field conditions.

Sorghum plants, which are hosts to shootfly (*Atherigona soccata*) or stemborer (*Chilo partellus*), also have greatly increased concentrations of phenolics (Woodhead, 1981). There is good evidence that such increases in some of the secondary plant products enhance the resistance of at least some plants to pathogens and predators, but nobody has investigated the possibility that such increases in allelochemics may increase the allelopathic effects of the infected or infested plants.

X. CONCLUSIONS

It is obvious that a moderate amount of information is available concerning the factors affecting concentrations of phenolics in plants. Moreover, a little research has been completed concerning the factors affecting concentrations of alkaloids and terpenoids; however, little or no information is available on factors affecting concentrations of other types of allelopathic agents in plants. Therefore, research is urgently needed in this area.

12

Evidence for Movement of Allelopathic Compounds from Plants and Absorption and Translocation by Other Plants

I. MOVEMENT FROM PLANTS

Evidence indicates that allelopathic compounds get out of plants by volatization, exudation from roots, leaching from plants by rain, or decomposition of residues. Much indirect evidence and some direct evidence was presented in previous chapters concerning these methods of egress of allelochemics. Some of the more direct evidence is reviewed here concerning the movement of known allelopathic agents, or closely related compounds, from plants and the uptake of these by neighboring plants. The possibility is also discussed that such compounds can move from plant to plant without intering the substrate.

A. Volatilization

Considerable research has been done on the emanation of volatile inhibitors from plants since Elmer (1932) demonstrated that a volatile substance from apples and pears inhibited the growth of potato sprouts. Molisch (1937) also became interested in the volatile substance produced by apple fruits and demonstrated that it has marked growth effects on many plants. He subsequently identified the compound as ethylene. It was his interest in this volatile allelopathic agent that caused him to coin the term *allelopathy*.

Muller *et al.* (1964) reported that *Salvia leucophylla, S. mellifera,* and *S. apiana* produce volatile inhibitors of other higher plants, and Muller and Muller (1964) identified six inhibitory terpenes in ether extracts of *Salvia* leaves. C. H.

Muller (1965) identified the two most inhibitory terpenes, camphor and cineole, in air around *Salvia* plants in the field and in the greenhouse. The volatile terpenes were adsorbed on soil and retained their inhibitory activity in the soil for several months.

Artemisia californica produces volatile compounds that are inhibitory to several species in the California annual grassland (Muller *et al.*, 1964; Halligan, 1973, 1975, 1976). Halligan (1975) identified five major terpenoids and three minor ones in leaves of this species. Camphor and 1,8-cineole were the most toxic of those identified. Many of the same volatile terpenoids were present in the soil and the litter under, and immediately adjacent to, the *Artemisia* stand.

The genus *Eucalyptus* has long been known to produce several volatile terpenes and several of these have been shown to be very toxic to seed germination and seedling growth of numerous species of plants (del Moral and Muller, 1970). Baker (1966) demonstrated that volatile growth inhibitors are produced by *E. globulus* and that they are more inhibitory to growth of *Cucumis* roots than to *Eucalyptus* roots and hypocotyls. Del Moral and Muller (1970) found that fresh leaves of *E. camaldulensis* produce large amounts of four volatile terpenes, which are toxic to plant growth, and two of these, α-pinene and 1,8-cineole, were found in large amounts in soil under the *Eucalyptus*. Concentrations of the compounds in the bare zone adjacent to the *Eucalyptus* stand were considerably lower but easily detectable, whereas no terpenes were detected in the grassland soil.

Numerous other species of plants have been shown to produce volatile inhibitors of other plants or of microorganisms, but most of the allelopathic agents have not been identified. It appears that volatile allelopathic compounds may be most significant ecologically under arid and semiarid conditions.

B. Exudation from Roots

Since the comprehensive investigation of Lyon and Wilson (1921) demonstrated that roots of several crop plants exuded large amounts of organic compounds even under sterile conditions, numerous investigators have found that many kinds of organic compounds are exuded by living roots of many species.

Lundegårdh and Stenlid (1944) reported that adenosine monophosphoric acid was the chief compound that exuded from sterile, intact roots (2–4 days old) of pea (*Pisum*) and wheat seedlings. A flavanone was also found to be exuded by wheat seedlings. The amount of nucleotide exuded in 5 hours corresponded to 0.5–1% of the total dry weight of the roots and 2–4% of the dry weight of the root tips. Fries and Forsman (1951) found that adenine, quanine, cytidine, and uridine were also exuded from sterile, intact pea roots. The weight of the compounds exuded in 5 hours amounted to 0.096% of the dry weight of the roots. These studies are pertinent because several allelopathic compounds have been identified as purines or nucleosides (see Chapter 10).

Bonner and Galston (1944) reported that cinnamic acid was one of two strong inhibitors of guayule seedling growth found in the leachate of living guayule roots. This compound is known to occur in the free form in the guayule plant, so these researchers stated that it most likely came from the roots and not from microbial activity. Many cinnamic acid derivatives have been identified as allelopathic agents, so it is likely that they can be exuded from roots of plants that produce them.

Phenylacetic acid and several of its derivatives have been identified as allelopathic compounds (Chou and Patrick, 1976; Chou and Lin, 1976; Tang and Young, 1982). It is noteworthy, therefore, that several studies have demonstrated clearly that some derivatives of phenylacetic acid are readily exuded from plants and absorbed by neighboring plants. α-Methoxyphenylacetic acid (MOPA) is a compound known to have marked plant growth-modifying properties; therefore, Preston *et al.* (1954) decided to determine if this compound moved out of roots of plants treated with it and into roots of untreated plants. They demonstrated first that it is not volatile enough to cause growth defects in plants enclosed in airtight bags with an appreciable amount of MOPA present, but not in contact with the plants. Next, they applied 100 μg of MOPA mixed with 1 part of Tween 20 and 4 parts of lanolin as a narrow band around the stem of one bean plant growing in a pot with another untreated bean plant. This was replicated several times and in each case, both treated and untreated plants in the same pot showed the typical growth effects of MOPA in 2 days. When treated plants were grown in aerated water for 1 week, then removed and replaced with untreated plants, the untreated plants developed the typical MOPA growth effects in a few days.

In another experiment, snap bean, sunflower, cucumber, buckwheat, cotton, and corn were planted together in pots containing soil. In any one pot, MOPA was applied as a band to the stem of all plants of a given species, but not to the others. This was repeated in other pots with other species serving as the treated species. On the basis of the formative effects induced, MOPA moved from the bean plants to sunflower, cucumber, buckwheat, and cotton; from cucumber to bean, sunflower, buckwheat, and cotton; from sunflower to bean; from buckwheat to bean; and from cotton to bean.

Linder *et al.* (1957) used a more direct test to demonstrate that MOPA exudes from the roots of treated plants. They prepared ^{14}C-carboxyl-tagged MOPA and applied 5 μg in water to the primary leaves of Pinto bean plants, the roots of which were emersed in water or nutrient solution. The presence of the ^{14}C-tagged MOPA was first detected in the water or nutrient solution about 5 hours after application of the MOPA to the leaves and reached a maximum concentration in slightly over 100 hours. The amount of MOPA exuded from the roots was proportional to the amount applied to the leaves over the range of doses used. When the roots were not aerated, the amount of MOPA exuded was reduced by almost 80%. Thus, it is clear that α-methoxyphenylacetic acid is readily exuded from the roots of at least several species of plants. It is noteworthy that Tang and Young (1982) found α-methoxyphenylacetic acid to be one of the allelopathic compounds exuded from the roots of bigalta limpograss (*Hemarthria altissima*).

Mitchell *et al.* (1959) reported that *m*-chloro-α-methoxyphenylacetic, *m*-fluoro-α-methoxyphenylacetic, and *p*-fluoro-α-methoxyphenylacetic acids are also easily translocated through Pinto bean plants and exuded from the roots in readily detectable amounts.

Benzoic acid and several of its derivatives have been identified as allelopathic agents in many plant species (Chapter 10). *p*-Hydroxybenzoic acid and vanillic acid are common allelopathic agents reported in soils under allelopathic plants. It is significant, therefore, that several benzoic acid derivatives have been found to be exuded readily from roots of several species of plants. Linder *et al.* (1958) reported that 2,3,6-trichlorobenzoic and 2,3,5,6-tetrachlorobenzoic acids applied to stems or leaves of bean plants were absorbed, translocated, and exuded from the roots into the substrate. They were taken up by neighboring plants and caused growth defects in these untreated plants. They were also found to move from bean to sunflower and cucumber, from sunflower to cucumber and bean, from cucumber to bean and sunflower, from barley to bean, and from corn to bean. Tests were run that demonstrated that the effects were not due to evaporation and diffusion of the chemicals to the neighboring plants.

Foy *et al.* (1971) used techniques similar to those described above and found that 2-methoxy-3,6-dichlorobenzoic (*dicamba*), 2-methoxy-3,5-dichlorobenzoic, and 2,3,6-trichlorobenzoic acids were readily translocated to the roots of bean plants, exuded from the roots of treated plants, and absorbed by the roots of adjacent untreated plants. The smallest amount of dicamba that could be applied and consistently produce symptoms in the neighboring untreated plants was about 2.2 µg. However, a growth response was occasionally observed in untreated plants when as little as 0.2 µg was applied to the top of the treated plant. The transfer of these compounds was demonstrated in soil, sand, and nutrient solution.

Autoradiographs (^{14}C-tagged compounds) and bioassay data indicated that these compounds were not exuded from the roots of treated plants when the phloem was girdled in the hypocotyl with a jet of steam. This demonstrated that the growth regulators were being translocated into the roots through the phloem in ungirdled donor plants. Aeration of the roots enhanced exudation of these compounds from donor roots, and killing of the roots by steaming prevented exudation. Thus, an intact, living root system with moderately good aeration appears to be essential for exudation. This suggests that exudation is an active process.

Martin (1957) reported that scopoletin (6-methoxy-7-hydroxycoumarin) is exuded in appreciable amounts from young intact sterile roots of oat plants. In 72 hours at 30°C, 45.8 μg of scopoletin were exuded per gram dry weight of root. Coumarin and several of its derivatives (see Chapter 10) have been identified as allelopathic compounds, which are produced by several species of plants. Fay and Duke (1977) found that the allelopathic effects of several accessions of oats against crunchweed appeared to be related to the relative amounts of scopoletin exuded from the roots (Chapter 2).

Winter (1961) contributed a great deal to our knowledge concerning the exudation and uptake of allelochemics by plants. He found that wheat and bean plants growing in a solution containing 200 μg/ml of hydroquinone take up this compound, synthesize its glycoside, arbutin, and exude the arbutin back into the solution. Normally, wheat and bean plants produce no arbutin. Winter was interested in the uptake of "alien metabolic products" from different plant species because this enabled him and his colleagues to demonstrate that selected allelopathic compounds could get out of plants (or residues) that produce them and into other species that do not normally produce them. Moreover, it often causes a species to produce compounds that it never produces without the uptake of the "alien" compounds, as was true of arbutin production by wheat and bean plants from hydroquinone.

Smith (1977) collected root exudates from the unsuberized tips of new woody roots of mature *Betula alleghaniensis, Fagus grandifolia,* and *Acer saccharum* in a northern hardwood forest. Exudates were fractionated into carbohydrates, amino acids/amides, organic acids, and nine inorganic ions. Organic acids were the most abundant component.

Van Staden (1976) reported that the natural cytokinins present in corn roots are released into the growing medium and suggested that they play a major role in controlling the germination of parasitic angiosperms. He pointed out further that cytokinins, which are similar to or identical with zeatin and zeatin riboside, can be extracted from soils supporting *Pinus, Carya,* and *Acacia* species. Cytokinins are released into the medium by cultures of tomato root tips also (Koda and Okazawa, 1978). These scientists found 8 times as much cytokinin activity in the medium after 7 days of growth as in the root tissues.

Tang and Young (1982) developed an excellent technique for collecting exudates from the roots of plants. This involved circulating the nutrient solution continuously through pots containing the root system and a column containing XAD-4 resin. Adsorbed organic compounds were eluted and identified by GC-Mass Spectrometry. Use of this technique enabled them to identify 16 allelochemics in the root exudate of bigalta limpograss (Chapter 2).

The evidence is thus quite strong that many identified allelopathic compounds and related compounds are exuded from roots of plants, taken up by neighboring plants, and cause growth changes in the receptor species. Readers who desire a review of plant root exudates are referred to Rovira (1965, 1969, 1971).

C. Leaching by Rain

At least some biologists have been aware of the leaching phenomenon since early in the eighteenth century, and many kinds of inorganic ions, elements, and compounds, and organic compounds have been identified in the leachate of plant foliage (Tukey, 1966, 1969, 1971). In fact, Lee and Monsi (1963) found a statement by Banzan Kumazawa in some ancient Japanese documents, about 300 years old, that rain water or dew from leaves of *Pinus densiflora* is harmful to crops underneath. There is no question, therefore, that plants are leaky systems even when alive and more so after death. Factors affecting the leaching of materials from leaves were discussed by Tukey (1969, 1971). In this discussion of the release of inhibitors by leaching, only some of the investigations in which the allelochemics were identified in the leachate of living plants or of plant residues and were tested for toxicity will be discussed. The fact that phytotoxins have been demonstrated in extracts of various plant parts does not mean that they will leach or exude from the plant. Of course, water soluble toxins, which are still present after death of a plant part, can leach out.

Winter (1961) demonstrated that several allelopathic compounds leached from plant residues into soil were taken up by wheat plants, which do not produce any of the compounds involved. Arbutin is an allelochemic produced by *Arctostaphylos uva ursi,* and when residues of this plant were placed on soil in which wheat was planted, arbutin was subsequently found in the soil and in the roots of the wheat plants. Phlorizin is a potent allelopathic compound produced in apple roots, and when Winter placed apple roots on the surface of soil that contained no phlorizin, this compound was soon found in the soil along with phloroglucinol, a decomposition product of phlorizin. Roots of wheat plants growing in the soil contained phlorin, a sugar ester of phloroglucinol.

Horse chestnut leaves and bark contain the allelopathic compound esculin and Winter found that placement of residues of this plant on soil in which wheat was planted resulted in the presence of esculin and esculetin (the aglycone) in the soil and in the wheat roots. Both compounds strongly inhibit the growth of wheat plants, and both are derivatives of coumarin. Sweetclover (*Melilotus alba*) leaves contain appreciable amounts of coumarin, and mulching soil with leaves of this species resulted in the presence of coumarin in the soil and of a coumarin derivative in wheat plants growing in the soil.

Ten phenolic inhibitors were found in leachates of the litter from *Eucalyptus camaldulensis,* and five were identified: caffeic, chlorogenic, *p*-coumaric, ferulic, and gallic acids (del Moral and Muller, 1970). Litter leachates of *E. baxteri* were inhibitory in bioassays and contained gentisic, gallic, sinapic, caffeic, and ellagic acids, several unidentified phenolic aglycones, numerous glycosides, and terpenoids (del Moral *et al.,* 1978).

AlSaadawi *et al.* (1983) identified nine fatty acids in residues of *Polygonum aviculare* and seven of these occurred in soil under the residues. All were allelopathic to bermudagrass and to several strains of nitrogen-fixing bacteria (see Chapter 10).

Numerous allelopathic compounds have been identified in rainfall (or artificial rain) and fog drip leachates of living tops of several species of plants. In many more instances, the leachates have been shown to inhibit plant growth but the allelopathic compounds were not identified. A suspected α-napthol derivative and scopolin were identified in the leachate of sunflower leaves (Wilson and Rice, 1968).

Fog drip from *Eucalyptus globulus* leaves contained chlorogenic, *p*-coumaryl-quinic, and gentisic acids (del Moral and Muller, 1969), and foliar leachates of *E. baxteri* contained gentisic and ellagic acids (del Moral *et al.,* 1978). All these compounds inhibited the growth of test plants.

Two inhibitors were identified in the leachate of fresh leaves of *Platanus occidentalis*—chlorogenic acid and scopolin (Al-Naib and Rice, 1971). Each was inhibitory to the growth of plants. Chlorogenic acid was found also in the leachate of leaves of *Arctostaphylos glandulosa* (Chou and Muller, 1972), along with arbutin, hydroquinone, gallic, protocatechuic, vanillic, and *p*-hy-droxybenzoic acids.

The evidence is clear that many allelopathic agents that occur in living plants or in residues of such plants can leach out in appreciable amounts due to rain or dew.

D. Decomposition of Residues

The problem of determining whether inhibitors already present are being released from plant material by decay is very difficult if not impossible to solve. There is always the possibility that microorganisms change nontoxic compounds to toxic ones, as in the case of amygdalin in peach residues (Patrick, 1955). It is also possible that microorganisms synthesize inhibitors, as in the production of patulin by *Penicillium urticae* growing on wheat straw residue (Norstadt and McCalla, 1963), the production of patulin and a phenolic inhibitor by *Penicillium expansum* growing on apple residue (Börner, 1963a,b), and the production of other inhibitors by microorganisms that decompose residues of various crop plants (Patrick and Koch, 1958; Patrick *et al.,* 1963). Moreover, water-soluble inhibitors should easily leach out of plant residues after death when the various membranes lose their differential permeability. There are, of course, many potent inhibitors such as most flavonoids (aglycones), which are only slightly soluble in water; these are probably released primarily by decomposition.

It should be understood that the change of nontoxic compounds to toxic ones by microorganisms or the synthesis of toxins from other materials in the residues adds a new dimension to the possible allelopathic effects of a plant.

In Chapters 2, 3, 5, and 7, a great many allelopathic species of plants in which inhibitors were shown to be released or produced from their residues during decomposition were discussed. All of those will not be repeated here, but I will discuss briefly those cases where the allelopathic agents were identified.

Chou and Lin (1976) found that aqueous extracts of decomposing rice residues in soil inhibited the radicle growth of rice and lettuce seedlings and the growth of rice plants. Five allelochemics were identified from decomposing rice straw, and several unknowns were isolated. Those identified were *p*-hydroxybenzoic, *p*-coumaric, vanillic, ferulic, and *o*-hydroxyphenylacetic acids. The same inhibitors were identified in paddy soil, and the amounts were higher in paddies in which rice stubble was left than in paddies from which the rice stubble was removed (Chou *et al.,* 1977).

Chou and Patrick (1976) identified eighteen allelopathic compounds in decomposing corn residues (whole plant): salicylaldehyde, *p*-hydroxybenzaldehyde, phloroglucinol, resorcinol, and butyric, phenylacetic, 4-phenylbutyric, benzoic, *p*-hydroxybenzoic, vanillic, ferulic, *o*-coumaric, *o*-hydroxyphenylacetic, salicylic, syringic, *p*-coumaric, *trans*-cinnamic, and caffeic acids.

Nine allelochemics were identified in decomposing rye residues (whole plant): vanillic, ferulic, phenylacetic, 4-phenylbutyric, *p*-coumaric, *p*-hydroxybenzoic, salicylic, and *o*-coumaric acids, and salicylaldehyde (Chou and Patrick, 1976). All of the compounds identified reduced the growth of lettuce seedlings significantly in minimum concentrations of 25–100 ppm.

Tang and Waiss (1978) identified the major allelopathic agents in decomposing wheat straw as salts of acetic, propionic, and butyric acids. However, they also found traces of isobutyric, pentanoic, and isopentanoic acids.

The research of AlSaadawi *et al.* (1983) was discussed in Section I,C in connection with the leaching of allelopathic compounds from plant residues. It is possible that the seven long-chain fatty acids were released from the residue of *Polygonum aviculare* by decomposition.

II. UPTAKE BY PLANTS

Several examples of the uptake by plants of allelopathic compounds, or closely related compounds, were discussed in Section I. A few more experiments, which clearly demonstrated the uptake of allelopathic compounds, or related substances, by different plants, will be discussed here.

When 2,4-dichlorophenoxyacetic acid (2,4-D), labeled in the carboxyl position with ^{14}C, was added to the culture solution in which *Lemna minor* was growing, the compound was taken up at the maximum rate in the first 20 minutes and then the rate fell slowly until it reached zero in 1 to 2 hours (Blackman *et al.*, 1959). 2,6-Dichlorophenoxyacetic, 2-chlorophenoxyacetic, and phenoxyacetic acids tagged in the carboxyl position with ^{14}C were also taken up by *Lemna minor*. Phenoxyacetic acid, which is very closely related to several known allelochemics, was taken up at a steady rate over a 24-hour period.

Sodium 2,2-dichloropropionate (labeled in the 2-position with ^{14}C) and sodium 2,2-dichloropropionate (labeled with ^{36}Cl) were absorbed through the leaves and roots of sorghum and cotton and translocated throughout the plants (Foy, 1961). Some derivatives of propionic acid have been identified as allelopathic compounds (Chapter 10).

Scopoletin is a common allelopathic agent, and Einhellig *et al.* (1970) found that it is readily absorbed by the roots of sunflower, pigweed (*Amaranthus*), and tobacco plants. Scopoletin treated plants had a mottled, greenish appearance in the roots after a few hours treatment. Treated plants fluoresced along the leaf veins under UV light by the fifth day of treatment. The plant tissue and culture solution were analyzed for scopoletin 11 to 14 days after its addition to the nutrient solution. No scopoletin remained in any of the nutrient solutions, and the treated plants had many times as much scopoletin in the roots and tops as did the controls. The treated plants also had many times as much scopolin (glycoside of scopoletin) as the controls. Scopoletin tagged with radioisotopes was absorbed by roots of plants also (personal communication from Tom Innerarity).

Glass and Bohm (1971) reported that ^{14}C-labeled arbutin (hydroquinone-β-D-glucoside) and hydroquinone were rapidly taken up by roots of barley. Kinetic studies indicated that there was an active uptake of arbutin, whereas hydroquinone entered the root by diffusion.

III. TRANSLOCATION

Several examples of movement of allelopathic compounds and of closely related substances through plants were discussed in Sections I and II of this chapter. The movement in several cases was shown to be due to translocation through the phloem. There is a large volume of literature on the translocation of various growth-regulating compounds through many different species of plants (Leopold, 1955; Crafts, 1961). Experimental evidence indicates that growth-regulating compounds applied to the tops of plants are translocated through the phloem. Hitchcock and Zimmerman (1935) showed, however, that growth regulators applied to the soil and absorbed by the roots were apparently moved upward through the plants in the transpiration stream. This was demonstrated subsequently by Weaver and DeRose (1946), who killed a section of a stem with a flame and showed that movement of 2,4-D from the soil to the growing part of the stem was unimpeded.

The rate of translocation of growth-regulating compounds through the phloem is affected by several environmental factors, such as temperature, light intensity, and availability of several mineral elements (Crafts, 1961). The rate of translocation of 2,4-D through bean seedlings was estimated to be between 10 to 100 cm/per hour (Day, 1952). The rate of movement from the roots up through the xylem can be much faster, but this rate obviously depends on the rate of transpiration.

Macleod and Pridham (1965) measured the rates of translocation of 13 phenols and flavonoids in stems of *Vicia faba*. Rates of movement down stem from an apical leaf varied from 12 to 108 cm/hour, with the movement of flavonoids being slowest. Tests with aphids demonstrated that the compounds moved through the phloem and retained the original chemical form. Moreover, virtually all of the compounds involved have been identified as allelopathic compounds.

IV. POSSIBLE PLANT–PLANT MOVEMENT THROUGH ROOT GRAFTS, FUNGAL BRIDGES, OR HAUSTORIA OF PARASITIC VASCULAR PLANTS

In the previous discussions of the exudation of allelochemics from roots, the subject was treated as though such compounds always move into the substrate around the roots and have to be absorbed from the substrate by roots of neighboring plants to bring about allelopathic responses in receptor plants. Extensive evidence, however, indicates that many vascular plants do not grow as distinct individuals (Atsatt, 1970). Bridges are formed between plants by natural root or stem grafts, mycorrhizal fungi, or haustorial connections of parasitic vascular plants.

Almost 200 species of angiosperms and gymnosperms were reported by Graham and Bormann (1966) to form natural root grafts. There are probably a great many more species that do this also. Probably many kinds of chemical compounds pass through the grafts because even spores of pathogenic fungi, such as the spores of the fungus which causes Dutch elm disease, can be transported to other individuals through root grafts (Verrall and Graham, 1935).

The flowering plant called pinesap or Indian pipe (*Monotropa hypopitys*) is well supplied with mycorrhizal fungi, and these fungi are shared with roots of pine and spruce trees (Björkman, 1960). Moreover, compounds pass from the trees to Indian pipe through the fungal bridges. Mycorrhizal sharing has also been demonstrated between several nongreen orchids and various forest trees (Atsatt, 1970).

Mycorrhizal fungi are apparently essential to the growth of the majority of the world's commercially important plants (Wilde and Lafond, 1967). Woods and Brock (1964) suggested, moreover, that mycorrhizal fungi may be mutually shared by the root systems of many forest trees. If this is so, allelopathic compounds could readily move from plant to plant through the fungal bridges. Woods and Brock applied either radioactive phosphorus (^{32}P) or radioactive calcium (^{45}Ca) in aqueous solution to freshly cut stumps of young red maple (*Acer rubrum*) trees in a mixed hardwood forest stand. They took foliage samples subsequently from all trees within 24 feet of the treated ones and analyzed them for the presence of the radioisotope. They found that the radioactive element was transferred into 19 different species of shrubs and trees surrounding the donor trees. They pointed out that the radioisotope could have moved from the donor plants to the others through root grafts, by root exudation and uptake, and through mutually shared mycorrhizal fungi. They suggested that the trees were too young to have root grafts based on previous studies of this phenomenon, so the results were probably due to exudation or fungal bridges. The exchange rates were such that within a period of several weeks, every woody plant in the local community would have acquired at least a few of the radioactive ions originally introduced to the donor.

Mycorrhizae are present in herbaceous plants also, but less is known about their overall significance in such plants. There is growing evidence, however, that mycorrizae are just as necessary for the satisfactory growth of many herbaceous species as they are for woody species.

V. CONCLUSIONS

It is obvious that there is a large body of indirect evidence, but only a relatively small body of direct evidence, concerning the movement of allelopathic compounds from plants which produce them and the uptake and translocation of these compounds in neighboring plants. This is no doubt the weakest link in our chain of information concerning the phenomenon of allelopathy. There is an urgent need, therefore, for careful research in this area. Potential allelopathic compounds need to be tagged in suspected allelopathic plants, and paths of the compounds should be traced out of the donor plant and into and through affected acceptor plants.

Mechanisms of Action of Allelopathic Agents

I. INTRODUCTION

This topic is a very important one in allelopathy, and it is gratifying to see an increase in the rate of research concerning it. In spite of this increase, the surface has just been scratched in determining the mechanisms by which the different kinds of allelopathic compounds exert their actions. One reason for this is that it is difficult to separate secondary effects from primary causes. In spite of the pitfalls, it is important that much more work be done in this area.

II. EFFECTS ON DIVISION, ELONGATION, AND ULTRASTRUCTURE OF THE CELL

The chief criteria used in determining the presence or relative effectiveness of allelopathic agents are changes in size and weight of test organisms. Appreciable increases in either of these require cell division and enlargement. A discussion of the effects of known toxins on these processes seems to be a logical starting point in a consideration of mechanisms of action.

A saturated aqueous solution of coumarin blocks all mitoses in onion and lily roots within 2 to 3 hours (Cornman, 1946). The initial effect is similar to that resulting from colchicine treatment, destruction of the spindle with the resultant interruption of anaphases and accumulation of metaphases. These interrupted mitoses form tetraploid nuclei or binucleate cells. Coumarin also prevents the entry of cells into mitosis. Parasorbic acid in a saturated aqueous solution causes an accumulation of metaphases in onion roots, but apparently only by slowing mitosis because the retarded metaphases eventually continue into normal anaphases and telophases (Cornman, 1946). Parasorbic acid also prevents the inception of mitosis.

Jensen and Welbourne (1962) found a marked decrease in numbers of root cells of *Pisum sativum* in mitosis 4 and 8 hours after treatment with an aqueous extract of *Juglans nigra* hulls or *trans*-cinnamic acid. Moreover, during treatment with the walnut hull extract, sizeable numbers of cells were found in metaphase at 8 and 12 hours after beginning the treatment, whereas no metaphases were found in the controls at those times. They concluded therefore that the walnut hull extract slows mitosis and also prevents inception of the process. They made no additional comments on *trans*-cinnamic acid, but their results suggested that it prevents inception of mitosis only.

Volatile terpenes from macerated leaves of *Salvia leucophylla* completely prevented mitosis in roots of *Cucumis sativus* seedlings (W. H. Muller, 1965). In addition, they prevented cells from elongating in roots and hypocotyls. The cell became wider than control cells; thus, they had a considerably different appearance. Terpenes found in the air near *S. leucophylla* plants in the field were cineole and camphor.

W. H. Muller (1965) found that volatiles from leaves of *Salvia leucophylla* inhibited growth (cell division) of 32 of 44 bacterial isolates obtained from soil in and around *Salvia* stands. Five were stimulated, and seven were not affected. He reported that a relatively small concentration of cineole inhibited growth of the 36 isolates against which it was tested.

Avers and Goodwin (1956) demonstrated that scopoletin and coumarin decreased mitosis in *Phleum pratense* roots and that scopoletin was most effective. Bukolova (1971) found that toxins from three weedy species, *Sonchus arvensis, Chenopodim album* and *Cirsium arvense,* reduced mitotic activity in roots of wheat, rye, and garden cress.

Jankay and Muller (1976) reported that umbelliferone decreased the cell elongation rate in cucumber roots but increased radial expansion. Previously, Hogetsu *et al.* (1974) demonstrated a retardation of cellulose synthesis in bean epicotyls by coumarin.

Exposure of cucumber roots to volatile terpenes from *Salvia leucophylla* leaves caused accumulation of globules (which appeared to be lipid) in the cytoplasm of root tip cells, a drastic reduction in number of several intact organelles, including mitochondria, and a disruption of membranes surrounding nuclei, mitochondria, and dictyosomes (Lorber and Muller, 1976). Disruption of the root meristems resulted in the death of the cucumber seedlings.

III. EFFECTS ON HORMONE-INDUCED GROWTH

Andreae (1952) reported that scopoletin inhibited oxidation of indoleacetic acid (IAA). Sondheimer and Griffin (1960) found that indoleacetic acid oxidase from etiolated pea epicotyls was inhibited by chlorogenic acid, isochlorogenic acid, neochlorogenic acid, band 510, and dihydrochlorogenic acid. All of these are polyphenols. They found that dihydro-*p*-coumaric and *p*-coumaric acids were strong activators of the enzyme.

Lee and Skoog (1965) reported that monohydroxybenzoic acids stimulated inactivation of IAA by a crude enzyme extract of tobacco callus and that the order of increasing effectiveness was as follows: 2-, 3-, and 4-hydroxybenzoic acid. 2,4-Dihydroxybenzoic acid stimulates IAA inactivation, but 3,4-dihydroxybenzoic acid inhibits it. In fact, at equimolar concentrations, 3,4-dihydroxybenzoic acid completely prevents the stimulatory effect of 4-hydroxybenzoic acid on IAA inactivation. *p*-Coumaric acid strongly accelerates IAA inactivation, whereas ferulic acid strongly inhibits it. 2-Hydroxyphenylacetic and 2-methoxyphenylacetic acids are moderate inhibitors of IAA inactivation. The same relative activities were found in the case of horseradish peroxidase breakdown of IAA.

Tomaszewski and Thimann (1966) supported the previous work and suggested that polyphenols synergize IAA-induced growth by counteracting IAA decarboxylation. They pointed out that ferulic acid and sinapic acid act like polyphenols. They also suggested that monophenols stimulate the decarboxylation of IAA under conditions where they depress growth and that this action is enhanced by Mn^{2+}. They hypothesized further that the major role of polyphenolase may be the control of hormone balance, because changing monophenols to polyphenols changes their action from auxin destroying to auxin preserving.

Stenlid (1968) reported that naringenin, 2′,4,4′-trihydroxychalcone, and phlorizin plus some related flavonoid glycosides are strong stimulators of IAA oxidase. This could explain the potent growth inhibitory effects of many flavonoids.

Kefeli and Turetskaya (1967) reported that phenolic growth inhibitors from *Salix rubra* and apple trees suppressed the activity of IAA and gibberellin (GA). Wurzburger and Leshem (1969) reported that the germination inhibitor in the glumes and hull of the grass species *Aegilops kotschyi* inhibited gibberellin-induced growth. Corcoran (1970) found that inhibitors from carob, *Ceratonia siliqua,* also inhibited gibberellin-induced growth but not IAA-induced growth.

Six chemically defined tannins inhibited hypocotyl growth induced by gibberellic acid in cucumber seedlings (Geissman and Phinney, 1972), but growth induced by IAA was not inhibited. Many chemically defined tannins inhibited gibberellin-induced growth in dwarfpea seedlings (Corcoran *et al.,* 1972; Green and Corcoran, 1975). Coumarin, cinnamic acid, and several phenolic compounds inhibited gibberellin-induced growth but less so than did tannins in similar concentrations. The tannins were particularly inhibitory to the activity of GA_4 and GA_{14}. Inhibition could be completely reversed in all cases by increasing the amount of gibberellin.

Stahl *et al.* (1973) found that T-2 toxin produced by *Fusarium tricinctum* inhibited auxin-promoted elongation in *Glycine max* var. Hawkeye 63. A 1-hour preincubation with 5 μM toxin prevented the induction of a faster rate of elongation by auxin. Inhibition of elongation by cytokinin was similar to that of the toxin, but the mode of action appeared to be different, because the inhibitory effects were additive. Toxin treatment did not diminish cytokinin-induced radial enlargement.

Victorin, the phytotoxin from the host-specific pathogen *Helminthosporium victoriae* completely inhibited the growth response of susceptible tissue to auxin, whereas it had no effect on the response of resistant tissue to auxin (Saftner and Evans, 1974).

When dormant apple, crabapple, and apricot buds were sprayed with benzyladenine, ethylene production by the buds increased significantly in 24 to 48 hours, and dormancy was broken (Zimmerman *et al.*, 1977). Application of the ethoxy analog of rhizobitoxine at the same time as the benzyladenine prevented the increase in production of ethylene and inhibited growth of the buds. These data suggest that inhibition of ethylene biosynthesis by the rhizobitoxine analog prevented bud growth. However, the authors pointed out that rhizobitoxine produces other metabolic effects, and the relationship may not be as direct as it seems.

Lee (1977) found that six phenolic compounds, including several commonly identified allelopathic agents, interfered with IAA-induced spectral changes in horseradish peroxidase (HRP). This prevented the HRP from reacting with the IAA and consequently prevented IAA degradation.

IV. EFFECTS ON MEMBRANE PERMEABILITY

Muller *et al.* (1969) stated that two volatile terpenes, cineole and dipentene, from leaves of *Salvia leucophylla* decreased the permeability of cell membranes. They did not, however, furnish any actual evidence for such a change. Levitan and Barker (1972) presented definite evidence, on the other hand, that several known allelopathic agents change membrane permeability. They found that salicylate, benzoate, cinnamate, 2-naphthoate and derivatives increased the permeability of neuronal membranes of the marine mollusk *Navanax inermis* to potassium and decreased permeability of the membrane to chloride. Moreover, the effectiveness in changing membrane permeability was closely correlated (positively) with the octanol–water partition coefficient and pK_a value. The effect, in other words, was related to the solubility of the compound in the membrane and the concentration of aromatic anions available.

Roshchina and Roshchina (1970) tested aqueous leachates of leaves of 32 woody species on permeability of beet root membranes to anthocyanin during a

15 to 20 hour exposure of beet slices at 26°C. Leaves of fourteen species increased the permeability of the membranes to anthocyanin. The strongest action was shown by *Acer campestre, Betula verrucosa, Cotinus coggygria, Padus racemosa, Populus nigra, Rhus typhina,* and *Sorbus aucuparia.*

Allelopathic extracts of leaf litter of *Tilia cordata, Albizzia julibrissin, Aesculus hippocastanum,* and *Pinus nigra* decreased the bioelectrical activity of seedlings of *Triticum vulgare* by 96, 93, 86, and 50%, respectively (Petrushenko *et al.,* 1974). These decreases corresponded to the allelopathic activity of these species.

Owens (1969) pointed out that several of the polypeptide antibiotics, polypeptide marasmins, and animal polypeptide toxins are postulated to exert their biological influence primarily by altering the permeability of certain membranes. The primary effect of victorin is thought to be a change in membrane permeability. When victorin-treated tissue from a susceptible oat variety is placed in a bathing solution the tissue begins to lose electrolytes into the solution within 5 minutes from the time of treatment (Owens, 1969). Tissues from resistant oat varieties are not affected, which shows that the effect is host-specific.

Four glycopeptide wilt toxins produced by *Corynebacterium* also change membrane permeability (Owens, 1969). Another marasmin, fusaric acid, produced by several species of *Fusarium* changes membrane permeability, as do α-picolinic acid and dehydrofusaric acid (Owens, 1969). Owens stated in fact that, with few exceptions, the most common early sign of marasmin damage to cells is an alteration in water or ion permeability of the cytoplasmic membrane. However, he pointed out further, that it is not known in many cases whether this is the primary effect of the toxins.

Keck and Hodges (1973) reported that the toxin (victorin) produced by *Helminthosporium victoriae* dramatically increased the permeability of both the plasma membrane and tonoplast of root cells in a suceptible cultivar of *Avena sativa,* but not in a resistant cultivar. The *H. maydis* race T toxin increased the permeability of plasma membranes of *Zea mays* leaf cells, but the effect was not host-specific. Moreover, this toxin did not affect the tonoplast permeability. Saftner and Evans (1974) substantiated the results of Keck and Hodges concerning effects of victorin on membrane permeability in resistant and susceptible host cells.

Aescin, a triterpene glycoside, induced leakage of ribonucleotide material, nucleosides, and pentose phosphate or pentose from hyphae of *Ophiobolus graminis* and *Neurospora crassa* (Olsen, 1975). Loss of viability seemed to be correlated with the loss of oligonucleotides from the cells. *Aspergillus niger* was relatively insensitive to the inhibitor, and aescin induced leakage of only pentose or pentose phosphate.

Scharff and Perry (1976) reported that under anaerobic conditions, at low pH and 30°C, commercial baker's yeast lost K^+ ions in the presence of salicylic acid, and glucose utilization was inhibited. Both effects were reversed by washing the cells free of salicylate. They concluded that a fundamental action of this compound in many organisms may be its ability to reduce the K^+ content of the cells.

The evidence is clear the changes in permeability of membranes represent an important mechanism of action of at least some allelopathic substances. This field needs further research, however.

V. EFFECTS ON MINERAL UPTAKE

Most persons working on allelopathy quickly become aware of the rapid and pronounced effects of various phytotoxins on the appearance of roots of test plants. This fact, together with the basic function of roots in mineral absorption, have stimulated considerable research on the effects of allelopathic compounds on mineral absorption. This activity is continuing and is now being extended to explain how phytotoxins affect mineral uptake.

Sugar beets alter the zinc status of the soil to the extent that zinc-sensitive crops, such as corn and beans, are severely zinc-deficient when they follow beets in the cropping sequence (Boawn, 1965). This has been observed in all parts of the United States where sugar beets are grown. Boawn found in field experiments that beets did not remove as much zinc as sorghum, which does not cause zinc deficiency in subsequent crops. Moreover, he found that sugar beets did not change the acid-extractable zinc, titratable alkalinity, or pH any more than did sorghum. His only conclusion was that excessive removal of zinc by sugar beets cannot be considered a factor contributing to zinc deficiency in subsequent crops. It appears, therefore, that sugar beets must be adding some toxin to the soil that interferes with the uptake of zinc by other crops.

Chambers and Holm (1965) measured the uptake of phosphorus (^{32}P) by bean plants (*Phaseolus vulgaris*) growing alone or in association with other bean plants, pigweed (*Amaranthus retroflexus*), or green foxtail (*Setaria viridis*). They found that one associated bean plant reduced the ^{32}P uptake by the test bean plant as much as two, three, or four associated bean plants. Moreover, the weed species caused less reduction in phosphorus uptake than did the associated bean plants, even though these particular weed species are noted for absorbing large quantities of the major elements. In fact, pigweed absorbed 7 times as much phosphorus as the bean plant and still had less effect on absorption of ^{32}P by the test bean plant than other bean plants did. These facts indicated that competition for limited phosphorus did not cause the reduced phosphorus uptake by bean plants. Consequently, the authors concluded that an allelopathic interrelationship was involved. Tillberg (1970) found that salicylic acid at concentrations of 10^{-6} to 10^{-3} M decreased phosphorus uptake by the alga, *Scenedesmus,* but *trans*-cinnamic and abscisic acids had no effect.

Lastuvka and Minarz (1970) reported that when maize and peas are grown together in solution culture, the removal of nitrogen, phosphorus, and potassium is almost always greater than when each crop is grown alone. Under the mixed condition, the migration of nutrients into the aboveground parts of maize is better than in pure culture, but it is worse in peas. This suggested that root exudates were affecting mineral uptake. In subsequent experiments, Lastuvka (1970) used differentially permeable membranes to separate macromolecular root secretions and found that the substances secreted affected ion absorption and accumulation by test plants.

Olmsted and Rice (1970) found that the total uptake of K^+ and Ca^{2+} by *Amaranthus retroflexus* seedlings from a culture solution was significantly reduced by both chlorogenic and tannic acids. There was no reduction in uptake, however, based on the amount of each ion absorbed per gram dry weight of *Amaranthus* tissue.

Buchholtz (1971) observed that corn plants growing in areas infested with quackgrass, *Agropyron repens,* appeared to be suffering from a severe deficiency of mineral elements, particularly nitrogen and potassium. Analysis of corn stover from such areas demonstrated that it was very low in nitrogen and potassium compared with controls from non-quackgrass areas. Heavy fertilization with nitrogen and potassium in quackgrass areas did not improve the yield of corn greatly even though only a small fraction of the added elements was absorbed by the quackgrass.

In a subsequent experiment, a container was buried flush with the soil surface and filled with full-strength Hoagland's solution. Two seminal roots of a corn seedling were placed in the solution, and the remainder of the root system of the corn was allowed to grow outside the container in quackgrass-infested soil. In this experiment, the corn plants grew almost as well as controls without quackgrass even though most of the roots of the test corn plants were in the quackgrass-infested soil. Buchholtz concluded, therefore, that the allelopathic effect of quackgrass is predominately localized and not systemic and that reduced growth of corn plants associated with quackgrass is caused by a deficiency of minerals within the corn shoot tissue. Buchholtz suggested four possible reasons for the failure of shoots of corn plants growing in quackgrass-infested soil to obtain sufficient potassium and nitrogen for normal growth: (1) Nutrients in the soil are made unavailable in some way other than by depletion; (2) the corn root system may be reduced; (3) the absorptive capacity of the corn roots may be impaired; and (4) the nitrogen and potassium may be absorbed but are not transported to the shoots. From various bits of evidence, Buchholtz felt that numbers (2) and (4) are not very likely reasons. The problem is still unsolved.

Susceptible corn roots exposed to the host-specific toxin of *Helminthosporium carbonum* race 1 removed nitrate from solution and accumulated it in tissues twice as fast as did control roots (Yoder and Scheffer, 1973a). Uptake by resistant roots was stimulated also, provided approximately 100 times higher concentrations of toxins were used. The stimulation occurred in the presence of tungstate, which eliminates nitrate reductase activity, and the toxin did not cause leakage of nitrate from roots under the experimental conditions employed. Thus, the enhanced nitrate accumulation was caused by increased uptake rather than by decreased nitrate metabolism or decreased nitrate leakage. Subsequently, Yoder and Scheffer (1973b) found that the same toxin caused susceptible corn roots to take up and retain more Na^+, Cl^-, 3-O-methylglucose, and leucine in addition to NO_3^-. The toxin did not affect uptake of NO_2^-, K^+, Ca^{2+}, phosphate ion, SO_4^-, and glutamic acid. Thus, the data suggest that this toxin does not cause the general plasma membrane derangement caused by some other host-selective toxins. Frick *et al.* (1976, 1977) reported that the *Helminthosporium maydis* race T toxin inhibited K^+ uptake by corn roots and inhibition was more marked in highly susceptible lines.

The importance of phenolic acids as allelopathic compounds caused Glass to conduct a series of investigations on effects of these compounds on ion uptake by barley roots (Glass, 1973, 1974a,b; Glass and Dunlop, 1974; Glass, 1975). All phenolic acids tested inhibited ^{32}P-labeled phosphate uptake, and the degree of inhibition correlated well with the lipid solubility of the compounds. Potassium uptake was inhibited also by 12 different phenolic acids, and the degree of inhibition was strongly correlated with the octanol–water partition coefficients of the compounds under investigation. The membrane potentials of barley root cells were rapidly depolarized by several benzoic and cinnamic acid derivatives, and there was a strong positive correlation between the depolarization values for the benzoic acids and their lipid solubilities. Glass concluded that all the studies supported the hypothesis that the inhibition of ion uptake by phenolic acids is caused by a generalized increase in membrane permeability to inorganic ions. He did not, however, quantitate ion efflux in the presence of phenolics and, thus, did not actually measure changes in membrane permeability.

Danks *et al.* (1975a) examined the influence of ferulic acid on ion (Mg^{2+}, Ca^{2+}, K^+, P, Fe^{3+}, Mn^{2+}, Mo^{3+}) uptake in sterile cultures of Paul's Scarlet rose during a 14-day growth cycle. In general, rates of uptake were higher in test cells than control in older cells and less than control rates in cells 3 to 5 days old. The degree of inhibition of uptake of ^{86}Rb also varied with age. Uptake by young cells (4 to 5 days) was inhibited by about 50% at high concentrations of RbCl (system 2) and by approximately 25% at low concentrations (system 1). The rate of ^{86}Rb uptake in 10-day cells, however, was not significantly affected by ferulic acid. Hodges *et al.* (1971) reported a similar differential sensitivity with age to Rb uptake in oat root sections treated with the antibiotics gramicidin and nigericin.

Kolesnichenko and Aleikina (1976) reported that absorption of minerals from soil was lower in oak roots (*Quercus robur*) growing close to roots of ash (*Fraxinus excelsior*) than in oak roots growing near other oak roots. In the laboratory, the uptake of minerals by oak roots was inhibited by chemical compounds from ash.

Newman and Miller (1977) applied root exudates (leachates of pots) of *Anthoxanthum odoratum, Lolium perenne, Plantago lanceolata,* and *Trifolium repens* to pots containing each of the same four species (one species in each pot). They reported that some exudates stimulated ^{32}P uptake in certain receiver species, whereas others inhibited uptake. Results on uptake of nitrogen and potassium were inconclusive.

Kinetic analysis of phosphate (^{32}P) uptake rates of roots of 3-day-old seedlings of three varieties (Hill, Ogden, and Jackson) of soybean showed that 0.5 and 1.0 mM ferulic acid inhibited absorption of phosphate by each variety (McClure *et al.*, 1978). Inhibition was noncompetitive; the ferulic acid reduced the V_{max} values of the absorption systems but did not alter the values of the apparent K_m. There was no effect on translocation of absorbed phosphate in intact seedlings.

Although many investigators have demonstrated that numerous allelopathic compounds inhibit or stimulate ion uptake, few have tried to elucidate the mechanism of this effect. Obviously, active absorption requires energy, and several workers have reported marked effects of various phytotoxins on respiration and oxidative phosphorylation (see review below). Another, and more direct, way that allelochemics may inhibit ion absorption by plants is by inhibiting plasma membrane-bound ATPases that are involved in ion transport (Balke, 1977; Balke and Hodges, 1977). Evidence is now unequivocal that cation transport in animals is mediated by cation-activated ATPase (Leonard and Hotchkiss, 1976), and these investigators furnished additional evidence that this enzyme plays a similar role in ion transport in plants.

Balke (1977) surveyed various phenolic compounds for their effects on K^+ absorption by excised oat roots and on ATPase activity of plasma membrane vesicles isolated from oat roots. He found that the flavonoids were generally more inhibitory than phenolic acids at a $10^{-4}\ M$ concentration. The most inhibitory phenolic tested was juglone with a 79% inhibition of K^+ absorption and a 42% inhibition of ATPase activity. Ferulic and salicylic acids caused only 7 and 15% inhibition of absorption from solutions buffered at pH 6.5, and no inhibition of ATPase in 10 minutes. Absorption from unbuffered solutions, on the other hand, was inhibited 40% by ferulic acid and 64% by salicylic acid. Thus, either the pH or the composition of the buffer solution markedly affected the inhibitory influence of the phenolic acids.

Balke (1977) and Balke and Hodges (1977) found that the synthetic diphenolic compound diethylstilbestrol (DES) markedly inhibited K^+ and Cl^- absorption by excised oat roots and inhibited plasma membrane ATPase also. They determined the times required for DES to change ion absorption, membrane permeability, and oxidative phosphorylation, and the results caused them to conclude that inhibition of K^+ absorption was accounted for primarily by decreased ATPase activity but that inhibition of ATP production was a minor factor after about 10 minutes.

In subsequent work, Balke and Hodges (1979a,b,c) reported that $10^{-4}\ M$ DES inhibited $^{36}Cl^-$ absorption in 1 minute and K^+ (^{86}Rb) absorption in 1 to 2 minutes in oat roots. With a 10-minute incubation period, K^+ and Cl^- absorption were inhibited 50% by $1.1 \times 10^{-5}\ M$ and $8.4 \times 10^{-6}\ M$ DES, respectively. DES increased the efflux of ^{86}Rb from excised oat roots only after a 10-minute lag period. Comparison of efflux curves for roots loaded for 20 hours with ^{86}Rb, and those loaded for 15 minutes suggested that DES increased the permeability of the plasma membrane after 10 minutes and the permeability of the tonoplast after 10 to 20 minutes.

DES inhibited noncompetitively the ATPase in the plasma membrane fraction from oat (cv. Goodfield) roots when assayed in the presence of $MgSO_4$ or $MgSO_4$ plus KCl (Balke and Hodges, 1979b). In the presence of $MgSO_4$, $7.1 \times 10^{-5}\ M$ DES inhibited the enzyme 50%; whereas in the presence of $MgSO_4$ plus KCl, $1.3 \times 10^{-4}\ M$ DES was required for the same inhibition.

Oxidative phosphorylation by isolated mitochondria was inhibited 50% by $2.6 \times 10^{-5}\ M$ DES, and concentrations of $10^{-4}\ M$ or greater were completely inhibitory (Balke and Hodges, 1979c). After a lag period of 2 minutes, $10^{-4}\ M$ DES produced a linear decrease in the ATP content of excised oat roots. After 20 minutes, the ATP content was about 50% of the control and remained at that level after 30 minutes in DES. These results showed that DES can inhibit K^+ absorption by reducing mitochondrial ATP production in addition to inhibiting the plasma membrane ATPase. However, the rapid (less than 5 minutes) reduction of absorption is caused by direct inhibition of the ATPase rather than the ATP supply, according to Balke and Hodges. The ATP supply is lowered only slightly in that short time period whereas the ATPase activity is inhibited greatly. Results of the more detailed studies led therefore to essentially the same conclusions as the preliminary studies (Balke and Hodges, 1977).

Harper and Balke (1980) extended the work of Balke (1977) on the effects of ferulic and salicylic acids on K^+ uptake and efflux in oat roots. They found that pH was very important in determining the effects of the phenolic compounds. Salicylic acid was found again to have the greater effect on K^+ uptake. At a pH of 4.5, $5 \times 10^{-4}\ M$ salicylic acid inhibited uptake by 96%, whereas at a pH of 7.5, it inhibited uptake by only 19%. The same concentration of salicylic acid caused a 90% efflux of K^+ at a pH of 4.5 in 4 hours, whereas at pH 6.5 and 7.5, it had no effect on efflux.

Ion uptake is of basic importance in the growth and reproduction of organisms of all levels of complexity, and evidence is accumulating that many types of allelopathic agents affect the rate of ion uptake. Therefore, this is a very important mechanism of action of many such agents and much more research is needed in this area at all levels of complexity, from one-celled organisms to intact higher plants.

VI. EFFECTS ON EASILY AVAILABLE PHOSPHORUS AND POTASSIUM IN SOILS

Gajić (1977) and Gajić et al. (1977) reported that addition of 0.8 to 3 g/ha of agrostemmin, an allelochemic from *Agrostemma githago,* significantly increased amounts of easily available phosphorus and potassium in manured, eroded Chernozem and hydromorphic black soil in Yugoslavia. Results in unmanured soils were variable, but at least some concentrations of agrostemmin increased amounts of available phosphorus and potassium.

VII. EFFECTS ON STOMATAL OPENING AND PHOTOSYNTHESIS

Einhellig et al. (1970) found that scopoletin markedly inhibited the photosynthetic rate of intact plants of Russian Mammoth sunflower, tobacco (var. One Sucker), and pigweed (*Amaranthus retroflexus*). There was a large reduction in net photosynthesis in tobacco plants the second day after treatment with $10^{-3} M$ and $5 \times 10^{-4} M$ concentrations. This reduction reached a low point by the fourth day (34% of the control rate) and was followed by a gradual recovery phase. The degree of reduction correlated well with the concentration of scopoletin used in treatment and leaf area expansion also correlated well with the concentration of scopoletin. Moreover, a calculation of CO_2 fixed per hour of illumination in the daily photosynthesis of tobacco showed a striking relationship with reduced CO_2 fixation in the treated plants (Fig. 37). By the end of the experiment the $10^{-3} M$ scopoletin treated tobacco plants fixed only 51% as much CO_2 as the controls. The dark respiration rate was not significantly affected by scopoletin treatment. The results with sunflower were very similar to those with tobacco; the photosynthetic and dark respiration rates in pigweed were measured only on the fifth day after treatment, and results were similar on that day to the results obtained with sunflower and tobacco.

Einhellig et al. (1970) found that the inhibitory activity of scopoletin on total plant growth correlated well with decreases in net photosynthesis, but as previously stated, no differences were found in rates of respiration. The evidence indicated that the effect of scopoletin on net photosynthesis was the cause of growth reduction, but they pointed out that the evidence did not eliminate other possible causes of growth inhibition.

Einhellig et al. (1970) found a loss in turgor pressure in tobacco plants treated

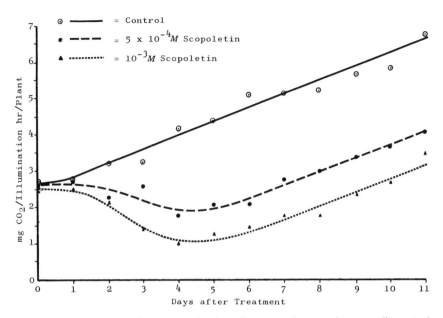

Fig. 37. Effects of scopoletin on CO_2 fixed per illumination hour in tobacco seedlings. Each point computed from mean leaf area and net photosynthesis of four plants. (From Einhellig et al., 1970.)

with 10^{-3} M scopoletin in addition to the reduction in photosynthesis. They suggested, therefore, that scopoletin may operate through its effect on the stomata, because stomata are the portals through which carbon dioxide enters the interior of the leaf during photosynthesis and through which large quantities of water vapor diffuse in transpiration. Moreover, Shimshi (1963a) found that spraying phenylmercuric acetate on tobacco and sunflower plants reduced stomatal apertures, affected transpiration, and reduced growth. Similar treatment reduced photosynthesis in maize (Shimshi, 1963b). In addition, Zelitch (1967) reported that leaf disks of tobacco floated on a 10^{-3} M chlorogenic acid solution had 50% stomatal closure.

Einhellig and Kuan (1971) followed up on the suggestion of Einhellig et al. (1970) and found that whole plants of tobacco and sunflower treated with 10^{-3} M and 5×10^{-4} M scopoletin and chlorogenic acid (root immersion) exhibited stomatal closure for several days after treatment. On the other hand, 10^{-4} M concentrations of both phenolics stimulated opening of the stomates of both test species. They concluded that inhibition of stomatal opening by scopoletin correlates well with growth reduction under similar conditions. They also pointed out that a comparison of the photosynthetic curves of Einhellig et al. (1970) with the curves of the effects of scopoletin on stomatal aperture indicated that the closing and partial closing of stomata were closely related to photosynthetic reductions. They cautioned, however, that it was not clear whether stomatal closure induced by scopoletin caused a reduction in photosynthesis or whether the latter induced the former. They pointed out that photosynthesis is one of the processes that has been implicated in light-induced stomatal opening.

A 10^{-3} M tannic acid solution caused significant reductions in stomatal apertures of tobacco plants for about 5 days after treatment, but a 10^{-4} M tannic acid solution did not affect stomatal opening of tobacco even though it significantly reduced plant growth (Einhellig, 1971). Surprisingly, the stomata of the 10^{-3} M tannic acid treated plants returned to normal aperture at a time when the plants appeared to be in a deteriorated condition. Thus, stomatal aperature changes cannot account for the pronounced growth effects of tannic acid on tobacco plants.

Vikherkova (1970) reported that extracts of fresh rhizomes of *Agropyron repens* added to soil in which *Linum usitatissimum* was growing caused decreased transpiration, water content, and osmotic pressure of the cell sap and reduced the opening of the stomates. He felt that these changes were secondary, however and that they were caused by a water deficiency, which could have resulted from a limitation of water inflow into the roots of the plant.

Turner (1972) found that oat plants treated with the marasmin victorin had reduced stomatal apertures and a decreased rate of transpiration. On the other hand, fusicoccin, which is also a marasmin, stimulated the opening of stomata and increased transpiration (Turner, 1972). A great many of the marasmins produced by pathogenic microorganisms cause plants to wilt. Thus, it would be interesting to determine if many of them act like fusicoccin and increase transpiration.

The evidence on the effects of allelopathic agents on stomatal opening is confusing at this point in time. It appears that the effects may be secondary ones generally, but much more research is needed before definite conclusions can be drawn.

Sikka *et al.* (1972) found that three quinones inhibited CO_2 fixation by isolated chloroplasts. The quinones were 2,3-dichloro-1,4-naphthoquinone (Dichlone), 2-amino-3-chloro-1,4-naphthoquinone (06K-Quinone), and 2,3,5,6-tetrachloro-1,4-benzoquinone (Chloranil). Dichlone was a strong inhibitor of both photosystems I and II, photosystem I was more sensitive to 06K-Quinone than was photosystem II, but the reverse was true in the case of Chloranil. These quinones are commercially produced and, to my knowledge, have not been shown to be produced by plants. Nevertheless, they do indicate that quinones can inhibit photosynthesis and suggest that tests should be run using quinones that have been identified as allelopathic agents.

Roshchina (1973) investigated the effects of water infusions of 20 woody plant species against dark reduction of 2,6-dichlorophenolindophenol (DCPI) and against photochemical reactions of photosynthesis in isolated chloroplasts. Infusions of *Phellodendron amurense, Juglans regia, Betula verrucosa,* and *Corylus avellana* contained polyphenols that caused dark reduction of DCPI and inhibited photochemical reactions in the isolated chloroplasts.

Boiled aqueous leaf extracts of *Celtis laevigata,* a known allelopathic species, significantly inhibited net CO_2 uptake by three grass species in light when added to the nutrient solution in which the grasses were growing (Lodhi and Nickell, 1973). This effect appeared to be independent of effects on water content.

Arntzen *et al.* (1974) found that kaempferol, a known allelopathic flavonol, inhibited coupled electron transport and both cyclic and noncyclic photophosphorylation in isolated pea (*Pisum sativum*) chloroplasts. There was no effect on basal or uncoupled electron flow or light-induced proton accumulation by isolated thylakoids over a concentration range that markedly inhibited ATP synthesis. They suggested, therefore, that kaempferol acts as an energy transfer inhibitor.

The phytotoxin produced by *Helminthosporium maydis* caused a rapid inhibition of photosynthesis in whole leaves of *Zea mays* having Texas male-sterile cytoplasm but not in leaves having normal cytoplasm (Arntzen *et al.,* 1973). Electron transport, phosphorylation, and proton uptake activities of isolated chloroplast lamellae were not affected by the toxin in either genetic type. The toxin however, was found to have a direct effect on stomatal functioning. The toxin inhibited light-induced K^+ uptake by guard cells in the Texas strain, and thus the stomates did not open.

Phlorizin (10^{-3} M) inhibited the rate of photoreduction of $NADP^+$ by isolated pea chloroplasts by 40%, electron transport via cytochrome f by 100% and via plastocyanin by 50% (Roshchina and Akulova, 1978). Crossover experiments demonstrated that phlorizin inhibited ADP-induced photoreduction of cytochrome f but had no effect on plastocyanin under identical conditions.

Treatment of pea chloroplasts with a low molecular weight (1000 to 1500) allelopathic agent produced by the green alga *Pandorina morum* resulted in a strong inhibition of uncoupled electron flow with methylviologen as electron acceptor and either water or hydroquinone as electron donor (Patterson *et al,* 1979). Photosystem II electron flow (water to dimethylquinone plus dibromothymoquinone) was markedly inhibited, whereas photosystem I electron flow (diaminodurene to methylviologen) was not affected.

Roshchina *et al.* (1979) reported that aqueous extracts of the vegetative organs of *Cicuta virosa* inhibited chloroplast movement in *Elodea canadensis.* Moreover, they strongly inhibited photophosphorylation and $NADP^+$ photoreduction in isolated pea chloroplasts, and slightly influenced cytochrome f reactions.

Cicutotoxin, the purified allelochemic from *Cicuta virosa* rhizomes, inhibited

NADP$^+$ photoreduction in isolated pea chloroplasts by 50% and electron transport at the cytochrome f level by 25 to 30%, even in low concentrations of 10^{-11} to 10^{-9} M (Roshchina et al., 1980).

Following a 6-day treatment cycle, dry weight increases of soybean seedlings were reduced by 10^{-3} M and 5×10^{-4} M concentrations of ferulic, p-coumaric, and vanillic acids (Einhellig and Rasmussen, 1979). Soybean weight reductions in each case were paralleled by a significant reduction in the concentration of chlorophylls a and b and total chlorophyll in the unifoliate leaves. This no doubt affected the rate of photosynthesis, although no such measurements were made. The growth of sorghum seedlings was also reduced by each of the compounds at the 5×10^{-4} M concentration, but leaf chlorophyll concentration was not reduced.

Patterson (1981) reported that 10^{-3} M concentrations of caffeic, trans-cinnamic, p-coumaric, ferulic, gallic acid, and vanillic acid caused marked reductions in concentrations of chlorophyll in leaves of soybean plants growing in aerated nutrient solutions. The same treatments severely reduced the net photosynthetic rate and stomatal conductance of single, fully expanded leaves.

Velvetleaf is a potent allelopathic plant as previously indicated, and aqueous extracts of its leaves increased diffusive resistance in leaves of treated soybean plants, suggesting partial stomatal closure (Colton and Einhellig, 1980). Moreover, the chlorophyll concentration of leaves of treated plants was reduced. Both effects would likely reduce the photosynthetic rate, but no measurements were made. The evidence indicates that many allelopathic agents inhibit the growth of affected plants through an indirect or a direct effect on the rate of photosynthesis.

VIII. EFFECTS ON RESPIRATION

Patrick (1955) found that water extracts of soil in which peach root residues were decomposing inhibited respiration in excised peach roots. Patrick and his colleagues demonstrated later that water extracts of soils in which several different crop residues were decomposing inhibited respiration in excised tobacco roots (Patrick and Koch, 1958; Patrick et al., 1964).

Hulme and Jones (1963) tested a large number of simple phenols, derivatives of benzoic acid, digallic and ellagic acids, tannic acid, cinnamic acid derivatives, esculetin and flavonoids against oxidation of succinate by succinate dehydrogenase in suspensions of isolated apple peel mitochondria. Most of the compounds inhibited succinate dehydrogenase at least slightly; many were very inhibitory to that enzyme. Two flavonoids and tannic and chlorogenic acids were tested against decarboxylation of malate by the malic enzyme in apple peel mitochondria and all of them inhibited the malic enzyme except chlorogenic acid, which stimulated the enzyme after the first hour.

4-Hydroxybenzoic acid stimulated oxidation of NADH by enzymes from tobacco leaves (Lee, 1966), but a shift of the hydroxyl group from the 4- to the 3- position and from the 3- to the 2- position decreased the activity. In the presence of 4-hydroxybenzoic acid, 2,4-dihydroxybenzoic and 2,5-dihydroxybenzoic acids further stimulated the oxidation of NADH. On the other hand, 3,4-dihydroxybenzoic, 2,3-dihydroxybenzoic, and 2,6-dihydroxybenzoic acids inhibited oxidation of NADH by the same enzymes.

Stenlid (1968) found that the flavonoids, naringenin and 2',4,4'-trihydroxychalcone, inhibited oxidative phosphorylation in higher plants and gave a distinct uncoupling effect. He found, however, that phlorizin and related glycosides were less active in that respect.

Muller *et al.* (1969) reported that two volatile terpenes, cineole and dipentene, which emanate from leaves of *Salvia leucophylla,* markedly reduced oxygen uptake by suspensions of mitochondria from *Avena fatua* or *Cucumis sativus.* The inhibition appeared to be localized in that part of the Kreb's cycle where succinate is converted to fumarate, or fumarate to malate.

Van Sumere *et al.* (1971) reported that several quinones, aldehydes, benzoic and cinnamic acid derivatives, and coumarins affected the rate of oxygen uptake by yeast (*Saccharomyces cerevisiae*) cells. Most of the compounds stimulated uptake of oxygen but a few inhibited it, notably the quinones. The same compounds have very similar effects on oxygen uptake by lettuce seeds as they do on uptake by yeast cells. Oxygen uptake by barley seeds was affected by these compounds, but there were many more cases of reduction in uptake, particularly on a short-term basis (10 minutes).

Several of the compounds that stimulated oxygen uptake by yeast cells lowered the ADP/O ratio resulting from the oxidation of NADH by suspensions of isolated yeast mitochondria and thus proved to be uncouplers of oxidative phosphorylation. Compounds included in this list were 2-methyl-1,4-naphthoquinone, salicylaldehyde, β-resorcylaldehyde, cinnamaldehyde, cinnamic acid, *o*-coumaric acid, and scopoletin. An additional compound, caffeic acid, also was an uncoupler of oxidative phosphorylation in yeast respiration.

Dedonder and Van Sumere (1971) demonstrated that almost all of 40 phenolics tested stimulated respiration (dark uptake of O_2) in *Chlorella vulgaris* in at least some combination of pH and concentration. The enhancement of respiration generally occurred at the concentrations of the phenolics that caused growth inhibition of the alga.

Koeppe (1972) found that 500 μM concentrations of juglone inhibited oxygen uptake by excised corn roots by more than 90% after a 1-hour treatment. Lesser reductions occurred with 50 and 250 μM concentrations of juglone. In the absence of inorganic phosphate (P_i), juglone stimulated the rate of oxygen uptake by isolated corn mitochondria oxidizing NADH, succinate, or malate + pyruvate. However, in the presence of P_i, juglone concentrations of 3 μM and greater inhibited the state-3 oxidation rates of succinate and malate + pyruvate and lowered respiratory control and ADP/O ratios obtained from the oxidation of NADH, malate + pyruvate, or succinate. Juglone + P_i also reduced the coupled deposition of calcium phosphate within isolated mitochondria, driven by the oxidation of malate + pyruvate. Koeppe concluded that this uncoupling of ATP production correlates well with the growth inhibition that results when tissues or organisms are treated with juglone.

Lodhi and Nickell (1973) reported that addition of aqueous leaf extract of *Celtis laevigata* to nutrient solution in which either of three species of grasses was growing stimulated dark CO_2 production up to 6 days, the time depending on the concentration of the *Celtis* extract. After the early stimulation, the rate of respiration returned to that of controls.

Macerated leaves and sequiterpene lactones obtained from *Artemisia tridentata* ssp. *vaseyana,* and other species of sagebrush, inhibited the growth and stimulated the respiration of *Cucumis sativus* (McCahon *et al.,* 1973). Arbusculin A, achillin deacetoxymatricarin, viscidulin C, and viscidulin B were the most active sesquiterpene lactones isolated and were effective at $10^4 M$ or higher concentrations. Weaver and Klarich (1977) found that vapors from *A. tridentata* reduced respiration rates of germinating seeds of wheat and cucumber and of newly emerged leaves of wheat and barley. On the other hand, the vapors stimulated respiration in mature leaves of wheat and oats, two species of *Juniperus,* four species of *Pinus,* douglas fir, two species of *Picea,* and two species of *Abies.* Wheat plants exposed to *Artemisia* volatiles in the field in naturally occurring concentrations also had higher respiration rates than controls.

Stenlid (1970) surveyed a large number of flavonoids for the effects on ATP production by mitochondria isolated from several plant species. He found that minor changes in the hydroxyl pattern affect the degree of inhibition of ATP formation, but he did not demonstrate the site of the flavonoid effect. Koeppe and Miller (1974) investigated the point of action of kaempferol on respiration in isolated corn mitochondria and found that it inhibited the rate of state-3 substrate oxidation, but not the state-4 rate. Moreover, kaempferol inhibited substrate-driven calcium phosphate deposition. These actions provided evidence that this flavonoid acts specifically on the phosphorylation mechanism and not on electron transfer. Koeppe and Miller concluded that kaempferol probably acts before formation of the phosphorylated high energy intermediate, but not as an uncoupler in the traditional 2,4-dinitrophenol mode.

Four of seven flavones and flavone glycosides inhibited mitochondrial ATPase (Lang and Racker, 1974). Quercetin, in low concentrations, inhibited both soluble and particulate mitochondrial ATPase, but had no effect on oxidative phosphorylation in submitochondrial particles. Quercetin also retarded the ATP-dependent reduction of NAD^+ by succinate in fully reconstituted submitochondrial particles. It was concluded that hydroxyl groups at the 3' and perhaps 3 position are important for the inhibition of ATPase activity by flavones.

Demos *et al.* (1975) tested 10 phenolic compounds against hypocotyl growth and mitochondrial metabolism of mung bean. They found that some compounds inhibited both hypocotyl growth and respiration or coupling responses in mitochondria, some did not affect either, and others reduced hypocotyl growth but did not affect metabolism of isolated mitochondria. All compounds that inhibited mitochondrial metabolism also retarded hypocotyl growth; therefore, they suggested that inhibition of mitochondrial metabolism may represent an important mechanism of action of these compounds.

The triterpeneglycoside, aescin, in phosphate buffer reduced the production of $^{14}CO_2$ from uniformly labelled glucose in mycelia of *Ophiobolus graminis* and *Neurospora crassa*, whereas aescin in succinate buffer had no effect (Olsen, 1974). The production of $^{14}CO_2$ from glucose or sucrose in *Aspergillus niger* was not affected, however. Johansson and Hägerby (1974) reported that phenolic compounds added to the medium in which *Fomes annosus* was growing stimulated laccase activity, generally in combination with a decrease in the ATP level.

Patterson *et al.* (1979) found that the allelopathic agent produced by the green alga *Pandorina morum* markedly inhibited electron transport in mitochondria isolated from white potato tubers (*Solanum tuberosum*). The inhibition occurred at a point between sites II and III.

The fact that respiratory activity by isolated respiratory enzymes, isolated mitochondria, one-celled organisms, and organs of plants has been shown to be adversely affected by many identified allelopathic agents indicates that respiratory effects probably represent an important mechanism of action of at least some inhibitors.

IX. INHIBITION OF PROTEIN SYNTHESIS AND CHANGES IN LIPID AND ORGANIC ACID METABOLISM

Krylov (1970) reported that potatoes produce toxic substances that inhibit tree growth when the potatoes are cultivated between rows of young apple trees. The toxins also decrease the total nitrogen content in the branches and roots of the apple trees, change the composition of proteins in the bark of the branches, increase the amount of soluble albumins, and decrease the amount of residual proteins.

Van Sumere *et al.* (1971) investigated the effects of many known allelopathic agents on incorporation of [1−^{14}C]phenylalanine into protein in yeast cells; most of the inhibitors reduced incorporation of the labeled phenylalanine. In other experiments, they tested the effects of ferulic acid and coumarin on the incorporation of [1-^{14}C] phenylalanine into protein in lettuce seeds and in barley seeds and embryos. They found that both inhibitors markedly reduced the incorporation of labeled phenylalanine in all cases. These workers also found that the uptake of labeled phenylalanine by yeast and seeds was generally reduced by the allelopathic compounds. They concluded, therefore, that some allelopathic agents may reduce growth by inhibiting the transport of amino acids and the formation of proteins.

Zweig *et al.* (1972) investigated the effects of certain quinones on the photosynthetic incorporation of $^{14}CO_2$ by the alga *Chlorella.* They found that the quinones caused an increase in the proportion of ^{14}C in sucrose and glycine accompanied by reduction of ^{14}C in lipids and glutamic acid. They suggested, therefore, that the quinones inactivate coenzyme A and cause a shortage of NADPH.

Incubation of Paul's Scarlet rose cells with [UL-^{14}C]glucose in the presence of ferulic acid resulted in increased incorporation of ^{14}C into the soluble lipid fraction along with a decreased incorporation of ^{14}C into protein, organic acids, and soluble amino acids (Danks *et al,* 1975b). Treatment of the cells with cinnamic acid resulted in a significant decrease in incorporation of ^{14}C into protein, but incorporation of ^{14}C into soluble amino acids was significantly increased. There was also a decreased incorporation of ^{14}C into protein amino acids. They concluded that the reduction in protein synthesis resulting from ferulic or cinnamic acid treatment would lead to a reduction in growth of the cultures. It was also inferred that ferulic and cinnamic acids affect protein synthesis in different ways. Cinnamic acid appeared to inhibit the mechanism of protein synthesis, whereas ferulic acid caused a diversion of [^{14}C] acetate into lipid synthesis rather than into Kreb's cycle and subsequent pathways leading to synthesis of amino acids and proteins.

Kolesnichenko and Aleikina (1976) found that the rate of protein biosynthesis was lower in roots of oak, *Quercus robur,* growing close to the roots of ash, *Fraxinus excelsior,* than in roots of oak growing with other oak trees. In laboratory experiments, protein synthesis in oak roots was inhibited by chemical compounds from ash roots.

Cinnamic acid in 0.05 mM and 0.5 mM concentrations and ferulic acid in a 0.5mM concentration significantly inhibited protein synthesis in lettuce seedlings (Black Seeded Simpson), when added from the beginning of the germination period or when added for a short period of time to seedlings that had already germinated under control conditions (Cameron and Julian, 1980). The same concentrations of the allelopathic agents also significantly inhibited root growth of the seedlings.

It is noteworthy that two commonly identified allelochemics, cinnamic and ferulic acids, have been shown to inhibit protein synthesis in one-celled organisms, cell cultures, seeds, and intact plants. It appears, therefore, that this is one of the important mechanisms of action of many allelopathic compounds. Certainly, this is a fruitful area for future research.

X. POSSIBLE INHIBITION OF PORPHYRIN SYNTHESIS

There is much evidence that several plants, which have been found to inhibit *Rhizobium,* reduce nodule numbers on heavily inoculated legumes and hemoglobin content of the nodules (Chapters 7 and 9). Legume nodules have to contain hemoglobin in order to be effective in nitrogen-fixation; thus, it is obvious that reduction of nodulation and of hemoglobin could reduce the growth of legumes in areas low in nitrogen.

Heavily inoculated bean plants growing in soil with plants known to produce inhibitors of *Rhizobium* have unusual patterns of chlorosis. In addition to a general appearance of nitrogen deficiency, the primary leaves are often completely devoid of chlorophyll in certain sections. Often the entire blade on one side of the midvein is completely devoid of chlorophyll. Considerable evidence was cited in Section VII of this chapter that allelopathic compounds often reduce chlorophyll concentration in affected plants.

Hemoglobin and chlorophyll are both porphyrin containing compounds; thus, it appears that some allelochemics interfere with porphyrin synthesis. This is another fertile area for research.

XI. INHIBITION OR STIMULATION OF SPECIFIC ENZYMES

A. Pectolytic Enzymes

The ability of pathogens to penetrate host cells depends strongly on the effectiveness of the pectolytic enzymes produced by the pathogens. Knowledge of effects of plant-produced substances on pectolytic enzymes is of considerable importance, therefore, in an understanding of resistance of plants to diseases. According to Williams (1963), Cole demonstrated in 1958 that apple juice inhibited pectolytic enzymes of *Sclerotinia fructigena,* which causes the fungal Brown Rot of fruit. Williams investigated the effects of known simple phenols, chlorogenic acid, and flavonoids (unoxidized and enzymatically oxidized) present in the apple fruit against the tissue-macerating and polygalacturonase activity of culture filtrates of *S. fructigena.* None of the unoxidized compounds had an appreciable effect on either the macerating or polygalacturonase activity. All the oxidized compounds were effective against both types of activity, with oxidized chlorogenic acid being least effective. In other experiments, he found that tannic acid is an extremely effective inhibitor of both types of activity, and the larger molecules of the gallotannins are even more effective. He concluded that the minimum molecular size for inhibitory activity at concentrations up to 0.2% is about 500. Benoit and Starkey (1968b) found that wattle tannin, a condensed tannin, also markedly inhibited the activity of a purified commercial polygalacturonase.

It is evident therefore, that some well-known allelopathic agents inhibit activity of pectolytic enzymes, at least after oxidation of the compounds. This effect of some phytoncides produced by higher plants on the activity of pathogenic microorganisms probably has significant ecological importance.

B. Cellulase

Cellulase is extremely important ecologically because of its role in decomposition and because of the large amount of cellulose in plants. Benoit and Starkey (1968b) demonstrated that wattle tannin strongly inhibited the action of cellulase. They previously demonstrated that wattle tannin slowed the decomposition of hemicellulose and cellulose by cultures of microorganisms obtained from fresh barnyard soil (Benoit and Starkey, 1968a). Although they did not specifically study the effect of tannin on hemicellulase, it appears that tannin also inhibits the action of that enzyme.

The role of allelopathic substances in decomposition probably is very important ecologically; thus, this phase of allelopathic research has not had the attention it merits.

C. Catalase and Peroxidase

The information is very limited concerning effects of inhibitors on these enzymes. Dzyubenko and Petrenko (1971) reported that root secretions of *Lupinus albus* and *Zea mays* inhibited growth and catalase and peroxidase activity of two weed species, *Chenopodium album* and *Amaranthus retroflexus*.

In Section III, I discussed the evidence of Lee and Skoog (1965) concerning the effects of several phytotoxins on horseradish peroxidase breakdown of IAA. Possibly many more types of inhibitors can affect plant growth through their effects on peroxidase and thus on IAA inactivation. Benoit and Starkey (1968b) stated that tannins inactivate peroxidase and catalase.

Jankay and Muller (1976) found that umbelliferone caused a swelling response in cucumber roots and that increased peroxidase levels appeared coincident with the swelling response. They suggested, therefore, that the higher peroxidase levels might be involved in the mediation of the radial expansion.

D. Phosphorylases

Schwimmer (1958) found that potato tubers contain polyphenols that inhibit the activity of potato phosphorylase. He tested the activity of chlorogenic acid, caffeic acid, and catechol individually against activity of the enzyme and found that all were very inhibitory. He decided that chlorogenic acid was the most important inhibitor of phosphorylase in the potato peel, however. Sondheimer (1962) stated that the concentration of polyphenols in the peel is probably sufficiently high to inhibit the phosphorylase completely.

E. β-Cystathionase (Cystathionine β-Lyase)

Giovanelli *et al.* (1971) found that rhizobitoxine irreversibly inactivated β-cystathionase isolated from spinach. They later found that it only partially inhibits β-cystathionase of spinach and corn seedlings *in vivo* (Giovanelli *et al.*, 1972). Rhizobitoxine-treated and control corn seedlings were allowed to assimilate $^{35}SO_4^{2-}$ for 3 or 6 hours, and the radioactivity incorporated into sulfur amino acids was determined. The most striking effect of the rhizobitoxine was an increase in radioactive cystathionine up to 22-fold. Accumulation of radioactivity in methionine was only slightly inhibited. It was concluded that the trans-sulfuration pathway contributes to methionine biosynthesis and that metabolism via this pathway is impaired but not entirely eliminated by rhizobitoxine.

F. Phenylalanine Ammonia-Lyase

Riov *et al.* (1969) reported that the induction of phenylalanine ammonia-lyase (PAL) in citrus fruit peel is controlled by ethylene.

Ethylene increased the activity of PAL and the total protein content of carrot root disks, until the maximum activity was reached after 36 to 48 hours of incubation (Chalutz, 1973). Cycloheximide or actinomycin D inhibited the ethylene-induced activity in disks that had not previously been exposed to ethylene. The results supported the hypothesis that the mode of action of ethylene involves both *de novo* synthesis of the enzyme protein and protection or regulation of activity of the induced enzyme. This enzyme is involved directly in the biosynthesis of phenolic compounds in plants. Marigo and Boudet (1975) demonstrated that exogenously supplied precursors of phenolics caused important increases in the phenolic content of young tomato plants and reduced their growth. Phenylalanine was one of the compounds supplied.

G. Sucrase (Invertase)

The triterpeneglycoside, aescin, completely inhibited the enzymatic hydrolysis of sucrose in mycelia of *Ophiobolus graminis* and *Neurospora crassa*

(Olsen, 1974). The concentration of aescin employed in experiments with *Ophiobolus* was 100 mg/liter, and with *Neurospora* was 300 mg/liter. Tannins are very effective inactivators of invertase also (Benoit and Starkey, 1968b).

H. Other Enzymes

There is extensive literature on factors and substances that regulate the activity of many enzymes, but most of the evidence does not relate to the role of established allelopathic agents. Benoit and Starkey (1968b) stated that tannins inactivate the following enzymes not mentioned previously: amylase, myrosinase, pepsin, proteinase, dehydrogenases, decarboxylases, phosphatases, β-glucosidase, aldolase, polyphenol oxidase, lipase, urease, trypsin, and chymotrypsin.

XII. EFFECTS ON CORKING AND CLOGGING OF XYLEM ELEMENTS, STEM CONDUCTANCE OF WATER, AND INTERNAL WATER RELATIONS

Bogdan and Grodzinsky worked for several years on the nonspecific effect of allelopathic compounds in causing the corking and clogging of xylem elements (Bogdan, 1971, 1977; Grodzinsky and Bogdan, 1972, 1973; Bogdan and Grodzinsky, 1974). They observed that aqueous extracts of many allelopathic plant species caused browning, corking, and clogging of xylem vessels in numerous test species. Histochemical tests indicated that the brown substance, which impregnated the walls and filled the cavities of the vessels, was a complex of pectins, lignin, suberin, melanins, and many unidentified substances. It was found subsequently that allelopathic agents caused increases in the reduced form of ascorbic acid in the xylem vessels, and histochemical evidence indicated that the reduced ascorbic acid played a role in the formation of the brown substance. Evidence was also obtained that the thiol groups of substances flowing to the area of corking and browning took part in the formation of the brown mass. It was concluded that intensification of redox processes in tissues of plants subjected to allelopathically active substances probably favors formation of the mechanical barriers to toxin penetration of the plant.

In spite of the results of Bogdan and Grodzinsky, very little research has been done on the effects of allelopathic agents on stem conductance of water and on general water relations of plants. This is probably due to the fact that few persons outside the U.S.S.R. are aware of their research. Most persons doing research in allelopathy have observed the wilting effects of allelopathic agents, and the wilting effects of Dutch elm disease are known to many persons, including nonscientists. Van Alfen and Turner (1975a) reported that, after 4 hours in a 200 μg/ml solution of water-soluble glycopeptide toxins from cultures of *Ceratocystis ulmi,* stem conductance of water in *Ulmus americana* seedlings was reduced by 79%. Moreover, leaf water potential was reduced by 3 bars to the point at which the seedlings wilted, the stomates closed, and transpiration ceased.

Van Alfen and Turner (1975b) found also that even 2 μg of glycopeptide toxin produced by *Corynebacterium insidiosum,* which causes wilt of alfalfa, significantly reduced water conductance through 15-cm long stems of alfalfa. They stated that the decrease in stem conductance best explains the rapid decrease in transpiration and stomatal conductance, and the resultant wilting, after alfalfa cuttings have been in 200 μg/ml toxin for 2 hours. Membrane damage resulting in water leakage was ruled out as a factor in wilting during the 2-hour period, and it was postulated that the toxin acts by interfering with water movement through pit membranes.

Phenol, resorcinol, phloroglucinol, pyrogallol, salicylic acid, and cinnamic acid in concentrations of 10^{-4} to 10^{-3} M caused marked changes in the water content of cells of the storage tissue of beets (Roshchina, 1972b). Compounds with an orthoplacement of hydroxyl groups had the greatest effects.

Lodhi and Nickell (1973) reported that very dilute aqueous leaf extracts of the allelopathic species *Celtis laevigata* greatly reduced the shoot water content (% of dry wt) of three test grasses. The initial extract had a water potential of -0.5 bar, and this was diluted 1:5 or 1:10 (v/v, extract/Hoagland nutrient solution) before treatment of the grasses.

Soybean plants, which were reduced in growth by aqueous velvetleaf extracts, had markedly reduced leaf water potentials and reduced water content as compared with controls (Colton and Einhellig, 1980). The lowest water potential of any control plant was -10.4 bars, whereas test plants had leaf water potentials as low as -32.4 bars with many below -20 bars.

Even though only a small amount of research has been done on this subject, it is obvious that some allelopathic agents have pronounced effects on water relations of affected plants. Thus, this is a promising area for research.

XIII. MISCELLANEOUS MECHANISMS

Hauschka *et al.* (1945) found that cysteine and glutathione inactivated parasorbic acid; however, cystine, glycine, and glutamic acid did not affect activity of that allelopathic compound. They suggested, therefore, that parasorbic acid and related unsaturated lactones, such as patulin and penicillic acid, may interfere with cellular proliferation because of their reactivity with SH groups essential to enzyme function. Cavallito and Haskell (1945) reported the same phenomenon about the same time. They found that the antibiotic properties of penicillin and several widely different bacteriostatic substances, all of which were unsaturated lactones, were inactivated by compounds having SH groups. They determined the details of how the compounds combine and the end products in most cases. They concluded that unsaturated lactone antibiotics may inhibit enzyme activity by uniting with SH and possibly amino groups of enzyme proteins. This could be a basic mechanism of action of all unsaturated lactone inhibitors including the coumarins, protoanemonin, strophanthidin, and digitoxigenin, in addition to others named above.

As was pointed out in Chapter 4, allelopathic substances sometimes promote the infection of plants by pathogens. In fact, they sometimes make species susceptible to certain diseases to which they are normally resistant (Patrick and Koch, 1963).

Adams *et al.* (1970) reported that water-repellent soils are present under several shrub species in southeastern California and that these are particularly pronounced under *Larrea divaricata, Prosopis juliflora,* and *Cercidium floridum.* The outward extension of the water-repelllent soils generally coincides with the extension of the crown, indicating that substances leached from the aboveground parts of the plants are responsible. Fire increases the thickness of the hydrophobic layer of soil, and no annual plants start for several years after a fire, even if the crowns are removed. Hummocks under the shrubs are nearly devoid of annual vegetation even without fire, whereas the surrounding soil is densely populated with annuals. This is clearly a case of allelopathy because the effect is due to substances added to the environment by the shrubs, but it certainly represents an unusual mechanism of action. In this case, available soil moisture is simply decreased under the shrubs.

As stated at the beginning of this chapter, the mechanisms of action of allelopathic agents have not been adequately researched. The future will no doubt reveal many important mechanisms that are unknown at present; hopefully, this brief coverage will stimulate more research in this important area of allelopathy.

14

Factors Determining Effectiveness of Allelopathic Agents after Egression from Producing Organisms

This topic is of great importance in the overall phenomenon of allelopathy, but only a relatively small amount of research has been done on it. Consequently, I have discussed this topic last in order to emphasize its significance and the critical need for more research on the subject.

I. CHEMICAL UNION OF SOME ALLELOCHEMICS WITH ORGANIC MATTER IN SOIL

Wang *et al.* (1967b) found that amounts of several phenolic acids extractable with alcoholic NaOH at pH 11 from several Taiwan soils increased markedly when any of seven volatile fatty acids was added to the soils. They suggested that these phenolic acids might have been derived from humic acids which decomposed as the volatile acids were added.

Wang *et al.* (1971) added known amounts of five different phenolic acids (one at a time) to 10 g of each of two common agricultural soils in Taiwan, TSES mudstone and Yuehmei latosol. They permitted the mixtures to stand for 3 hours after which they extracted the soils with alcoholic NaOH at pH 11 and determined the amounts of the various phenolic acids extracted. Between 60 and 80% of the added *p*-hydroxybenzoic, *p*-coumaric, and vanillic acids was bound by TSES mudstone and not recovered, whereas between 40 and 80% of these phenolic acids was bound by Yuehmei latosol. Syringic and ferulic acids were fixed in enormous quantities by the two soils, 60–98% of the applied amounts being bound.

Wang *et al.* pointed out that the fraction that was fixed could not have been retained by adsorption on soil colloids, nor could it have been made insoluble by the presence of polyvalent cations in the soil, since both forms should have been made soluble by the extractant employed. They pointed out further that the amount that was not recovered could not have been decomposed by microorganisms, because experiments conducted in alcohol and in water gave similar results. The time involved (3 hours) was not sufficient either to allow breakdown by microorganisms. Therefore, they suggested that the phenolic acids were probably tied up by the humic acids in the soil.

It is noteworthy that when a relatively small amount (0.04 µmole/10 g soil) of *p*-hydroxybenzoic, *p*-coumaric, or vanillic acid was added to one of the soils previously named, the total amount of the added acid could not be recovered (Wang *et al.,* 1971). A greater amount of the other phenolic acids could be recovered, however, than was recovered without the addition of a phenolic acid to the soil. Apparently some kind of exchange reaction occurred.

Addition of ^{14}C-labeled *p*-coumaric or ferulic acid to soil confirmed the idea that those parts of the coumaric and ferulic acids that had been added to the soil, and which were unrecoverable with an alcoholic NaOH solution of pH 11, had been fixed by the humic material. They could be recovered with 0.15 M sodium pyrophosphate solution at pH 11, which is known to hydrolyze the humic acids.

Rhus copallina produces large amount of tannic acid, which is a potent allelochemic (Blum and Rice, 1969). The litter under a stand of this species in an Oklahoma prairie contained approximately 6000 to 46,000 ppm of tannic acid throughout the year, and the top 5 cm of soil contained 600 to 800 ppm. Moreover, tannic acid was found to a depth of 75 cm in the soil with a zone of accumulation at the 45 to 55 cm level. Soxhlet extraction with acetone proved to be the best technique and solvent system to extract the tannic acid from soil.

When known amounts of tannic acid were added to soil obtained from areas adjacent to *R. copallina,* which contained no tannic acid extractable with acetone, it was found that a minimum of 400 ppm had to be added before any could be recovered immediately. It is noteworthy, therefore, that as small a concentration as 30 ppm added to the same soil reduced the nodule number of heavily inoculated legumes growing in that soil. Thus, at least some of the bound tannic acid is still biologically active.

Attempts by Rice and Pancholy (1973) to extract condensed tannins from soil by Soxhlet extraction with various solvent systems or by base exchange reactions were unsuccessful. They decided that the tannins are probably held by chemical union with peptide linkages in the proteins of the humus and other organic matter in a manner analogous to that in the tanning of leather. Therefore, they tested a number of different concentrations of sodium hydroxide and autoclaving times to extract the tannins. They found that autoclaving for 10 minutes at 20-lb pressure with $1N$ NaOH gave the best and most consistent yields of condensed tannins in the soil types investigated by them.

The concentration of condensed tannins in the top 15 cm of soil in an intermediate successional stage and the climax in three different vegetation types ranged from 8 to 93 ppm. Nevertheless, as little as 2 ppm of condensed tannins added to soils from the different research areas completely inhibited oxidation of NH_4 to NO_2 for 3 weeks (the duration of the tests) in all but one area. It appears, therefore, that at least part of the bound condensed tannins may be biologically active also.

Kaminsky and Muller (1977) recommended a method for the extraction of phytotoxins from soil using a neutral EDTA solution and tested it against some phenolic acids reported to be produced by certain shrubs. The procedure would probably be useful in many instances. Unfortunately, the authors made some statements in their paper, which have been shown to be questionable. One such statement concerns the covalent bonding of phytotoxins to organic matter in the soil. They stated that "Since such compounds cannot be absorbed by plants, they should be considered to be unavailable. The extraction of soil with procedures that would liberate such compounds, (e.g., alkaline procedures) would yield data of questionable ecological importance, and should, therefore, be avoided. . . ." In contrast to their statement, the evidence discussed in the previous paragraphs suggests that at least part of some bound allelopathic compounds are biologically active.

Lodhi (1978a) found that decaying litter of sycamore and northern red oak inhibited the growth of test species in January more than in April or August (see Chapter 3, Section I,A). On the other hand, hackberry and white oak were most inhibitory in April. He found that northern red oak and sycamore leaves contained phytotoxins that are primarily in the free form, whereas white oak and hackberry leaf litter released large numbers of toxins only after hydrolysis.

Another important point that was overlooked by Kaminsky and Muller is that phytotoxins can have very important effects on higher plants, indirectly, through their effects on microorganisms in the soil, which are important to the growth of higher plants. Thus, phototoxins do not have to be absorbed by higher plants in order for the phytotoxins to have important effects on such plants. There are several known instances also where phytotoxins are effective only in fine textured soils where they are adsorbed and concentrated to toxic proportions.

At present, our knowledge is limited in relation to which allelopathic compounds are bound to soil and which bound compounds retain at least some biological activity. The mechanisms of binding are certainly not clear either in most cases. Much detailed research is urgently needed in this important area, and tagged molecules should be helpful in clarifying some of the gray areas.

II. SOIL TEXTURE AND ACCUMULATION OF ALLELOCHEMICS TO PHYSIOLOGICALLY ACTIVE CONCENTRATIONS

Decomposing shoots of buffalobur nightshade (*Solanum rostratum*) inhibited the growth of both tomato and buffalobur nightshade plants in soil during a 30-day period; they either stimulated growth or had no effect in sand (Ahshapanek, 1962). Decomposing roots of buffalobur nightshade had similar effects on tomato plants. The plants were watered periodically during the 30-day growth period with nutrient solution and at other times with cistern water when necessary. It seems likely therefore that the inhibitory compounds were accumulated in the rooting zones of the plants in soil, whereas they were leached below the root zone in sand.

Muller noted that patterns of zonation around certain shrubs in the California chaparral were most pronounced in fine textured soils and particularly on heavy Zaca clay soils, which are widely distributed in the Santa Ynez Valley where he did most of his research (see Chapter 5, Section III,A). Muller and del Moral (1966) demonstrated that volatile terpenes from *Salvia* and other allelopathic shrubs are adsorbed on the colloidal particles in soil, remain in an active state for a prolonged period, and migrate from the surfaces of the soil particles to the sites of inhibition within the plants (see Figs. 16 and 17 in Chapter 5).

Eucalyptus camaldulensis is very inhibitory to annual herbs (Chapter 5, Section III,B) in many areas in California, but del Moral and Muller (1970) found that it fails to inhibit these plants on sand. They found that the volatile terpene inhibitors were adsorbed on the colloidal particles in the soil just as were the volatile inhibitors from allelopathic shrubs. Moreover, they placed 100 g samples of Oakley sand and Milpitas loam in Büchner funnels and flooded each 2 times with 100 ml of an aqueous extract of *E. camaldulensis* litter. Similar samples of each soil treated with distilled water served as controls. All soils were air-dried and radicle growth of *Bromus rigidis* was determined on all samples. Radicle growth in treated sand was reduced to 78% of the control, compared with 42% in the loam. Thus, this experiment indicated that the greater retention capacity of the loam for the water-soluble phenolic compounds in the litter was important in the attainment of physiologically active concentrations of these allelochemics in the substrate also.

The rather meager evidence available suggests that soil texture may be very important in determining the impact of a species with allelopathic potential. Much more research needs to be done, however, on this important subject.

III. DURATION OF ALLELOPATHIC ACTIVITY

There is little doubt that all plants and microorganisms produce compounds, which, if present in appropriate concentrations, can inhibit or stimulate the growth of other plants and microorganisms. It is also clear from evidence presented previously that many factors, both genetic and environmental, affect the amounts of potential phytotoxins produced. Once the phytotoxins are produced and escape into the environment, they begin to be decomposed either by microorganisms or by chemical action not involving microorganisms. Thus, potential allelopathic effects depend basically on the relative rates of addition of the allelochemics to the environment and decomposition, or inactivation. This was shown very clearly by Burgos-Leon (1976) and Burgos-Leon et al. (1980) in their work on allelopathic effects of sorghum residues in different soils in Africa (see Chapter 2, Section II). In many cases, an accumulation of an allelochemic may occur for only a short time, but this can be very important if it happens in a critical stage of a susceptible plant or microorganism. Even a brief period of inhibition or stimulation could affect a susceptible organism's competitive ability in relation to a nonsusceptible organism.

Guenzi et al. (1967) found that wheat, oat, corn, and sorghum residues collected at harvest time contained water-soluble compounds that inhibited the growth of wheat seedlings. Wheat and oat residues contained virtually no water-soluble toxic components after 8 weeks of decomposition under field conditions. On the other hand, corn and sorghum residues remained toxic for 22 to 28 weeks under the same conditions.

Blum and Rice found over 500 ppm of tannic acid, a potent allelopathic compound, throughout the year in the top 5 cm of soil under a stand of *Rhus copallina*. They found this compound to a depth of 75 cm with a zone of accumulation at the 45 to 55 cm level. Thus, this allelochemic is very stable and persists in the soil. The large amounts of condensed tannins, which occur in soils of several different ecosystems, and the persistence of these compounds were discussed previously (Chapter 9, Section III).

Kimber (1973) reported that germination of wheat was depressed for a period of 18 days by wheat straw applied to the surface of pots. Tang and Waiss (1978) found that toxicity of decomposing wheat straw in water increased gradually up to 12 days and that the amounts of salts of acetic, propionic, and butyric acids increased during that time also. Chou and Lin (1976) reported that decomposing rice residues in rice paddies remained inhibitory to radicle growth of rice and lettuce seedlings for 4 months.

Burgos-Leon (1976) found that a crop of *Sorghum vulgare* following the same crop was a complete failure on sandy soils in Senegal (Africa) but did alright on clay soils high in montmorillonite. This resulted from the persistence of the phytotoxins in sandy soils, whereas they were decomposed rapidly in clay soils.

AlSaadawi and Rice (1982b) and AlSaadawi et al. (1983) found four phenolic and nine fatty acid phytotoxins in the residues of *Polygonum aviculare* 4 months after death of this annual plant in the field. All the phenolic compounds and seven of the fatty acids were present in the soil under the residues. No later sampling was done.

It is evident that some allelopathic compounds are very stable in soil and others are not. It is also clear that the stability of any given compound depends considerably on the type of compound involved and on several environmental factors. This topic still remains the one about which we know the least in allelopathy, and it is urgent that more research be done.

IV. DECOMPOSITION OF ALLELOCHEMICS

A. Nonmicrobial Decomposition

Tubbs (1973) found that the leachate of rapidly growing sugar maple roots lost its inhibitory activity against yellow birch seedlings after standing for just a few days at room temperature. He subsequently sterilized some of the leachate with micropore filters and let it stand in a sealed container for 5 days at 5°C. The filtrate lost virtually all its toxicity in that short time period, indicating that the decomposition was spontaneous and not due to microorganisms.

No doubt many types of allelopathic compounds undergo spontaneous decomposition under at least some environmental conditions, but there is very little information in the literature about this phenomenon.

B. Microbial Decomposition

1. Phenolic Acids.   Cinnamic acid is an allelopathic compound that is exuded from the roots of guayule (Bonner and Galston, 1944), and Bonner (1946) reported that this compound is decomposed in soil. When it was added at the rate of 0.1 mg per gram dry weight of Hanford sandy loam and the soil was kept at field capacity at 25°C, the toxicity disappeared in 2 days. The toxicity of soil containing 1 mg of cinnamic acid per gram disappeared in 2 weeks under the same conditions. When the soil was sterilized after addition of the cinnamic acid, no reduction in toxicity occurred during the period of incubation, indicating that microorganisms were responsible for the decomposition under nonsterile conditions.

Henderson and several colleagues studied the decomposition by soil microorganisms of numerous phenolics known to be allelopathic compounds. Henderson and Farmer (1955) isolated many fungi from soils under a variety of vegetation types and found several of them able to decompose p-hydroxybenzaldehyde, ferulic acid, syringaldehyde, and vanillin. These compounds were used as the sole source of carbon by the organisms tested. It was found that vanillin and ferulic acid were converted to vanillic acid before the breaking of the benzene ring, and syringaldehyde was converted to syringic acid.

Henderson (1956) used spore suspensions of *Haplographium* sp., *Hormo-*

dendrum sp., *Penicillium* sp., and *Spicaria* sp. to investigate the metabolism of the same four phenolics employed by Henderson and Farmer (1955). Henderson's results confirmed those of Henderson and Farmer and extended them. He found that an intermediate compound in the breakdown of *p*-hydroxybenzaldehyde was *p*-hydroxybenzoic acid. Henderson and Farmer used mycelia in their studies and *p*-hydroxybenzaldehyde was metabolized so rapidly that no intermediate could be detected.

All four fungal species tested by Henderson metabolized *p*-hydroxybenzaldehyde most rapidly, followed by ferulic acid and vanillin (about the same), and syringaldehyde was metabolized most slowly by all species. The stimulation by the four compounds of oxygen uptake followed the same sequence also. The intermediate products of the metabolism of those compounds, *p*-hydroxybenzoic, syringic, and vanillic acids, were attacked by adaptive enzymes produced by the four fungal species.

In chernozem soil enriched with vanillin as the carbon and energy souce, the bacterial count almost tripled after 3 days' incubation (Kunc, 1971). A marked increase (from 7 to 68%) also occurred in the proportion of bacteria capable of decomposing vanillin as the only carbon and energy source. Of the 21 strains of bacterial vanillin decomposers isolated from soil, 15 (six species) were members of the genus *Pseudomonas,* five (two species) belonged to the genus *Cellulomonas,* and one to the genus *Achromobacter.* In another series of 37 bacterial isolates capable of decomposing vanillin, 22 (59.5%) attacked cellulose also.

Measurement of oxygen consumption by soil suspensions showed more than one maximum in the rate of oxidation of vanillin. Moreover, examination of the course of oxidation of vanillin by cultures of different bacterial isolates showed that not all strains oxidized vanillin in the same manner. The pH of the culture medium also markedly influenced the course of oxidation in pure bacterial cultures. In chernozem soil suspensions, vanillin was decomposed via vanillic and protocatechuic acids before the aromatic ring opened.

Microorganisms capable of utilizing gallic acid as their sole carbon source are not readily isolated but Tack *et al.* (1972) isolated a strain of *Pseudomonas putida* from soil in St. Paul, Minnesota, which readily oxidized gallate after growth at the expense of syringic acid. Cell-free extracts of this strain of *P. putida,* grown with syringic acid as the carbon source, catalyzed the oxidation of 1 mole of gallate by 1 mole of oxygen to give 2 moles of pyruvate and 1 mole of CO_2.

Ferulic, sinapic and *p*-coumaric acids are important in the synthesis of lignin and these same acids have been shown to be products of lignin degradation. It is not surprising, therefore, that ferulic and *p*-coumaric acids are the most common cinnamic acid derivatives found in soil (Chapter 10). Moreover, they are potent inhibitors of the growth of plants and microorganisms as indicated in several previous chapters. Numerous scientists have investigated the microbial decomposition of ferulic acid because it is usually produced in greatest abundance during lignin decomposition.

High amounts of ferulic acid are found in the leaves of *Celtis laevigata* (Lodhi and Rice, 1971), so Turner and Rice (1975) decided to follow the loss of ferulic acid from leaves of this species during the course of a year, and the concurrent change in the ferulic acid content of the soil under the leaves. Leaves were collected soon after leaf fall, oven dried, and aliquots were analyzed for ferulic acid concentration. The rest were sewn into nylon bags which were weighed and numbered. They were returned to the collection area and placed on the soil surface in five rings spaced 0.75 m apart around a *C. laevigata* tree. Each ring consisted of 12 bags to permit removal of one bag from each ring every month for a year. When the bags were collected, 300 g of soil were removed from the 0- to 15-cm layer beneath each one and the ferulic acid concentrations of the leaves and of the soil were determined.

Over 65% of the extractable ferulic acid was lost from the leaves in the first 100 days, and by day 300, less than 0.1% remained in the leaves. Depite the rather rapid loss of ferulic acid from the leaves, the concentration of ferulic acid in the soil under the bags of decaying leaves usually remained relatively constant from month to month except for minor fluctuations, which were not statistically significant (Table 59). The only exception to this was the concentration at day 100, which was significantly higher than the concentration at the previous sampling period and that at the following sampling period. Turner and Rice were unable to determine whether the small amounts of ferulic acid that disappeared from the leaves each month actually leached into the soil under the bags, because of the rather large amount of soil (300 g) that was analyzed beneath each bag. The large amount of soil was necessary because of the low concentration of ferulic acid in the soil. They concluded that the minor fluctuations in ferulic acid concentration in the soil from month to month could have resulted in part from the leaching of the released material from the leaves and that it is possible that the ferulic acid concentration increased considerably in a thin layer of soil next to the leaves. On the other hand, they concluded that the very large increase in ferulic acid concentration in the soil to a depth of 15 cm in March could not have resulted from material leaching from the leaves; thus, it must have been exuded

TABLE 59. Concentration of Ferulic Acid in Soil under Bags of Decomposing *Celtis laevigata* Leaves[a]

Days[b]:	0	30	100	130	161	191	222	253	300	365
Ppm[c]:	25	26	47[d]	22	22	30	34	27	32	23

[a] From Turner and Rice (1975).
[b] Number of days from start of decomposition.
[c] Each figure is average of five analyses.
[d] Difference from concentrations at day 30 and day 130 significant at 0.05 level.

from living roots of the *C. laevigata* or must have been leached from dead or decomposing roots. Young roots of this species contained 2000 ppm of ferulic acid.

Another possible explanation for the decrease in concentration of ferulic acid in the decaying leaves is that some, or all of it, was decomposed by microorganisms before leaching from the leaves. The disappearance of ferulic acid from the decomposing leaves without any increase in concentration in the soil under the leaves over the period of a year indicated that this allelopathic compound was being decomposed in either the leaves or the soil, or both.

Soil from under a *C. laevigata* tree was either amended with 1000 ppm ferulic acid or perfused with a 1000 ppm aqueous solution, and microorganisms that survived were isolated and characterized, and some were identified. Nine morphologically distinct bacteria were isolated, of which eight were gram-negative and one gram-positive. Several organisms were keyed to the genus *Pseudomonas*. Two fungi, *Cephalosporium curtipes* and *Rhodotorula rubra,* were also isolated. Further growth tests in a liquid medium with ferulic acid as the sole carbon source resulted in the elimination of all of the original isolates except the two fungi.

The optimum pH for the utilization of ferulic acid by *Rhodotorula rubra* was approximately 7.0, whereas the optimum pH for *Cephalosporium curtipes* was approximately 6.0. *R. rubra* grew almost as well on vanillic, cinnamic, and *p*-hydroxybenzoic acids as on ferulic acid. *C. curtipes* utilized ferulic acid more rapidly than it did vanillic, cinnamic, or *p*-hydroxybenzoic acids.

Of four species of *Cephalosporium* (*C. furcatum, C. khandalense, C. nordinii, C. roseum*) obtained from the American Type Culture Collection, only *C. furcatum* was able to utilize ferulic acid as the sole carbon source. *Rhodotorula glutinis, R. rubra,* and *R. lactosa* were obtained from the same collection, and, of these, only the first two listed were able to utilize ferulic acid as the sole carbon source.

Haider and Martin (1975) labeled 10 phenolic acids in various positions with ^{14}C and followed the loss of the labeled carbon as $^{14}CO_2$ upon incubation at 100 and 1000 ppm in Greenfield sandy loam over a 12-week period. Within 1 week about 90% of the $^{14}COOH$ carbon of *p*-hydroxybenzoic, syringic, vanillic, and anisic acids had evolved as $^{14}CO_2$. After 12 weeks, this had increased to about 95%. Loss of the $^{14}COOH$ carbon of 6-methylsalicylic acid was about 75% at 1 week and 87% after 12 weeks. Approximately 42 to 85% of the side-chain carbons of *p*-hydroxycinnamic and caffeic acids evolved as CO_2 and from 66 to 85% of the ring carbon of protocatechuic, benzoic, vanillic, veratric, and caffeic acids was lost over the 12-week incubation period. Linkage into model phenolase polymers greatly retarded the decomposition rate. Loss of ring carbon varied from 3 to 6% and side chain and COOH carbons from 8 to 75%. It should be kept in mind that these losses occurred under optimum temperature, aeration, and moisture conditions for the decomposers; therefore, it is difficult to extrapolate to field conditions.

Black and Dix (1976) examined the ability of 21 species of fungi to use ferulic acid as a sole carbon source. Species saprophytic on *Angiosperm* leaf litters showed wide variation in their ability to use this compound. The majority of soil and conifer litter inhabiting species examined utilized ferulic acid readily, the notable exceptions being isolates of *Trichoderma viride, Mucor* spp., and *Mortierella ramanniana*. The ability of these fungi to utilize ferulic acid was not limited by poor spore germination. In fact, germination of *Mucor, Trichoderma,* and some other fungi was stimulated by ferulic acid.

2. Flavonoids.   The fungus *Pullularia fermentans* was shown to decompose rutin to phloroglucinol, protocatechuic, and 2-protocatechuoylphloroglucinol carboxylic acids (Hattori and Noguchi, 1959). Westlake *et al.* (1959) tested 280 cultures of bacteria, actinomycetes, and fungi for their ability to decompose rutin. Cultures from 33 genera were unable to decompose that compound, whereas cultures from 19 identified genera and several unidentified genera were able to do so. Active genera were distributed among the actinomycetes, bacteria, and fungi; the fungi appeared, however, to be most active. *Aspergillus flavus* and *A. niger* were particularly effective in decomposing rutin. The aspergilli, when grown on either rutin or quercetin, produced extracellular enzymes that degraded both rutin and quercetin, but not quercitrin. Rutinose, protocatechuic acid, phloroglucinol carboxylic acid, and a phloroglucinol carboxylic acid–protocatechuic acid ester were produced during the decomposition.

Padron *et al.* (1960) obtained a species of *Aspergillus* from a soil enrichment procedure using quercetin as the enrichment agent. They grew this organism subsequently for 72 hours in a basic medium with quercetin as the sole carbon source. They obtained an extracellular enzyme preparation from this culture, which was capable of degrading quercetin readily. This enzyme preparation was tested against 17 other flavonoids, and it was able to decompose six of them: rutin, isoquercitrin, rhamnetin, fisetin, kaempferol, and galangin.

Nine isoflavonoids are found in red clover, and it was thought initially that red clover soil sickness was due to these compounds that are very inhibitory to growth of this species (Chang *et al.*, 1969). Attempts to isolate isoflavonoids from farm soil where clover sickness was present were not successful. Several phenolic acids were found, however, and they strongly inhibited the growth of red clover. Phenolic acids identified were *p*-methoxybenzoic, *p*-hydroxybenzoic, 2,4-dihydroxybenzoic, and salicylic acids.

It was decided, therefore, to determine what compounds were present in the culture water in which red clover plants had grown. Two isoflavonoids, biochanin A and formononetin, were detected plus *p*-methoxybenzoic, *p*-methoxyphenylacetic, and cinnamic acids. The three phenolic acids do not occur in red clover plants, so they were apparently formed in the medium from the iso-flavonoids, which had been exuded into the medium.

Degradation of biochanin A in aqueous 0.1 N NaOH for 12 weeks resulted in the formation of phloroglucinol, p-methoxybenzoic acid, and p-methoxy-phenylacetic acids, plus several unidentified compounds. p-Hydroxybenzoic, p-hydroxyphenylacetic, and 2,4-dihydroxybenzoic acids, and resorcinol were detected during the degradation of daidzein. The breakdown of trifolirhizin resulted in the production of 2,4-dihydroxybenzoic acid and some unidentified phenolic compounds. It was concluded that clover sickness results from the accumulation in soils of phenolic compounds as well as isoflavonoids, which can be converted into the former acids before or after the death of the red clover plants.

Börner (1959) concluded that phloretin, p-hydroxyhydrocinnamic acid, phloroglucinol, and p-hydroxybenzoic acid are decomposition products of phlorizin and that the decomposition is accomplished by microorganisms (See Chapter 3, Section II). The three phenolics are very inhibitory to the growth of apple seedlings, as are phlorizin and phloretin.

Holowczak et al. (1960) reported that isolates of the three known races of Venturia inaequalis made poor growth on a medium containing 2×10^{-3} M phlorizin as the sole carbon source. All isolates were able to degrade phlorizin in the presence of glucose to phloretin and glucose, and phloretin to p-hydroxy-hydrocinnamic acid and phloroglucinol. A few of the isolates could degrade p-hydrocinnamic acid to p-hydroxybenzoic acid.

An inducible enzyme catalyzing the hydrolysis of phloretin to phloroglucinol and p-hydroxyhydrocinnamic acid was extracted from the acetone-dried powders of the mycelial felts of an Aspergillus niger strain grown in the presence of phlorizin (Minamikawa et al., 1970). Its activity on phloretin appeared to be maximal at about pH 9.6. The enzyme showed a rather broad substrate specificity, and some other C-acylated phenols related to phloretin (e.g., phloracetophenone) were hydrolyzed also.

3. Tannins.   Tannins are common and generally very inhibitory allelopathic compounds against most test organisms (see Chapter 10). Nevertheless, there are microorganisms that can decompose at least some of these highly diverse compounds. Tannins are transformed slowly by certain fungi, principal among which are Penicillium and Aspergillus (Lewis and Starkey, 1968, 1969). Species of Fusarium, Cylindrocarpon, Gliocladium, Ramigella, Spicaria, and Endothia have been implicated also in the decomposition of various tannins. There is little information concerning the decomposition of tannins by basidiomycetes, although they are suspect because of their ability to decompose lignins, which contain polyphenolic residues similar to those found in tannins. One bacterium, Achromobacter sp., was demonstrated by Lewis and Starkey (1969) to be able to use gallotannin or chestnut tannin as its sole carbon source. Hydrolyzable tannins appear to be more susceptible to microbial attack than condensed tannins. The result is that the latter accumulate in soil over a long period of time (Rice and Pancholy, 1973).

A purified preparation of wattle tannin (a condensed tannin) was decomposed very slowly by a mixed-soil preparation in culture solution (Benoit *et al.*, 1968). From 3 to 17% of the tannin carbon was released as CO_2 aerobically in 21 days, depending on the pH and tannin concentration. Under the same conditions two-thirds or more of the carbon of proteins, hemicelluloses, and polysaccharides was oxidized to CO_2.

Lewis and Starkey (1968) determined the relative rates of decomposition of three hydrolyzable tannins (tannic acid, gallotannin, chestnut tannin) and two condensed tannins (wattle tannin, canaigre tannin) by a hardwood forest soil at pH 4.2. From about 25 to 43% of the hydrolyzable tannins was decomposed in 60 days; whereas only about 7% of the condensed tannins was decomposed during that time.

Grant (1976) isolated a strain of *Penicillium adametzi* from enrichment cultures with condensed tannins as the carbon source. The fungus was originally isolated from a rotting *Pinus radiata* log. This penicillium grew well on the low-molecular weight condensed tannins, which indicates that it would decompose such compounds very slowly. According to the author, this is the first report of isolation of an organism capable of growth on condensed tannins as the sole carbon source under defined culture conditions.

Much more information is needed on rates of production of allelopathic compounds, rates of escape into the environment, rates of decomposition, and cycling of these compounds back into plants.

V. SYNERGISTIC ACTION OF ALLELOCHEMICS

There have been numerous suggestions in the literature that synergistic actions probably result from combinations of identified allelopathic compounds. Most authors have not, however, tested this hypothesis.

Evenari (1949) discussed the synergistic effect of combinations of organic acids in inhibiting plant growth. He pointed out that a mixture of 0.025% citric, 0.025% tartaric, and 0.01% salicylic acid (total acids 0.06%) exerts a much stronger effect than each acid alone in a concentration of 0.06%.

Monoterpenes exert a marked synergistic effect in combinations (Asplund, 1969). Pulegone alone reduced radish seed germination by 50% at a concentration of 1.5 μM/liter in ambient air and the parallel value for (−)camphor was 3.1 μM/liter. A combination of these at a total concentration of only 0.0338 μM/liter reduced germination by 80%. This was a decrease in concentration of almost 100-

fold. Similar increases in toxicity as a result of synergism occurred also with (+)camphor and (−)borneol in combination with the terpenes discussed above and with each other. It is noteworthy that, when used singly, (+)camphor and (−)camphor did not produce significantly different effects; however, their synergistic effect was highly significant. It is interesting also that the combination of (−)camphor with (+)puligone had approximately 4 times as great an effect on germination as the combination of (+)camphor with (+)pulegone.

Ferulic and p-coumaric acids are two of the most commonly identified allelopathic compounds in soil. It is significant, therefore, that these compounds have synergistic phytotoxic effects (Rasmussen and Einhellig, 1977). Equimolar mixtures of both acids reduced sorghum seed germination, shoot elongation, and total seedling growth more than did either compound alone. Mixtures containing $5 \times 10^{-3} M$ p-coumaric and $5 \times 10^{-4} M$ ferulic acids reduced seed germination to 34% of control germination after 24 hours and 59% by 48 hours. The same concentration of either phenol alone reduced germination to only 69 and 92% of control germination. The inhibitory effect of the combination approximated the inhibitory effect on sorghum seed germination of $10^{-2} M$ ferulic acid and was a greater reduction than that caused by a $10^{-2} M$ p-coumaric acid treatment. Seedling growth of sorghum was more sensitive to the phenolics than seed germination. An equimolar mixture of $2.5 \times 10^{-4} M$ p-coumaric and 2.5×10^{-4} ferulic acids reduced seedling growth more than did $2.5 \times 10^{-4} M$ p-coumaric or ferulic acid alone. A $1.25 \times 10^{-4} M$ concentration of either phenol stimulated sorghum seedling growth, whereas a mixture of these two inhibited growth.

Vanillic and p-hydroxybenzoic acids, two other commonly identified allelopathic compounds in soil, also exert synergistic action (Einhellig and Rasmussen, 1978). At threshold inhibitory concentrations, $2.5 \times 10^{-3} M$ vanillic acid reduced radish seed germination by 29% after 24 hours, and 2.5×10^{-3} M p-hydroxybenzoic acid reduced germination by 5%. A mixture of these compounds, each in a $2.5 \times 10^{-3} M$ concentration, reduced seed germination by 48% in 24 hours. Sorghum root and shoot elongation and total seedling growth were more sensitive than germination to vanillic and p-hydroxybenzoic acid treatments, and synergistic effects were present also. A combination of $5 \times 10^{-3} M$ vanillic acid and $5 \times 10^{-3} M$ p-hydroxybenzoic acid reduced root growth more than either did individually, and a mixture of $5 \times 10^{-4} M$ vanillic acid and $5 \times 10^{-4} M$ p-hydroxybenzoic acid reduced sorghum seedling growth to approximately that resulting from a $10^{-3} M$ concentration of either phenol alone.

Rasmussen and Einhellig (1979b) extended their previous work on the synergistic effects of ferulic, p-coumaric, and vanillic acids. They found that an equimolar combination of $3.3 \times 10^{-3} M$ of each of the three compounds showed a synergistic inhibition of sorghum seed germination. Moreover, that combination also depressed seed germination more than a combination of any two of those compounds.

The three allelopathic compounds, individually, had similar effects on seed germination and seedling growth of sorghum at 10^{-2} M and 3.3×10^{-3} M concentrations. However, the 3.3×10^{-3} M concentration of vanillic acid stimulated seedling growth. Equimolar concentrations (3.3×10^{-3} M) of the three compounds in combination did not exert synergistic action on seedling growth as they did on seed germination. This might have resulted from the stimulatory effect exerted by the 3.3×10^{-3} M concentration of vanillic acid on seedling growth. Differences in mechanisms of action of various allelopathic compounds have been demonstrated; thus, differences in activity are to be expected (see Chapter 13).

Williams and Hoagland (1982) tested twelve phenolic compounds, singly and in some two member combinations against seed germination of four crop and five weed species. The concentrations tested were 10^{-3} and 10^{-5} M. There was little effect of any compound at 10^{-5} M. At 10^{-3} M, coumarin, hydrocinnamic acid, juglone, and pyrocatechol strongly inhibited seed germination; p-hydroxybenzaldehyde and p-hydroxybenzoic acid were ineffective; and fumaric, p-coumaric, ferulic, caffeic, chlorogenic, and gallic acids had intermediate effects. Chlorogenic acid, p-hydroxybenzaldehyde, and pyrocatechol, each combined with coumarin markedly inhibited germination of some test species. The combination of coumarin and p-hydroxybenzaldehyde had an additive effect on germination of two weed species, inhibiting germination to a greater extent than either compound alone. This was true also of the combination of coumarin and pyrocatechol against one weed species.

Einhellig et al. (1982) investigated the possible synergistic effects on grain sorghum of four cinnamic acid derivatives, which are known allelopathic compounds. These were trans-cinnamic acid, caffeic acid, p-coumaric acid, and ferulic acid. They first determined threshold concentrations for inhibition of seed germination, radicle elongation, and seedling growth by each of the compounds. They determined a rate of germination index as follows:

$$\text{Index} = \text{at } \frac{\text{no. germ.}}{1} + \frac{\text{no. germ.}}{2} + \frac{\text{no. germ.}}{3} + \frac{\text{no. germ.}}{4} + \frac{\text{no. germ.}}{5}$$

Effects on root elongation were determined by measuring the length of the longest root of each seedling, 6 days after planting the seeds. Effects on seedling growth were determined by taking oven-dry weights of the seedlings 6 days after treatment with the selected compounds.

All of the original data were converted to percent-of-control values and Colby's (1967) method was used for testing the effects of the various combinations of allelopathic compounds. An expected (E) percent-of-control value was calculated for combinations of two as $E = XY/100$, where X and Y were percent-of-control values for single treatments. The expected response for higher combinations was obtained by calculating the product of the percent-of-control values for compounds applied alone and dividing by 10,000 with combinations of three and 1,000,000 with combinations of four. Observed values less than the expected indicate synergism, more than expected suggest antagonism, and when the observed value equals the expected, the combination is additive.

All combinations of two, three, or four of the compounds tested exerted synergistic effects against sorghum seed germination except one. The exception was the combination of ferulic and *trans*-cinnamic acids. All combinations except one had slight to marked synergistic effects on sorghum radicle elongation (Table 60). The exception was the combination of caffeic and *trans*-cinnamic acids. All combinations except two had synergistic effects on sorghum seedling growth also. The exceptions were the combinations of caffeic and ferulic acids and ferulic and *trans*-cinnamic acids. In all cases, the combinations of three and four phenols exerted synergistic action on seed germination, radicle elongation, and seedling growth.

The five known allelopathic compounds present in decomposing rice straw (Chapters 2 and 9) were tested singly and in different combinations against three strains of *Rhizobium* (Rice *et al.*, 1981). The bacteria selected for tests included *R. leguminosarum*, American Type Culture (ATC) strain 10314; *Rhizobium* sp., ATC strain 10703; and *R. japonicum*, Taiwan strain WSM isolated from soybean nodules. Each of the five compounds inhibited growth of all three rhizobial strains even in concentrations as low as $10^{-4} M$. When combinations of equimolar concentrations ($10^{-3} M$) of the phenolics were tested, synergistic effects resulted in several instances. The combination of *p*-coumaric and vanillic acids had a synergistic effect on *Rhizobium* strains 10314 and 10703, the combination of *p*-hydroxybenzoic and *o*-hydroxyphenylacetic acids had synergistic effects against all three strains, the combination of ferulic and vanillic acids had synergistic effects on strains 10314 and 10703, and the combination of all five phenolics had a synergistic effect against strain 10314.

Thus, evidence is strong that combinations of allelopathic compounds often exert additive or synergistic actions against the growth of plants and microorganisms. This synergistic action is of considerable significance in natural and agronomic communities, because considerable discussion has centered around the realistic concentrations necessary for inhibitory action of allelopathic compounds under field conditions. Seemingly insignificant concentrations of individual allelochemics become significant in view of the synergistic capabilities demonstrated.

It is revealing to consider microbial decomposition of allelopathic compounds in relation to synergism. Isoflavonoids are decomposed to several phenolic com-

TABLE 60. Effects of Combinations of Phenolic Acids on Sorghum Radicle Elongation[a]

Phenolic treatment (mM)				Radicle length (mm ± SE)[c]	% of Control (observed)	% Expected in combination[d]	Observed, minus expected[e]
CA	pCA	FA	tCnA[b]				
—	—	—	—	20.7 ± 1.9a			
1.0	—	—	—	14.3 ± 1.6b	69		
—	1.0	—	—	6.8 ± 1.2c	33		
—	—	1.0	—	17.8 ± 2.0ab	86		
—	—	—	0.25	19.3 ± 1.8a	93		
Combinations of two:							
1.0	—	1.0	—	6.3 ± 1.4cd	30	(59)	−29
1.0	1.0	—	—	4.3 ± 0.8cde	21	(23)	− 2
—	1.0	1.0	—	2.9 ± 1.0cde	14	(28)	−14
—	—	1.0	0.25	6.2 ± 1.1cd	30	(80)	−50
—	1.0	—	0.25	5.8 ± 1.2cd	28	(31)	− 3
1.0	—	—	0.25	17.5 ± 1.7ab	85	(64)	+21
Combination of three:							
1.0	1.0	1.0	—	1.3 ± 0.5e	6	(20)	−14
1.0	—	1.0	0.25	2.6 ± 1.0de	13	(55)	−42
1.0	1.0	—	0.25	2.2 ± 0.9de	11	(21)	−10
—	1.0	1.0	0.25	0.5 ± 0.3e	2	(26)	−24
Combination of four:							
1.0	1.0	1.0	0.25	1.3 ± 0.3e	6	(18)	−12

[a] Modified from Einhellic et al. (1982).
[b] Symbols: CA, caffeic acid; pCA, p-coumaric acid; FA, ferulic acid; tCnA, trans-cinnamic acid.
[c] Means not followed by the same letter are significantly different, $P < 0.05$, by Duncan's Multiple Range Test.
[d] See formula in text.
[e] Difference values with a minus sign indicate synergism.

pounds; vanillin and ferulic acid are converted to vanillic acid; *p*-hydroxybenzaldehyde is converted to *p*-hydroxybenzoic acid; most other allelochemics are decomposed through a series of intermediate compounds that also have allelopathic potential. It is noteworthy, therefore, that the partial decomposition of one compound may result in the presence of several allelopathic compounds, which may exert synergistic allelopathic effects. Thus, partial decomposition could increase allelopathic action, rather than decrease it.

VI. ENHANCEMENT OF ALLELOPATHIC ACTIVITY BY OTHER STRESS FACTORS

Many factors that determine the degree of allelopathic activity of a given plant or microorganism have been discussed. I pointed out in Chapter 11 that genes and many environmental factors affect concentrations of allelochemics produced in the organisms. Various genetic and environmental factors are known also to regulate the movement of allelochemics from donor plants. In previous sections of this chapter, it was noted that numerous factors affect the potential availability and activity of allelochemics in soil. One would expect, therefore, that various genetic and environmental factors would regulate the response of a receptor plant or microorganism to allelochemics that enter it. Unfortunately, very little research has been done on this phase of allelopathy. Probably all scientists who have done considerable research in allelopathy have observed that all individuals of a given test species (even of a rather genetically uniform crop cultivar) do not respond uniformly to a given allelopathic compound. This is true even under controlled environmental conditions. Such results are no doubt caused partially by genetic differences in individuals, but probably many factors are involved.

Probably stress factors affect the response of a plant to a specific allelochemic, just as many stress factors affect the production of allelochemics by the donor plants. Unfortunately, little research has been done on this subject. Einhellig and Eckrich (1983) observed that, even under greenhouse conditions with adequate water and nutrients, there was considerable variation in response of plants to treatment with an allelochemic. They hypothesized that a stress factor, such as high temperature, might determine the degree of response of the receptor plant to the alielochemic when the chemical is near its threshold concentration for inhibition of seedling growth. They treated grain sorghum and soybean seedlings with various concentrations of ferulic acid and divided plants of each species into two groups. One group of sorghum plants was kept in a temperature range of 33.5° to 41.2°C during daylight hours and 18.5° to 22.6°C at night. The other group was kept in a range of 26.3° to 32.3°C during daylight hours and 15.6° to 19.4°C at night. These conditions were maintained in separate greenhouses. The day/night relative humidities for the hot and cool environments were 24/56% and 41/64%, respectively.

Both groups of soybeans were kept in growth chambers using a 16/8 hour light/dark cycle with a light intensity of 18,000 lux. One chamber was kept at a temperature of 34.4°C during the light period and 28.9°C during the dark period, with an average light/dark relative humidity of 40/53%. The cool chamber was maintained at 23.3°C during the light period and 14.4°C at night, with an average light/dark relative humidity of 63/73%.

Seedling growth of sorghum was reduced by both 0.2 mM and 0.4 mM concentrations of ferulic acid under the high temperature regime but only by the 0.4 mM concentration under the cool regime (Table 61). Abaxial leaf resistance was increased above the control value by the 0.2 mM and 0.4 mM concentrations in the high regime and by the 0.1 mM, 0.2 mM, and 0.4 mM concentrations in the low temperature regime.

Seedling growth of soybean was significantly reduced by both 0.1mM and 0.25 mM concentrations of ferulic acid in the high temperature regime but only by the 0.25 mM concentration in the low regime (Table 62). Abaxial leaf resistance was reduced below the value in the control by the 0.25 mM concentration in the high temperature regime, but was not changed significantly by any ferulic acid concentration tested in the low regime.

The combined effects of temperature and ferulic acid treatment on seedling growth were highly significant in both species (Tables 61 and 62). This was true

TABLE 61.   Interaction of Temperature and Ferulic Acid (FA) Stress on Sorghum Seedlings[a]

Treatment	Abaxial resistance[b] (s cm^{-1})	Plant dry weight[b] (mg ± SE)
Hot[c] (37.0°/20.3°C)		
Control	2.1 ± 0.2a	225.8 ± 12.1a
0.1 mM FA	2.7 ± 0.3ab	231.9 ± 24.0a
0.2 mM FA	3.6 ± 0.4b	150.6 ± 9.3b
0.4 mM FA	3.7 ± 0.9b	75.2 ± 6.7c
Cool[c] (29.3°/17.9°C)		
Control	1.3 ± 0.2a	226.2 ± 8.9a
0.1 mM FA	2.5 ± 0.2b	220.2 ± 14.5a
0.2 mM FA	2.1 ± 0.1bc	225.9 ± 13.4a
0.4 mM FA	2.9 ± 0.2c	151.9 ± 8.6b
Two-way ANOVA F Value and (probability)		
Temperature	17.4 (0.0001)	13.7 (0.0004)
FA	11.9 (0.0001)	31.5 (0.0001)
Temp-FA	2.4 (0.0766)	6.7 (0.0005)

[a] Data from Einhellig and Eckrich (1983).
[b] Column means within a temperature regime not followed by the same letter are significantly different $P < 0.05$, ANOVA with Duncan's Multiple Range Test.
[c] Day/Night means, see text for details.

TABLE 62.   Interaction of Temperature and Ferulic Acid (FA) Stress on Soybean Seedlings[a]

Treatment	Abaxial resistance[b] (s cm^{-1})	Plant dry weight[b] (mg ± SE)
Hot[c] (34.4°/28.9°C)		
Control	3.5 ± 0.5a	410.0 ± 7.7a
0.1 mM FA	2.7 ± 0.2ab	349.0 ± 11.8b
0.25 mM FA	2.1 ± 0.1b	259.4 ± 8.4c
Cool[c] (23.3°/14.4°C)		
Control	1.3 ± 0.1a	393.5 ± 11.8a
0.1 mM FA	1.4 ± 0.1a	392.5 ± 12.7a
0.25 mM FA	1.4 ± 0.1a	328.1 ± 14.8b
Two-way ANOVA F Value and (probability)		
Temperature	59.6 (0.0001)	11.7 (0.001)
FA	4.4 (0.0175)	47.2 (0.0001)
Temp-FA	6.6 (0.0028)	7.3 (0.0012)

[a] Data from Einhellig and Eckrich (1983).
[b] Column means within a temperature regime that are not followed by the same letter are significantly different, $P < 0.05$, ANOVA with Duncan's Multiple Range Test.
[c] Day/Night means, see text for details.

also for the combined effect on abaxial leaf resistance in soybean. The amount and direction of response varied, however, both with the characteristic being considered and the plant species.

There are obviously many stress factors which could affect response of a plant or microorganism to a given allelopathic compound or combination of compounds. The striking results of Einhellig and Eckrich indicate that these interactions should be investigated.

BIBLIOGRAPHY

Abdul-Wahab, A. S. (1964). The toxicity of Johnson grass excretions: A mechanism of root competition. M.S. Thesis, Louisiana State Univ., Baton Rouge.

Abdul-Wahab, A. S., and Al-Naib, F. A. G. (1972). Inhibitional effects of *Imperata cylindrica* (L.) P.B. *Bull. Iraq Nat. Hist. Mus.* **5**, 17–24.

Abdul-Wahab, A. S., and Rice, E. L. (1967). Plant inhibition by Johnson grass and its possible significance in old-field succession. *Bull. Torrey Bot. Club* **94**, 486–497.

Adams, R. P. (1979). Diurnal variation in the terpenoids of *Juniperus scopulorum* (Cupressaceae)—summer versus winter. *Am. J. Bot.* **66**, 986–988.

Adams, R. P., and Hagerman, A. (1977). Diurnal variation in the volatile terpenoids of *Juniperus scopulorum* (Cupressaceae). *Am. J. Bot.* **64**, 278–285.

Adams, S., Strain, B. R., and Adams, M. S. (1970). Water-repellent soils, fire, and annual plant cover in a desert scrub community of southeastern California. *Ecology* **51**, 696–700.

Addoms, R. M. (1937). Nutritional studies of loblolly pine. *Plant Physiol.* **12**, 199–205.

Agnihothruda, B. (1955). Incidence of fungistatic organisms in the rhizosphere of pigeon-pea (*Cajanus cajan*) in relation to resistance and susceptibility to wilt caused by *Fusarium udum* Butler. *Naturwissenschaften* **42**, 373.

Agnihotri, V. P., and Vaartaja, O. (1968). Seed exudates from *Pinus resinosa* and their effects on growth and zoospore germination of *Pythium afertile*. *Can. J. Bot.* **46**, 1135–1141.

Ahlgren, H. L., and Aamodt, O. S. (1939). Harmful root interactions as a possible explanation for effects noted between various species of grasses and legumes. *J. Am. Soc. Agron.* **31**, 982–985.

Ahshapanek, D. C. (1962). Ecological studies on plant inhibition by *Solanum rostratum*. Ph.D. Dissertation, Univ. of Oklahoma, Norman.

Akehurst, S. C. (1931). Observations on pond life, with special reference to the possible causation of swarming of phytoplankton. *R. Microsc. Soc. J.* **51**, 237–265.

Alexander, M. (1977). "Introduction to Soil Microbiology," 2nd ed. Wiley, New York.

Alexander, M., and Clark, F. E. (1965). Nitrifying bacteria. *In* "Methods of Soil Analysis" (C. A. Black, D. D. Evans, J. L. White, L. E. Ensminger, and F. E. Clark, eds.), Vol. 2, pp. 1477–1483. Am. Soc. Agron., Madison, Wisconsin.

Allen, R. N., and Newhook, F. J. (1974). Suppression by ethanol of spontaneous turning activity in zoospores of *Phytophthora cinnamomi*. *Trans. Br. Mycol. Soc.* **63**, 383–385.

Allison, F. E. (1931). Forms of nitrogen assimilated by plants. *Q. Rev. Biol.* **6**, 313–321.

Allison, F. E. (1955). The enigma of soil nitrogen balance sheets. *Adv. Agron.* **7**, 213–250.

Al-Mousawi, A. H., and Al-Naib, F. A. G. (1975). Allelopathic effects of *Eucalyptus microtheca* F. Muell. *J. Univ. Kuwait, Sci.* **2,** 59–66.

Al-Mousawi, A. H., and Al-Naib, F. A. G. (1976). Volatile growth inhibitors produced by *Eucalyptus microtheca. Bull. Biol. Res. Centre* **7,** 17–23.

Al-Naib, F. A. (1968). Allelopathic effects of *Platanus occidentalis.* M.S. Thesis, Univ. of Oklahoma, Norman.

Al-Naib, F. A., and Al-Mousawi, A. H. (1976). Allelopathic effects of *Eucalyptus microtheca:* Identification and characterization of the phenolic compounds in *Eucalyptus microtheca. J. Univ. Kuwait, Sci.* **3,** 83–88.

Al-Naib, F. A., and Rice, E. L. (1971). Allelopathic effects of *Platanus occidentalis. Bull. Torrey Bot. Club* **98,** 75–82.

AlSaadawi, I. S., and Rice, E. L. (1982a). Allelopathic effects of *Polygonum aviculare* L. I. Vegetational patterning. *J. Chem. Ecol.* **8,** 993–1009.

AlSaadawi, I. S., and Rice, E. L. (1982b). Allelopathic effects of *Polygonum aviculare* L. II. Isolation, characterization, and biological activities of phytotoxins. *J. Chem. Ecol.* **8,** 1011–1023.

AlSaadawi, I. S., Rice, E. L., and Karns, T. K. B. (1983). Allelopathic effects of *Polygonum aviculare* L. III. Isolation, characterization, and biological activities of phytotoxins other than phenols. *J. Chem. Ecol.* **9,** 761–774.

Altiera, M. A., and Doll, J. D. (1978). The potential of allelopathy as a tool for weed management in crops. *PANS* **24,** 495–502.

Anaya, A. L., and Del Amo, S. (1978). Allelopathic potential of *Ambrosia cumanensis* H.B.K. (Compositae) in a tropical zone of Mexico. *J. Chem. Ecol.* **4,** 289–304.

Anaya, A. L., and Gómez-Pompa, A. (1971). Inhibicion del crecimiento producida por el "piru" (*Schinus molle* L.). *Rev. Soc. Mex. Hist. Nat.* **32,** 99–109.

Anaya, A. L., Del Amo, S., Ruy-Ocotla, G., and Ortiz, L. M. (1978). Allelopathic potential of a coffee plantation. *Int. Cong. Ecol., 2nd, Jerusalem, Israel,* p. 8. (Abstr.)

Anaya, A. L., Ruy-Ocotla, G., Ortiz, L. M., and Ramos, L. (1982). Potencial alelopatico de las principales plantas de un cafetal. *In* "Estudios Ecologicos en el Agroecosistema Cafetalero" (E. Jiménez Avila and A. Gómez-Pompa, eds.), pp. 85–94. Instituto Nacional de Investigaciones sobre Recursos Bioticos, Xalapa, Veracruz.

Anderson, R. C., Katz, A. J., and Anderson, M. R. (1978). Allelopathy as a factor in the success of *Helianthus mollis* Lam. *J. Chem. Ecol.* **4,** 9–16.

Anderson, R. C., Liberta, A. E., Packheiser, J., and Neville, M. E. (1980). Inhibition of selected fungi by bacterial isolates from *Tripsacum dactyloides* L. *Plant Soil* **56,** 149–152.

Anderson-Prouty, A. J., and Albersheim, P. (1975). Host-pathogen interactions. VIII. Isolation of a pathogen-synthesized fraction rich in glucan that elicits a defense response in the pathogen's host. *Plant Physiol.* **56,** 286–291.

Andreae, W. A. (1952). Effects of scopoletin on indoleacetic acid metabolism. *Nature (London)* **170,** 83–84.

Anonymous (1963). Mosses inhibit growth of bacteria and fungi. *Nat. Conservancy News* **13,** 12.

Anonymous (1969). Natural weed killer. *Sci. Am.* **221,** 54.

Anonymous (1980). Metabolites of plant pathogens may prove useful in weed control. *Crops Soils* **32,** 26.

Armstrong, G. M., Rohrbaugh, L. M., Rice, E. L., and Wender, S. H. (1970). The effect of nitrogen deficiency on the concentration of caffeoylquinic acids and scopolin in tobacco. *Phytochemistry* **9,** 945–948.

Armstrong, G. M., Rohrbaugh, L. M., Rice, E. L., and Wender, S. H. (1971). Preliminary studies on the effect of deficiency in potassium or magnesium on concentration of chlorogenic acid and scopolin in tobacco. *Proc. Okla. Acad. Sci.* **51,** 41–43.

Arnold, J. F. (1964). Zonation of understory vegetation around a juniper tree. *J. Range Manage.* **17**, 41–42.

Arntzen, C. J., Haugh, M. F., and Bobick, S. (1973). Induction of stomatal closure by *Helminthosporium maydis* pathotoxin. *Plant Physiol.* **52**, 569–574.

Arntzen, C. J., Falkenthal, S. V., and Bobick, S. (1974). Inhibition of photophosphorylation by kaempferol. *Plant Physiol.* **53**, 304–306.

Ashraf, N., and Sen, D. N. (1978). Allelopathic potential of *Celosia argentea* in arid land crop fields. *Oecol. Plant.* **13**, 331–338.

Asian Vegetable Research and Development Center (1978). Soybean Report for 1976. AVRDC, Shanhua, Taiwan, R.O.C.

Asplund, R. O. (1968). Monoterpenes: Relationship between structure and inhibition of germination. *Phytochemistry* **7**, 1995–1997.

Asplund, R. O. (1969). Some quantitative aspects of the phytotoxicity of monoterpenes. *Weed Sci.* **17**, 454–455.

Atsatt, P. R. (1970). Biochemical bridges between vascular plants. *Proc. 29th Annu. Biol Colloq.* [Oreg. State Univ.] pp. 53–68.

Aubert, M., Pesando, D., and Gauthier, M. (1970). Phénomènes d'antibiose d'origine phytoplanktonique in milieu marine. Substances antibactérienne produites par un diatomée *Asterionella japonica* Cleve. *Rev. Int. Océanogr. Méd.* **18/19**, 69–76.

Avers, C. J., and Goodwin, R. H. (1956). Studies on roots. IV. Effects of coumarin and scopoletin on the standard root growth pattern of *Phleum pratense. Am. J. Bot.* **43**, 612–620.

Ayers, A. R., Ebel, J., Finelli, F., Berger, N., and Albersheim, P. (1976a). Host-pathogen interactions. IX. Quantitative assays of elicitor activity and characterization of the elicitor present in the extracellular medium of cultures of *Phytophthora megasperma* var. *sojae. Plant Physiol.* **57**, 751–759.

Ayers, A. R., Ebel, J., Valent, B., and Albersheim, P. (1976b). Host-pathogen interactions. X. Fractionation and biological activity of an elicitor isolated from the mycelial walls of *Phytophthora megasperma* var. *sojae. Plant Physiol.* **57**, 760–765.

Ayers, A. R., Valent, B., Ebel, J., and Albersheim, P. (1976c). Host-pathogen interactions. XI. Composition and structure of wall-released elicitor fractions. *Plant Physiol.* **57**, 766–774.

Bailey, L. H., Bailey, E. Z., and Staff of Liberty Hyde Bailey Hortorium (1976). "Hortus Third: A Concise Dictionary of Plants Cultivated in the United States and Canada." Macmillan, New York.

Baker, H. G. (1966). Volatile growth inhibitors produced by *Eucalyptus globulus. Madroño* **18**, 207–210.

Baker, K. F., and Cook, R. J. (1974). "Biological Control of Plant Pathogens." Freeman, San Francisco, California.

Baker, K. F., and Snyder, W. C., eds. (1965). "Ecology of Soil-Borne Plant Pathogens." Univ. of California Press, Berkeley.

Balke, N. E. (1977). Inhibition of ion absorption in *Avena sativa* L. roots by diethylstilbestrol and other phenolic compounds. Ph.D. Thesis, Purdue Univ., W. Lafayette, Indiana. *Diss. Abstr.* No. 7813025.

Balke, N. E., and Hodges, T. K. (1977). Inhibition of ion absorption in oat roots: Comparison of diethylstilbestrol and oligomycin. *Plant Sci. Lett.* **10**, 319–325.

Balke, N. E., and Hodges, T. K. (1979a). Effect of diethylstilbestrol on ion fluxes in oat roots. *Plant Physiol.* **63**, 42–47.

Balke, N. E., and Hodges, T. K. (1979b). Inhibition of adenosine triphosphatase activity of the plasma membrane fraction of oat roots by diethylstilbestrol. *Plant Physiol.* **63**, 48–52.

Balke, N. E., and Hodges, T. K. (1979c). Comparison of reductions in adenosine triphosphate content, plasma membrane-associated adenosine triphosphatase activity, and potassium absorption in oat roots by diethylstilbestrol. *Plant Physiol.* **63**, 53–56.

Ballester, A., and Vieitez, E. (1971). Estudio de sustancias de crecimiento aisladas de *Erica cinerea* L. *Acta Cien. Compostelana* **8,** 79–84.

Ballester, A., Arines, J., and Vieitez, E. (1972). Compuestos fenolicos en suelos de brezal. *An. Edafol. Agrobiol.* **31,** 359–366.

Ballester, A., Albo, J. M., and Vieitez, E. (1977). The allelopathic potential of *Erica scoparia* L. *Oecologia* **30,** 55–61.

Ballester, A., Vieitez, A. M., and Vieitez, E. (1979). The allelopathic potential of *Erica australis* L. and *E. arborea* L. *Bot. Gaz. (Chicago)* **140,** 433–436.

Bandeen, J. D., and Buchholtz, K. P. (1967). Competitive effects of quackgrass upon corn as modified by fertilizer. *Weeds* **15,** 220–224.

Baranetsky, G. G. (1973). On the chemical nature of the biologically active water-soluble substances in fallen ash and lime leaves. *In* "Physiological-Biochemical Basis of Plant Interactions in Phytocenoses" (A. M. Grodzinsky, ed.), Vol. 4, pp. 85–88. Naukova Dumka, Kiev. (In Russian, English summary.)

Basaraba, J. (1964). Influence of vegetable tannins on nitrification in soil. *Plant Soil* **21,** 8–16.

Baskin, J. M., Ludlow, C. J., Harris, T. M., and Wolf, F. T. (1967). Psoralen, an inhibitor in the seeds of *Psoralea subacaulis* (Leguminosae). *Phytochemistry* **6,** 1209–1213.

Bate, G. C., and Heelas, B. V. (1975). Studies on the nitrate nutrition of two indigenous Rhodesian grasses. *J. Appl. Ecol.* **12,** 941–952.

Bate-Smith, E. C., and Metcalfe, C. R. (1957). The nature and systematic distribution of tannins in dicotyledonous plants. *Bot. J. Linn. Soc.* **55,** 669–705.

Battle, J. P., and Whittington, W. J. (1969). The relation between inhibitory substances and variability in time to germination of sugar beet clusters. *J. Agric. Sci.* **73,** 337–346.

Bazzaz, F. A. (1975). Plant species diversity in old-field successional ecosystems in southern Illinois. *Ecology* **56,** 485–488.

Becker, Y., and Guyot, L. (1951). Sur les toxines racinaires des sols incultes. *C. R. Acad. Sci.* **232,** 105–107.

Becker, Y., Guyot, L., and Montegut, J. (1951). Sur quelques incidences phytosociologiques du problème des excrétions racinaires. *C. R. Acad. Sci.* **232,** 2472–2474.

Becking, J. H. (1970). Frankiaceae Fam. Nov. (Actinomycetales) with one new combination and six new species of the genus *Frankia* Brunchorst 1886, 174. *Int. J. Syst. Bacteriol.* **20,** 201–220.

Beggs, J. P. (1964). Spectacular clover establishment with formalin treatment suggest growth inhibitor in soil. *N. Z. J. Agric.* **108,** 529–535.

Bell, A. A. (1974). Biochemical bases of resistance of plants to pathogens. *In* "Biological Control of Plant Insects and Diseases" (F. G. Maxwell and F. S. Harris, eds.), pp. 403–461. Univ. Press of Mississippi, Jackson.

Bell, A. A. (1977). Plant pathology as influenced by allelopathy. *In* "Report of the Research Planning Conference on the Role of Secondary Compounds in Plant Interactions (Allelopathy)" (C. G. McWhorter, A. C. Thompson, and E. W. Hauser, eds.), pp. 64–99. USDA, Agricultural Research Service, Tifton, Georgia.

Bell, A. R., and Nalewaja, J. D. (1968a). Competitive effects of wild oat in flax. *Weed Sci.* **16,** 501–504.

Bell, A. R., and Nalewaja, J. D. (1968b). Competition of wild oat in wheat and barley. *Weed Sci.* **16,** 505–508.

Bell, A. R., and Nalewaja, J. D. (1968c). Effect of duration of wild oat competition in flax. *Weed Sci.* **16,** 509–512.

Bell, D. T., and Koeppe, D. E. (1972). Noncompetitive effects of giant foxtail on the growth of corn. *Agron. J.* **64,** 321–325.

Bell, D. T., and Muller, C. H. (1973). Dominance of California annual grasslands by *Brassica nigra*. *Am. Midl. Nat.* **90,** 277–299.

Bell, S., and Klikoff, L. G. (1979). Allelopathic and autopathic relationships among the ferns *Polystichum acrostichoides, Polypodium vulgare* and *Onoclea sensibilis. Am. Midl. Nat.* **102**, 168–171.

Bendall, G. M. (1975). The allelopathic activity of California thistle (*Cirsium arvense* (L.) Scop.) in Tasmania. *Weed Res.* **15**, 77–81.

Benedict, H. M. (1941). The inhibitory effect of dead roots on the growth of brome-grass. *J. Am. Soc. Agron.* **33**, 1108–1109.

Bennett, E. L., and Bonner, J. (1953). Isolation of plant growth inhibitors from *Thamnosma montana. Am. J. Bot.* **40**, 29–33.

Benoit, R. E., and Starkey, R. L. (1968a). Enzyme inactivation as a factor in the inhibition of decomposition of organic matter by tannins. *Soil Sci.* **105**, 203–208.

Benoit, R. E., and Starkey, R. L. (1968b). Inhibition of decomposition of cellulose and some other carbohydrates by tannin. *Soil Sci.* **105**, 291–296.

Benoit, R. E., Starkey, R. L., and Basaraba, J. (1968). Effect of purified plant tannin on decomposition of some organic compounds and plant materials. *Soil Sci.* **105**, 153–158.

Bentley, H. R., Cunningham, K. G., and Spring, F. S. (1951). Cordycepin, a metabolic product from cultures of *Cordyceps militaris* (Linn.) Link. Part II. The structure of cordycepin. *J. Chem. Soc., London* pp. 2301–2305.

Beobachter, Home Correspondence (1845). The highland pine. *Gard. Chron.* **5**, 69–70.

Berestetsky, O. A. (1970). On the role of decomposition products of root residue for garden soil toxicity. *In* "Physiological-Biochemical Basis of Plant Interactions in Phytocenoses" (A. M. Grodzinsky, ed.), Vol. 1, pp. 113–118. Naukova Dumka, Kiev. (In Russian, English summary.)

Berestetsky, O. A. (1972). Formation of phytotoxic substances by soil microorganisms on root residues of fruit trees. *In* "Physiological-Biochemical Basis of Plant Interactions in Phytocenoses" (A. M. Grodzinsky, ed.), Vol. 3, pp. 121–124. Naukova Dumka, Kiev. (In Russian, English summary.)

Berglund, H. (1969). Stimulation of growth of two marine algae by organic substances excreted by *Enteromorpha linza* in unialgal and axenic cultures. *Physiol. Plant.* **22**, 1069–1073.

Berlier, Y., Dabin, B., and Leneuf, N. (1956). Comparaison physique, chimique et microbiologique entre les sols de foret et de savane sur les sables tertiaires de la Basse Côte d'Ivoire. *Trans., 6th Int. Congr. Soil Sci. E* pp. 499–502.

Bevege, D. I. (1968). Inhibition of seedling hoop pine (*Araucaria cunninghamii* Ait.) on forest soils by phytotoxic substances from the root zones of *Pinus*, Araucaria, and *Flindersia. Plant Soil* **29**, 263–273.

Bhakuni, D. S., and Silva, M. (1974). Biodynamic substances from marine flora. *Bot. Mar.* **17**, 40–51.

Bhandari, M. C., and Sen, D. N. (1971). Effect of *Citrullus colocynthis* (Linn.) Schrad. on the seedling growth of *Pennisetum typhoideum* Rich. *Z. Pflanzenphysiol.* **64**, 466–469.

Bhandari, M. C., and Sen, D. N. (1972). Growth regulation specificity exhibited by substances present in the fruit pulp of *Citrullus lanatus* (Thunb.) Mansf. *Z. Naturforsch.* **27**, 72–75.

Bhowmik, P. C., and Doll, J. D. (1979). Evaluation of allelopathic effects of selected weed species on corn and soybeans. *Proc. North Cent. Weed Control Conf.* **34**, 43–45.

Bhowmik, P. C., and Doll, J. D. (1982). Corn and soybean response to allelopathic effects of weed and crop residues. *Agron. J.* **74**, 601–606.

Bieber, G. L., and Hoveland, C. S. (1968). Phytotoxicity of plant materials on seed germination of crownvetch, *Coronilla varia* L. *Agron. J.* **60**, 185–188.

Björkman, E. (1960). *Monotropa hypopitys* L.—an epiparasite on tree roots. *Physiol. Plant.* **13**, 308–327.

Black, R. L. B., and Dix, N. J. (1976). Utilization of ferulic acid by microfungi from litter and soil. *Trans. Br. Mycol. Soc.* **66**, 313–317.

Blackman, G. E., Sen, G., Birch, W. R., and Powell, R. G. (1959). The uptake of growth substances. I. Factors controlling the uptake of phenoxyacetic acids by *Lemna minor*. *J. Exp. Bot.* **10**, 33–54.

Blum, U., and Rice, E. L. (1969). Inhibition of symbiotic nitrogen-fixation by gallic and tannic acid, and possible roles in old-field succession. *Bull. Torrey Bot. Club* **96**, 531–544.

Boawn, L. C. (1965). Sugar beet induced zinc deficiency. *Agron. J.* **57**, 509.

Bode, H. R. (1940). Über die Blattausscheidungen des Wermuts und ihre Wirkung auf andere Pflanzen. *Planta* **30**, 567–589.

Bogdan, G. P. (1971). Anatomical study of effects in the plant conducting system of allelopathic substances. *Ukr. Bot. Zh.* **28**, 703–707.

Bogdan, G. P. (1977). Mutual effect of couch grass and cultivated plants in phytocenoses. *In* "Interactions of Plants and Microorganisms in Phytocenoses" (A. M. Grodzinsky, ed.), pp. 36–43. Naukova Dumka, Kiev. (In Russian, English summary.)

Bogdan, G. P., and Grodzinsky, A. M. (1974). Role of sulfhydryl groups in protective reactions of plants during allelopathic damage. *Ukr. Bot. Zh.* **30**, 771–778. (In Ukrainian, Russian and English summaries.)

Boiko, M. F. (1973). Allelopathic peculiarities of epiphytic mosses, lichens and algae of floodplain forests of the Berkul River. *In* "Physiological-Biochemical Basis of Plant Interactions in Phytocenoses" (A. M. Grodzinsky, ed.), Vol. 4, pp. 103–107. Naukova Dumka, Kiev. (In Russian, English summary.)

Bokhari, U. G. (1978). Allelopathy among prairie grasses and its possible ecological significance. *Ann. Bot. (London)* **42**, 127–136.

Bold, H. C. (1949). The morphology of *Chlamydomonas chlamydogama* sp. nov. *Bull. Torrey Bot. Club* **76**, 101–108.

Bold, H. C. (1957). "Morphology of Plants." Harper, New York.

Bonasera, J., Lynch, J., and Leck, M. A. (1979). Comparison of the allelopathic potential of four marsh species. *Bull. Torrey Bot. Club* **106**, 217–222.

Bonner, J. (1946). Further investigation of toxic substances which arise from guayule plants: Relation of toxic substances to the growth of guayule in soil. *Bot. Gaz. (Chicago)* **107**, 343–351.

Bonner, J., and Galston, A. W. (1944). Toxic substances from the culture media of guayule which may inhibit growth. *Bot. Gaz. (Chicago)* **106**, 185–198.

Booth, W. E. (1941a). Revegetation of abandoned fields in Kansas and Oklahoma. *Am. J. Bot.* **28**, 415–422.

Booth, W. E. (1941b). Algae as pioneers in plant succession and their importance in erosion control. *Ecology* **22**, 38–46.

Börner, H. (1959). The apple replant problem. I. The excretion of phlorizin from apple root residues. *Contrib. Boyce Thompson Inst.* **20**, 39–56.

Börner, H. (1963a). Untersuchungen über die Bildung antiphytotischer und antimikrobieller Substanzen durch Mikroorganismen im Boden und ihre mögliche Bedeutung für die Bodenmüdigkeit beim Apfel (*Pirus malus* L.). I. Bildung von Patulin und einer phenolischen Verbindung durch *Penicillium expansum* auf Wurzel-und Blattrückstanden des Apfel. *Phytopathol. Z.* **48**, 370–396.

Börner, H. (1963b). II. Der Einflutz verschiedener Factoren auf die Bildung von Patulin und einer phenolischen Verbindung durch *Penicillium expansum* auf Blatt-und Wurzebrückstanden des Apfels. *Phytopathol. Z.* **49**, 1–28.

Boswell, F. C., and Anderson, O. E. (1974). Nitrification inhibitor studies of soil in field-buried polyethylene bags. *Soil Sci. Soc. Am. Proc.* **38**, 851–852.

Boughey, A. S., Munro, P. E., Meiklejohn, J., Strang, R. M., and Swift, M. J. (1964). Antibiotic reactions between African savanna species. *Nature (London)* **203,** 1302–1303.

Bould, C., and Hewitt, E. J. (1963). Mineral nutrition of plants in soils and in culture media. *In* "Plant Physiology: A Treatise" (F. C. Steward, ed.), Vol. III, pp. 15–133. Academic Press, New York.

Bowen, G. D. (1961). The toxicity of legume seed diffusates toward Rhizobia and other bacteria. *Plant Soil.* **15,** 155–165.

Bracken, A. F., and Greaves, J. E. (1941). Losses of nitrogen and organic matter from dry-land soils. *Soil Sci.* **51,** 1–15.

Brady, N. C. (1974). "The Nature and Properties of Soils," 8th ed. Macmillan, New York.

Brown, D. D. (1968). The possible ecological significance of inhibition by *Euphorbia supina.* Unpublished M.S. Thesis, Univ. of Oklahoma, Norman.

Brown, R. T. (1967). Influence of naturally occurring compounds on germination and growth of jack pine. *Ecology* **48,** 542–546.

Brown, R. T., and Mikola, P. (1974). The influence of fruticose soil lichens upon the mycorrhizae and seedling growth of forest trees. *Acta For. Fenn.* **141,** 1–22.

Brown, S. A. (1964). Lignin and tannin biosynthesis. *In* "Biochemistry of Phenolic Compounds" (J. B. Harborne, ed.), pp. 361–398. Academic Press, New York.

Bruehl, G. W., ed. (1975). "Biology and Control of Soil-Borne Plant Pathogens." Am. Phytopathol. Soc., St. Paul, Minnesota.

Buchholtz, K. P. (1971). The influence of allelopathy on mineral nutrition. *In* "Biochemical Interactions among Plants" (Environ. Physiol. Subcomm., U. S. Natl. Comm. for IBP, eds.), pp. 86–89. Natl. Acad. Sci., Washington, D. C.

Bukolova, T. P. (1971). A study of the mechanism of action of water-soluble substances of weeds on cultivated plants. *In* "Physiological-Biochemical Basis of Plant Interactions in Phytocenoses" (A. M. Grodzinsky, ed.), Vol. 2, pp. 66–69. Naukova Dumka, Kiev. (In Russian, English summary.)

Bundy, L. G., and Bremner, J. M. (1974). Inhibition of nitrification in soils. *Soil Sci. Soc. Am. Proc.* **37,** 396–398.

Burbott, A. J., and Loomis, W. D. (1967). Effects of light and temperature on the monoterpenes of peppermint. *Plant Physiol.* **42,** 20–28.

Burger, W. P. (1981). Allelopathy in citrus orchards. Doctoral Thesis, Univ. of Port Elizabeth, South Africa.

Burgos-Leon, W. (1976). Phytotoxicité induite par les résidus de récolte de *Sorghum vulgare* dans les sols sableux de l'ouest Africain. Thèse pour Doctorat, Université de Nancy, France.

Burgos-Leon, W., Gaury, F., Nicou, R., Chopart, J. L., and Dommergues, Y. (1980). Études et travaux: un cas de fatigue des sols induite par la culture du sorgho. *Agron. Trop. (Paris)* **35,** 319–334.

Burkholder, P. R., Burkholder, L. M., and Almodóvar, L. R. (1960). Antibiotic activity of some marine algae of Puerto Rico. *Bot. Mar.* **2,** 149–156.

Byrde, R. J. W., Fielding, A. H., and Williams, A. H. (1960). The roles of oxidized polyphenols in the varietal resistance of apples to brown rot. *In* "Phenolics in Plants in Health and Disease" (J. B. Pridham, ed.), pp. 95–99. Pergamon, Oxford.

Cadman, C. H. (1959). Some properties of an inhibitor of virus infection from leaves of raspberry. *J. Gen. Microbiol.* **20,** 113–128.

Cain, J. C. (1952). A comparison of ammonium and nitrate nitrogen for blueberries. *Am. Soc. Hortic. Sci. Proc.* **59,** 161–166.

Callaham, D., Del Tredici, P., and Torrey, J. G. (1978). Isolation and cultivation in vitro of the actinomycete causing nodulation in *Comptonia. Science* **199,** 899–902.

Cameron, H. J., and Julian, G. R. (1980). Inhibition of protein synthesis in lettuce (*Lactuca sativa* L.) by allelopthic compounds. *J. Chem. Ecol.* **6,** 989–995.

Camm, L. E., Chi-Kit, W., and Towers, G. H. N. (1976). An assessment of the roles of furanocoumarins in *Heracleum lanatum*. *Can. J. Bot.* **54**, 2562–2566.

Campbell, G., Lambert, J. D. H., Arnason, T., and Towers, G. H. N. (1982). Allelopathic properties of α-terthienyl and phenylheptatriyne, naturally occurring compounds from species of Asteraceae. *J. Chem. Ecol.* **8**, 961–972.

Campbell, H. (1964). Notes on viability of honey locust seeds in relation to age. *Turtox News* **42**, 134–135.

Carballeira, A. (1980). Phenolic inhibitors in *Erica australis* L. and the associated soil. *J. Chem. Ecol.* **6**, 593–596.

Carnahan, G., and Hull, A. C. (1962). The inhibition of seeded plants by tarweed. *Weeds* **10**, 87–90.

Cavallito, C. J., and Haskell, T. H. (1945). The mechanism of antibiotics. The reaction of unsaturated lactones with cysteine and related compounds. *J. Am. Chem. Soc.* **67**, 1991–1994.

Cavallito, C. J., Buck, J. S., and Suter, C. M. (1944). Allicin, the antibacterial principle of *Allium sativum*. II. Determination of the chemical structure. *J. Am. Chem. Soc.* **66**, 1952–1954.

Chaffin, W. A. (no date). Soil improvement program for Oklahoma. Circular 412 of Extension Service, Oklahoma State University, Stillwater, Oklahoma.

Chalutz, E. (1973). Ethylene-induced phenylalanine ammonia-lyase activity in carrot roots. *Plant Physiol.* **51**, 1033–1036.

Chambers, E. E., and Holm, L. G. (1965). Phosphorus uptake as influenced by associated plants. *Weeds* **13**, 312–314.

Chan, E. C. S., Basavanand, P., and Liivak, T. (1970). The growth inhibition of *Azotobacter chroococcum* by *Pseudomonas* sp. *Can. J. Microbiol.* **16**, 9–16.

Chandler, J. M. (1977). Competition of spurred anoda, velvetleaf, prickly sida, and Venice mallow in cotton. *Weed Sci.* **25**, 151–158.

Chandramohan, D., Purushothaman, D., and Kothandaraman, R. (1973). Soil phenolics and plant growth inhibition. *Plant Soil* **39**, 303–308.

Chang, C. F., Suzuki, A., Kumai, S., and Tamura, S. (1969). Chemical studies on "clover sickness." Part II. Biological functions of isoflavonoids and their related compounds. *Agric. Biol. Chem.* **33**, 398–408.

Char, M. B. S. (1977). Pollen allelopathy. *Naturewissenshaften* **64**, 489–490.

Chaumont, J. P., and Simeray, J. (1982). Les propriétés antifongiques de 225 Basidiomycetes et Ascomycetes vis-à-vis de 7 champignons pathogenes cultivés *in vitro*. *Cryptog. Mycol.* **3**, 249–259.

Chou, C. H. (1977). Phytotoxic substances in twelve subtropical grasses. I. Additional evidences of phytotoxicity in the aqueous fractions of grass extracts. *Bot. Bull. Acad. Sin.* **18**, 131–141.

Chou, C. H. (1980). Allelopathic researches in the subtopical vegetation in Taiwan. *Comp. Physiol. Ecol.* **5**, 222–234.

Chou, C. H., and Chiou, S. J. (1979). Autointoxication mechanism of *Oryza sativa*. II. Effects of culture treatments on the chemical nature of paddy soil and on rice productivity. *J. Chem. Ecol.* **5**, 839–859.

Chou, C. H., and Chung, Y. T. (1974). The allelopathic potential of *Miscanthus floridulus*. *Bot. Bull. Acad. Sin.* **15**, 14–27.

Chou, C. H., and Hou, M. H. (1981). Allelopthic researches of subtropical vegetations in Taiwan. I. Evaluation of allelopathic potential of bamboo vegetation. *In* "Annual Report for July 1980–June 1981," pp. 10–11. Institute of Botany, Academia Sinica, Taipei, Taiwan.

Chou, C. H., and Lin, H. J. (1976). Autointoxication mechanisms of *Oryza sativa*. I. Phytotoxic effects of decomposing rice residues in soil. *J. Chem. Ecol.* **2**, 353–367.

Chou, C. H., and Muller, C. H. (1972). Allelopathic mechanisms of *Arctostaphylos glandulosa* var. *zacaensis*. *Am. Midl. Nat.* **88**, 324–347.

Chou, C. H., and Patrick, Z. A. (1976). Identification and phytotoxic activity of compounds produced during decomposition of corn and rye residues in soil. *J. Chem. Ecol.* **2**, 369–387.

Chou, C. H., and Waller, G. R. (1980a). Possible allelopathic constituents of *Coffea arabica*. *J. Chem. Ecol.* **6**, 643–654.

Chou, C. H., and Waller, G. R. (1980b). Isolation and identification by mass spectrometry of phytotoxins in *Coffea arabica*. *Bot. Bull. Acad. Sin.* **21**, 25–34.

Chou, C. H., and Young, C. C. (1975). Phytotoxic substances in twelve subtropical grasses. *J. Chem. Ecol.* **1**, 183–193.

Chou, C. H., Lin, T. J., and Kao, C. I. (1977). Phytotoxins produced during decomposition of rice stubbles in paddy soil and their effect on leachable nitrogen. *Bot. Bull. Acad. Sin.* **18**, 45–60.

Chou, C. H., Chiang, Y. C., and Cheng, H. H. (1981). Autointoxication mechanism of *Oryza sativa*. III. Effect of temperature on phytotoxin production during rice straw decomposition in soil. *J. Chem. Ecol.* **7**, 741–752.

Chouteau, J., and Loche, J. (1965). Incidence de la nutrition azotée de la plante de tabac sur l'accumulation des composés phénoliques dans les feuilles. *C. R. Acad. Sci.* **260**, 4568–4588.

Christensen, N. L., and Muller, C. H. (1975). Effects of fire on factors controlling plant growth in *Adenostoma* chaparral. *Ecol. Monogr.* **45**, 29–55.

Christersson, L. (1972). The influence of urea and other nitrogen sources on growth rate of Scots pine seedlings. *Physiol. Plant.* **27**, 83–88.

Chu-Chou, M. (1978). Effects of root residues on growth of *Pinus radiata* seedlings and a mycorrhizal fungus. *Ann. Appl. Biol.* **90**, 407–416.

Chumakov, V. V., and Aleikina, M. M. (1977). On qualitative changes in root exudates of woody plants growing together. *In* "Interactions of Plants and Microorganisms in Phytocenoses" (A. M. Grodzinsky, ed.), pp. 147–156. Naukova Dumka, Kiev. (In Russian, English summary.)

Clark, R. S., Kuć, J., Henze, R. E., and Quackenbush, F. W. (1959). The nature and fungitoxicity of an amino acid addition product of chlorogenic acid. *Phytopathology* **49**, 594–597.

Cobb, E. W. J., Krstic, M., Zavarin, E., and Barbe, H. W., Jr. (1968). Inhibitory effects of volatile oleoresin component on *Fomes annosus* and four *Ceratocystis* species (disease reistance). *Phytopathology* **58**, 1327–1335.

Coble, H. D., and Ritter, R. L. (1978). Pennsylvania smartweed (*Polygonum pennsylvanicum*) interference in soybeans (*Glycine max*). *Weed Sci.* **26**, 556–559.

Cochrane, V. W. (1948). The role of plant residues in the etiology of root rot. *Phytopathology* **38**, 185–196.

Colby, S. R. (1967). Calculating synergistic and antagonistic responses of herbicide combinations. *Weeds* **15**, 20–22.

Colton, C. E., and Einhellig, F. A. (1980). Allelopathic mechanisms of velvetleaf (*Abutilon theophrasti* Medic., Malvaceae) on soybean. *Am. J. Bot.* **67**, 1407–1413.

Conn, E. E., and Akazawa, T. (1958). Biosynthesis of *p*-hydroxybenzaldehyde. *Fed. Proc., Fed. Am. Soc. Exp. Biol.* **17**, 205.

Conover, J. J., and Sieburth, J. M. (1964). Effect of *Sargassum* distribution on epibiota and antibacterial activity. *Bot. Mar.* **6**, 147–157.

Cook, M. T. (1921). Wilting caused by walnut trees. *Phytopathology* **11**, 346.

Cook, M. T., and Taubenhaus, J. J. (1911). The rleation of parasitic fungi to the contents of the cells of the host plants. I. The toxicity of tannin. *Del. Agric. Exp. Stn., Bull.* No. 91, pp. 3–77.

Cooper, W. S., and Stoesz, A. D. (1931). The subterranean organs of *Helianthus scaberrimus*. *Bull. Torrey Bot. Club* **58**, 67–72.

Corcoran, M. R. (1970). Inhibitors from Carob (*Ceratonia siliqua* L. II. Effect on growth induced by indoleacetic acid or gibberellins A_1, A_4, A_5, and A_7. *Plant Physiol.* **46**, 531–534.

Corcoran, M. R., Geissman, T. A., and Phinney, B. O. (1972). Tannins as gibberellin antagonists. *Plant Physiol.* **49**, 323–330.

Cornman, I. (1946). Alteration of mitosis by coumarin and parasorbic acid. *Am. J. Bot.* **33**, 217.

Corse, J., Lundin, R. E., and Waiss, A. C., Jr. (1965). Identification of several components of isochlorogenic acid. *Phytochemistry* **4**, 527–529.

Cowles, H. C. (1911). The causes of vegetative cycles. *Bot. Gaz. (Chicago)* **51**, 161–183.

Crafts, A. S. (1961). "Translocation in Plants." Holt, New York.

Craigie, J. S., and McLachlan, J. (1964). Excretion of colored ultraviolet absorbing substances by marine algae. *Can. J. Bot.* **42**, 23–33.

Cramer, M., and Myers, J. (1948). Nitrate reduction and assimilation in *Chlorella. J. Gen. Physiol.* **32**, 92–102.

Crocker, R. L., and Major, J. (1955). Soil development in relation to vegetation and surface age at Glacier Bay, Alaska. *J. Ecol.* **43**, 427–448.

Cruickshank, I. A. M., and Perrin, D. R. (1964). Pathological function of phenolic compounds in plants. *In* "Biochemistry of Phenolic Compounds" (J. B. Harborne, ed.), pp. 511–544. Academic Press, New York.

Culpeper, N. (1633). "English Physitian and Complete Herball." Foulsham, London. (Reprinted 1955.)

Curtis, J. T. (1959). "The Vegetation of Wisconsin." Univ. of Wisconsin Press, Madison.

Curtis, J. T., and Cottam, G. (1950). Antibiotic and autotoxic effects in prairie sunflower. *Bull. Torrey Bot. Club* **77**, 187–191.

Dadykin, V. P., Stepanov, L. N., and Ryzhkova, B. E. (1970). On the importance of volatile plant secretions under the development of closed systems. *In* "Physiological-Biochemical Basis of Plant Interactions in Phytocenoses" (A. M. Grodzinsky, ed.), Vol. 1, pp. 118–124. Naukova Dumka, Kiev. (In Russian.)

Daniel, H. A., and Langham, W. H. (1936). The effect of wind erosion and cultivation on the total nitrogen and organic matter content of soil in the southern high plains. *J. Am. Soc. Agron.* **28**, 587–596.

Danks, M. L., Fletcher, J. S., and Rice, E. L. (1975a). Influence of ferulic acid on mineral depletion and uptake of ^{86}Rb by Paul's scarlet rose cell-suspension cultures. *Am. J. Bot.* **62**, 749–755.

Danks, M. L., Fletcher, J. S., and Rice, E. L. (1975b). Effects of phenolic inhibitors on growth and metabolism of glucose-UL-^{14}C in Paul's Scarlet Rose cell-suspension cultures. *Am. J. Bot.* **62**, 311–317.

Datta, S. C., and Chakrabarti, S. D. (1978). Germination and growth inhibitors in *Clerodendrum viscosum* Vent., a perennial weed. *Proc. 6th Asian-Pac. Weed Sci Soc. Conf.* pp. 240–245.

Datta, S. C., and Chatterjee, A. K. (1978). Some characteristics of an inhibitory factor in *Polygonum orientale* L. *Indian J. Weed Sci.* **10**, 23–33.

Datta, S. C., and Chatterjee, A. K. (1980a). Allelopathy in *Polygonum orientale:* Inhibition of seed germination and seedling growth of mustard. *Comp. Physiol. Ecol.* **5**, 54–59.

Datta, S. C., and Chatterjee, A. K. (1980b). Allelopathic potential of *Polygonum orientale* L. in relation to germination and seedling growth of weeds. *Flora* **169**, 456–465.

Datta, S. C., and Sinha-Roy, S. P. (1975). Phytotoxic effects of *Croton bonplandianum* Baill. on weedy associates. *Vegetatio* **30**, 157–163.

Davidonis, G. H., and Ruddat, M. (1973). Allelopathic compounds, thelypterin A and B in the fern *Thelypteris normalis. Planta* **111**, 23–32.

Davidonis, G. H., and Ruddat, M. (1974). Growth inhibition in gametophytes and oat coleoptiles by thelypterin A and B released from roots of the fern *Thelypteris normalis. Am. J. Bot.* **61**, 925–930.

Davis, E. F. (1928). The toxic principle of *Juglans nigra* as identified with synthetic juglone and its toxic effects on tomato and alfalfa plants. *Am. J. Bot.* **15**, 620.

Dawes, D. S., and Maravolo, N. C. (1973). Isolation and characteristics of a possible allelopathic factor supporting the dominant role of *Hieracium aurantiacum* in the bracken-grasslands of northern Wisconsin. *Trans. Wis. Acad. Sci., Arts Lett.* **61**, 235–251.

Dawson, J. H. (1964). Competition between irrigated field beans and annual weeds. *Weeds* **12**, 206–208.

Dawson, J. H. (1965). Competition between irrigated sugar beets and annual weeds. *Weeds* **13**, 245–249.

Dawson, J. H. (1977). Competition of late-emerging weeds with sugar beets. *Weed Sci.* **25**, 168–170.

Day, B. E. (1952). The absorption and translocation of 2,4-D by bean plants. *Plant Physiol.* **27**, 143–152.

De, P. K., and Mandal, L. N. (1956). Fixation of nitrogen by algae in rice soils. *Soil Sci.* **81**, 453–458.

De, P. K., and Sulaïman, M. (1950. Influence of algal growth in the rice fields on the yield of crops. *India J. Agric. Sci.* **20**, 327–342.

Dear, J., and Aronoff, S. (1965). Relative kinetics of chlorogenic and caffeic acids during the onset of boron deficiency in sunflower. *Plant Physiol.* **40**, 458–459.

De Bell, D. S. (1971). Phytotoxic effects of cherry-bark oak. *For. Sci.* **17**, 180–185.

De Bell, D. S., and Radwan, M. A. (1979). Growth and nitrogen relations of coppiced black cottonwood and red alder in pure and mixed planting. *Bot. Gaz. (Chicago)* **140**, (Suppl.), S97–S101.

DeCandolle, M. A.-P. (1832). "Physiologie Végétale," Tome III, pp. 1474–1475. Béchet Jeune, Lib, Fac. Méd., Paris.

Dedonder, A., and Van Sumere, C. F. (1971). The effect of phenolics and related compounds on the growth and the respiration of *Chlorella vulgaris*. *Z. Pflanzenphysiol.* **65**, 70–80.

Del Amo, S., and Anaya, A. L. (1978). Effect of some sesquiterpenic lactones on the growth of certain secondary tropical species. *J. Chem. Ecol.* **4**, 305–313.

Deleuil, M. G. (1950). Mise en évidence de substances toxiques pour les thérophytes dans les associations du Rosmarino-Ericion. *C. R. Acad. Sci.* **230**, 1362–1364.

Deleuil, M. G. (1951a). Origine des substances toxiques du sol des associations sans thérophytes du Romarino-Ericion. *C. R. Acad. Sci.* **232**, 2038–2039.

Deleuil, M. G. (1951b). Explication de la présence de certains thérophytes rencontrés parfois dans les associations du Rosmarino-Ericion. *C. R. Acad. Sci.* **232**, 2476–2477.

del Moral, R. (1972). On the variability of chlorogenic acid concentration. *Oecologia* **9**, 289–300.

del Moral, R., and Cates, R. G. (1971). Allelopathic potential of the dominant vegetation of western Washington. *Ecology* **52**, 1030–1037.

del Moral, R., and Muller, C. H. (1969). Fog drip: A mechanism of toxin transport from *Eucalyptus globulus*. *Bull. Torrey Bot. Club* **96**, 467–475.

del Moral, R., and Muller, C. H. (1970). The allelopathic effects of *Eucalyptus camaldulensis*. *Am. Midl. Nat.* **83**, 254–282.

del Moral, R., Willis, R. J., and Ashton, D. H. (1978). Suppression of coastal heath vegetation by *Eucalyptus baxteri*. *Aust. J. Bot.* **26**, 203–219.

Demos, E. K., Woolwine, M., Wilson, R. H., and McMillan, C. (1975). The effects of ten phenolic compounds on hypocotyl growth and mitochondrial metabolism of mung bean. *Am. J. Bot.* **62**, 97–102.

Denffer, D. (1948). Über einen Wachstumhemmstoff in alternden Diatomeenkulturen. *Biol. Zentralb.* **67**, 7–13.

Dieterman, L. J., Lin, C.-Y., Rohrbaugh, L., Thiesfeld, V., and Wender, S. H. (1964a). Identification and quantitative determination of scopolin and scopoletin in tobacco plants treated with 2,4-dichlorophenoxyacetic acid. *Anal. Biochem.* **9**, 139–145.

Dieterman, L. J., Lin, C.-Y., Rohrbaugh, L. M., and Wender, S. H. (1964b). Accumulation of ayapin and scopolin in sunflower plants treated with 2,4-dichlorophenoxyacetic acid. *Arch. Biochem. Biophys.* **106**, 275–279.

Dommergues, Y. (1954). Biology of forest soils of central and eastern Madagascar. *Trans. 5th Int. Congr. Soil Sci.* pp. 3, 24–28.

Dommergues, Y. (1956). Study of the biology of soils of dry tropical forests and their evolution after clearing. *Trans. 6th Int. Congr. Soil Sci. E,* pp. 605–610.

Dormaar, J. F. (1970). Seasonal pattern of water-soluble constituents from leaves of *Populus* x 'Northwest' (Hort.). *J. Soil Sci.* **21,** 105–110.

Drew, W. B. (1942). The revegetation of abandoned cropland in the Cedar Creek area, Boone and Callaway counties, Missouri. *Univ. M. Exp. Stn., Bull.* No. 344.

Drost, D. C., and Doll, J. D. (1980). The allelopathic effect of yellow nutsedge (*Cyperus esculentus*) on corn (*Zea mays*) and soybeans (*Glycine max*). *Weed Sci.* **28,** 229–233.

Dube, V. P., Singhal, V. P., and Tyagi, S. (1979). *Parthenium hysterophorus:* Allelopathic effects on vegetable crops. *Bot. Prog.* **2,** 62–69.

Dzyubenko, N. N., and Petrenko, N. I. (1971). On biochemical interaction of cultivated plants and weeds. *In* ''Physiological-Biochemical Basis of Plant Interactions in Phytocenoses'' (A. M. Grodzinsky, ed.), Vol. 2, pp. 60–66. Naukova Dumka, Kiev. (In Russian, English summary.)

Dzyubenko, N. N., Krupa, L. I., and Boiko, P. I. (1977). Dynamics of inhibitor accumulation in the soil under continuous and crop rotation culture. *In* ''Interactions of Plants and Microorganisms in Phytocenoses'' (A. M. Grodzinsky, ed.), pp. 70–77. Naukova, Dumka, Kiev. (In Russian, English summary.)

Eaton, B. J., Russ, O. G., and Feltner, K. C. (1976). Competition of velvetleaf, prickly sida, and venice mallow in soybeans. *Weed Sci.* **24,** 224–228.

Ebel, J., Ayers, A. R., and Albersheim, P. (1976). Host-pathogen interactions. XII. Response of suspension-cultured soybean cells to the elicitor isolated from *Phytophthora megasperma* var. *sojae,* a fungal pathogen of soybeans. *Plant Physiol.* **57,** 775–779.

Eden, T. (1951). Some agricultural properties of Ceylon montane tea soils. *J. Soil Sci.* **2,** 43–49.

Einhellig, F. A. (1971). Effects of tannic acid on growth and stomatal aperture in tobacco. *Proc. S. D. Acad. Sci.* **50,** 205–209.

Einhellig, F. A., and Eckrich, P. C. (1983). Interaction of temperature and ferulic acid stress on grain sorghum and soybeans. *J. Chem. Ecol.* In press.

Einhellig, F. A., and Kuan, L. (1971). Effects of scopoletin and chlorogenic acid on stomatal aperture in tobacco and sunflower. *Bull. Torrey Bot. Club* **98,** 155–162.

Einhellig, F. A., and Rasmussen, J. A. (1973). Allelopathic effects of *Rumex crispus* on *Amaranthus retroflexus,* grain sorghum and field corn. *Am. Midl. Nat.* **90,** 79–86.

Einhellig, F. A., and Rasmussen, J. A. (1978). Synergistic inhibitory effects of vanillic and p-hydroxybenzoic acids on radish and grain sorghum. *J. Chem. Ecol.* **4,** 425–436.

Einhellig, F. A., and Rasmussen, J. A. (1979). Effects of three phenolic acids on chlorophyll content and growth of soybean and grain sorghum seedlings. *J. Chem. Ecol.* **5,** 815–824.

Einhellig, F. A., and Schon, M. K. (1982). Noncompetitive effects of *Kochia scoparia* on grain sorghum and soybeans. *Can. J. Bot.* **60,** 2923–2930.

Einhellig, F. A., Rice, E. L., Risser, P. G., and Wender, S. H. (1970). Effects of scopoletin on growth, CO_2 exchange rates, and concentration of scopoletin, scopolin, and chlorogenic acids in tobacco, sunflower, and pigweed. *Bull. Torrey Bot. Club* **97,** 22–33.

Einhellig, F. A., Schon, M. K., and Rasmussen, J. A. (1982). Synergistic effects of four cinnamic acid compounds on grain sorghum. *J. Plant Growth Reg.* **1,** 251–258.

Elkan, G. H. (1961). A nodulation-inhibiting root excretion from a non-nodulating soybean strain. *Can. J. Microbiol.* **7,** 851–856.

Ellis, J. R., and McCalla, T. M. (1973). Effects of patulin and method of application on growth stages of wheat. *Appl. Microbiol.* **25,** 562–566.

Ellison, L., and Houston, W. R. (1958). Production of herbaceous vegetation in openings and under canopies of western aspen. *Ecology* **39,** 337–345.

Elliston, J., Kuć, J., and Williams, E. B. (1976). Protection of *Phaseolus vulgaris* against anthracnose by *Colletotrichum* species nonpathogenic to bean. *Phytopathol. Z.* **86,** 117–126.

Elmer, O. H. (1932). Growth inhibition of potato sprouts by the volatile products of apples. *Science* **75,** 193.

Elmore, C. D. (1980). Inhibition of turnip (*Brassica rapa*) seed germination by velvetleaf (*Abutilon theophrasti*) seed. *Weed Sci.* **28,** 658–660.

Eussen, J. H. H. (1978). Isolation of growth inhibiting substances from alang-alang [*Imperata cylindrica* (L.) Beauv.]. *In* "Studies on the Tropical Weed *Imperata cylindrica* (L.) Beauv. var. *major*" (J. H. H. Eussen, ed.), Paper No. 7. Drukkerij Elinkwijk BV, Utrecht.

Eussen, J. H. H., and Soerjani, M. (1978). Allelopathic activity of alang-alang [*Imperata cylindrica* (L.) Beauv.], isolation of growth regulating substances from leaves. *In* "Studies on the Tropical Weed *Imperata cylindrica* (L.) Beauv. var. *major*" (J. H. H. Eussen, ed.), Paper No. 6. Drukkerij Elinkwijk BV, Utrecht.

Evenari, M. (1949). Germination inhibitors. *Bot. Rev.* **15,** 153–194.

Evetts, L. L. (1971). Ecological studies with common milkweed. M.S. Thesis, Univ. of Nebraska, Lincoln.

Evetts, L. L., and Burnside, O. C. (1972). Germination and seedling development of common milkweed and other species. *Weed Sci.* **20,** 371–378.

Evetts, L. L., and Burnside, O. C. (1973). Competition of common milkweed with sorghum. *Agron. J.* **65,** 931–932.

Fales, S. L., and Wakefield, R. C. (1981). Effects of turfgrass on the establishment of woody plants. *Agron. J.* **73,** 605–610.

Farkas, G. L., and Kiraly, Z. (1962). Role of phenolic compounds in the physiology of plant diseases and disease resistance. *Phytopathol. Z.* **44,** 105–150.

Farnsworth, R. B., and Clawson, M. A. (1972). Nitrogen-fixation by *Artemisia ludoviciana* determined by acetylene-ethylene gas assay. *Agron. Abstr., 1972 Ann. Meet., Miami Beach, Florida.*

Fay, P. K., and Duke, W. B. (1977). An assessment of allelopathic potential in *Avena* germplasm. *Weed Sci.* **25,** 224–228.

Feenstra, W. J. (1960). The genetic control of the formation of phenolic compounds in the seedcoat of *Phaseolus vulgaris* L. *In* "Phenolics in Plants in Health and Disease" (J. B. Pridham, ed.), pp. 127–131. Pergamon, Oxford.

Feltner, K. C., Hurst, H. R., and Anderson, L. E. (1969). Yellow foxtail competition in grain sorghum. *Weed Sci.* **17,** 211–213.

Fenical, W. (1975). Halogenation in the Rhodophyta: A review. *J. Phycol.* **11,** 245–259.

Ferenczy, L. (1956). Occurrence of antibacterial compounds in seeds and fruits. *Acta Biol. Acad. Sci. Hung.* **6,** 317–323.

Ferguson, A. R., and Bollard, E. G. (1969). Nitrogen metabolism of *Spirodela oligorrhiza*. I. Utilization of ammonium, nitrate, and nitrite. *Planta* **88,** 344–352.

Fernald, M. L. (1950). "Gray's Manual of Botany," 8th ed. American Book Co., New York.

Finnell, H. H. (1933). The economy of soil nitrogen under semi-arid conditions. *Okla. Agric. Exp. Stn., Bull.* No. 215, 22 p.

Fisher, R. F. (1978). Juglone inhibits pine growth under certain moisture regimes. *Soil Sci. Soc. Am. J.* **42,** 801–803.

Fisher, R. F., Woods, R. A., and Glavicic, M. R. (1978). Allelopathic effects of goldenrod and aster on young sugar maple. *Can. J. For. Res.* **8,** 1–9.

Fitzgerald, G. P. (1969). Some factors in competition or antagonism among bacteria, algae, and aquatic weeds. *J. Phycol.* **5,** 351–359.

Fletcher, R. A., and Renney, A. J. (1963). A growth inhibitor found in *Centaurea* spp. *Can. J. Plant Sci.* **43,** 475–481.

Floyd, G. L., and Rice, E. L. (1967). Inhibition of higher plants by three bacterial inhibitors. *Bull. Torrey Bot. Club* **94**, 125–129.

Fogg, G. E., and Boalch, G. T. (1958). Extracellular products in pure cultures of a brown alga. *Nature (London)* **181**, 789–790

Fogg, G. E., and Westlake, D. F. (1955). The importance of extracellular products of algae in freshwater. *Int. Ver. Theor. Angew. Limnol. Mitt.* **12**, 219–232.

Fogg, G. E., Nalewajko, C., and Watt, W. D. (1965). Extracellular products of phytoplankton photosynthesis. *Proc. R. Soc. London, Ser. B* **162**, 517–534.

Fomenko, B. S. (1968). Effect of ionizing radiation on the metabolism of some phenols in the shoots of plants differing in their radiosentivity. *Biol. Nauk.* **11**, 45–50.

Fottrell, P. F., O'Connor, S., and Masterson, C. L. (1964). Identification of the flavonol myricetin in legume seeds and its toxicity to nodule bacteria. *Ir. J. Agric. Res.* **3**, 246–249.

Foy, C. L. (1961). Absorption, distribution, and metabolism of 2,2-dichloropropionic acid in relation to phytotoxicity. I. Penetration and translocation of Cl^{36}—and C^{14}—labeled Dalapon. *Plant Physiol.* **36**, 688–697.

Foy, C. L., Hurtt, W., and Hale, M. G. (1971). Root exudation of plant-growth regulators. In "Biochemical Interactions among Plants" (U. S. Natl. Comm. for IBP, eds.), pp. 75–85. Natl. Acad. Sci., Washington, D. C.

Fraenkel, G. S. (1959). The raison d'être of secondary plant substances. *Science* **129**, 1466–1470.

Frank, P. A., and Dechoretz, N. (1980). Allelopathy in dwarf spikerush (*Eleocharis coloradoensis*). *Weed Sci.* **28**, 499–505.

Franz, E. H., and Haines, B. L. (1977). Nitrate reductase activities of vascular plants in a terrestrial sere: Relationship of nitrate to uptake and the cybernetics of biogeochemical cycles. *Bull. Ecol. Soc. Am.* **58**, 62.

Frei, I. K., Sister, and Dodson, C. H. (1972). The chemical effect of certain bark substrates on the germination and early growth of epiphytic orchids. *Bull. Torrey Bot. Club* **99**, 301–307.

Frey-Wyssling, A., and Bäbler, S. (1957). Zur Biochemie des Gewachshaustabaks. *Experientia* **13**, 399–400.

Frick, H., Nicholson, R. L., Hodges, T. K., and Bauman, L. F. (1976). Influence of *Helminthosporium maydis*, Race T, toxin on potassium uptake in maize roots. *Plant Physiol.* **57**, 171–174.

Frick, H., Bauman, L. F., Nicholson, R. L., and Hodges, T. K. (1977). Influence of *Helminthosporium maydis*, Race T, toxin on potassium uptake in maize roots. II. Sensitivity of development of the augmented uptake potential to toxin and inhibitors of protein synthesis. *Plant Physiol.* **59**, 103–106.

Friedman, J., and Orshan, G. (1975). The distribution, emergence and survival of seedlings of *Artemisia herba alba* Asso in the Negev desert of Israel in relation to distance from the adult plants. *J. Ecol.* **63**, 627–632.

Friedman, J., Orshan, G., and Ziger-Cfir, Y. (1977). Suppression of annuals by *Artemisia herba-alba* in the Negev desert of Israel. *J. Ecol.* **65**, 413–426.

Friedman, J., Rushkin, E., and Waller, G. R. (1982). Highly potent germination inhibitors in aqueous eluate of fruits of Bishop's weed (*Ammi majus* L.) and avoidance of autoinhibition. *J. Chem. Ecol.* **8**, 55–65.

Friedman, T., and Horowitz, M. (1971). Biologically active substances in subterranean parts of purple nutsedge. *Weed Sci.* **19**, 398–401.

Fries, N., and Forsman, B. (1951). Quantitative determination of certain nucleic acid derivatives in pea root exudate. *Physiol. Plant.* **4**, 410–420.

Friesen, H. A. (1961). Some factors affecting the control of wild oats with Barban. *Weeds* **9**, 185–194.

Funke, G. L. (1941). Essai de phytosociologie experimentale. *Bull. Soc. Hist. Nat. Toulouse* **76**, 19–21.

Funke, G. L. (1943). The influence of *Artemisia absinthium* on neighboring plants. *Blumea* **5**, 281–293.

Gabor, W. E., and Veatch, C. (1981). Isolation of a phytotoxin from quackgrass (*Agropyron repens*) rhizomes. *Weed Sci.* **29**, 155–159.

Gabriel, W. J. (1975). Allelopathic effects of black walnut on white birches. *J. For.* **73**, 234–237.

Gaidamak, V. M. (1971). Biologically active substances in nutrient solutions after cucumbers and tomatoes were grown on pure and multiple used broken brick. *In* "Physiological-Biochemical Basis of Plant Interactions in Phytocenoses" (A. M. Grodzinsky, ed.), Vol. 2, pp. 55–60. Naukova Dumka, Kiev. (In Russian, English summary.)

Gajić, D. (1966). Interaction between wheat and corn cockle on brown soil and smonitsa. *J. Sci. Agric. Res.* **19**, 63–96.

Gajić, D. (1969). Effect of substances x on wheat yield. *Savrem. Poljopr.* **17**, 351–358.

Gajić, D. (1973). The effect of agrostemins as a means of improvement of the quality and quantity of the grass-cover of the Zlatibor—as a preventive measure against the weeds. Yugoslav Symposium on Weed Control in Hilly and Mountainous Areas. Sarajevo.

Gajić, D. (1977). Effect of agrostemin as an exometabolite on strengthening of ecological metabolism, in particular on the increase of phosphorus content and productivity. *In* "Interactions of Plants and Microorganisms in Phytocenoses" (A. M. Grodzinsky, ed.), pp. 114–116. Naukova Dumka, Kiev. (In Russian, English summary.)

Gajić, D., and Nikočević, G. (1973). Chemical allelopathic effect of *Agrostemma githago* upon wheat. *Fragm. Herb. Jugoslavica* **18**, 1–5.

Gajić, D., and Vrbaški, M. (1972). Identification of the effect of bioregulators from *Agrostemma githago* upon wheat in heterotrophic feeding, with special respect to agrostemmin and allantoin. *Fragm. Herb. Croatica* **7**, 1–6.

Gajić, D., Malenčić, S., Vrbaški, M., and Vrbaški, S. (1976). Study of the quantitative and qualitative improvement of wheat yield through agrostemin as an allelopathic factor. *Fragm. Herb. Jugoslavica* **63**, 121–141.

Gajić, D., Vrbaški, M., and Vrbaški, S. (1977). Investigations of allelopathic effects of agrostemin on the dynamics of phosphorus (P_2O_5) and potassium (K_2O) in soil of manured and unmanured smonitsa and chernozem. *Fragm. Herb. Jugoslavica* **2**, 5–16.

Galston, A. W. (1975). The water fern-rice connection. *Nat. Hist.* **12**, 10.

Gamborg, O. L., and Shyluk, J. P. (1970). The culture of plant cells with ammonium salts as the sole nitrogen source. *Plant Physiol.* **45**, 598–600.

Gant, R. E., and Clebsch, E. E. C. (1975). The allelopathic influences of *Sassafras albidum* in old-field succession in Tennessee. *Ecology* **56**, 604–615.

Gardner, R. L., and Payne, M. G. (1964). Identification of a phenolic substance in sugar beets responsible for resistance to *Cercospora beticola*. *J. Colo.-Wyo. Acad. Sci.* **5**, 42. (Abstr.)

Gayed, S. K., and Rosa, N. (1975). Levels of chlorogenic acid in tobacco cultivars; healthy and infected with *Thielaviopsis basicola*. *Phytopathology* **65**, 1049–1053.

Geissman, T. A., and Phinney, B. O. (1972). Tannins as gibberellin antagonists. *Plant Physiol.* **49**, 323–330.

Geissman, T. A., Griffin, S., Waddell, T. G., and Chen, H. H. (1969). Sesquiterpene lactones. Some new constitutents of *Ambrosia* species: *A. psilostachya* and *A. acanthicarpa*. *Phytochemistry* **8**, 145–150.

Gerrettson-Cornell, L., and Humphreys, F. R. (1978). Results of an experiment on the effects of *Pinus radiata* bark on the formation of sporangia in *Phytophthora cinnamomi* Rands. ΦΥΤΟΝ **36**, 15–17.

Gigon, A., and Rorison, I. H. (1972). The response of some ecologically distinct plant species to nitrate- and to ammonium-nitrogen. *J. Ecol.* **60**, 93–102.

Gilmore, A. R. (1977). Effects of soil moisture stress on monoterpenes in loblolly pine. *J. Chem. Ecol.* **3**, 667–676.

Gilmore, A. R. (1980). Phytotoxic effects of giant foxtail on loblolly pine seedlings. *Comp. Physiol. Ecol.* **5**, 183–192.

Gilmore, A. R., and Boggess, W. R. (1963). The effect of past agricultural soiling practices on the survival and growth of planted pine and hardwood seedlings in southern Illinois. *Soil Sci. Soc. Am. Proc.* **27**, 98–102.

Giovanelli, J., Owens, L. D., and Mudd, S. H. (1971). Mechanism of inhibition of spinach β-cystathionase by rhizobitoxine. *Biochim. Biophys. Acta* **227**, 671–684.

Giovanelli, J., Owens, L. D., and Mudd, S. H. (1972). β-cystathionase—in vivo inactivation by rhizobitoxine and role of the enzyme in methionine biosynthesis in corn seedlings. *Plant Physiol.* **51**, 492–503.

Glass, A. D. M. (1973). Influence of phenolic acids on ion uptake. I. Inhibition of phosphate uptake. *Plant Physiol.* **51**, 1037–1041.

Glass, A. D. M. (1974a). Influence of phenolic acids upon ion uptake. II. A structure-activity study of the inhibition of phosphate uptake by benzoic acid derivatives. *Bull. R. Soc. N. Z.* No. 12, pp. 159–164.

Glass, A. D. M. (1974b). Influence of phenolic acids upon ion uptake. III. Inhibition of potassium absorption. *J. Exp. Bot.* **25**, 1104–1113.

Glass, A. D. M. (1975). Inhibition of phosphate uptake in barley roots by hydroxybenzoic acids. *Phytochemistry* **14**, 2127–2130.

Glass, A. D. M. (1976). The allelopathic potential of phenolic acids associated with the rhizosphere of *Pteridium aquilinum*. *Can. J. Bot.* **54**, 2440–2444.

Glass, A. D. M., and Bohm, B. A. (1969). The accumulation of cinnamic and benzoic acid derivatives in *Pteridium aquilinum* and *Athyrium felix-femina*. *Phytochemistry* **8**, 371–377.

Glass, A. D. M., and Bohm, B. A. (1971). The uptake of simple phenols by barley roots. *Planta* **100**, 93–105.

Glass, A. D. M., and Dunlop, J. (1974). Influence of phenolic acids on ion uptake. IV. Depolarization of membrane potentials. *Plant Physiol.* **54**, 855–858.

Gliessman, S. R. (1976). Allelopathy in a broad spectrum of environments as illustrated by bracken. *J. Linn. Soc., Bot.* **73**, 95–104.

Gliessman, S. R. (1978). Allelopathy as a potential mechanism of dominance in the humid tropics. *Trop. Ecol.* **19**, 200–208.

Gliessman, S. R., and Muller, C. H. (1972). The phytotoxic potential of bracken, *Pteridium aquilinum* (L.) Kuhn. *Madroño* **21**, 299–304.

Gliessman, S. R., and Muller, C. H. (1978). The allelopathic mechanisms of dominance in bracken (*Pteridium aquilinum*) in southern California. *J. Chem. Ecol.* **4**, 337–362.

González de la Parra, M., Anaya, A. L., Espinosa, F., Jiménez, M., and Castillo, R. (1981). Allelopathic potential of *Piqueria trinerva* (Compositae) and piquerols A and B. *J. Chem. Ecol.* **7**, 509–515.

Goodwin, R. H., and Kavanagh, F. (1949). The isolation of scopoletin, a blue-fluorescing compound of oat roots. *Bull. Torrey Bot. Club* **76**, 255–265.

Gordon, J. C., and Dawson, J. O. (1979). Potential uses of nitrogen-fixing trees and shrubs in commercial forestry. *Bot. Gaz. (Chicago)* **140** (Suppl.), S88–S90.

Graham, B. F., Jr., and Bormann, F. H. (1966). Natural root grafts. *Bot. Rev.* **32**, 255–292.

Grand, L. F., and Ward, W. W. (1969). An antibiotic detected in conifer foliage and its relation to *Cenococcum graniforme* mycorrhizae. *For. Sci.* **15**, 286–288.

Grant, E. A., and Sallans, W. G. (1964). Influence of plant extracts on germination and growth of eight forage species. *J. Br. Grass. Soc.* **19**, 191–197.

Grant, W. D. (1976). Microbial degradation of condensed tannins. *Science* **193**, 1137–1138.

Gray, F., and Galloway, H. M. (1959). Soils of Oklahoma. *Okla., Agric. Exp. Stn., [Misc. Publ.]* MP–56.

Gray, F., and Stahnke, C. (1970). Classification of soils in the savanna-forest transition in eastern Oklahoma. *Okla., Agr. Exp. Stn., Bull.* B–672.

Gray, R., and Bonner, J. (1948a). An inhibitor of plant growth from the leaves of *Encelia farinosa. Am. J. Bot.* **35**, 52–57.

Gray, R., and Bonner, J. (1948b). Structure determination and synthesis of a plant growth inhibitor, 3-acetyl-6-methoxybenzaldehyde, found in the leaves of *Encelia farinosa. J. Am. Chem. Soc.* **70**, 1249–1253.

Green, F. B., and Corcoran, M. R. (1975). Inhibitory action of five tannins on growth induced by several gibberellins. *Plant Physiol.* **56**, 801–806.

Greenland, D. J. (1958). Nitrate fluctuations in tropical soils. *J. Afric. Sci.* **50**, 82–92.

Gressel, J. B., and Holm, L. G. (1964). Chemical inhibition of crop germination by weed seeds and the nature of inhibition by *Abutilon theophrasti. Weed Res.* **4**, 44–53.

Grodzinsky, A. M., and Bogdan, G. P. (1972). Histochemical study of pectins, lignin, suberin and melanins in plants treated with allelopathically active substances. *Ukr. Bot. Zh.* **29**, 137–143. (In Ukrainian, Russian and English summaries.)

Grodzinsky, A. M., and Bogdan, G. P. (1973). Role of ascorbic acid in formation of the brown mass in xylem of plants under the effect of allelopathic factors. *Ukr. Bot. Zh.* **30**, 28–35. (In Ukrainian, Russian and English summaries.)

Grodzinsky, A. M., and Gaidamak, V. M. (1971). Allelopathic influence of woody plants on herbaceous ones in the Ukrainian forest-steppe region. *In* "Physiological-Biochemical Basis of Plant Interactions in Phytocenoses" (A. M. Grodzinsky, ed.), Vol. 2, pp. 3–11. Naukova Dumka, Kiev. (In Russian, English summary.)

Grodzinsky, A. M., and Panchuk, M. A. (1974). Allelopathic properties of crop residues of wheat-wheat grass hybrids. *In* "Physiological-Biochemical Basis of Plant Interactions in Phytocenoses" (A. M. Grodzinsky, ed.), Vol. 5, pp. 51–55. Naukova Dumka, Kiev. (In Russian, English summary.)

Groner, M. G. (1974). Intraspecific allelopathy in *Kalanchoe daigremontiana. Bot. Gaz. (Chicago)* **135**, 73–79.

Groner, M. G. (1975). Allelopathic influence of *Kalanchoe daigremontiana* on other species of plants. *Bot. Gaz. (Chicago)* **136**, 207–211.

Grosjean, J. (1950). Substances with fungicidal activity in the bark of deciduous trees. *Nature (London)* **165**, 853–854.

Grümmer, G. (1955). "Die gegenseitige Beeinflussung höherer Pflanzen-Allelopathie." Fischer, Jena.

Grümmer, G. (1961). The role of toxic substances in the interrelationships between higher plants. *In* "Mechanisms in Biological Competition" (F. L. Milthorpe, ed.), pp. 219–228. Academic Press, New York.

Grümmer, G., and Beyer, H. (1960). The influence exerted by species of *Camelina* on flax by means of toxic substances. *Symp. Br. Ecol. Soc.* **1**, 153–157.

Guenzi, W., and McCalla, T. (1962). Inhibition of germination and seedling development by crop residues. *Soil Sci. Soc. Am. Proc.* **26**, 456–458.

Guenzi, W. D., and McCalla, T. M. (1966a). Phenolic acids in oats, wheat, sorghum, and corn residues and their phytotoxicity. *Agron. J.* **58**, 303–304.

Guenzi, W. D., and McCalla, T. M. (1966b). Phytotoxic substances extracted from soil. *Soil Sci. Soc. Am. Proc.* **30**, 214–216.

Guenzi, W. D., McCalla, T. M., and Norstadt, F. A. (1967). Presence and persistence of phytotoxic substances in wheat, oat, corn, and sorghum residues. *Agron. J.* **59**, 163–165.

Guillard, R. R. L., and Hellebust, J. A. (1971). Growth and the production of extracellular substances by two strains of *Phaeocystis poucheti. J. Phycol.* **7**, 330–338.

Gupta, N. N., and Houdeshell, J. (1976). A differential-difference equations model of a dynamic aquatic ecosystem. *Int. J. Syst. Sci.* **7,** 481–492.

Guyot, A. L. (1957). Les microassociations végétales au sein du Brometum Erecti. *Vegetatio* **7,** 321–354.

Hadwiger, L. A. (1972). Induction of phenylalanine ammonia lyase and pisatin by photosensitive psoralen compounds. *Plant Physiol.* **49,** 779–782.

Hagood, E. S., Jr., Bauman, T. T., Williams, J. L., Jr., and Schreiber, M. M. (1980). Growth analysis of soybeans (*Glycine max*) in competition with velvetleaf (*Abutilon theophrasti*). *Weed Sci.* **28,** 729–734.

Haider, K., and Martin, J. P. (1975). Decomposition of specifically carbon-14 labeled benzoic and cinnamic acid derivatives in soil. *Soil Sci. Soc. Am. Proc.* **39,** 657–662.

Haines, B. L. (1977). Nitrogen uptake: Apparent pattern during old-field succession in southeastern United States. *Oecologia* **26,** 295–303.

Hall, A. B., Blum, U., and Fites, R. C. (1982). Stress modification of allelopathy of *Helianthus annuus* L. debris on seed germination. *Am. J. Bot.* **69,** 776–783.

Halligan, J. P. (1973). Bare areas associated with shrub stands in grasslands: The case of *Artemisia californica*. *BioScience* **23,** 429–432.

Halligan, J. P. (1975). Toxic terpenes from *Artemisia californica*. *Ecology* **56,** 999–1003.

Halligan, J. P. (1976). Toxicity of *Artemisia californica* to four associated herb species. *Am. Midl. Nat.* **95,** 406–421.

Handley, W. R. C. (1963). Mycorrhizal associations and *Calluna* heathland afforestation. *Bull. For. Comm., London,* No. 36.

Hansen, J. A. (1973). Antibiotic activity of the chrysophyte *Ochromonas malhamensis Physiol. Plant.* **29,** 234–238.

Harborne, J. B. (1964). Phenolic glycosides and their natural distribution. *In* "Biochemistry of Phenolic Compounds" (J. B. Harborne, ed.), pp. 129–169. Academic Press, New York.

Harborne, J. B., and Simmonds, N. W. (1964). The natural distribution of the phenolic aglycones. *In* "Biochemistry of Phenolic Compounds" (J. B. Harborne, ed.), pp. 77–127. Academic Press, New York.

Harder, R. (1917). Ernahrungsphysiologische Untersuchungen au Cyanophyceen, hauptsachlich dem endophytischen *Nostoc punctiforme*. *Z. Bot.* **9,** 145–242.

Harmsen, G. W., and Kolenbrander, G. J. (1965). Soil inorganic nitrogen. *In* "Soil Nitrogen" (W. V. Bartholomew and F. E. Clark, eds.), pp. 43–92. Am. Soc. Agron., Madison, Wisconsin.

Harper, H. J. (1932). Easily soluble phosphorus in Oklahoma soils. *Okla. Agr. Exp. Stn. Bull.* No. 205, 24 p.

Harper, J. R., and Balke, N. E. (1980). Inhibition of potassium absorption in excised oat roots by phenolic acids. *In* "Plant Membrane Transport: Current Conceptual Issues" (R. M. Spanswick, W. J. Lucas, and J. Dainty, eds.), pp. 399–400. Elsevier/North-Holland, New York.

Harris, D. O. (1970). An autoinhibitory substance produced by *Platydorina caudata* Kofoid. *Plant Physiol.* **45,** 210–214.

Harris, D. O. (1971a). Growth inhibitors produced by the green algae (Volvocaceae). *Arch. Mikrobiol.* **76,** 47–50.

Harris, D. O. (1971b). A model system for the study of algal growth inhibitors. *Arch. Protistenkd.* **113,** 230–234.

Harris, D. O. (1971c). Inhibition of oxygen evolution in *Volvox globator* by culture filtrates from *Pandorina morum*. *Microbios* **3,** 73–75.

Harris, D. O., and Caldwell, C. D. (1974). Possible mode of action of a photosynthetic inhibitor produced by *Pandorina morum*. *Arch. Microbiol.* **95,** 193–204.

Harris, D. O., and Parekh, M. C. (1974). Further observations on an algicide produced by *Pandorina morum*, a colonial green flagellate. *Microbios* **9,** 259–265.

Harris, H. B., and Burns, R. E. (1970). Influence of tannin content on preharvest seed germination in sorghum. *Agron. J.* **62**, 835–836.

Harris, H. B., and Burns, R. E. (1972). Inhibiting effects of tannin in sorghum grain on preharvest seed molding. *Agron. Abstr., 1972, Annu. Meet., Miami Beach, Florida.*

Harris, T. H., Hay, J. V., and Quarterman, E. (1973). Isolation of 2-(4-hydroxybenzyl) malic acid from *Petalostemon gattingeri. J. Org. Chem.* **38**, 4457–4459.

Hartman, R. T. (1960). Algae and metabolites of natural waters. *In* "The Ecology of Algae" (C. A. Tryon, Jr., and R. T. Hartman, eds.), pp. 38–55, Special Publ. No. 2, Pymatuning Laboratory of Field Biology. University of Pittsburgh, Pittsburgh, Pennsylvania.

Harvey, R. G., and Linscott, J. J. (1978). Ethylene production in soil containing quack grass rhizomes and other plant materials. *Soil Sci. Soc. Am. J.* **42**, 721–724.

Harwood, R. R. (1979). Natural weed controls are looking good. *New Farm* **1**(6), 56–58.

Hattingh, M. J., and Louw, H. A. (1969a). The influence of antagonistic rhizoplane bacteria on the clover: *Rhizobium* symbiosis. *Phytophylactica* **1**, 205–208.

Hattingh, M. J., and Louw, H. A. (1969b). Clover rhizoplane bacteria antagonistic to *Rhizobium trifolii. Can. J. Bot.* **15**, 361–364.

Hattori, S., and Noguchi, I. (1959). Microbial degradation of rutin. *Nature (London)* **184**, 1145–1146.

Hauschka, T., Toennies, G., and Swain, A. P. (1945). The mechanism of growth inhibition by hexenolactone. *Science* **101**, 383–385.

Havill, D. C., Lee, J. A., and Stewart, G. R. (1974). Nitrate utilization by species from acid and calcareous soils. *New Phytol.* **73**, 1221–1231.

Havis, L., and Gilkeson, A. L. (1947). Toxicity of peach roots. *Proc. Am. Soc. Hort. Sci.* **50**, 203–205.

Hayes, L. E. (1947). Survey of higher plants for presence of antibacterial substances. *Bot. Gaz. (Chicago)* **108**, 408–414.

Heisey, R. M., Delwiche, C. C., Virginia, R. A., Wrona, A. F., and Bryan, B. A. (1980). A new nitrogen-fixing non-legume: *Chamaebatia foliolosa* (Rosaceae). *Am. J. Bot.* **67**, 429–431.

Hellebust, J. A. (1974). Extracellular products. *In* "Algal Physiology and Biochemistry" (W. D. P. Stewart, ed.), pp. 838–863. Univ. of California Press, Berkeley.

Henderson, M. E. K. (1956). A study of the metabolism of phenolic compounds by soil fungi using spore suspensions. *J. Gen. Microbiol.* **14**, 684–691.

Henderson, M. E. K., and Farmer, V. C. (1955). Utilization by soil fungi of *p*-hydroxybenzaldehyde, ferulic acid, syringaldehyde and vanillin. *J. Gen. Microbiol.* **12**, 37–46.

Henis, Y., Tagari, H., and Volcani, R. (1964). Effect of water extract of carob pods, tannic acid, and their derivatives on the morphology and growth of microorganisms. *Appl. Microbiol.* **12**, 204–209.

Hennequin, J. R., and Juste, C. (1967). Présence d'acides phénols libres dans le sol. Étude de leur influence sur la germination et la croissance des végétaux. *Ann. Agron.* **18**, 545–569.

Hill, L. V., and Santelmann, P. W. (1969). Competitive effects of annual weeds on Spanish peanuts. *Weed Sci.* **17**, 1–2.

Hilton, J. L. (1979). Research on the physiology and biology of weeds. *Weeds Today* **10**(4), 5–6.

Hitchcock, A. E., and Zimmerman, P. W. (1935). Absorption and movement of synthetic growth substances from soil as indicated by responses of aerial parts. *Contrib. Boyce Thompson Inst.* **7**, 447–476.

Hoagland, D. R., and Arnon, D. I. (1950). The water-culture method for growing plants without soil. *Calif. Agric. Exp. Sta. Cir.* No. 347.

Hodges, T. K., Darding, R. L., and Weidner, T. (1971). Gramicidin-D-stimulated influx of monovalent cations into plant roots. *Planta* **97**, 245–256.

Hodgson, J. M. (1958). Canada thistle (*Cirsium arvense* Scop.) control with cultivation, cropping, and chemical sprays. *Weeds* **6**, 1–11.

Hoffman, G. R., and Hazlett, D. L. (1977). Effects of aqueous *Artemisia* extracts and volatile substances on germination of selected species. *J. Range Manage.* **30,** 134–137.

Hogetsu, T., Shibaoka, H., and Shimokoriyama, M. (1974). Involvement of cellulose in actions of gibberellin and kinetin-coumarin interactions on stem elongation. *Plant Cell Physiol.* **15,** 265–272.

Holm, L. (1969). Weed problems in developing countries. *Weed Sci.* **17,** 113–118.

Holm, L., Pancho, J. V., Herberger, J. P., and Plucknett, D. L. (1979). "A Geographical Atlas of World Weeds." Wiley, New York.

Holowczak, J., Kuć, J., and Williams, E. G. (1960). Metabolism *in vitro* of phloridzin and other host compounds by *Venturia inaequalis. Phytopathology* **50,** 640.

Hook, D. D., and Stubbs, J. (1967). An observation of understory growth retardation under three species of oaks. *U. S. For. Ser. Res. Note* SE-70.

Horowitz, M., and Friedman, T. (1971). Biological activity of subterranean residues of *Cynodon dactylon* L., *Sorghum halepense* L., and *Cyperus rotundus* L. *Weed Res.* **11,** 88–93.

Horsley, S. B. (1977a). Allelopathic inhibition of black cherry by fern, grass, goldenrod, and aster. *Can. J. For. Res.* **7,** 205–216.

Horsley, S. B. (1977b). Allelopathic inhibition of black cherry. II. Inhibition by woodland grass, ferns, and club moss. *Can. J. For. Res.* **7,** 515–519.

Horton, J. S., and Kraebel, C. J. (1955). Development of vegetation after fire in the chamise chaparral of southern California. *Ecology* **36,** 244–262.

Huang, C. Y. (1978). Effects of nitrogen fixing activity of blue-green algae on the yield of rice plants. *Bot. Bull. Acad. Sin.* **19,** 41–52.

Hubbes, M. (1962). Inhibition of *Hypoxylon pruinatum* by pyrocatechol isolated from bark of aspen. *Science* **136,** 156.

Huber, D.M., Warren, H. L., Nelson, D. W., and Tsai, C. Y. (1977). Nitrification inhibitors: New tools for food production. *BioScience* **27,** 523–529.

Hughes, J. C., and Swain, T. (1960). Scopolin production in potato tubers infected with *Phytophthora infestans. Phytopathology* **50,** 398–400.

Hull, J. C., and Muller, C. H. (1977). The potential for dominance by *Stipa pulchra* in a California grassland. *Am. Midl. Nat.* **97,** 147–175.

Hulme, A. C., and Jones, J. D. (1963). Tannin inhibition of plant mitochondria. *In* "Enzyme Chemistry of Phenolic Compounds" (J. B. Pridham, ed.), pp. 97–120. Macmillan, New York.

Hunter, B. T. (1971). "Gardening without Poisons." 2nd ed. Houghton, Boston, Massachusetts.

Huntsman, S. A., and Barber, R. T. (1975). Modification of phytoplankton growth by excreted compounds in low-density populations. *J. Phycol.* **11,** 10–13.

Hussain, A., and Mallik, M. A. B. (1972). Study of rhizosphere microflora of berseem (*Trifolium alexandrinum* L.) and their effect on *Rhizobium trifolii. J. Sci.* **1,** 139–145.

Ingham, J. (1972). Phytoalexins and other natural products as factors in plant disease resistance. *Bot. Rev.* **38,** 343–424.

Isleib, D. R. (1960). Quackgrass control in potato production. *Weeds* **8,** 631–635.

Iuzhina, Z. I. (1958). Relationship between toxic properties of soil of Kola Peninsula and number of bacterial antagonists of *Azotobacter. Microbiology (Engl. Transl.)* **27,** 452–456. (Transl. from Russian.)

Iverson, L. R., and Wali, M. K. (1982). Reclamation of coal mined land: The role of *Kochia scoparia* and other pioneers in early succession. *Reclam. Reveg. Res.* **1,** 123–160.

Jackson, J. R., and Willemsen, R. W. (1976). Allelopathy in the first stages of secondary succession on the piedmont of New Jersey. *Am. J. Bot.* **63,** 1015–1023.

Jackson, R. M. (1965). Antibiosis and fungistasis of soil microorganisms. *In* "Ecology of Soil-Borne Plant Pathogens" (K. F. Baker and W. C. Snyder, eds.), pp. 363–369. Univ. of California Press, Berkeley.

Jacquemin, H., and Berlier, Y. (1956). Evolution du pouvoir nitrifiant d'un sol de basse Côte d'Ivoire sous l'action du climat et de la végétation. *Trans. 6th Int. Congr. Soil Sci. C* pp. 343–347.

Jaffe, M. J., and Isenberg, F. M. R. (1969). Red light photoenhancement of the synthesis of phenolic compounds and lignin in potato tuber tissue. *ΦYTON* **26**, 51–67.

Jakob, H. (1954). Compatibilitiés et antagonismes entre algues du sol. *C. R. Acad. Sci.* **238**, 928–930.

Jameson, D. A. (1961). Growth inhibitors in native plants of northern Arizona. *Res. Note No. 61 Rocky M. For. Range Exp. Stn., USDA.*

Jameson, D. A. (1966). Pinyon-juniper litter reduces growth of blue grama. *J. Range Manage.* **9**, 214–217.

Jankay, P., and Muller, W. H. (1976). The relationships among umbelliferone, growth, and peroxidase levels in cucumber roots. *Am. J. Bot.* **63**, 126–132.

Jensen, T. E., and Welbourne, F. (1962). The cytological effects of growth inhibitors on excised roots of *Vicia faba* and *Pisum sativum. Proc. S. D. Acad. Sci.* **41**, 131–136.

Jobidon, R., and Thibault, J. R. (1981). Allelopathic effects of balsam poplar on green alder germination. *Bull. Torrey Bot. Club* **108**, 413–418.

Jobidon, R., and Thibault, J. R. (1982). Allelopathic growth inhibition of nodulated and unnodulated *Alnus crispa* seedlings by *Populus balsamifera. Am. J. Bot.* **69**, 1213–1223.

Johansson, M., and Hägerby, E. (1974). Influence of growth conditions, metabolic inhibitors, and phenolic compounds on the ATP pool in *Fomes annosus. Physiol. Plant.* **32**, 23–32.

Johnson, G., and Schaal, L. A. (1952). Relation of chlorogenic acid to scab resistance in potatoes. *Science* **115**, 627–629.

Jones, J. M., and Richards, B. N. (1977). Effect of reforestation on turnover of ^{15}N-labelled nitrate and ammonium in relation to changes in soil microflora. *Soil Biol. Biochem.* **9**, 383–392.

Jørgensen, E. (1956). Growth inhibiting substances formed by algae. *Physiol. Plant.* **9**, 712–726.

Jørgensen, E. (1962). Antibiotic substances from cells and culture solutions of unicellular algae with special reference to some chlorophyll derivatives. *Physiol. Plant.* **15**, 530–545.

Jørgensen, E., and Steemann-Nielsen, E. (1961). Effect of filtrates from cultures of unicellular algae on the growth of *Staphylococcus aureus. Physiol. Plant.* **14**, 896–908.

June, S. R. (1976). Investigations on allelopathy in a red beech forest. *Mauri Ora* **4**, 87–91.

Junttila, O. (1975). Allelopathy in *Heracleum laciniatum:* inhibition of lettuce seed germination and root growth. *Physiol. Plant.* **33**, 22–27.

Junttila, O. (1976). Allelopathic inhibitors in seeds of *Heracleum laciniatum. Physiol. Plant.* **36**, 374–378.

Kaminsky, R., and Muller, W. H. (1977). The extraction of soil phytotoxins using a neutral EDTA solution. *Soil Sci.* **124**, 205–210.

Kanchan, S. D., and Jayachandra (1979a). Allelopathic effects of *Parthenium hysterophorus* L. I. Exudation of inhibitors through roots. *Plant Soil* **53**, 27–35.

Kanchan, S. D., and Jayachandra (1979b). Allelopathic effects of *Parthenium hysterophorus* L. III. Inhibitory effects of the weed residue. *Plant Soil* **53**, 37–47.

Kanchan, S., and Jayachandra (1980). Pollen allelopathy: A new phenomenon. *New Phytol.* **84**, 739–746.

Kapusta, G., and Strieker, C. F. (1975). Selective control of downy brome in alfalfa. *Weed Sci.* **23**, 202–206.

Kapustka, L. A., and Moleski, F. L. (1976). Changes in community structure in Oklahoma old field succession. *Bot. Gaz. (Chicago)* **137**, 7–10.

Kapustka, L. A., and Rice, E. L. (1976). Acetylene reduction (N_2-fixation) in soil and old field succession in central Oklahoma. *Soil Biol. Biochem.* **8**, 497–503.

Kapustka, L. A., and Rice, E. L. (1978a). Acetylene reduction (N$_2$-fixation) of glucose-amended soils from central Oklahoma old field succession plots. *Southwest. Nat.* **23**, 389–396.

Kapustka, L. A., and Rice, E. L. (1978b). Symbiotic and asymbiotic N$_2$-fixation in a tall grass prairie. *Soil Biol. Biochem.* **10**, 553–554.

Katayama, T. (1962). Volatile constituents. *In* "Physiology and Biochemistry of Algae" (R. A. Lewin, ed.), pp. 467–473. Academic Press, New York.

Katznelson, J. (1972). Studies in clover soil sickness. I. The phenomenon of soil sickness in berseem and Persian clover. *Plant Soil* **36**, 379–393.

Kaurov, I. A. (1970). Interaction of bird's foot and yellow lupine in pure and mixed cultures. *In* "Physiological-Biochemical Basis of Plant Interactions in Phytocenoses" (A. M. Grodzinsky, ed.), Vol. 1, pp. 66–71. Naukova Dumka, Kiev. (In Russian, English summary.)

Keating, K. I. (1977). Allelopathic influence on blue-green bloom sequence in a eutrophic lake. *Science* **196**, 885–887.

Keating, K. I. (1978). Blue-green algal inhibition of diatom growth: Transition from mesotrophic to eutrophic community structure. *Science* **199**, 971–973.

Keck, R. W., and Hodges, T. K. (1973). Membrane permeability in plants: Changes induced by host-specific pathotoxins. *Phytopathology* **63**, 226–230.

Keeley, P. E., and Thullen, R. J. (1975). Influence of yellow nutsedge competition on furrow-irrigated cotton. *Weed Sci.* **23**, 171–175.

Kefeli, V. I., and Turetskaya, R. K. (1967). Comparative effect of natural growth inhibitors, narcotics, and antibiotics on plant growth. *Fiziol. Rast. (Moscow)* **14**, 796–803. (In Russian, English summary.)

Kerr, A. (1972). Biological control of crown gall: Seed inoculation *J. Appl. Bacteriol.* **35**, 493–497.

Keynes, G., ed. (1929). "The Works of Sir Thomas Browne. IV. Hydriotaphia, Brampton Urns, The Garden of Cyrus." Faber and Gwyer Ltd., London.

Khailov, K. M. (1971). "Ecological Metabolism in the Sea." Naukova Dumka, Kiev. (In Russian.)

Khailov, K. M., ed.(1974). "Biochemical Trophodynamics in Marine Coastal Ecosystems." Naukova Dumka, Kiev. (In Russian.)

Kil, B. S. (1981). Allelopathic effect of *Pinus densiflora* on the floristic compositions of undergrowth in pine forest. Doctoral Dissertation, Department of Biology, Chung-Ang Univ., Iri, Korea.

Kimber, R. W. L. (1973). Phytotoxicity from plant residues. III. The relative effect of toxins and nitrogen immobilisation on the germination and growth of wheat. *Plant Soil* **38**, 543–555.

King, J. E., and Coley-Smith, J. R. (1968). Effects of volatile products of *Allium* species and their extracts on germination of sclerotia of *Sclerotium cepivorum* Berk. *Ann. Appl. Biol.* **61**, 407–414.

Klaus, H. (1939). Das Problem der Bodenmüdikeit unter Berücksichtigung des Obstbaues. *Landwirtsch. Jahrb.* **89**, 413–459.

Kleifeld, Y. (1970). Combined effect of trifluralin and MSMA on Johnsongrass control in cotton. *Weed Sci.* **18**, 16–18.

Klöpping, H. L., and van der Kerk, G. J. M. (1951). Antifungal agents from the bark of *Populus candicans*. *Nature (London)* **167**, 996–997.

Knake, E. L., and Slife, F. W. (1962). Competition of *Setaria faberii* with corn and soybeans. *Weeds* **10**, 26–29.

Knake, E. L., and Slife, F. W. (1965). Giant foxtail seeded at various times in corn and soybeans. *Weeds* **13**, 331–334.

Knudsen, L. (1913). Tannic acid fermentation. II. Effect of nutrition on the production of the enzyme tannase. *J. Biol. Chem.* **14**, 185–202.

Kobayashi, A., Morimoto, S., Shibata, Y., Yamashita, K., and Numata, M. (1980). C_{10}-poly-acetylenes as allelopathic substances in dominants in early stages of secondary succession. *J. Chem. Ecol.* **6**, 119–131.

Koch, L. W. (1955). The peach replant problem in Ontario. I. Symptomatology and distribution. *Can. J. Bot.* **33**, 450–460.

Kochhar, M., Blum, U., and Reinert, R. A. (1980). Effects of O_3 and (or) fescue on ladino clover: Interactions. *Can. J. Bot.* **58**, 241–249.

Koda, Y., and Okazawa, Y. (1978). Cytokinin production by tomato root: Occurrence of cytokinins in staled medium of root culture. *Physiol. Plant.* **44**, 412–416.

Koeppe, D. E. (1972). Some reactions of isolated corn mitochondria influenced by juglone. *Physiol. Plant.* **27**, 89–94.

Koeppe, D. E., and Miller, R. J. (1974). Kaempferol inhibitions of corn mitochondrial phosphorylation. *Plant Physiol.* **54**, 374–378.

Koeppe, D. E., Rohrbaugh, L. M., and Wender, S. H. (1969). The effect of varying U.V. intensities on the concentration of scopolin and caffeoylquinic acids in tobacco and sunflower. *Phytochemistry* **8**, 889–896.

Koeppe, D. E., Rohrbaugh, L. M., Rice, E. L., and Wender, S. H. (1970a). The effect of x-radiation on the concentration of scopolin and caffeoylquinic acids in tobacco. *Radia. Bot.* **10**, 261–265.

Koeppe, D. E., Rohrbaugh, L. M., Rice, E. L., and Wender, S. H. (1970b). The effect of age and chilling temperatures on the concentration of scopolin and caffeoylquinic acids in tobacco. *Physiol. Plant.* **23**, 258–266.

Koeppe, D. E., Rohrbaugh, L. M., Rice, E. L., and Wender, S. H. (1970c). Tissue age and caffeoylquinic acid concentration in sunflower. *Phytochemistry* **9**, 297–301.

Koeppe, D. E., Southwick, L. M., and Bittell, J. E. (1976). The relationship of tissue chlorogenic acid concentrations and leaching of phenolics from sunflowers grown under varying phosphate nutrient conditions. *Can. J. Bot.* **54**, 593–599.

Kogan, S. B., Kareva, M. A., and Kozyritskaya, B. E. (1973). Formation of B vitamins by actinomycetes in sterile soil. *In* "Physiological-Biochemical Basis of Plant Interactions in Phytocenoses" (A. M. Grodzinsky, ed.), Vol. 4, pp. 118–121. Naukova Dumka, Kiev. (In Russian, English summary.)

Kohlmuenzer, S. (1965a). Botanical and chemical studies of the collective species *Galium mullugo* with reference to karyotypes growing in Poland. V. Phytochemical studies. *Diss. Pharm.* **17**, 357–367. (Via CA 64: 10085 g.)

Kohlmuenzer, S. (1965b). Botanical and chemical studies of the collective species *Galium mollugo* with reference to karyotypes growing in Poland. VI. Effect of extracts and some other chemical components of *Galium mollugo* on the germination of seeds and growth of selected plants. *Diss. Pharm.* **17**, 369–379. (Via CA 64: 10085–10086 g.)

Kokino, N. A., Podtelok, M. P., and Prutenskaya, N. I. (1973). Dynamics of biologically active volatile and water-soluble substances from fallen maple leaves. *In* "Physiological-Biochemical Basis of Plant Interactions in Phytocenoses" (A. M. Grodzinsky, ed.), Vol. 4, pp. 94–100. Naukova Dumka, Kiev. (In Russian, English summary.)

Kolesnichenko, M. V., and Aleikina, M. M. (1976). The rate of protein biosynthesis and absorption of mineral substances by the roots of oak and ash growing together in the forest. *Fiziol. Rast. (Moscow)* **23**, 127–131. (In Russian, English summary.)

Kommedahl, T., Kotheimer, J. B., and Bernardini, J. V. (1959). The effects of quackgrass on germination and seedling development of certain crop plants. *Weeds* **7**, 1–12.

Konishi, K. (1931). Effect of soil bacteria on the growth of the root nodule bacteria. *Mem. Col. Agric., Kyoto Imp. Univ.* No. 16, pp. 1–17.

Kossanel, J. P., Martin, J., Annelle, P., Peinot, M., Vallet, J. K., and Kurnej, K. (1977). Inhibition of growth of young radicles of maize by exudations in culture solutions and extracts of ground roots of *Chenopodium album* L. *In* "Interactions of Plants and Microorganisms in Phytocenoses" (A. M. Grodzinsky, ed.), pp. 77–86. Naukova Dumka, Kiev. (In Russian, English summary.)

Kozel, P. C., and Tukey, H. B., Jr. (1968). Loss of gibberellins by leaching from stems and foliage of *Chrysanthemum morifolium* 'Princess Anne.' *Am. J. Bot.* **55**, 1184–1189.

Kranz, E., and Jacob, F. (1977a). Zur Mineralstoff-Konkurrenz zwischen *Linum* und *Camelina* I. Aufnahme von ^{35}S-Sulfat. *Flora* **166**, 491–503.

Kranz, E., and Jacob, F. (1977b). Zur Mineralstoff-Konkurrenz zwischen *Linum* und *Camelina* II. Aufnahme von ^{32}P-Phosphat und ^{86}Rubidium. *Flora* **166**, 505–516.

Krogstad, O., and Solbraa, K. (1975). Effects of extracts of crude and composted bark from spruce on some selected biological systems. *Acta Agric. Scand.* **25**, 306–312.

Krupa, S. V. (1974). Biological significance of polyacetylenes and terpenoids in fungi and higher plants. *Phytochem. Bull.* **7**, 9–16.

Krylov, Y. V. (1970). Influence of potatoes on an apple tree and its photosynthesis. *In* "Physiological-Biochemical Basis of Plant Interactions in Phytocenoses" (A. M. Grodzinsky, ed.), Vol. 1, pp. 128–134. Naukova Dumka, Kiev. (In Russian, English summary.)

Krywolap, G. N., and Casida, L. E., Jr. (1964). An antibiotic produced by the mychorrhizal fungus *Cenococcum graniforme*. *Can. J. Microbiol.* **10**, 365–370.

Krywolap, G. N., Grand, L. F., and Casida, L. E., Jr. (1964). Natural occurrence of an antibiotic in the mycorrhizal fungus *Cenococcum graniforme*. *Can. J. Microbiol.* **10**, 323–328.

Kuć, J. (1972). Phytoalexins. *Annu. Rev. Phytopathol.* **10**, 207–232.

Kuć, J., Henze, R. E., Ulstrup, A. J., and Quackenbush, F. W. (1956). Chlorogenic and caffeic acids as fungistatic agents produced by potatoes in response to inoculation with *Helminthosporium carbonum*. *J. Am. Chem. Soc.* **78**, 3123–3125.

Kuhn, R., Jerchel, D., Moewus, F., and Moeller, E. F. (1943). Über die chemische Natur der Blastokoline und ihre Einwirkung auf keimende Samen, Pollenkörner, Hefen, Bakterien, Epithelgewebe und Fibroblasten. *Naturwissenschaften* **31**, 468.

Kunc, F. (1971). Decomposition of vanillin by soil microorganisms. *Folia Microbiol. (Prague)* **16**, 41–50.

Kushnir, G. P. (1973a). Microflora of *Crambe cordifolia* Stev. and *Heracleum sosnoroskyi* Manden rhizosphere. *In* "Physiological-Biochemical Basis of Plant Interactions in Phytocenoses" (A. M. Grodzinsky, ed.), Vol. 4, pp. 112–115. Naukova Dumka, Kiev. (In Russian, English summary.)

Kushnir, G. P. (1973b). Biological activity of rhizosphere organisms of some plants from steppe phytocenoses. *In* "Physiological-Biochemical Basis of Plant Interactions in Phytocenoses" (A. M. Grodainsky, ed.), Vol. 4, pp. 115–118. Naukova Dumka, Kiev. (In Russian, English summary.)

Kushnir, G. P. (1977). Actinomycetes as antagonists in the rhizosphere of crysanthemum. *In* "Interactions of Plants and Microorganisms in Phytocenoses" (A. M. Grodzinsky, ed.), pp. 157–161. Naukova Dumka, Kiev. (In Russian, English summary.)

Kushnir, G. P., and Shrol, T. S. (1974). On metabolites of fungi in rhizospheres of some steppe plants. *In* "Physiological-Biochemical Basis of Plant Interactions in Phytocenoses" (A. M. Grodzinsky, ed.), Vol. 5, pp. 92–94. Naukova Dumka, Kiev. (In Russian, English summary.)

Lakhtanova, L. I. (1977). Effect of *Lupinus polyphyllus* Lindl. root exudates on certain physiological processes in *Picea excelsa* L. *In* "Interactions of Plants and Microorganisms in Phytocenoses" (A. M. Grodzinsky, ed.), pp. 86–91. Naukova Dumka, Kiev. (In Russian, English summary.)

Lalonde, M., and Quispel, A. (1977). Ultrastructural and immunological demonstration of the nodulation of the European *Alnus glutinosa* (L.) Gaertn. host plant by the North American *Alnus crispa* var. *mollis* Fern. root nodule endophyte. *Can. J. Microbiol.* **23**, 1529–1547.

Lane, F. E. (1965). Dormancy and germination in fruits of the sunflower. Ph.D. Dissertation, Univ. of Oklahoma, Norman.

Lang, D. R., and Racker, E. (1974). Effects of quercetin and F_1 inhibitor on mitochondrial ATPase and energy-linked reactions in submitochondrial particles. *Biochim. Biophys. Acta* **333**, 180–186.

Larson, M. M., and Schwarz, E. L. (1980). Allelopathic inhibition of black locust, red clover, and black alder by six common herbaceous species. *For. Sci.* **26**, 511–520.

Lastuvka, Z. (1970). Allelopathy and the processes of ion absorption and accumulation. *In* "Physiological-Biochemical Basis of Plant Interactions in Phytocenoses" (A. M. Grodzinsky, ed.), Vol. 1, pp. 37–40. Naukova Dumka, Kiev. (In Russian, English summary.)

Lastuvka, Z., and Minarz, I. (1970). Mutual effect of maize and pea in water cultures with additional nutrition. *In* "Physiological-Biochemical Basis of Plant Interactions in Phytocenoses" (A. M. Grodzinsky, ed.), Vol. 1, pp. 55–59. Naukova Dumka, Kiev. (In Russian, English summary.)

Lazauskas, P., and Balinevichiute, Z. (1972). Influence of the excretions from *Vicia villosa* Roth seeds on germination and primary growth of some crops and weeds. *In* "Physiological-Biochemical Basis of Plant Interactions in Phytocenoses" (A. M. Grodzinsky, ed.), Vol. 3, pp. 76–79. Naukova Dumka, Kiev. (In Russian, English summary.)

Leather, G. R. (1982). Weed control using allelopathic crop plants. *North Am. Symp. Allelopathy Nov. 14–17, Urbana-Champaign, Illinois.* (Abstr.)

Leather, G. R. (1983). Sunflowers (*Helianthus annuus*) are allelopathic to weeds. *Weed Sci.* **31**, 37–42.

Leather, J. W. (1911). Records of drainage in India. *Mem. Dep. Agric. India, Chem. Ser.* **2**, 63–140.

Lee, I. K., and Monsi, M. (1963). Ecological studies on *Pinus densiflora* forest 1. Effects of plant substances on the floristic composition of the undergrowth. *Bot. Mag.* **76**, 400–413.

Lee, I. K., Im, O. A., and Bark, M. S. (1967). Studies on the sick soil phenomena of *Setaria italica* and *Solanum melongena*. *Nuc. Eng. Stud. Ser.* **7**(No. 1, Pt.2), 39–44. (In Korean, English summary.)

Lee, T. T. (1966). Effects of hydroxybenzoic acids on oxidation of reduced nicotinamide adenine dinucleotide by enzymes from tobacco leaves. *Physiol. Plant.* **19**, 660–671.

Lee, T. T. (1977). Role of phenolic inhibitors in peroxidase-mediated degradation of indole-3-acetic acid. *Plant Physiol.* **59**, 372–375.

Lee, T. T., and Skoog, F. (1965). Effects of hydroxybenzoic acids on indoleacetic acid inactivation by tobacco callus extracts. *Physiol. Plant.* **18**, 577–585.

Lefevre, M. (1950). Compatibilités et antagonismes entre algues d'eau douce dans les collections d'eau naturelles. *Proc. Int. Assoc. Limnol.* **11**, 224–229.

Lefevre, M., and Nisbet, M. (1948). Sur la sécrétion par certaines espèces d'Algues de substances inhibitrices d'autres espèces d'Algues. *C. R. Acad. Sci.* **226**, 107–109.

Lefevre, M., Nisbet, M., and Jakob, H. (1949). Action des substances excretées, en culture, par certaines espèces d'algues, sur le metabolisme d'autres espèces d'algues. *Verh. Int. Ver. Limnol.* **10**, 259–264.

Lefevre, M., Jakob, H., and Nisbet, M. (1950). Sur la secretion, par certaines Cyanophytes, de substances algostatiques dans les collections d'eau naturelles. *C. R. Acad. Sci.* **230**, 2226–2227.

Lefevre, M., Jakob, H., and Nisbet, M. (1952). Auto- et heteroantagonisme chez les algues d'eau douce. *Ann. Stn. Cent. Hydrobiol. Appl.* **4**, 5–197.

Lehman, R. H., and Rice, E. L. (1972). Effect of deficiencies of nitrogen, potassium and sulfur on chlorogenic acids and scopolin in sunflower. *Am. Midl. Nat.* **87**, 71–80.

Lellinger, D. B. (1976). Incompatible companion plants. *Bull. Am. Fern Soc.* **3**, 1.

Leonard, R. T., and Hotchkiss, C. W. (1976). Cation-stimulated adenosine triphosphatase activity and cation absorption in corn roots. *Plant Physiol.* **58**, 331–335.

Leopold, A. C. (1955). "Auxins and Plant Growth." Univ. of California Press, Berkeley.

Le Tourneau, D., and Heggeness, H. G. (1957). Germination and growth inhibitors in leafy spurge foliage and quackgrass rhizomes. *Weeds* **5**, 12–19.

Le Tourneau, D., Failes, G. D., and Heggeness, H. G. (1956). The effect of aqueous extracts of plant tissue on germination of seeds and growth of seedlings. *Weeds* **4**, 363–368.

Leuck, E. E., II, and Rice, E. L. (1976). Inhibition of *Rhizobium* and *Azotobacter* by rhizosphere bacteria of *Aristida oligantha*. *Bot. Gaz. (Chicago)* **137**, 160–164.

Levin, D. A. (1971). Plant phenolics: An ecological perspective. *Am. Nat.* **105**, 157–181.

Levin, D. A. (1972). The role of phenolics in plant defense. *In* "A Symposium on Ecosystematics" (R. T. Allen and F. C. James, eds.), pp. 165–193. University of Arkansas Museum Occasional Paper No. 4, Fayetteville.

Levitan, H., and Barker, J. L. (1972). Membrane permeability: Cation selectivity reversibly altered by salicylate. *Science* **178**, 63–64.

Levy, G. F. (1970). The phytosociology of northern Wisconsin upland openings. *Am. Midl. Nat.* **83**, 213–237.

Lewis, J. A., and Starkey, R. L. (1968). Vegetable tannins, their decomposition and effects on decomposition of some organic compounds. *Soil Sci.* **106**, 241–247.

Lewis, J. A., and Starkey, R. L. (1969). Decomposition of plant tannins by some soil microorganisms. *Soil Sci.* **107**, 235–241.

Li, C. Y. (1974). Phenolic compounds in understory of alder, conifer, and mixed alder-conifer stands of coastal Oregon. *Lloydia* **37**, 603–607.

Li, C. Y. (1977). Conversion of *p*-coumaric acid to caffeic acid and of *p*-hydroxyphenylacetic acid to 3,4-dihydroxyphenylacetic acid by *Alnus rubra*. *Lloydia* **40**, 298–300.

Li, C. Y., Lu, K. C., Trappe, J. M., and Bollen, W. B. (1969a). A simple, quantitative method of assaying soil for inhibitory fungi. *U. S. For. Serv. Res. Note* PNW–108.

Li, C. Y., Lu, K. C., Nelson, E. E., Bollen, W. B., and Trappe, J. M. (1969b). Effect of phenolic and other compounds on growth of *Poria weirii in vitro*. *Microbios* **1**, 305–311.

Li, C. Y., Lu, K. C., Trappe, J. M., and Bollen, W. B. (1970). Separation of phenolic compounds in alkali hydrolysates of a forest soil by thin-layer chromatography. *Can. J. Soil Sci.* **50**, 458–460.

Li, C. Y., Lu, K. C., Trappe, J. M., and Bollen, W. B. (1972). *Poria weirii*-inhibiting and other phenolic compounds in roots of red alder and Douglas fir. *Microbios* **5**, 65–68.

Li, C. Y., Lu, K. C., Trappe, J. M., and Bollen, W. B. (1973). Formation of *p*-hydroxybenzoic acid from phenylacetic acid by *Poria weirii*. *Can. J. Bot.* **51**, 827–828.

Lieth, H. (1960). Patterns of change within grassland communities. *Symp. Br. Ecol. Soc.* **1**, 27–39.

Likens, G. E., Bormann, F. H., and Johnson, N. M. (1969). Nitrification: Importance to nutrient losses from a cutover forested ecosystem. *Science* **163**, 1205–1206.

Lill, R. E., and McWha, J. A. (1976). Production of ethylene by incubated litter of *Pinus radiata*. *Soil Biol. Biochem.* **8**, 61–63.

Lill, R. E., and Waid, J. S. (1975). Volatile phytotoxic substances formed by litter of *Pinus radiata*. *N. Z. J. For. Sci.* **5**, 165–170.

Lill, R. E., McWha, J. A., and Cole, A. L. J. (1979). The influence of volatile substances from incubated litter of *Pinus radiata* on seed germination. *Ann. Bot.* **43**, 81–85.

Linder, P. J., Craig, J. C., Jr., and Walton, T. R. (1957). Movement of C^{14} tagged *alpha*-methoxyphenylacetic acid out of roots. *Plant Physiol.* **32**, 572–575.

Linder, P. J., Craig, J. C., Jr., Cooper, F. E., and Mitchell, J. W. (1958). Movement of 2,3,6-trichlorobenzoic acid from one plant to another through their root systems. *J. Agric. Food Chem.* **6**, 356–357.

Lingappa, B. T., Prasad, M., and Lingappa, Y. (1969). Phenethyl alcohol and tryptophol: Autoantibiotics produced by the fungus *Candida albicans*. *Science* **163**, 192–194.

Litav, M., and Isti, D. (1974). Root competition between two strains of *Spinacia oleracea*. II. Effects of nutrient supply and non-simultaneous emergence. *J. Appl. Ecol.* **11**, 1017–1025.

Livingston, B. E. (1905). Physiological properties of bog water. *Bot. Gaz. (Chicago)* **39**, 348–355.

Loche, J., and Chouteau, J. (1963). Incidences des carences en Ca, Mg or P sur l'accumulation des polyphenol dans la feuille de tabac. *C. R. Hebd. Seances Acad. Agric. Fr.* **49**, 1017–1026.

Lockwood, J. L. (1959). *Streptomyces* spp. as a cause of natural fungitoxicity in soil. *Phytopathology* **49**, 327–334.

Lodhi, M. A. K. (1975a). Allelopathic effects of hackberry in a bottomland forest community. *J. Chem. Ecol.* **1**, 171–182.

Lodhi, M. A. K. (1975b). Soil-plant phytotoxicity and its possible significance in patterning of herbaceous vegetation in a bottomland forest. *Am. J. Bot.* **62**, 618–622.

Lodhi, M. A. K. (1976). Role of allelopathy as expressed by dominating trees in a lowland forest in controlling productivity and pattern of herbaceous growth. *Am. J. Bot.* **63**, 1–8.

Lodhi, M. A. K. (1977). The influence and comparison of individual forest trees on soil properties and possible inhibition of nitrification due to intact vegetation. *Am. J. Bot.* **64**, 260–264.

Lodhi, M. A. K. (1978a). Allelopathic effects of decaying litter of dominant trees and their associated soil in a lowland forest community. *Am. J. Bot.* **65**, 340–344.

Lodhi, M. A. K. (1978b). Comparative inhibition of nitrifiers and nitrification in a forest community as a result of the allelopathic nature of various tree species. *Am. J. Bot.* **65**, 1135–1137.

Lodhi, M. A. K. (1979a). Germination and decreased growth of *Kochia scoparia* in relation to its autoallelopathy. *Can. J. Bot.* **57**, 1083–1088.

Lodhi, M. A. K. (1979b). Allelopathic potential of *Salsola kali* L. and its possible role in rapid disappearance of weedy stage during revegetation. *J. Chem. Ecol.* **5**, 429–437.

Lodhi, M. A. K. (1979c). Inhibition of nitrifying bacteria, nitrification, and mineralization in spoil soils as related to their successional stages. *Bull. Torrey Bot. Club* **106**, 284–289.

Lodhi, M. A. K. (1981). Accelerated soil mineralization, nitrification, and revegetation of abandoned fields due to the removal of crop-soil phytotoxicity. *J. Chem. Ecol.* **7**, 685–694.

Lodhi, M. A. K., and Killingbeck, K. T. (1980). Allelopathic inhibition of nitrification and nitrifying bacteria in a ponderosa pine (*Pinus ponderosa* Dougl.) community. *Am. J. Bot.* **67**, 1423–1429.

Lodhi, M. A. K., and Nickell, G. L. (1973). Effects of leaf extracts of *Celtis laevigata* on growth, water content, and carbon dioxide exchange rates of three grass species. *Bull. Torrey Bot. Club* **100**, 159–165.

Lodhi, M. A. K., and Rice, E. L. (1971). Allelopathic effects of *Celtis laevigata*. *Bull. Torrey Bot. Club* **98**, 83–89.

Löfgren, N., Lüning, B., and Hedström, H. (1954). The isolation of nebularine and the determination of its structure. *Acta. Chem. Scand.* **8**, 670–680.

Lorber, P., and Muller, W. H. (1976). Volatile growth inhibitors produced by *Salvia leucophylla:* Effects on seedling root tip ultrastructure. *Am. J. Bot.* **63**, 196–200.

Lott, H. V. (1960). Über den Einfluss der kurzwelligen Strahlung auf die Biosynthese der pflanzlichen Polyphenole. *Planta* **55**, 480–495.

Lovett, J. V., and Duffield, A. M. (1981). Allelochemicals of *Camelina sativa*. *J. Appl. Ecol.* **18**, 283–290.

Lovett, J. V., and Jackson, H. F. (1980). Allelopathic activity of *Camelina sativa* (L.) Crantz in relation to its phylloshere bacteria. *New Phytol.* **86**, 273–277.

Lovett, J. V., and Sagar, G. R. (1978). Influence of bacteria in the phyllosphere of *Camelina sativa* (L.) Crantz on germination of *Linum usitatissimum* L. *New Phytol.* **81**, 617–625.

Lucena, J. M., and Doll, J. (1976). Efectos inhibidores de crecimiento del coquito (*Cyperus rotundus* L.) sobre sorgo y soya. *Rev. Comalfi* **3**, 241–256.

Ludwig, R. A. (1957). Toxin production by *Helminthosporium sativum* P. K. & B. and its significance in disease development. *Can. J. Bot.* **35**, 291–303.

Lundegårdh, H., and Stenlid, G. (1944). On the exudation of nucleotides and flavanone from living roots. *Ark. Bot.* **31A**, 1–27.

Lykvar, D. F., and Nazarova, N. S. (1970). On importance of legume varieties in mixed cultures with maize. *In* "Physiological-Biochemical Basis of Plant Interactions in Phytocenoses" (A. M. Grodzinsky, ed.), Vol. 1, pp. 83–88. Naukova Dumka, Kiev. (In Russian, English summary.)

Lyon, T. L., and Wilson, J. K. (1921). Liberation of organic matter by roots of growing plants. *Cornell Univ. Agric. Exp. Stn. Mem.* No. 40.

Lyon, T. L., Bizzell, J. A., and Wilson, B. D. (1923). Depressive influence of certain higher plants on the accumulation of nitrates in the soil. *J. Am. Soc. Agron.* **15,** 457–467.

McCahon, C. B., Kelsey, R. G., Sheridan, P. P., and Shafizadeh, F. (1973). Physiological effects of compounds extracted from sagebrush. *Bull. Torrey Bot. Club* **100,** 23–28.

McCalla, T. M., and Duley, F. L. (1948). Stubble mulch studies: Effect of sweetclover extract on corn germination. *Science* **108,** 163.

McCalla, T. M., and Duley, F. L. (1949). Stubble mulch studies. III. Influence of soil microorganisms and crop residues on the germination, growth, and direction of root growth of corn seedlings. *Soil Sci. Soc. Am. Proc.* **14,** 196–199.

McCalla, T. M., and Haskins, F. A. (1964). Phytotoxic substances from soil microorganisms and crop residues. *Bacteriol. Rev.* **28,** 181–207.

McClure, P. R., Gross, H. D., and Jackson, W. A. (1978). Phosphate absorption by soybean varieties: The influence of ferulic acid. *Can. J. Bot.* **56,** 764–767.

McCracken, M. D., Middaugh, R. E., and Middaugh, R. S. (1980). A chemical characterization of an algal inhibitor obtained from *Chlamydomonas. Hydrobiologia* **70,** 271–276.

MacDaniels, L. H., and Pinnow, D. L. (1976). Walnut toxicity, an unsolved problem. *67th Annu. Rep. North. Nut Grow. Assoc.* pp. 114–122.

McFee, W. W., and Stone, E. L., Jr. (1968). Ammonium and nitrate as nitrogen sources for *Pinus radiata* and *Picea glauca. Soil Sci. Soc. Am. Proc.* **32,** 879–884.

McGrath, W. T. (1972). Biological control of *Fomes annosus:* A new possibility in the United States. *Consultant* **17,** 94–96.

McKey, D., Waterman, P. G., Mbi, C. N., Gartlan, J. S., and Struhsaker, T. T. (1978). Phenolic content of vegetation in two African rain forests: Ecological implications. *Science* **202,** 61–64.

McKnight, R. S., and Lindegren, C. C. (1936). Bacteriocidal effects of vapors from crushed garlic on *Mycobacterium cepae. Proc. Soc. Exp. Biol. Med.* **35,** 477–479.

Macleod, N. J., and Pridham, J. B. 1965). Observations on the translocation of phenolic compounds. *Phytochemistry* **5,** 777–781.

McNaughton, S. J. (1968). Autotoxic feedback in relation to germination and seedling growth in *Typha latifolia. Ecology* **49,** 367–369.

McPherson, J. K., and Muller, C. H. (1969). Allelopathic effects of *Adenostoma fasciculatum,* "chamise," in the California chaparral. *Ecol. Monogr.* **39,** 177–198.

McPherson, J. K., and Thompson, G. L. (1972). Competitive and allelopathic suppression of understory by Oklahoma oak forest. *Bull. Torrey Bot. Club* **99,** 293–300.

McWhorter, C. G., and Hartwig, E. E. (1972). Competition of johnsongrass and cocklebur with six soybean varieties. *Weed Sci.* **20,** 56–59.

Maksimova, I. V., and Pimenova, M. N. (1969). Liberation of organic acids by green unicellular algae. *Microbiology* **38,** 64–70. (Transl. of *Mikrobiologiya.*)

Mallik, M. A. B., and Hussain, A. (1972). Effect of rhizosphere microflora of *Melilotus alba* on *Rhizobium meliloti. J. Sci.* **1,** 133–138.

Marigo, G., and Boudet, A. M. (1975). Rôle des polyphénols dans la croissance. Définition d'un modèle expérimental chez *Lycopersicum esculentum. Physiol. Plant.* **34,** 51–55.

Markova, S. A. (1972). Experimental investigations of the influence of oats on growth and development of *Erysimum cheiranthoides* L. *In* "Physiological-Biochemical Basis of Plant Interactions in Phytocenoses" (A. M. Grodzinsky, ed.), Vol. 3, pp. 66–68. Naukova Dumka, Kiev. (In Russian, English summary.)

Martin, D. F., Kutt, E. C., and Kim, Y. S. (1974). Use of a multiple diffusion chamber unit in culture studies. Application to *Gomphosphaeria aponina*. *Environ. Lett.* **7**(1), 39–46.

Martin, J. P. (1948). Fungus flora of some California soils in relation to slow decline of citrus trees. *Soil Sci. Soc. Am. Proc.* **12**, 209–214.

Martin, J. P. (1950). Effects of soil fungi on germination of sweet orange seeds and development of the young seedlings. *Soil Sci. Soc. Am. Proc.* **14**, 184–188.

Martin, J. P., and Ervin, J. O. (1958). Greenhouse studies on the influence of other crops and of organic materials on growth of orange seedlings in old citrus soils. *Soil Sci.* **85**, 141–147.

Martin, J. P., Aldrich, D. G., Murphy, W. S., and Bradford, G. R. (1953). Effects of soil fumigation on growth and chemical composition of citrus plants. *Soil Sci.* **75**, 137–151.

Martin, J. P., Klotz, L. J., DeWolfe, T. A., and Ervins, J. O. (1956). Influence of some common soil fungi on growth of citrus seedlings. *Soil Sci.* **81**, 259–267.

Martin, P. (1957). Die Abgabe von organischen Verbindungen insbesondere von Scopoletin aus den Keimwurzeln des Hafers. *Z. Bot.* **45**, 475–506.

Martin, P., and Rademacher, B. (1960). Studies on the mutual influences of weeds and crops. *Symp. Br. Ecol. Soc.* **1**, 143–152.

Marx, D. H. (1969a). The influence of ectotrophic mycorrhizal fungi on the resistance of pine roots to pathogenic infections. I. Antagonism of mycorrhizal fungi to root pathogenic fungi and soil bacteria. *Phytopathology* **59**, 153–163.

Marx, D. H. (1969b). The influence of ectotrophic mycorrhizal fungi on the resistance of pine roots to pathogenic infections. II. Production, identification, and biological activity of antibiotics produced by *Leucopaxillus cerealis* var. *piceina*. *Phytopathology* **59**, 411–417.

Marx, D. H. (1972). Ectomycorrhizae as biological deterrents to pathogenic root infections. *Annu. Rev. Phytopathol.* **10**, 429–454.

Marx, D. H.,and Davey, C. B. (1969a). The influence of ectotrophic mycorrhizal fungi on the resistance of pine roots to pathogenic infections. III. Resistance of aseptically formed mycorrhizae to infection by *Phytophthora cinnamomi*. *Phytopathology* **59**, 549–558.

Marx, D. H., and Davey, C. B. (1969b). The influence of ectotrophic mycorrhizal fungi on the resistance of pine roots to pathogenic infections. IV. Resistance of naturally occurring mycorrhizae to infections by *Phytophthora cinnamomi*. *Phytopathology* **59**, 559–565.

Mason, C. P., and Gleason, F. K. (1981). An antibiotic from *Scytonema hofmanni* (Cyanophyta). *J. Phycol.(Suppl.)* **17**, 8.

Massey, A. B. (1925). Antagonism of the walnuts (*Juglans nigra* L. and *J. cinera* L.) in certain plant associations. *Phytopathology* **15**, 773–784.

Mathes, M. C., Helton, E. D., and Fisher, K. D. (1971). The production of microbial-regulatory materials by isolated aspen tissue. *Plant Cell Physiol.* **12**, 593–601.

Matveev, N. M., Krisanov, G. N., and Lyzhenko, I. I. (1975). Role of plant excretions in the formation of the grass stand in black locust and smoketree plantings of the steppe zone. *Biol. Nauki (Moscow)* **18**, 80–84. (In Russian.)

Mautner, H. G., Gardner, G. M., and Pratt, R. (1953). Antibiotic activity of seaweed extracts. II. *Rhodomela larix*. *J. Am. Pharm. Assoc., Sci. Ed.* **42**, 294–296.

Meiklejohn, J. (1962). Microbiology of the nitrogen cycle in some Ghana soils. *Emp. J. Exp. Agric.* **30**, 115–126.

Meiklejohn, J. (1968). Numbers of nitrifying bacteria in some Rhodesian soils under natural grass and improved pasture. *J. Appl. Ecol.* **5**, 291–300.

Meissner, R., Nel, P. C., and Smit, N. S. H. (1979). Influence of red nutgrass (*Cyperus rotundus*) on growth and development of some crop plants. *Proc. 3rd Natl. Weed Conf. S. Afr.* pp. 39–52.

Melin, E. (1963). Some effects of forest tree roots on mycorrhizal Basidiomycetes. *In* "Symbiotic Associations" (P. S. Nutman and B. Mosse, eds.), pp. 125–145. Cambridge Univ. Press, London and New York.

Mensah, K. O. A. (1972). Allelopathy as expressed by sugar maple on yellow birch. *Diss. Abstr. B.* **33**(5), 1877.

Menzies, J. D., and Gilbert, R. G. (1967). Responses of soil microflora to volatile components in plant residues. *Soil Sci. Soc. Am. Proc.* **31**, 495–496.

Mergen, F. (1959). A toxic principle in the leaves of *Ailanthus. Bot. Gaz. (Chicago)* **121**, 32–36.

Miller, D. A. (1982). Allelopathic effects of alfalfa. *North Am. Symp. Allelopathy, Nov. 14–17, Urbana-Champaign, Illinois.* (Abstr.)

Miller, H. E., Mabry, T. J., Turner, B. L., and Payne, W. W. (1968). Infraspecific variation of sesquiterpene lactones in *Ambrosia psilostachya* (Compositae). *Am. J. Bot.* **55**, 316–324.

Millhollon, R. W. (1970). MSMA for Johnsongrass control in sugarcane. *Weed Sci.* **18**, 333–336.

Mills, W. R. (1953). Nitrate accumulation in Uganda soils. *East Afr. Agric. J.* **19**, 53–54.

Minamikawa, T., Akazawa, T., and Uritani, I. (1963). Analytical study of umbelliferone and scopoletin synthesis in sweet potato roots infected by *Ceratocystis fimbriata. Plant Physiol.* **38**, 493–497.

Minamikawa, T., Jayasankar, N. P., Bohm, B. A., Taylor, I. E. P., and Towers, G. H. N. (1970). An inducible hydrolase from *Aspergillus niger,* acting on carbon-carbon bonds, for phlorrhizin and other C-acylated phenols. *Biochem. J.* **116**, 889–897.

Minar, J. (1974). The effect of couch grass on the growth and mineral uptake of wheat. *Folia Fac. Sci. Nat. Univ. Purkynianae Brun.* **15**, 1–84.

Mishra, R. R., and Kanaujia, R. S. (1972). Investigations into rhizosphere mycoflora XIII. Effect of foliar application of certain plant extracts on *Pennisetum typhoides* f. *burm* Stapf and Hubb. *Isr. J. Agric. Res.* **22**, 3–9.

Mishra, R. R., and Pandey, K. K. (1974). Studies on soil fungistasis. V. Effect of temperature, moisture content, and incubation period. *Indian Phytopathol.* **27**, 475–479.

Mishra, R. R., and Pandey, K. K. (1975). Studies on soil fungistasis. IV. Effect of physico-chemical characters and soil fungal flora on fungistasis. *Ann. Edafol. Agrobiol.* **34**, 423–428.

Mitchell, J. W., Smale, B. C., and Preston, W. H., Jr. (1959). New plant regulators that exude from roots. *J. Agric. Food Chem.* **7**, 841–843.

Mitin, V. V. (1970). On study of chemical nature of growth inhibitors in the dead leaves of hornbean and beech. *In* "Physiological-Biochemical Basis of Plant Interactions in Phytocenoses"(A. M. Grodzinsky, ed.), Vol. 1, pp. 177–181. Naukova Dumka, Kiev. (In Russian, English summary.)

Mitin, V. V. (1971). The water-soluble inhibitors of seed germination from beech (*Fagus silvatica* L.) autumn leaves. *In* "Physiological-Biochemical Basis of Plant Interactions in Phytocenoses" (A. M. Grodzinsky, ed.), Vol. 2, pp. 22–25. Naukova Dumka, Kiev. (In Russian, English summary.)

Moebus, K. (1972a). Seasonal changes in antibacterial activity of North Sea water. *Mar. Biol.* **13**, 1–13.

Moebus, K. (1972b). Bacteriocidal properties of natural and synthetic sea water as influenced by addition of low amounts of organic matter. *Mar. Biol.* **15**, 81–88.

Mohnot, K., and Soni, S. (1976). Observations on the allelochemic factor in air-dried leaves of *Salvadora oleoides. Comp. Physiol. Ecol.* **1**, 125–128.

Mohnot, K., and Soni, S. (1977). Ecophysiological studies of desert plants. II. Growth retarding factor in air-dried stem of *Solanum surattense* Burm. F. *Comp. Physiol. Ecol.* **2**, 97–100.

Moleski, F. L. (1976). Condensed tannins in soil: Inputs and effects on microbial populations. Ph.D. Diss., Univ. of Oklahoma, Norman. *Diss. Abstr.* 76-15816.

Molisch, H. (1937). "Der Einfluss einer Pflanze auf die andere-Allelopathie." Fischer, Jena.

Monahan, T. J., and Trainor, F. R. (1970). Stimulatory properties of filtrate from the green alga *Hormotila blennista.* I. Description. *J. Phycol.* **6**, 263–269.

Monahan, T. J., and Trainor, F. R. (1971). Stimulatory properties of filtrate from the green alga *Hormotila blennista*. II. Fractionation of filtrate. *J. Phycol.* **7**, 170–176.

Moore, C. W. E., and Keraitis, K. (1971). Effect of nitrogen source on growth of eucalypts in sand culture. *Aust. J. Bot.* **19**, 125–141.

Moore, D. R. E., and Waid, J. S. (1971). The influence of washings of living roots on nitrification. *Soil Biol. Biochem.* **3**, 69–83.

Moore, L. W., and Warren, G. (1979). *Agrobacterium radiobacter* strain 84 and biological control of crown gall. *Annu. Rev. Phytopathol.* **17**, 163–179.

Mountain, W. B., and Boyce, H. R. (1958). The peach replant problem in Ontario. VI. The relation of *Pratylenchus penetrans* to the growth of young peach trees. *Can. J. Bot.* **36**, 135–151.

Mountain, W. B., and Patrick, Z. A. (1959). The peach replant problem in Ontario. VII. The pathogenicity of *Pratylenchus penetrans* (Cobb, 1917) Filip. and Stek, 1944. *Can. J. Bot.* **37**, 459–470.

Muir, A. D., and Majak, W. (1981). Knapweed allelopathy project: Progress report, Sept., 1981. *Res. Stn. Agric. Can., Kamloops, B.C.*

Mulder, E. G. (1954). Molybdenum in relation to growth of higher plants and microorganisms. *Plant Soil* **5**, 368–415.

Muller, C. H. (1965).Inhibitory terpenes volatilized from *Salvia* shrubs. *Bull. Torrey Bot. Club* **92**, 38–45.

Muller, C. H. (1966). The role of chemical inhibition (allelopathy) in vegetational composition. *Bull. Torrey Bot. Club* **93**, 332–351.

Muller, C. H. (1969). Allelopathy as a factor in ecological process. *Vegetatio* **18**, 348–357.

Muller, C. H., and del Moral, R. (1966). Soil toxicity induced by terpenes from *Salvia leucophylla*. *Bull. Torrey Bot. Club* **93**, 130–137.

Muller, C. H., Muller, W. H., and Haines, B. L. (1964). Volatile growth inhibitors produced by shrubs. *Science* **143**, 471–473.

Muller, C. H., Hanawalt, R. B., and McPherson, J. K. (1968). Allelopathic control of herb growth in the fire cycle of California chaparral. *Bull. Torrey Bot. Club* **95**, 225–231.

Muller, W. H. (1965). Volatile materials produced by *Salvia leucophylla:* effects on seedling growth and soil bacteria. *Bot. Gaz. (Chicago)* **126**, 195–200.

Muller, W. H., and Muller, C. H. (1964). Volatile growth inhibitors produced by *Salvia* species. *Bull. Torrey Bot. Club* **91**, 327–330.

Muller, W. H., Lorber, P., Haley, B., and Johnson, K. (1969). Volatile growth inhibitors produced by *Salvia leucophylla:* Effect on oxygen uptake by mitochondrial suspensions. *Bull. Torrey Bot. Club* **96**, 89–96.

Munro, P. E. (1966a). Inhibition of nitrite-oxidizers by roots of grass. *J. Appl. Ecol.* **3**, 227–229.

Munro, P. E. (1966b). Inhibition of nitrifiers by grass root extracts. *J. Appl. Ecol.* **3**, 231–238.

Murphy, T. P., Lean, D. R. S., and Nalewajko, C. (1976). Blue-green algae: Their excretion of Fe-selective chelators enables them to dominate other algae. *Science* **192**, 900.

Murthy, M. S., and Nagodra, T. (1977). Allelopathic effects of *Aristida adscensionis* on *Rhizobium*. *J. Appl. Ecol.* **14**, 279–282.

Murthy, M. S., and Ravindra, R. (1974). Inhibition of nodulation of *Indifogera cordifolia* by *Aristida adscensionis*. *Oecologia* **16**, 257–258.

Murthy, M. S., and Ravindra, R. (1975). Allelopathic effects of *Aristida adscensionis*. *Oecologia* **18**, 243–250.

Nadkernichnyi, S. P. (1974). On the problem of distribution of toxin producing microscopic fungi in soddy-medium podzolic soil under some farm crops. *In* "Physiological-Biochemical Basis of Plant Interactions in Phytocenoses" (A. M. Grodzinsky, ed.), Vol. 5, pp. 97–100. Naukova Dumka, Kiev. (In Russian, English summary.)

Nalewajko, C., and Lean, D. R. S. (1972). Growth and excretion in planktonic algae and bacteria. *J. Phycol.* **8**, 361–366.

Naqvi, H. H. (1972). Preliminary studies of interference exhibited by Italian ryegrass. *Biologia (Lahore)* **18**, 201–210.

Naqvi, H. H., and Muller, C. H. (1975). Biochemical inhibition (allelopathy) exhibited by Italian ryegrass (*Lolium multiflorum* L.). *Pak. J. Bot.* **7**, 139–147.

Neal, J. L., Jr. (1969). Inhibition of nitrifying bacteria by grass and forb root extracts. *Can. J. Bot.* **15**, 633–635.

Neill, R. L., and Rice, E. L. (1971). Possible role of *Ambrosia psilostachya* on patterning and succession in old-fields. *Am. Midl. Natr.* **86**, 344–357.

Neish, A. C. (1964). Major pathways of biosynthesis of phenols. *In* "Biochemistry of Phenolic Compounds" (J. B. Harborne, ed.), pp. 295–359. Academic Press, New York.

Neustruyeva, S. N., and Dobretsova, T. N. (1972). Influence of some summer crops on white goosefoot. *In* "Physiological-Biochemical Basis of Plant Interactions in Phytocenoses" (A. M. Grodzinsky, ed.), Vol. 3, pp. 68–73. Naukova Dumka, Kiev. (In Russian, English summary.)

New, P. B., and Kerr, A. (1972). Biological control of crown gall: Field measurements and glass-house experiments. *J. Appl. Bacteriol.* **35**, 279–287.

Newman, E. I., and Miller, M. H. (1977). Allelopathy among some British grassland species II. Influence of root exudates on phosphorus uptake. *J. Ecol.* **65**, 399–401.

Newman, E. I., and Rovira, A. D. (1975). Allelopathy among some British grassland species. *J. Ecol.* **63**, 727–737.

Nickell, L. G. (1960). Antimicrobial activity of vascular plants. *Econ. Bot.* **13**, 281–318.

Nielsen, K. F., and Cunningham, R. K. (1964). The effects of soil temperature and form and level of nitrogen on growth and chemical composition of Italian rye-grass. *Soil Sci. Soc. Am. Proc.* **28**, 213–218.

Nielsen, K. F., Cuddy, T., and Woods, W. (1960). The influence of the extract of some crops and soil residues on germination and growth. *Can. J. Plant. Sci.* **40**, 188–197.

Nierenstein, M. (1934). "The Natural Organic Tannins." Churchill, London.

Norby, R. J., and Kozlowski, T. T. (1980). Allelopathic potential of ground cover species on *Pinus resinosa* seedlings. *Plant Soil* **57**, 363–374.

Norstadt, F. A., and McCalla, T. M. (1963). Phytotoxic substance from a species of *Pencillium*. *Science* **140**, 410–411.

Nozawa, K. (1968). The effect of *Peridinium* toxin on other algae. *Bull. Misaki Mar. Biol. Inst., Kyoto Univ.* No. 12, pp. 21–24.

Numata, M., Kobayashi, A., and Ohga, N. (1973). Studies on allelopathic substances concerning the formation of the urban flora. *In* "Fundamental Studies in the Characteristics of Urban Eco-systems" (M. Numata, ed.), pp. 59–64.

Numata, M., Kobayashi, A., and Ohga, N. (1974). Studies on allelopathic substances concerning the formation of the urban flora. *In* "Studies in Urban Ecosystems" (M. Numata, ed.), pp. 22–25.

Numata, M., Kobayashi, A., and Ohga, N. (1975). Studies on the role of allelopathic substances. *In* "Studies in Urban Ecosystems" (M. Numata, ed.), pp. 38–41.

Nye, R. H., and Greenland, D. J. (1960). The soil under shifting cultivation. *Tech. Commun. Commonw. Bur. Soils* No. 51, pp. 1–156.

Oborn, E. T., Moran, W. T., Greene, K. T., and Bartley, T. R. (1954). Weed control investigations on some important aquatic plants which impede the flow of western irrigation waters. Joint Lab. Rept. SI-2, U. S. Dep. Inter. Bur. of Reclam., U. S. Dept. Agric., Agric. Res. Serv., Denver, Colorado. 84 pp.

Odunfa, V. S. A. (1978). Root exudation in cowpea and sorghum and the effect on spore germination and growth of some soil fusaria. *New Phytol.* **80**, 607–612.

Oelke, E. A., and Morse, M. D. (1968). Propanil and molinate for control of barnyardgrass in water seeded rice. *Weed Sci.* **16**, 235–239.

Oertli, J. J. (1963). Effect of the form of nitrogen and pH on growth of blueberry plants. *Agron. J.* **55,** 305–307.

Ohman, J., and Kommedahl, T. (1960). Relative toxicity of extracts from vegetative organs of quackgrass (*Agropyron repens* L.) to alfalfa. *Weeds* **8,** 666–670.

Ohman, J., and Kommedahl, T. (1964). Plant extracts, residues, and soil minerals in relation to competition of quackgrass with oats and alfalfa. *Weeds* **12,** 222–231.

Okafor, L. I., and DeDatta, S. K. (1976). Competition between upland rice and purple nutsedge for nitrogen, moisture, and light. *Weed Sci.* **24,** 43–46.

Oleksevich, V. M. (1970). On the allelopathic activity of trees and shrubs used for landscape gardening. *In* "Physiological-Biochemical Basis of Plant Interactions in Phytocenoses" (A. M. Grodzinsky, ed.), Vol. 1, pp. 186–190. Naukova Dumka, Kiev. (In Russian, English summary.)

Oliver, L. R. (1979). Influence of soybean (*Glycine max*) planting data on velvetleaf (*Abutilon theophrasti*) competition. *Weed Sci.* **27,** 183–188.

Olmsted, C. E., III, and Rice, E. L. (1970). Relative effects of known plant inhibitors on species from first two stages of old-field succession. *Southwest. Nat.* **15,** 165–173.

Olney, H. O. (1968). Growth substances from *Veratrum tenuipetalum*. *Plant Physiol.* **43,** 293–302.

Olsen, R. A. (1973a). Triterpeneglycosides as inhibitors of fungal growth and metabolism. V. Role of the sterol contents of some fungi. *Physiol. Plant.* **28,** 507–515.

Olsen, R. A. (1973b). Triterpenglycosides as inhibitors of fungal growth and metabolism. VI. The effect of aescin on fungi with reduced sterol contents. *Physiol. Plant.* **29,** 145–149.

Olsen, R. A. (1974). Triterpeneglycosides as inhibitors of fungal growth and metabolism. VII. The effect of aescin on the utilization of glucose and sucrose. *Physiol. Plant.* **30,** 279–282.

Olsen, R. A. (1975). Triterpeneglycosides as inhibitors of fungal growth and metabolism. VIII. Induced leakage of nucleotide materials. *Physiol. Plant.* **33,** 75–82.

Olsen, R. A., Odham, G., and Lindberg, G. (1971). Aromatic substances in leaves of *Populus tremula* as inhibitors of mycorrhizal fungi. *Physiol. Plant.* **25,** 122–129.

Overland, L. (1966). The role of allelopathic substances in the "smother crop" barley. *Am. J. Bot.* **53,** 423–432.

Owens, L. D. (1969). Toxins in plant disease: Structure and mode of action. *Science* **165,** 18–25.

Owens, L. D. (1973). Herbicidal potential of rhizobitoxine. *Weed Sci.* **21,** 63–66.

Owens, L. D., Lieberman, M., and Kunishi, A. (1971). Inhibition of ethylene production by rhizobitoxine. *Plant Physiol.* **48,** 1–4.

Owens, L. D., Thompson, J. F., and Fennessey, P. V. (1972). Dihydrorhizobitoxine, a new ether amino-acid from *Rhizobium japonicum*. *J. Chem. Soc., Chem. Commun.* p. 715.

Padron, J., Grist, K. L., Clark, J. B., and Wender, S. H. (1960). Specificity studies on an extracellular enzyme preparation obtained from quercetin grown cells of *Aspergillus*. *Biochem. Biophy. Res. Commun.* **3,** 412–416.

Panchuk, M. A., and Prutenskaya, N. I. (1973). On the problem of the presence of allelopathic properties in wheat-wheat grass hybrids and their initial forms. *In* "Physiological-Biochemical Basis of Plant Interactions in Phytocenoses" (A. M. Grodzinsky, ed.), Vol. 4, pp. 44–47. Naukova Dumka, Kiev. (In Russian, English summary.)

Pandya, S. M. (1975). Effect of *Celosia argentea* extracts on root and shoot growth of bajra seedlings. *Geobios (Jodhpur)* **2,** 175–178.

Pandya, S. M. (1976). Effect of *Celosia argentea* Linn. extracts on dry weight of bajra seedlings. *Geobios (Jodhpur)* **3,** 137–138.

Pandya, S. M. (1977). On the relative nature of the inhibiting effects of *Celosia argentea* Linn. of different ages. *Sci. Culture* **43,** 343–344.

Pandya, S. M., and Pota, K. B. (1978). On the allelopathic potentials of root exudates from different ages of *Celosia argentea* Linn., a weed. *Natl. Acad. Sci. Lett.* **1,** 56–58.

Panova, L. S. (1977). Allelopathic activity of plants in the steppe phytocenoses of the Kamennyje Mogily reservation. I. Activity of root exudates. *In* "Interactions of Plants and Microorganisms in Phytocenoses" (A. M. Grodzinsky, ed.), pp. 131–137. Naukova Dumka, Kiev. (In Russian, English summary.)

Pareek, R. P., and Gaur, A. C. (1973). Organic acids in the rhizosphere of *Zea mays* and *Phaseolus aureus* plants. *Plant Soil* **39**, 441–444.

Parenti, R. L., and Rice, E. L. (1969). Inhibitional effects of *Digitaria sanguinalis* and possible role in old-field succession. *Bull. Torrey Bot. Club* **96**, 70–78.

Parker, V. T., and Muller, C. H. (1979). Allelopathic dominance by a tree-associated herb in a California annual grassland. *Oecologia* **37**, 315–320.

Parks, J. M., and Rice, E. L. (1969). Effects of certain plants of old-field succession on the growth of blue-green algae. *Bull. Torrey Bot. Club* **96**, 345–360.

Parpiev, Y. P. (1971). Influence of excretions of seeds and fall of some tree shrubby species of middle Asia deserts on seed germination of subcrown plants. *In* "Physiological-Biochemical Basis of Plant Interactions in Phytocenoses" (A. M. Grodzinsky, ed.), Vol. 2, pp. 46–51. Naukova Dumka, Kiev. (In Russian, English summary.)

Patrick, Z. A. (1955). The peach replant problem in Ontario. II. Toxic substances from microbial decomposition products of peach root residues. *Can. J. Bot.* **33**, 461–486.

Patrick, Z. A. (1971). Phytotoxic substances associated with the decomposition in soil of plant residues. *Soil Sci.* **111**, 13–18.

Patrick, Z. A., and Koch, L. W. (1958). Inhibition of respiration, germination, and growth by substances arising during the decomposition of certain plant residues in soil. *Can. J. Bot.* **36**, 621–647.

Patrick, Z. A., and Koch, L. W. (1963). The adverse influence of phytotoxic substances from decomposing plant residues on resistance of tobacco to black root rot. *Can. J. Bot.* **41**, 747–758.

Patrick, Z. A., Toussoun, T. A., and Snyder, W. C. (1963). Phytotoxic substances in arable soils associated with decomposition of plant residues. *Phytopathology* **53**, 152–161.

Patrick, Z. A., Toussoun, T. A., and Koch, L. W. (1964). Effect of crop residue decomposition products on plant roots. *Annu. Rev. Phytopathol.* **2**, 267–292.

Patterson, D. T. (1981). Effects of allelopathic chemicals on growth and physiological responses of soybean (*Glycine max*). *Weed Sci.* **29**, 53–59.

Patterson, G. M. L., Harris, D. O., and Cohen, W. S. (1979). Inhibition of photosynthetic and mitochondrial electron transport by a toxic substance isolated from the alga *Pandorina morum*. *Plant Sci. Lett.* **15**, 293–300.

Patterson, M. G., Buchanan, G. A., Street, J. E., and Crowley, R. H. (1980). Yellow nutsedge (*Cyperus esculentus*) competition with cotton (*Gossypium hirsutum*). *Weed Sci.* **28**, 327–329.

Paul, E. A., Myers, R. J. K., and Price, W. A. (1971). Nitrogen fixation in grassland and associated cultivated ecosystems. *In* "Biological Nitrogen Fixation in Natural and Agricultural Habitats" (T. A. Lie and E. G. Mulder, eds.), pp. 495–507. Nijhoff, The Hague.

Peet, M., Anderson, R., and Adams, M. S. (1975). Effect of fire on big bluestem production. *Am. Midl. Nat.* **94**, 15–26.

Perry, G. S. (1932). Some tree antagonisms. *Proc. Pa. Acad. Sci.* **6**, 136–141.

Persidsky, D. J., Loewenstein, H., and Wilde, S. A. (1965). Effect of extracts of prairie soils and prairie grass roots on the respiration of ectotrophic mycorrhizae. *Agron. J.* **57**, 311–312.

Peters, E. J. (1968). Toxicity of tall fescue to rape and birdsfoot trefoil seeds and seedlings. *Crop Sci.* **8**, 650–653.

Peterson, E. B. (1965). Inhibition of black spruce primary roots by a water-soluble substance in *Kalmia angustifolia*. *For. Sci.* **11**, 473–479.

Peterson, G. B. (1972). Determination of the presence, location, and allelopathic effects of substances produced by *Juniperus scopulorum* Sarg. *Diss. Abstr. B* **32**(7), 3811–3812.

Petranka, J. W., and McPherson, J. K. (1979). The role of *Rhus copallina* in the dynamics of the forest-prairie ecotone in north-central Oklahoma. *Ecology* **60**, 956–965.

Petrova, A. G. (1977). Effect of phytoncides from soybean, gram chick-pea and bean on the uptake of phosphorus by maize. *In* "Interactions of Plants and Microorganisms in Phytocenoses" (A. M. Grodzinsky, ed.), pp. 91–97. Naukova Dumka, Kiev. (In Russian, English summary.)

Petrushenko, V. V., Kovalenko, S. G., and Kostuchek, M. N. (1974). On changes in electrophysiological parameters of plants under the influence of allelopathic water-soluble exudates. *In* "Physiological-Biochemical Basis of Plant Interactions in Phytocenoses" (A. M. Grodzinsky, ed.), Vol. 5, pp. 32–36. Naukova Dumka, Kiev. (In Russian, English summary.)

Pharis, R. P., Barnes, R. L., and Naylor, A. W. (1964). Effects of nitrogen level, calcium level, and nitrogen source upon growth and composition of *Pinus taeda* L. *Physiol. Plant.* **17**, 560–572.

Pickering, S. V. (1917). The effect of one plant on another. *Ann. Bot.* **31**, 181–187.

Pickering, S. V. (1919). The action of one crop on another. *J. R. Hort. Soc.* **43**, 372–380.

Plinius Secundus, C. (1 A.D.) "Natural History," 10 Vols. Engl. transl. by H. Rackam, W. H. S. Jones, and D. E. Eichholz, Harvard Univ. Press, Cambridge, Massachusetts, 1938–1963.

Pratt, D. M. (1966). Competition between *Skeletonema costatum* and *Olisthodiscus luteus* in Narragansett Bay and in culture. *Limnol. Oceanogr.* **11**, 447–455.

Pratt, R. (1940). Influence of the size of the inoculum on the growth of *Chlorella vulgaris* in freshly prepared culture medium. *Am. J. Bot.* **27**, 52–56.

Pratt, R. (1942). Studies on *Chlorella vulgaris*. V. Some properties of the growth inhibitor formed by *Chlorella* cells. *Am. J. Bot.* **29**, 142–148.

Pratt, R. (1944). Studies on *Chlorella vulgaris*. IX. Influence on the growth of *Chlorella* of continuous removal of chlorellin from the solution. *Am. J. Bot.* **31**, 418–421.

Pratt, R. (1948). Studies on *Chlorella vulgaris*. XI. Relation between surface tension and accumulation of chlorellin. *Am. J. Bot.* **35**, 634–637.

Pratt, R., and Fong, J. (1940). Studies on *Chlorella vulgaris*. II. Further evidence that *Chlorella* cells form a growth-inhibiting substance. *Am. J. Bot.* **27**, 431–436.

Preston, W. H., Jr., Mitchell, J. W., and Reeve, W. (1954). Movement of *alpha*-methoxyphenylacetic acid from one plant to another through their root systems. *Science* **119**, 437–438.

Priester, D. S., and Pennington, M. T. (1978). Inhibitory effects of broomsedge extracts on the growth of young loblolly pine seedlings. *U. S. For. Ser. Res. Pap.* SE-182.

Proctor, V. W. (1957a). Some controlling factors in the distribution of *Haematococcus pluvialis*. *Ecology* **38**, 457–462.

Proctor, V. W. (1957b). Studies of algal antibiosis using *Haematococcus* and *Chlamydomonas*. *Limnol. Oceanogr.* **2**, 125–139.

Proebsting, E. L. (1950). A case history of a "peach replant" situation. *Proc. Am. Soc. Hort. Sci.* **56**, 46–48.

Proebsting, E. L., and Gilmore, A. E. (1941). The relation of peach root toxicity to the re-establishing of peach orchards. *Proc. Am. Soc. Hort. Sci.* **38**, 21–26.

Pronin, V. A., and Yakovlev, A. A. (1970). Influence of nutrition conditions and rhizospheric microorganisms on the interrelations of maize and fodder beans in mixed culture. *In* "Physiological-Biochemical Basis of Plant Interactions in Phytocenoses" (A. M. Grodzinsky, ed.), Vol. 1, pp. 93–101. Naukova Dumka, Kiev. (In Russian, English summary.)

Prutenskaya, N. I. (1972). Presence of inhibitors and stimulators of *Sinapis arvensis* L. in germinating seeds of cultivated plants. *In* "Physiological-Biochemical Basis of Plant Interactions in Phytocenoses" (A. M. Grodzinsky, ed.), Vol. 3, pp. 73–75. Naukova Dumka, Kiev. (In Russian, English summary.)

Prutenskaya, N. I. (1974). Peculiarities of interaction between *Sinapis arvensis* L. and cultivated plants. *In* ''Physiological-Biochemical Basis of Plant Interactions in Phytocenoses'' (A. M. Grodzinsky, ed.), Vol. 5, pp. 66–68. Naukova Dumka, Kiev. (In Russian, English summary.)

Prutenskaya, N. I., Yurchak, L. D., and Soroka, M. A. (1970). Physiologically active substances of microorganisms and decomposing plant residue *In* ''Physiological-Biochemical Basis of Plant Interactions in Phytocenoses'' (A. M. Grodzinsky, ed.), Vol. 1, pp. 218–222. Naukova Dumka, Kiev. (In Russian, English summary.)

Purchase, B. S. (1974). The influence of phosphate deficiency on nitrification. *Plant Soil* **41**, 541–547.

Putnam, A. R., and DeFrank, J. (1979). Use of cover crops to inhibit weeds. *Proc. IX Int. Cong. Plant Prot.* pp. 580–582.

Putnam, A. R., and DeFrank, J. (1983). Use of phytotoxic plant residues for selective weed control. *Crop Prot.* **2**, 173–181.

Putnam, A. R., and Duke, W. B. (1974). Biological suppression of weeds: evidence for allelopathy in accessions of cucumber. *Science* **185**, 370–372.

Putnam, A. R., and Duke, W. B. (1978). Allelopathy in agroecosystems. *Annu. Rev. Phytopathol.* **16**, 431–451.

Putnam, A. R., Hassett, K., Fobes, J. F., and Hodupp, R. M. (1983). Evidence for allelopathy and autotoxicity in *Asparagus officinalis* L. *Plant Sci. Lett.* In press.

Quarterman, E. (1973). Allelopathy in cedar glade plant communities. *J. Tenn. Acad. Sci.* **48**, 147–150.

Quinn, J. A. (1974). *Convolvulus sepium* in old field succession on the New Jersey Piedmont. *Bull. Torrey Bot. Club* **101**, 89–95.

Ragan, M. A., and Craigie, J. S. (1978). Phenolic compounds in brown and red algae. *In* ''Handbook of Phycological Methods: Physiological and Biochemical Methods'' (J. A. Hellebust and J. S. Craigie, eds.), pp. 157–179. Cambridge Univ. Press, London and New York.

Rajan, L. (1973). Growth inhibitor(s) from *Parthenium hysterophorus* L. *Curr. Sci.* **42**, 729–730.

Rakhteenko, I. N., Kaurov, I. A., and Minko, I. T. (1973a). Effect of water-soluble metabolites of a series of crops on some physiological processes. *In* ''Physiological-Biochemical Basis of Plant Interactions in Phytocenoses'' (A. M. Grodzinsky, ed.), Vol. 4, 23–26. Naukova Dumka, Kiev. (In Russian, English summary.)

Rakhteenko, I. N., Kaurov, I. A., and Minko, I. F. (1973b). On the problem of exchange of root excretions in some agricultural plants in agrophytocenoses. *In* ''Physiological-Biochemical Basis of Plant Interactions in Phytocenoses'' (A. M. Grodzinsky, ed.), Vol. 4, pp. 16–19. Naukova Dumka, Kiev. (In Russian, English summary.)

Rao, V. R., Rao, N. S. S., and Mukerji, K. G. (1973). Inhibition of *Rhizobium in vitro* by non-nodulating legume roots and root extracts. *Plant Soil* **39**, 449–452.

Rasmussen, J. A., and Einhellig, F. A. (1975). Non-competitive effects of common milkweed, *Asclepias syriaca* L., on germination and growth of sorghum. *Am. Midl. Nat.* **94**, 478–483.

Rasmussen, J. A., and Einhellig, F. A. (1977). Synergistic inhibitory effects of *p*-coumaric and ferulic acids on germination and growth of grain sorghum. *J. Chem. Ecol.* **3**, 197–205.

Rasmussen, J. A., and Einhellig, F. A. (1979a). Allelochemic effects of leaf extracts of *Ambrosia trifida* (Compositae). *Southwest. Nat.* **24**, 637–644.

Rasmussen, J. A., and Einhellig, F. A. (1979b). Inhibitory effects of combinations of three phenolic acids on grain sorghum germination. *Plant Sci. Lett.* **14**, 69–74.

Rasmussen, J. A., and Rice, E. L. (1971). Allelopathic effects of *Sporobolus pyramidatus* on vegetational patterning. *Am. Midl. Nat.* **86**, 309–326.

Rice, E. L. (1964). Inhibition of nitrogen-fixing and nitrifying bacteria by seed plants. I. *Ecology* **45**, 824–837.

Rice, E. L. (1965a). Inhibition of nitrogen-fixing and nitrifying bacteria by seed plants. II. Characterization and identification of inhibitors. *Physiol. Plant.* **18**, 255–268.

Rice, E. L. (1965b). Inhibition of nitrogen-fixing and nitrifying bacteria by seed plants. III. Comparison of three species of *Euphorbia. Proc. Okla. Acad. Sci.* **45**, 43–44.

Rice, E. L. (1965c). Inhibition of nitrogen-fixing and nitrifying bacteria by seed plants. IV. The inhibitors produced by *Ambrosia elatior* L. and *Ambrosia psilostachya* DC. *Southwest. Nat.* **10**, 248–255.

Rice, E. L. (1968). Inhibition of nodulation of inoculated legumes by pioneer plant species from abandoned fields. *Bull. Torrey Bot. Club* **95**, 346–358.

Rice, E. L. (1969). Inhibition of nitrogen-fixing and nitrifying bacteria by seed plants. VI. Inhibitors from *Euphorbia supina* Raf. *Physiol. Plant.* **22**, 1175–1183.

Rice, E. L. (1971a). Some possible roles of inhibitors in old-field succession. *In* "Biochemical Interactions among Plants" (Environ. Physiol. Subcomm., U. S. Natl. Comm. for IBP, eds.), pp. 128–132. Natl. Acad. Sci., Washington, D. C.

Rice, E. L. (1971b). Inhibition of nodulation of inoculated legumes by leaf leachates from pioneer plant species from abandoned fields. *Am. J. Bot.* **58**, 368–371.

Rice, E. L. (1972). Allelopathic effects of *Andropogon virginicus* and its persistence in old fields. *Am. J. Bot.* **59**, 752–755.

Rice, E. L. (1974). "Allelopathy." Academic Press, New York.

Rice, E. L. (1976). Allelopathy and grassland improvement. *In* "The Grasses and Grasslands of Oklahoma" (J. R. Estes and R. J. Tyrl, eds.), pp. 90–111. Noble Foundation, Ardmore, Oklahoma.

Rice, E. L., and Pancholy, S. K. (1972). Inhibition of nitrification by climax ecosystems. *Am. J. Bot.* **59**, 1033–1040.

Rice, E. L., and Pancholy, S. K. (1973). Inhibition of nitrification by climax ecosystems. II. Additional evidence and possible role of tannins. *Am. J. Bot.* **60**, 691–702.

Rice, E. L., and Pancholy, S. K. (1974). Inhibition of nitrification by climax ecosystems. III. Inhibitors other than tannins. *Am. J. Bot.* **61**, 1095–1103.

Rice, E. L., and Parenti, R. L. (1967). Inhibition of nitrogen-fixing and nitrifying bacteria by seed plants. V. Inhibitors produced by *Bromus japonicus* Thunb. *Southwest. Nat.* **12**, 97–103.

Rice, E. L., and Parenti, R. L. (1978). Causes of decreases in productivity in undisturbed tall grass prairie. *Am. J. Bot.* **65**, 1091–1097.

Rice, E. L., Penfound, W. T., and Rohrbaugh, L. M. (1960). Seed dispersal and mineral nutrition in succession in abandoned fields in central Oklahoma. *Ecology* **41**, 224–228.

Rice, E. L., Lin, C. Y., and Huang, C. Y. (1980). Effects of decaying rice straw on growth and nitrogen fixation of a bluegreen alga. *Bot. Bull. Acad. Sin.* **21**, 111–117.

Rice, E. L., Lin, C. Y., and Huang, C. Y. (1981). Effects of decomposing rice straw on growth of and nitrogen fixation by *Rhizobium. J. Chem. Ecol.* **7**, 333–344.

Rice, T. R. (1954). Biotic influences affecting population growth of planktonic algae. *U. S., Fish Wildl. Serv., Fish Bull.* **54**, 227–245.

Richardson, H. L. (1935). The nitrogen cycle in grassland soils. *Trans. 3rd Int. Congr. Soil Sci.* **1**, 219–221.

Richardson, H. L. (1938). Nitrification in grassland soils: with especial reference to the Rothamsted Park Grass experiment. *J. Agric. Sci., Camb.* **28**, 73–121.

Rietveld, W. J. (1975). Phytotoxic grass residues reduce germination and initial root growth of ponderosa pine. *USDA For. Serv. Res. Pap.* RM–153.

Riov, J., Monselise, S. P., and Kahan, R. S. (1969). Ethylene-controlled induction of phenylalanine ammonia-lyase in citrus fruit peel. *Plant Physiol.* **44**, 631–635.

Rizvi, S. J. H., Mukerji, D., and Mathur, S. N. (1980). A new report on a possible source of natural herbicide. *Indian J. Exp. Biol.* **18**, 777–778.

Rizvi, S. J. H., Mukerji, D., and Mathur, S. N. (1981). Selective phyto-toxicity of 1,3,7-trimethyl-xanthine between *Phaseolus mungo* and some weeds. *Agric. Biol. Chem.* **45,** 1255–1256.

Robertson, G. P.,and Vitousek, P. M. (1981). Nitrification potentials in primary and secondary succession. *Ecology* **62,** 376–386.

Robinson, E. L. (1976). Effect of weed species and placement on seed cotton yields. *Weed Sci.* **24,** 353–355.

Robinson, J. B. (1963). Nitrification in a New Zealand grassland soil. *Plant Soil* **19,** 173–183.

Robinson, R. K. (1972). The production by roots of *Calluna vulgaris* of a factor inhibitory to growth of some mycorrhizal fungi. *J. Ecol.* **60,** 219–224.

Robinson, T. (1983). "The Organic Constituents of Higher Plants." 5th ed. Cordus Press, North Amherst, Massachusetts.

Rodhe, W. (1948). Environmental requirements of fresh-water plankton algae. *Symb. Bot. Ups.* **10,** 1–149.

Roeth, F. W. (1973). Johnsongrass control in corn with soil incorporated herbicides. *Weed Sci.* **21,** 474–476.

Roshchina, V. D. (1972a). The use of gas chromatography in allelopathic investigations. *For. J.* **3,** 29–31. (In Russian.)

Roshchina, V. D. (1972b). The study of the state of intracellular water treated with phenols. *Agric. Biol.* **7,** 554–558. (In Russian, English summary.)

Roshchina, V. D. (1973). Reducing power of woody plant leaf infusions and inhibition of the Hill reaction with 2,6-dichlorophenol by them. *In* "Physiological-Biochemical Basis of Plant Interactions in Phytocenoses" (A. M. Grodzinsky, ed.), Vol. 4, pp. 10–15. Naukova Dumka, Kiev. (In Russian, English summary.)

Roshchina, V. D. (1974). Volatile and water-soluble metabolites of woody plant leaves. *In* "Physiological-Biochemical Basis of Plant Interactions in Phytocenoses" (A. M. Grodzinsky, ed.), Vol. 5, pp. 36–40. Naukova Dumka, Kiev. (In Russian, English summary.)

Roshchina, V. D., and Roshchina, V. V. (1970). Influence of water-soluble secretions from the woody species leaves on cytoplasm permeability for anthocyanin. *In* "Physiological-Biochemical Basis of Plant Interactions in Phytocenoses" (A. M. Grodzinsky, ed.), Vol. 1, pp. 257–262. Naukova Dumka, Kiev. (In Russian, English summary.)

Roshchina, V. D., Roshchina, V. V., and Kotova, I. N. (1979). The effect of extracts from *Cicuta virosa* on chloroplast movement and on some photosynthetic reactions. *Plant Physiol. (USSR)* **26,** 147–152. (In Russian, English summary.)

Roshchina, V. V., and Akulova, E. A. (1978). Effect of phloridzin on photosynthetic electron transport. *Biochemistry* **43,** 899–903. (In Russian, English summary.)

Roshchina, V. V., Solomatkin, V. P., and Roshchina, V. D. (1980). Cicutotoxin as an inhibitor of electron transport in photosynthesis. *Plant Physiol. (USSR)* **27,** 704–709. (In Russian, English summary.)

Rovira, A. D. (1965). Plant root exudates and their influence upon soil microorganisms. *In* "Ecology of Soil-Borne Plant Pathogens-Prelude to Biological Control" (K. F. Baker and W. C. Snyder, eds.), pp. 170–184. Univ. of California Press, Berkeley.

Rovira, A. D. (1969). Plant root exudates. *Bot. Rev.* **35,** 35–59.

Rovira, A. D. (1971). Plant root exudates. *In* "Biochemical Interactions among Plants" (U. S. Natl. Committee, IBP, eds.), pp. 19–24. Natl. Acad. Sci., Washington, D. C.

Russell, E. J. (1914). The nature and amount of the fluctuations in nitrate contents of arable soils. *J. Agric. Sci.* **6,** 50–53.

Russell, E. J., and Russell, E. W. (1961). "Soil Conditions and Plant Growth," 9th ed. Wiley, New York.

Rychert, R. C., and Skujins, J. (1974). Nitrogen fixation by blue-green algae-lichen crusts in the Great Basin Desert. *Soil Sci. Soc. Am. Proc.* **38,** 768–771.

Rydrych, D. J. (1974). Competition between winter wheat and downy brome. *Weed Sci.* **22**, 211–214.

Rydrych, D. J., and Muzik, T. J. (1968). Downy brome competition and control in dryland wheat. *Agron. J.* **60**, 279–280.

Saftner, R. A., and Evans, M. L. (1974). Selective effects of victorin on growth and the auxin response in *Avena. Plant Physiol.* **53**, 382–387.

Saiki, H., and Yoneda, K. (1981). Possible dual roles of an allelopathic compound, *cis*-dehydromatricaria ester. *J. Chem. Ecol.* **8**, 185–193.

Sajise, P. E., and Lales, J. S. (1975). Allelopathy in a mixture of cogon (*Imperata cylindrica*) and *Stylosanthes guyanesis. Kalikasan Philipp. J. Biol.* **4**, 155–164.

Salas, M. C., and Vieitez, E. (1972). Activated de crecimiento de Ericaceas. *An. Edafol. Agrobiol.* **31**, 1001–1009.

Santoro, T., and Casida, L. E., Jr. (1962). Elaboration of antibiotics by *Boletus luteus* and certain other mycorrhizal fungi. *Can. J. Microbiol.* **8**, 43–48.

Sarkar, S. K., and Phan, C. T. (1974). Effect of ethylene on the qualitative and quantitative composition of the phenol content of carrot roots. *Physiol. Plant.* **30**, 72–76.

Sarma, K. K. V. (1974a). Allelopathic potential of *Digera arvensis* Forsk. on *Pennisetum typhoides* Stapf. et Hubb. *Geobios (Jodhpur)* **1**, 137.

Sarma, K. K. V. (1974b). Allelopathic potential of *Echinops echinatus* and *Solanum surattense* on seed germination of *Argemone mexicana. Trop. Ecol.* **15**, 156–157.

Sarma, K. K. V., Giri, G. S., and Subramanyam, K. (1976). Allelopathic potential of *Parthenium hysterophorus* Linn. on seed germination and dry matter production in *Arachis hypogea* Willd., *Crotalaria juncea* Linn. and *Phaseolus mungo* Linn. *Trop. Ecol.* **17**, 76–77.

Savage, D. A., and Runyon, H. E. (1937). Natural revegetation of abandoned farm land in the central and southern Great Plains. *Int. Grass. Congr.* Aberystwyth, Great Britain, Rep. Sect. 1 (Grassland Ecology), pp. 178–182.

Schaal, L. A., and Johnson, G. (1955). The inhibitory effect of phenolic compounds on the growth of *Streptomyces scabies* as related to the mechanism of scab resistance. *Phytopathology* **45**, 626–628.

Scharff, T. G., and Perry, A. C. (1976). The effects of salicylic acid on metabolism and potassium ion content in yeast. *Proc. Soc. Exp. Biol.* **151**, 72–77.

Schenck, S., and Stotzky, G. (1975). Effect on microorganisms of volatile compounds released from germinating seeds. *Can. J. Microbiol.* **21**, 1622–1634.

Schlatterer, E. F., and Tisdale, E. W. (1969). Effects of litter of *Artemisia, Chrysothamnus,* and *Tortula* on germination and growth of three perennial grasses. *Ecology* **50**, 869–873.

Schneiderhan, F. J. (1927). The black walnut (*Juglans nigra* L.) as a cause of the death of apple trees. *Phytopathology* **17**, 529–540.

Schreiber, M. M. (1977). Longevity of foxtail taxa in undisturbed sites. *Weed Sci.* **25**, 66–72.

Schreiber, M. M., and Williams, J. L., Jr. (1967). Toxicity of root residues of weed grass species. *Weeds* **15**, 80–81.

Schreiner, O., and Lathrop, E. C. (1911). Examination of soils for organic constituents. *U. S. Dept. Agric., Bur. Soils Bull.* No. 80.

Schreiner, O., and Reed, H. S. (1907a). Certain organic constituents of soil in relation to soil fertility. *U. S. Dept. Agric., Bur. Soils Bull.* No. 47.

Schreiner, O., and Reed, H. S. (1907b). The production of deleterious excretions by roots. *Bull. Torrey Bot. Club* **34**, 279–303.

Schreiner, O., and Reed, H. S. (1908). The toxic action of certain organic plant constituents. *Bot. Gaz. (Chicago)* **45**, 73–102.

Schreiner, O., and Shorey, E. D. (1909). The isolation of harmful organic substances from soils. *U. S. Dept. Agric., Bur. Soils Bull.* No. 53.

Schreiner, O., and Sullivan, M. X. (1909). Soil fatigue caused by organic compounds. *J. Biol. Chem.* **6**, 39–50.

Schumacher, W. J., Thill, D. C., and Lee, G. A. (1982). The allelopathic potential of wild oat (*Avena fatua* L.) on spring wheat (*Triticum aestivum* L.) growth. *North Am. Symp. Allelopathy, Nov. 14–17, 1982, Urbana-Champaign, Illinois* (Abstr.)

Schwimmer, S. (1958). Influence of polyphenols and potato components on potato phosphorylase. *J. Biol. Chem.* **232**, 715–721.

Schwinghamer, E. A. (1964). Association between antibiotic resistance and ineffectiveness in mutant strains of *Rhizobium* spp. *Can. J. Microbiol.* **10**, 221–233.

Schwinghamer, E. A. (1967). Effectiveness of *Rhizobium* as modified by mutation for resistance to antibiotics. *Antonie von Leeuwenoek* **33**, 121–136.

Selleck, G. W. (1972). The antibiotic effects of plants in laboratory and field. *Weed Sci.* **20**, 189–194.

Sen, D. N. (1976). Ecophysiological studies on weeds of cultivated fields with special reference to bajra (*Pennisetum typhoideum* Rich.) and til (*Sesamum indicum* Linn.) crops. *Second Prog. Rept. Proj. No. A7-CR-425,* Lab. of Plant Ecology, Univ. of Jodhpur, Jodhpur, India.

Sevilla-Santos, P., Encinas, C. J., and Leus-Palo, S. (1964). The antibacterial activities of aqueous extracts from Philippine Basidiomycetes. *Philipp. J. Sci.* **93**, 479–498.

Shafizadeh, F., and Bhadane, N. R. (1972a). Badgerin, a new germacranolide from *Artemisia arbuscula* ssp. *arbuscula. J. Org. Chem.* **37**, 274–277.

Shafizadeh, F., and Bhadane, N. R. (1972b). Sesquiterpene lactones of sagebrush. New guaianolides from *Artemisia cana* ssp. *viscidula. J. Org. Chem.* **37**, 3168–3173.

Shafizadeh, F., and Melnikoff, A. B. (1970). Coumarins of *Artemisia tridentata* ssp. *vaseyana. Phytochemistry* **9**, 1311–1316.

Shafizadeh, F., Bhadane, N. R., Morris, M. S., Kelsey, R. G., and Khanna, S. N. (1971). Sesquiterpene lactones of big sagebrush. *Phytochemistry* **10**, 2745–2754.

Sharma, K. D., and Sen, D. N. (1971). Growth regulators in the fruit pulp of *Solanum surattense* L. *Z. Pflanzenphysiol.* **65**, 458–460.

Sharp, J. H., Underhill, P. A., and Hughes, D. J. (1979). Interaction (allelopathy) between marine diatoms: *Thalassiosira pseudonana* and *Phaeodactylum tricornutum. J. Phycol.* **15**, 353–362.

Shen, T. C. (1969). The induction of nitrate reductase and the preferential assimilation of ammonium in germinating rice seedlings. *Plant Physiol.* **44**, 1650–1655.

Shields, L. M., and Durrell, L. W. (1964). Algae in relation to soil fertility. *Bot. Rev.* **30**, 92–128.

Shimshi, D. (1963a). Effect of chemical closure of stomata on transpiration in varied soil and atomospheric environments. *Plant Physiol.* **38**, 709–712.

Shimshi, D. (1963b). Effect of soil moisture and phenylmercuric acetate upon stomatal aperture, transpiration, and photosynthesis. *Plant Physiol.* **38**, 713–721.

Shukla, A. N., Arora, D. K., and Dwivedi, R. S. (1977). Effect of microbial culture filtrates on the growth of sal (*Shorea robusta* Gaertn.) leaf litter fungi. *Soil Biol. Biochem.* **9**, 217–219.

Sieburth, J. M. (1959). Antibacterial activity of Anarctic marine phytoplankton. *Limnol. Oceanogr.* **4**, 419–424.

Sieburth, J. M. (1960). Acrylic acid, an ''antibiotic'' principle in *Phaeocystis* blooms in Anarctic waters. *Science* **132**, 676–677.

Sigmund, W. (1924). Über die Einwirkung von Stoffwechselendprodukten auf die Pflanzen. III. Einwirkung N-freier pflanzlicher Stoffwechselendprodukte auf die Keimung von Samen (Aetherische Oele, Terpene u.a.). *Biochem. Z.* **146**, 389–419.

Sikka, H. C., Shimabukuro, R. H., and Zweig, G. (1972). Studies on effect of certain quinones. I. Electron transport, photophosphorylation, and CO_2 fixation in isolated chloroplasts. *Plant Physiol.* **49**, 381–384.

Singh, P. N. (1977). Effect of root exudates and extracts of *Solanum nigrum* and *Argemone mexicana* seedlings on rhizosphere mycoflora. *Acta Bot. Indica* **5**, 123–127.

Singh, R. N. (1961). "Role of Blue Green Algae in Nitrogen Economy of Indian Agriculture." Indian Council of Agric. Res., New Delhi.

Singh, S. P. (1968). Presence of a growth inhibitor in the tubers of nutgrass (*Cyperus rotundus* L.) *Proc. Indian Acad. Sci.* **67**, 18–23.

Sinha-Roy, S. P., and Chakraborty, D. P. (1976). Psoralen a powerful germination inhibitor. *Phytochemistry* **15**, 2000–2006.

Smale, B. C., Wilson, R. A., and Keil, H. L. (1964). A survey of green plants for antimicrobial substances. Abstract. *Phytopathology* **54**, 748.

Smith, J. L., and Rice, E. L. (1983). Differences in nitrate reductase activity between species of different stages in old field succession. *Oecologia* **57**, 43–48.

Smith, R. J., Jr. (1960). Chemical control of barnyardgrass in rice. *Weeds* **8**, 256–267.

Smith, R. J., Jr. (1968). Weed competition in rice. *Weed Sci.* **16**, 252–255.

Smith, R. J., Jr. (1974). Competition of barnyardgrass with rice cultivars. *Weed Sci.* **22**, 423–426.

Smith, W. H. (1976). Character and significance of forest tree root exudates. *Ecology* **57**, 324–331.

Smith, W. H., Bormann, F. H., and Likens, G. E. (1968). Response of chemoautotrophic nitrifiers to forest cutting. *Soil Sci.* **106**, 471–473.

Society of American Bacteriologists (1957). "Manual of Microbiological Methods." McGraw-Hill, New York.

Solbraa, K. (1979). Composting of bark. IV. Potential growth-reducing compounds and elements in bark. Rept. 34.16 of the Norwegian Forest Research Institute, Box 61, 1432 ÅS-NLH, Norway.

Somers, T. C., and Harrison, A. F. (1967). Wood tannins-isolation and significance in host resistance to *Verticillium* wilt disease. *Aust. J. Biol. Sci.* **20**, 475–479.

Sondheimer, E. (1962). The chlorogenic acids and related compounds. *In* "Plant Phenolics and their Industrial Significance" (V. C. Runeckles, ed.), pp. 15–37. Symposium of Plant Phenolics Group of North America. Imperial Tobacco Co. of Canada, Montreal, Quebec.

Sondheimer, E., and Griffin, D. H. (1960). Activation and inhibition of indoleacetic acid oxidase activity from peas. *Science* **131**, 672.

Spencer, K. C. (1979). Chemical constituents of the Hepaticae. *Phytochem. Bull.* **12**, 4–19.

Spoehr, H. A., Smith, J. H. C., Strain, H. H., Milner, H. W., and Hardin, G. J. (1949). *Carnegie Inst. Washington Publ.* No. 586.

Stachon, W. J., and Zimdahl, R. L. (1980). Allelopathic activity of Canada thistle (*Cirsium arvense*) in Colorado. *Weed Sci.* **28**, 83–86.

Stahl, C., Vanderhoef, L. N., Siegel, N., and Helgeson, J. P. (1973). *Fusarium tricinctum* T-2 toxin inhibits auxin-promoted elongation in soybean hypocotyl. *Plant Physiol.* **52**, 663–666.

Staniforth, D. W. (1957). Effect of annual grass weeds on the yield of corn. *Agron. J.* **49**, 551–555.

Staniforth, D. W. (1958). Soybean-foxtail competition under varying soil moisture conditions. *Agron. J.* **50**, 13–15.

Staniforth, D. W. (1961). Responses of corn hybrids to yellow foxtail competition. *Weeds* **9**, 132–136.

Staniforth, D. W. (1965). Competitive effects of three foxtail species on soybeans. *Weeds* **13**, 191–193.

Steenhagen, D. A., and Zimdahl, R. L. (1979). Allelopathy of leafy spurge (*Euphorbia esula*). *Weed Sci.* **27**, 1–3.

Stenlid, G. (1968). On the physiological effects of phloridzin, phloretin and some related substances upon higher plants. *Physiol. Plant.* **21**, 882–894.

Stenlid, G. (1970). Flavonoids as inhibitors of the formation of adenosine triphosphate in plant mitochondria. *Phytochemistry* **9**, 2251–2256.

Stepanov, E. V. (1977). Volatile substances of conifer root systems as a factor of the forest community environment. *In* "Interactions of Plants and Microorganisms in Phytocenoses" (A. M. Grodzinsky, ed.), pp. 58–65. Naukova Dumka, Kiev. (In Russian, English summary.)

Stevenson, F. J. (1967). Organic acids in soil. *In* "Soil Biochemistry" (A. D. McLaren and G. H. Peterson, eds.), pp. 119–142. Dekker, New York.

Stewart, J. R., and Brown, R. M., Jr. (1969). *Cytophaga* that kills or lyses algae. *Science* **164,** 1523–1524.

Stewart, R. E. (1975). Allelopathic potential of western bracken *J. Chem. Ecol.* **1,** 161–169.

Stewart, W. D. (1966). "Nitrogen Fixation in Plants." Oxford Univ. Press, London and New York.

Steyn, P. L., and Delwiche, C. C. (1970). Nitrogen fixation by nonsymbiotic microorganisms in some California soils. *Environ. Sci. Technol.* **4,** 1122–1128.

Stickney, J. S., and Hoy, P. R. (1881). Toxic action of black walnut. *Trans. Wis. State Hort. Soc.* **11,** 166–167.

Stillwell, M. A. (1966). A growth inhibitor produced by *Cryptosporiopsis* sp., an imperfect fungus isolated from yellow-birch, *Betula alleghaniensis* Britt. *Can. J. Bot.* **44,** 259–267.

Stiven, G. (1952). Production of antibiotic substances by the roots of a grass [*Trachypogon plumosus* (H.B.K.) Nees] and of *Pentanisia variabilis* (E. Mey.) Harv. (Rubiaceae). *Nature (London)* **170,** 712–713.

Stoller, E. W., Wax, L. M., and Slife, F. W. (1979). Yellow nutsedge (*Cyperus esculentus*) competition and control in corn (*Zea mays*). *Weed Sci.* **27,** 32–37.

Stotzky, G., and Schenck, S. (1976). Observations on organic volatiles from germinating seeds and seedlings. *Am. J. Bot.* **63,** 798–805.

Stowe, L. G. (1979). Allelopathy and its influence on the distribution of plants in an Illinois old-field. *J. Ecol.* **67,** 1065–1085.

Strobel, G. A. (1974). Phytotoxins produced by plant parasites. *Annu. Rev. Plant Physiol.* **25,** 541–566.

Swain, T. (1965). The tannins. *In* "Plant Biochemistry" (J. Bonner and J. E. Varner, eds.), pp. 552–580. Academic Press, New York.

Swain, T. (1977). Secondary compounds as protective agents. *Annu. Rev. Plant Physiol.* **28,** 479–501.

Swan, H. S. D. (1960). The mineral nutrition of Canadian pulpwood species. I. The influence of nitrogen, phosphorous, potassium, and magnesium deficiencies on the growth and development of white spruce, black spruce, jack pine, and western hemlock seedlings grown in a controlled environment. *Pulp Pap. Inst. Can., Tech. Rep.* No. 168.

Szczepańska, W. (1971). Allelopathy among the aquatic plants. *Pol. Arch. Hydrobiol.* **18,** 17–30.

Szczepański, A. (1971). Allelopathy and other factors controlling the macrophytes production. *Hidrobiologia* **12,** 193–197.

T.A., Home Correspondence (1845). Rotation of crops. *Gard. Chron.* **5,** 159.

Tack, B. F., Chapman, P. J., and Dagley, S. (1972). Metabolism of gallic and syringic acids by *Pseudomonas putida. J. Biol. Chem.* **247,** 6438–6443.

Takatori, F., and Souther, F. (1978). "Asparagus Workshop Proceedings." Department of Plant Science, Univ of California, Riverside, California.

Tam, R. K., and Clark, H. E. (1943). Effects of chloropicrin and other soil disinfectants on the nitrogen nutrition of the pineapple plant. *Soil Sci.* **56,** 245–261.

Tames, R. S., Gesto, M. D. V., and Vieitez, E. (1973). Growth substances isolated from tubers of *Cyperus esculentus* var. *aureus. Physiol. Plant.* **28,** 195–200.

Tamura, S., Chang, C., Suzuki, A., and Kumai, S. (1967). Isolation and structure of a novel isoflavone derivative in red clover. *Agric. Biol. Chem.* **31,** 1108–1109.

Tamura, S., Chang, C., Suzuki, A., and Kumai, S. (1969). Chemical studies on "clover sickness."

Part I. Isolation and structural elucidation of two new isoflavonoids in red clover. *Agric. Biol. Chem.* **33**, 391–397.

Tang, C. S., and Waiss, A. C., Jr. (1978). Short-chain fatty acids as growth inhibitors in decomposing wheat straw. *J. Chem. Ecol.* **4**, 225–232.

Tang, C. S., and Young, C. C. (1982). Collection and identification of allelopathic compounds from the undisturbed root system of bigalta limpograss (*Hemarthria altissima*). *Plant Physiol.* **69**, 155–160.

Tang, C. S., Bhothipaksa, K., and Frank, H. A. (1972). Bacterial degradation of benzyl isothiocyanate. *Appl. Microbiol.* **23**, 1145–1148.

Tarrant, R. F., and Trappe, J. M. (1971). The role of *Alnus* in improving the forest environment. *In* "Biological Nitrogen Fixation in Natural and Agricultural Habitats" (T. A. Lie and E. G. Mulder, eds.), pp. 335–348. Nijhoff, The Hague.

Taylor, A. O. (1965). Some effects of photoperiod on the biosynthesis of phenylpropane devivatives in *Xanthium. Plant Physiol.* **40**, 273–280.

Templeton, G. E., TeBeest, D. O., and Smith, R. J. (1979). Biological weed control with mycoherbicides. *Annu. Rev. Phytopathol.* **17**, 301–310.

Theodorou, C., and Bowen, G. D. (1971). Effects of non-host plants on growth of mycorrhizal fungi of radiata pine. *Aust. For.* **35**, 17–22.

Theophrastus (ca 300 B.C.). "Enquiry into Plants and Minor Works on Odours and Weather Signs." 2 Vols., transl. to English by A. Hort., W. Heinemann, London, 1916.

Theron, J. J. (1951). The influence of plants on the mineralization of nitrogen and the maintenance of organic matter in the soil. *J. Agric. Sci., Camb.* **41**, 289–296.

Thibault, J. R., Fortin, J. A., and Smirnoff, W. A. (1982). *In vitro* allelopathic inhibition of nitrification by balsam poplar and balsam fir. *Am. J. Bot.* **69**, 676–679.

Thomas, A. S., Jr. (1974). The effect of aqueous extracts of blue spruce leaves on seed germination and seedling growth of several plant species. (Abstr.) *Phytopathology* **64**(5), 587.

Thomas, G., and Thelfall, D. R. (1974). Incorporation of shikimate and 4-(2'-carboxyphenyl)-4-oxobutyrate into phylloquinone. *Phytochemistry* **13**, 807–813.

Thorne, D. W., and Brown, P. E. (1937). The growth and respiration of some soil bacteria in juices of leguminous and non-leguminous plants. *J. Bacteriol.* **34**, 567–580.

Tillberg, J. E. (1970). Effects of abscisic acid, salicylic acid and *trans*-cinnamic acid on phosphate uptake, ATP-level and oxygen evolution in *Scenedesmus. Physiol. Plant.* **23**, 647–653.

Tinnin, R., and Muller, C. (1971). The allelopathic potential of *Avena fatua*: Influence on herb distribution. *Bull. Torrey Bot. Club* **98**, 243–250.

Toai, T. V., and Linscott, D. L. (1979). Phytotoxic effect of decaying quackgrass (*Agropyron repens*) residues. *Weed Sci.* **27**, 595–598.

Todd, R. L., Swank, W. T., Douglass, J. E., Kerr, P. C., Brockway, D. L., and Monk, C. D. (1975). The relationship between nitrate concentration in the southern Appalachian mountain streams and terrestrial nitrifiers. *Agro-Ecosystems* **2**, 127–132.

Tomanek, G. W., Albertson, F. W., and Riegel, A. (1955). Natural revegetation on a field abandoned for thirty-three years in central Kansas. *Ecology* **36**, 407–412.

Tomaszewski, M., and Thimann, K. V. (1966). Interactions of phenolic acids, metallic ions and chelating agents on auxin-induced growth. *Plant Physiol.* **41**, 1443–1454.

Torrey, J. G. (1978). Nitrogen fixation by actinomycete-nodulated angiosperms. *BioScience* **28**, 586–592.

Toussoun, T. A., and Patrick, Z. A. (1963). Effect of phytotoxic substances from decomposing plant residues on root rot of bean. *Phytopathology* **53**, 265–270.

Trappe, J. M., Li, C. Y., Lu, K. C., and Bollen, W. B. (1973). Differential response of *Poria weirii* to phenolic acids from Douglas-fir and red alder roots. *For. Sci.* **19**, 191.

Tso, T. C., Sorokin, T. P., Engelhaupt, M. E., Anderson, R. A., Bortner, C. E., Chaplin, J. F., Miles, J. D., Nichols, B. C., Shaw, L., and Street, O. E. (1967). Nitrogenous and phenolic compounds of *Nicotiana* plants. I. Field and greenhouse grown plants. *Tob. Sci.* **11**, 133–136.

Tso, T. C., Kasperbauer, M. J., and Sorokin, T. P. (1970). Effect of photoperiod and end-of-day light quality on alkaloids and phenolic compounds of tobacco. *Plant Physiol.* **45**, 330–333.

Tubbs, C. H. (1973). Allelopathic relationships between yellow birch and sugar maple seedlings. *For. Sci.* **19**, 139–145.

Tukey, H. B., Jr. (1966). Leaching of metabolites from above-ground plant parts and its implications. *Bull. Torrey Bot. Club* **93**, 385–401.

Tukey, H. B., Jr. (1969). Implications of allelopathy in agricultural plant science. *Bot. Rev.* **35**, 1–16.

Tukey, H. B., Jr. (1971). Leaching of substances from plants. *In* "Biochemical Interactions among Plants" (U. S. Natl. Comm. for IBP, eds.), pp. 25–32. Natl. Acad. Sci., Washington, D. C.

Turner, B. H., and Quarterman, E. (1975). Allelochemic effects of *Petalostemon gattingeri* on the distribution of *Arenaria patula* in cedar glades. *Ecology* **56**, 924–932.

Turner, J. A., and Rice, E. L. (1975). Microbial decomposition of ferulic acid in soil. *J. Chem. Ecol.* **1**, 41–58.

Turner, N. C. (1972). Stomatal behavior of *Avena sativa* treated with two phytotoxins, victorin and fusicoccin. *Am. J. Bot.* **59**, 133–136.

Turner, W. B. (1971). "Fungal Metabolites." Academic Press, New York.

Tyson, B. J., Dement, W. A., and Mooney, H. A. (1974). Volatilisation of terpenes from *Salvia mellifera*. *Nature (London)* **252**, 119–120.

Ukeles, R., and Bishop, J. (1975). Enhancement of phytoplankton growth by marine bacteria. *J. Phycol.* **11**, 142–149.

U. S. Forest Service (1963). Annual Rept. U. S. Dept. Agric., Rocky Mountain Forest and Range Exp. Stn.

Van Alfen, N. K., and Turner, N. C. (1975a). Influence of a *Ceratocystis ulmi* toxin on water relations of elm (*Ulmus americana*). *Plant Physiol.* **55**, 312–316.

Van Alfen, N. K., and Turner, N. C. (1975b). Changes in alfalfa stem conductance induced by *Corynebacterium insidiosum* toxin. *Plant Physiol.* **55**, 559–561.

Van der Merwe, K. J., Van Jaarsveld, P. P., and Hattingh, M. J. (1967). The isolation of 2,4-diacetyl-phloroglucinol from a *Pseudomonas* sp. *S. Afr. Med. J.* **41**, 1110.

Van der Valk, A. G., and Davis, C. B. (1976). The seed banks of prairie glacial marshes. *Can. J.Bot.* **54**, 1832–1838.

Van Staden, J. (1976). The release of cytokinins by maize roots. *Plant Sci. Lett.* **7**, 279–283.

Van Sumere, C. F., and Massart, L. (1959). Natural substances in relation to germination. *Proc. Int. Congr. Biochem. 4th*, **5**, 20–32.

Van Sumere, C. F., Cottenie, J., De Greef, J., and Kint, J. (1971). Biochemical studies in relation to the possible germination regulatory role of naturally occurring coumarin and phenolics. *Recent Adv. Phytochem.* **4**, 165–221.

Varga, M., and Köves, E. (1959). Phenolic acids as growth and germination inhibitors in dry fruits. *Nature (London)* **183**, 401.

Vartia, K. O. (1973). Antibiotics in lichens. *In* "The Lichens" (V. Ahmadjian and M. E. Hale, eds.), pp. 547–561. Academic Press, New York.

Verrall, A. F., and Graham, T. W. (1935). The transmission of *Ceratostomella ulmi* through root grafts. *Phytopathology* **25**, 1039–1040.

Vieitez, E., and Ballester, A. (1972). Compuestos fenólicos y cumáricos en *Erica cinerea* L. *An. Inst. Bot. A. J. Cavanilles* **29**, 129–142.

Vikherkova, M. (1970). Influence of active substances from rhizome of wheatgrass on growth and water balance of flax. *In* "Physiological-Biochemical Basis of Plant Interactions in Phytocenoses" (A. M. Grodzinsky, ed.), Vol. 1, pp. 135–140. Naukova Dumka, Kiev. (In Russian, English summary.)

Viro, P. J. (1963). Factorial experiments on forest humus decomposition. *Soil Sci.* **95**, 24–30.

Virtanen, A. I., Erkama, J., and Linkola, H. (1947). On the relation between nitrogen fixation and leghaemoglobin content of leguminous root nodules. II. *Acta Chem. Scand.* **1**, 861–870.

Visona, L., and Pesce, E. (1963). Action de quelques antibiotiques sur la symbiose du trèfle rouge (*Trifolium pratense* L.). *Ann. Inst. Pasteur, Paris* **105**, 368–382.

Visona, L., and Tardieux, P. (1964). Antagonistes des *Rhizobium* dans la rhizosphère du trèfle et de la luzerne. *Ann. Pasteur Inst. Paris* Suppl., No. 3, pp. 297–302.

Vitousek, P. M. (1977). The regulation of element concentrations in mountain streams in the northeastern United States. *Ecol. Monogr.* **47**, 65–87.

Vitousek, P. M., and Reiners, W. A. (1975). Ecosystem succession and nutrient retention: a hypothesis. *BioScience* **25**, 376–381.

Vlassak, K., Paul, E. A., and Harris, R. E. (1973). Assessment of biological nitrogen fixation in grassland and associated sites. *Plant Soil* **38**, 637–649.

Vogl, R. J. (1974). Effects of fire on grasslands. *In* "Fire and Ecosystems" (T. T. Kozlowski and C. E. Ahlgren, eds.), pp. 139–194. Academic Press, New York.

von Glombitza, K. W., and Stoffelen, H. (1972). 2,3-Dibromo-5-hydroxybenzyl-1,4-disulfat (dikaliumsalz) aus Rhodomelaceen. *Planta Med.* **22**, 391–395.

von Glombitza, K. W., Stoffelen, H., Murawski, U., Bielaczek, J., and Egge, H. (1974). Antibiotica aus Algen. 9. Mitt. Bromphenole aus Rhodomelaceae. *Planta Med.* **25**, 105–114.

Vrbaški, M. M., Grujić-Injac, B., and Gajić, D. (1977). Preparation, identification, and biological activity of substance "A" from seeds of *Agrostemma githago*. *Biochem. Physiol. Pflanz.* **171**, 69–74.

Waks, C. (1936). The influence of extract from *Robinia pseudoacacia* on the growth of barley. *Publ. Fac. Sci. Univ. Charles, Prague* **150**, 84–85.

Waksman, S. A. (1937). Soil deterioration and soil conservation from the viewpoint of soil microbiology. *J. Am. Soc. Agron.* **29**, 113–122.

Waksman, S. A. (1947). "Microbial Antagonisms and Antibiotic Substances." Commonwealth Fund, New York.

Wali, M. K., and Iverson, L. R. (1978). Revegetation of coal mine spoils and autoallelopathy in *Kochia scoparia*. *Abst., 144th Natl. Am. Assoc. Adv. Sci. Meet., Washington, D. C.* pp. 121–122.

Waller, C. W., Patrick, J. B., Fulmor, W., and Meyer, W. E. (1957). The structure of nucleocidin. I. *J. Am. Chem. Soc.* **79**, 1011–1012.

Walters, D. T., and Gilmore, A. R. (1976). Allelopathic effects of fescue on the growth of sweetgum. *J. Chem. Ecol.* **2**, 469–479.

Walton, S. (1980). Biocontrol agents prey on pests and pathogens. *BioScience* **30**, 445–447.

Wang, T. S. C., Cheng, S. Y., and Tung, H. (1967a). Extraction and analysis of soil organic acids. *Soil Sci.* **103**, 360–366.

Wang, T. S. C., Yang, T., and Chuang, T. (1967b). Soil phenolic acids as plant growth inhibitors. *Soil Sci.* **103**, 239–246.

Wang, T. S. C., Yeh, K. L., Cheng, S. Y., and Yang, T. K. (1971). Behavior of soil phenolic acids. *In* "Biochemical Interactions among Plants" (U. S. Natl. Comm. for IBP, eds.), pp. 113–120. Natl. Acad. Sci., Washington, D. C.

Ward, G. M., and Durkee, A. B. (1956). The peach replant problem in Ontario. III. Amygdalin content of peach tree tissues. *Can. J. Bot.* **34**, 419–422.

Warren, M. (1965). A study of soil-nutritional and other factors operating in secondary succession in highveld grassland in the neighborhood of Johannesburg. Ph.D. Thesis, Univ. of Witwatersrand, Johannesburg, South Africa.

Watanabe, R., McIlrath, W. J., Skok, J., Chorney, W., and Wender, S. H. (1961). Accumulation of scopoletin glucoside in boron-deficient tobacco leaves. *Arch. Biochem. Biophys.* **94**, 241–243.

Weatherspoon, D. M., and Schweizer, E. E. (1969). Competition between *Kochia* and sugarbeets. *Weed Sci.* **17**, 464–467.

Weaver, R. J., and DeRose, H. R. (1946). Absorption and translocation of 2,4-D. *Bot. Gaz. (Chicago)* **107**, 509–521.

Weaver, T. W., and Klarich, D. (1977). Allelopathic effects of volatile substances from *Artemisia tridentata* Nutt. *Am. Midl. Nat.* **97**, 508–512.

Webb, L. J., Tracey, J. G., and Haydock, K. P. (1967). A factor toxic to seedlings of the same species associated with living roots of the non-gregarious subtropical rain forest tree, *Grevillea robusta*. *J. Appl. Ecol.* **4**, 13–25.

Weetman, G. F. (1961). The nitrogen cycle in temperate forest stands (Literature Review). *Pulp Pap. Res. Inst. Can., Res. Note* No. 4.

Weissman, G. S. (1972). Influence of ammonium and nitrate nutrition on enzymatic activity in soybean and sunflower. *Plant Physiol.* **49**, 138–141.

Welbank, P. J. (1961). A study of the nitrogen and water factors in competition with *Agropyron repens* (L.) Beauv. *Ann. Bot. N.S.* **25**, 116–137.

Welbank, P. J. (1964). Competition for nitrogen and potassium in *Agropyron repens*. *Ann. Bot. N.S.* **28**, 1–16.

Wellhausen, E. J. (1962). Weeds and man in Latin America. *Weeds* **10**, 200–209.

Went, F. W. (1942). The dependence of certain annual plants on shrubs in southern California deserts. *Bull. Torrey Bot. Club* **69**, 100–114.

Werner, P. A. (1975). The effects of plant litter on germination in teasel, *Dipsacus sylvestris* Huds. *Am. Midl. Nat.* **94**, 470–476.

Westlake, D. W. S., Talbot, G., Blakley, E. R., and Simpson, F. J. (1959). Microbial decomposition of rutin. *Can. J. Microbiol.* **5**, 621–629.

White, G. A., and Starratt, A. N. (1967). The production of a phytotoxic substance by *Alternaria zinniae*. *Can. J. Bot.* **45**, 2087–2090.

Whitehead, D. C. (1964). Identification of *p*-hydroxybenzoic, vanillic, *p*-coumaric, and ferulic acids in soils. *Nature (London)* **202**, 417–418.

Whittaker, R. H., and Feeney, P. P. (1971). Allelochemics: Chemical interactions between species. *Science* **171**, 757–770.

Widera, M. (1978). Competition between *Hieracium pilosella* L. and *Festuca rubra* L. under natural conditions. *Ekol. Pol.* **26**, 359–390.

Wiechers, B. G., and Rovalo-Merino, M. (1982). Potencial alelopatico y microbicida de *Helietta parvifolia*. *Biotica* **7**, 405–416.

Wilde, S. A., and Lafond, A. (1967). Symbiotrophy of Lignophytes and fungi: its terminological and conceptual deficiencies. *Bot. Rev.* **33**, 99–104.

Wilkins, W. H., and Harris, G. C. M. (1944). Investigations into the production of bacteriostatic substances by fungi. VI. Examination of the larger Basiodiomycetes. *Ann. Appl. Biol.* **31**, 261–270.

William, R. D., and Warren, G. F. (1975). Competition between purple nutsedge and vegetables. *Weed Sci.* **23**, 317–323.

Williams, A. H. (1960). The distribution of phenolic compounds in apple and pear trees. *In* "Phenolics in Plants in Health and Disease" (J. B. Pridham, ed.), pp. 3–7. Pergamon, New York.

Williams, A. H. (1963). Enzyme inhibition by phenolic compounds. *In* "Enzyme Chemistry of Phenolic Compounds" (J. B. Pridham, ed.), pp. 87–96. Macmillan, New York.

Williams, J. T. (1964). A study of the competitive ability of *Chenopodium album* L. *Weed Res.* **4,** 283–295.

Williams, L. (1971). The role of heteroinhibition in the development of *Anabaena flos-aquae* waterblooms. Diss., Rutgers Univ. *Diss. Abstr. Int.* 72-9689.

Williams, R. D., and Hoagland, R. E. (1982). The effects of naturally occurring phenolic compounds on seed germination. *Weed Sci.* **30,** 206–212.

Wilson, R. E., and Rice, E. L. (1968). Allelopathy as expressed by *Helianthus annuus* and its role in old-field succession. *Bull. Torrey Bot. Club* **95,** 432–448.

Wilson, R. G., Jr. (1981). Effect of Canada thistle (*Cirsium arvense*) residue on growth of some crops. *Weed Sci.* **29,** 159–164.

Wiltshire, G. H. (1973). Response of grasses to nitrogen source. *J. Appl. Ecol.* **10,** 429–435.

Winkler, B. C. (1967). Quantitative analysis of coumarins by thin layer chromatography, related chromatographic studies, and the partial identification of a scopoletin glycoside present in tobacco tissue culture. Ph.D. Dissertation, Univ. of Oklahoma, Norman.

Winter, A. G. (1961). New physiological and biological aspects in the interrelationships between higher plants. *Symp. Soc. Exp. Biol.* **15,** 229–244.

Wolfe, J. M., and Rice, E. L. (1979). Allelopathic interactions between algae. *J. Chem. Ecol.* **5,** 533–542.

Wood, H. E. (1953). The occurrence and the problem of wild oats in the Great Plains region of North America. *Weeds* **2,** 292–294.

Wood, R. K. S., and Graniti, A., eds. (1976). "Specificity in Plant Diseases." Plenum, New York.

Wood, R. K. S., Ballio, A., and Graniti, A., eds.(1972). "Phytotoxins in Plant Disease." Academic Press, New York.

Woodhead, S. (1981). Environmental and biotic factors affecting the phenolic content of different cultivars of *Sorghum bicolor*. *J. Chem. Ecol.* **7,** 1035–1047.

Woods, F. W., and Brock, K. (1964). Interspecific transfer of Ca^{45} and P^{32} by root systems. *Ecology* **45,** 886–889.

Woodwell, G. M. (1974). Success, succession, and Adam Smith. *BioScience* **24,** 81–87.

Wright, J. M. (1956). The production of antibiotics in soil. IV. Production of antibiotics in coats of seeds sown in soil. *Ann. Appl. Biol.* **44,** 561–566.

Wurzburger, J., and Leshem, Y. (1969). Physiological action of the germination inhibitor in the husk of *Aegilops kotschyi* Boiss. *New Phytol.* **68,** 337–341.

Yakhontov, A. F. (1973). On the possibility of using the allelopathic action of some plants for controlling *Dactilospheara viticola* F. *In* "Physiological-Biochemical Basis of Plant Interactions in Phytocenoses" (A. M. Grodzinsky, ed.), Vol. 4, pp. 57–60. Naukova Dumka, Kiev. (In Russian, English summary.)

Yeo, R. R., and Fisher, T. W. (1970). Progress and potential for biological weed control with fish, pathogens, competitive plants, and snails. *In* "Tech. Papers FAO International Conference on Weed Control," pp. 450–463. Weed Sci. Soc. Am.

Yoder, O. C., and Scheffer, R. P. (1973a). Effects of *Helminthosporium carbonum* toxin on nitrate uptake and reduction by corn tissues. *Plant Physiol.* **52,** 513–517.

Yoder, O. C., and Scheffer, R. P. (1973b). Effects of *Helminthosporium carbonum* toxin on absorption of solutes by corn roots. *Plant Physiol.* **52,** 518–523.

Young, A. (1804). "The Farmers Calendar." London.

Young, B. R., Newhook, F. J., and Allen, R. N. (1977). Ethanol in the rhizosphere of seedlings of *Lupinus angustifolius* L. *N. Z. J. Bot.* **15,** 189–191.

Young, C. C., and Bartholomew, D. P. (1981). Allelopathy in a grass-legume association: I. Effects of *Hemarthria altissima* (Poir.) Stapf. and Hubb. root residues on the growth of *Desmodium intortum* (Mill.) Urb. and *Hemarthria altissima* in a tropical soil. *Crop. Sci.* **21,** 770–774.

Younger, P. D., Koch, R. G., and Kapustka, L. A. (1980). Allelochemic interference by quaking aspen leaf litter on selected herbaceous species. *For. Sci.* **26,** 429–434.

Yurchak, L. D. (1974). Active metabolites of microorganisms decomposing *Lupinus. In* "Physiological-Biochemical Basis of Plant Interactions in Phytocenoses" (A. M. Grodzinsky, ed.), Vol. 5, pp. 100–103. Naukova Dumka, Kiev. (In Russian, English summary.)

Zabyalyendzik, S. F. (1973). Allelopathic interaction of buckwheat and its components through root excretions. *Vyestsi Akad. Navuk BSSR Syer Biyal Navuk* **5,** 31–34. (In Belorussian, Russian summary.)

Zaikova, V. A. (1973). Study of the allelopathic regime in meadow phytocenoses of the Karelian ASSR. *Bot. Zh.* **58,** 1753–1760. (In Russian.)

Zelitch, I. (1967). Control of leaf stomata: Their role in transpiration and photosynthesis. *Am. Sci.* **55,** 472–486.

Zhamba, G. E. (1972). Allelopathic role of coumarin compounds on propagation of cow parsnip *Heracleum. Izv. Akad. Nauk Mold. SSR, Ser. Biol. Khim. Nauk* **1,** 86–87. (In Russian.)

Zimmerman, R. H., Lieberman, M., and O. C. Broome. (1977). Inhibitory effect of rhizobitoxine analog on bud growth after release from dormancy. *Plant Physiol.* **59,** 158–160.

Zinke, P. J. (1962). The pattern of influence of individual forest trees on soil properties. *Ecology* **43,** 130–133.

Zucker, M. (1963). The influence of light on synthesis of protein and of chlorogenic acid in potato tuber tissue. *Plant Physiol.* **38,** 575–580.

Zucker, M. (1969). Induction of phenylalanine ammonia-lyase in *Xanthium* leaf discs. Photosynthetic requirement and effect of daylength. *Plant Physiol.* **44,** 912–922.

Zucker, M., Nitsch, C., and Nitsch, J. P. (1965). The induction of flowering in *Nicotiana.* II. Photoperiodic alteration of the chlorogenic acid concentration. *Am. J. Bot.* **52,** 271–277.

Zweig, G., Carroll, J., Tamas, I., and Sikka, H. C. (1972). Studies on effects of certain quinones. II. Photosynthetic incorporation of $^{14}CO_2$ by *Chlorella. Plant Physiol.* **49,** 385–387.

INDEX

A

Abies, allelopathic potential, 88
Abutilon theophrasti, see Velvetleaf
Acer, allelopathic potential, 88–89, 324
Acer pseudoplatanus, see Sycamore maple
Acetylenes, halogenated, 271
Acorus calamus, 155
Acrylic acid, an allelochemic, 204
Actinomycetes, phytotoxin production, 46
Adenostoma fasciculatum, patterning effects, 148–150, 163–167
Aegilops kotschyi, 322
Aesculus hippocastanum, 324
Agrobacterium radiobacter, crown gall prevention by strain 84, 124
Agropyron repens, see Quackgrass
Agropyron smithii, see Western wheatgrass
Agrostemma githago, see Corn cockle
Ailanthus altissima, see Tree-of-heaven
Alang-alang, *see* Cogongrass
Albizzia julibrissin, 324
Alcohols, straight chain, 267–268
Aldehydes, aliphatic, 267–268
Alfalfa, allelopathic effects, 43, 62–63, 66
Algal-like fungi, 5
Algal succession, role of allelopathy, 189–202
Alkaloids, as allelochemics, 233, 285–287
Allelochemic efficacy
　duration of activity, 348–350
　enhancement by stress, 361–363
　microbial decomposition, 350–356
　nonmicrobial decomposition, 350
　soil adsorption to active concentrations, 348
　synergistic action of allelochemics, 356–361
　union with soil organic matter, 345–348
Allelochemic production, effects of
　age of plant organ, 305–306
　allelochemics, 303–305
　daylength, 296

　genetics, 306–307
　ionizing radiation, 293
　light intensity, 295–296
　light quality, 293–295
　mineral deficiency, 296–301
　pathogens, predators, 307–308
　temperature, 303
　water stress, 301–303
Allelochemics
　adsorption on soil, 177, 348
　effects on
　　catalase, 340
　　cell division, 320–321
　　cell elongation, 320–321
　　cellulase, 340
　　cell ultrastructure, 320–321
　　clogging of xylem, 342
　　β-cystathionase, 341
　　easily available phosphorus, 330
　　hormone-induced growth, 321–323
　　lipid metabolism, 337–339
　　membrane permeability, 323–325
　　mineral uptake, 325–330
　　organic acid metabolism, 337–339
　　pectolytic enzymes, 339–340
　　peroxidases, 340
　　phenylalanine-ammonia-lyase, 341
　　phosphorylases, 341
　　photosynthesis, 330–334
　　porphyrin synthesis, 339
　　protein synthesis, 337–339
　　respiration, 334–337
　　stomatal opening, 330–332
　　sucrase, 341–342
　　water conduction, 342–343
　　water relations, 342–343
　egress by plant decay, 315–316
　leaching by rain, 314–315
　from marine flora, 202–204
　plant–plant movement, through fungal bridges, 318–319

413

PHYSIOLOGICAL ECOLOGY

A Series of Monographs, Texts, and Treatises

EDITED BY

T. T. KOZLOWSKI

University of Wisconsin
Madison, Wisconsin

JAMES A. LARSEN. The Boreal Ecosystem, 1980

SIDNEY A. GAUTHREAUX, JR. (Ed.). Animal Migration, Orientation, and Navigation, 1981

F. JOHN VERNBERG AND WINONA B. VERNBERG (Eds.). Functional Adaptations of Marine Organisms, 1981

R. D. DURBIN (Ed.). Toxins in Plant Disease, 1981

CHARLES P. LYMAN, JOHN S. WILLIS, ANDRÉ MALAN, and LAWRENCE C. H. WANG. Hibernation and Torpor in Mammals and Birds, 1982

T. T. KOZLOWSKI (Ed.). Flooding and Plant Growth, 1984

ELROY L. RICE. Allelopathy, Second Edition, 1984